Springer Proceedings in Complexity

Springer Proceedings in Complexity publishes proceedings from scholarly meetings on all topics relating to the interdisciplinary studies of complex systems science. Springer welcomes book ideas from authors. The series is indexed in Scopus.

Proposals must include the following:

- name, place and date of the scientific meeting
- a link to the committees (local organization, international advisors etc.)
- scientific description of the meeting
- list of invited/plenary speakers
- an estimate of the planned proceedings book parameters (number of pages/articles, requested number of bulk copies, submission deadline)

Submit your proposals to: christoph.baumann@springer.com

More information about this series at http://www.springer.com/series/11637

Dan Braha · Marcus A. M. de Aguiar ·
Carlos Gershenson · Alfredo J. Morales ·
Les Kaufman · Elena N. Naumova · Ali A. Minai ·
Yaneer Bar-Yam
Editors

Unifying Themes in Complex Systems X

Proceedings of the Tenth International Conference on Complex Systems

Springer

Editors
Dan Braha
New England Complex Systems Institute
and University of Massachusetts
Cambridge, MA, USA

Carlos Gershenson
Universidad Nacional Autonoma de Mexico
Mexico, Distrito Federal, Mexico

Les Kaufman
Boston University
Boston, MA, USA

Ali A. Minai
University of Cincinnati
Cincinnati, OH, USA

Marcus A. M. de Aguiar
State University of Campinas
Sao Paulo, Brazil

Alfredo J. Morales
New England Complex Systems Institute
Cambridge, MA, USA

Elena N. Naumova
Tufts Medical Center
Boston, MA, USA

Yaneer Bar-Yam
New England Complex Systems Institute
Cambridge, MA, USA

ISSN 2213-8684 ISSN 2213-8692 (electronic)
Springer Proceedings in Complexity
ISBN 978-3-030-67317-8 ISBN 978-3-030-67318-5 (eBook)
https://doi.org/10.1007/978-3-030-67318-5

This Springer imprint is published by the registered company Springer Nature Switzerland AG
The registered company address is: Gewerbestrasse 11, 6330 Cham, Switzerland

Preface

The Tenth International Conference on Complex Systems (ICCS 2020) took place online during July 27–31, 2020, in the midst of the COVID-19 pandemic. The global situation shows that we face several challenges with a complex nature besides the pandemic: climate change, economic interdependencies, violent conflicts, social polarization, and technological change, just to name a few. Understanding complex systems is crucial for addressing all of these challenges and the ones yet to arise.

Since the first edition of the conference in 1997, the field and the community have matured, but its spirit has been maintained: By its very nature, the study of complexity requires a shift in thinking across many disciplines, including the physical, biological, and social sciences, engineering, medicine, etc. The traditional reductionist approach has been very useful but is limited as it tries to simplify and isolate in order to predict. Complex phenomena are characterized by relevant interactions that may generate novel information that limit predictability, so reductionism is inappropriate. New concepts and tools have been developed and applied in recent decades, increasing our ability to solve complex problems.

This volume contains selected contributions from ICCS 2020. The diversity of topics included highlights the prevalence of complexity across disciplines. The answers developed in recent decades are far from complete. There is much more to be done. Thus, the results presented in this book are provisional at best but should be seen as part of a greater effort to advance our understanding of complex systems and their properties. We will not be immune to complex problems, but we will be better prepared to face them.

It is interesting that the conference has attracted not only academics from all disciplines, but also people from the private, government, military, and social sectors. Complex challenges demand not only an interdisciplinary approach to face them, but also multisectorial collaboration. Academia, government, or private corporations alone will not solve complex problems . Coordination between all relevant actors is required. Events such as International Conference on Complex Systems provide a forum where non-traditional collaborations can take place, and the important topics presented at the conference can reach a broader audience through the publication of this volume.

Cambridge, USA	Dan Braha
Sao Paulo, Brazil	Marcus A. M. de Aguiar
Mexico, Mexico	Carlos Gershenson
Cambridge, USA	Alfredo J. Morales
Boston, USA	Les Kaufman
Boston, USA	Elena N. Naumova
Cincinnati, USA	Ali A. Minai
Cambridge, USA	Yaneer Bar-Yam

Contents

Contributors

Thomas Adams Sandia National Laboratories, Albuquerque, NM, USA

Sinan G. Aksoy Pacific Northwest National Laboratory, Seattle, WA, USA

John Bicknell More Cowbell Unlimited, Inc., Lake Oswego, OR, USA

Gabriel C. Birch Sandia National Laboratories, Albuquerque, NM, USA

Tiffany J. Callahan University of Colorado Anschutz Medical Campus, Denver, CO, USA

Susan Caskey Sandia National Laboratories, Albuquerque, NM, USA

Sandra C. Chapman University of Warwick, Coventry, UK

Adit Chawdhary Department of Electrical Engineering and Computer Science, University of Cincinnati, Cincinnati, OH, USA

Aleksei Chechkin Akhiezer Institute for Theoretical Physics, Kharkov, Ukraine;
Institute for Physics and Astronomy, University of Potsdam, Potsdam, Germany

Qinghua Chen School of Systems Science, Beijing Normal University, Beijing, China;
New England Complex Systems Institute, Cambridge, MA, USA;
Department of Chemistry, Brandeis University, Waltham, MA, USA

Ian Ford University College London, London, UK

Aidee Lashmi García-Kroepfly Laboratorio de Redes Neuronales, Departamento de Fisiología, Facultad de Medicina, Universidad Nacional Autónoma de México, Ciudad de México, México

Mario Giampietro Universitat Autònoma de Barcelona, Bellaterra, Spain;
ICREA, Barcelona, Spain

Virgil Griffith Ethereum Foundation, Zug, Switzerland;
Hong Kong Polytechnic University, Hong Kong, China

Thushara Gunda Sandia National Laboratories, Albuquerque, NM, USA

Hossein Haeri Francis College of Engineering, University of Massachusetts, Lowell, USA

Marcus Harris Portland State University, Portland, OR, USA

Harold M. Hastings Bard College at Simon's Rock, Great Barrington, USA

James K. Hazy Adelphi University, New York, USA

Jesús Hernández-Falcón Laboratorio de Redes Neuronales, Departamento de Fisiología, Facultad de Medicina, Universidad Nacional Autónoma de México, Ciudad de México, México

Lawrence E. Hunter University of Colorado Anschutz Medical Campus, Denver, CO, USA

Brett Jefferson Pacific Northwest National Laboratory, Seattle, WA, USA

Kshitij Jerath Francis College of Engineering, University of Massachusetts, Lowell, USA

Ziyan Jiang School of Government, Beijing Normal University, Beijing, China

Bonnie Johnson Naval Postgraduate School, Monterey, CA, USA

Cliff A. Joslyn Pacific Northwest National Laboratory, Seattle, WA, USA

Rainer Klages Queen Mary University of London, London, UK

Mieczyslaw M. Kokar Department of Electrical and Computer Engineering, Northeastern University, Boston, MA, USA

Werner Krebs Acculation, Inc., Los Angeles, CA, USA

Honggang Li School of Systems Science, Beijing Normal University, Beijing, China

Xiaohang Liu School of Systems Science, School of Government, Beijing Normal University, Beijing, China

A. H. Louie Ottawa, ON, Canada

David J. Lyle Princeton University, Princeton, NJ, USA;
Springer Heidelberg, Heidelberg, Germany

Kit Martin Northwestern University, Evanston, IL, USA

Karina Mendoza-Ángeles Laboratorio de Redes Neuronales, Departamento de Fisiología, Facultad de Medicina, Universidad Nacional Autónoma de México, Ciudad de México, México

Ali A. Minai Department of Electrical Engineering and Computer Science, University of Cincinnati, Cincinnati, OH, USA

Robin Wooyeong Na College of Liberal Studies, Seoul National University, Seoul, Republic of Korea

Iván Oliver-Domínguez Laboratorio de Redes Neuronales, Departamento de Fisiología, Facultad de Medicina, Universidad Nacional Autónoma de México, Ciudad de México, México

Mireya Osorio-Palacios Laboratorio de Redes Neuronales, Departamento de Fisiología, Facultad de Medicina, Universidad Nacional Autónoma de México, Ciudad de México, México

Kevin P. O'Keeffe SENSEable City Laboratory Massachusetts Institute of Technology, Cambridge, MA, USA;
Hong Kong Polytechnic University, Hong Kong, China

Brenda Praggastis Pacific Northwest National Laboratory, Seattle, WA, USA

Emilie Purvine Pacific Northwest National Laboratory, Seattle, WA, USA

Carlo Ratti SENSEable City Laboratory Massachusetts Institute of Technology, Cambridge, MA, USA;
Hong Kong Polytechnic University, Hong Kong, China

Ansel Renner Universitat Autònoma de Barcelona, Bellaterra, Spain

J. Rowan Scott Department of Psychiatry, University of Alberta, Edmonton, AB, Canada

Paolo Santi SENSEable City Laboratory Massachusetts Institute of Technology, Cambridge, MA, USA;
Hong Kong Polytechnic University, Hong Kong, China

Josef Schaff Naval Air Systems Command, Patuxent River, USA

J. Rowan Scott Department of Psychiatry, University of Alberta, Edmonton, AB, Canada

Pratim Sengupta Mind, Matter and Media Lab, University of Calgary, Calgary, AB, Canada

Shweta Singh Department of Electrical and Computer Engineering, Northeastern University, Boston, MA, USA

David A. Stainforth University of Warwick, Coventry, UK;
London School of Economics and Political Science, London, UK

Bongwon Suh Graduate School of Convergence Science and Technology, Seoul National University, Seoul, Republic of Korea

Durga Suresh Department of Electrical and Computer Engineering, Northeastern University, Boston, MA, USA

Predrag T. Tošić Independent AI and Computer Science Researcher, Spokane, WA, USA;
Department of Mathematics & Statistics (Adjunct Faculty), Washington State University, Pullman, WA, USA

Ignacio J. Tripodi University of Colorado Anschutz Medical Campus, Denver, CO, USA

Percy Venegas Economy Monitor, King's College London, London, UK

Nan Wang DeepMacro Inc, New York City, NY, USA

Thomas Wang Bard College at Simon's Rock, Great Barrington, USA

Wei Wang School of Mathematics, Renmin University of China, Beijing, China

Yan Wang Department of Mathematics, University of California, Los Angeles, USA

Nicholas W. Watkins University of Warwick, Coventry, UK;
London School of Economics and Political Science, London, UK;
The Open University, Milton Keynes, UK

Adam D. Williams Sandia National Laboratories, Albuquerque, NM, USA

Jamie Wingo Sandia National Laboratories, Albuquerque, NM, USA

Yang Xu SENSEable City Laboratory Massachusetts Institute of Technology, Cambridge, MA, USA;
Ethereum Foundation, Zug, Switzerland;
Hong Kong Polytechnic University, Hong Kong, China

Zhaohui Yang Francis College of Engineering, University of Massachusetts, Lowell, USA

Tai Young-Taft Bard College at Simon's Rock, Great Barrington, MA, USA;
Levy Economics Institute of Bard College, Annandale-on-Hudson, NY, USA

Martin Zwick Portland State University, Portland, OR, USA

Multi-agent Simulations of Intra-colony Violence in Ants

Kit Martin and Pratim Sengupta

Abstract This paper seeks to elucidate key aspects of a rarely-studied interaction in ant colonies—*intra-colony* violence—using multi-agent-based computational simulations. A central finding is that intra-colony violence is heritable, though not prevalent. Results from our simulations reveal specific conditions in which such infrequent forms of violence occur and can be inherited, which in turn helps us understand why *Atta cephalotes* may persist killing colony members, even though it dampens colony carrying capacity. We also discuss the concerns and implications of our work for modeling conflict and violence more broadly, which in turn raises questions about the ontological nature of the computational and evolutionary models.

Keywords Multi-agent simulations · Emergence · Ant colonies · Violence

1 Introduction

Intra-specific violence is a common phenomenon among ant colonies, as reported by biologists such as Whitehouse's [1] work on *Atta laevigata*. Nearly all observations of violence in ants are observations of violence among colonies within the same species [2] rather than within a colony, because generally, within ant colonies, sterile workers cooperate to rear colonial young [3]. Instances of intra-colony violence are present in *Atta cephalotes* colonies, as observed during fieldwork conducted by the first author (see Fig. 1). The goal of this paper is to present a set of computational

This research was supported by the US National Science Foundation under a CAREER Grant awarded to Pratim Sengupta (NSF CAREER 1150230). All opinions are the authors' and not endorsed by any organization.

K. Martin (✉)
Northwestern University, Evanston, IL 60202, USA
e-mail: kitmartin@u.northwestern.edu

P. Sengupta
Mind, Matter and Media Lab, University of Calgary, Calgary, AB T2N 1N4, Canada
e-mail: pratim.sengupta@ucalgary.ca

D. Braha et al. (eds.), *Unifying Themes in Complex Systems X*, Springer Proceedings in Complexity, https://doi.org/10.1007/978-3-030-67318-5_1

Fig. 1 Photo of two ants (*Atta cephalotes*) engaged in intra-colony conflict. Image credit: Kit Martin

experiments that shed light on the evolutionary stability of this significantly less studied form of violence within *Atta cephalotes* colonies. Nest-mates (i.e., members of the same colony) killing each other is an interesting case, because theoretically the benefits of cooperation should outweigh the benefits of intra-colony competition because of the associated net loss of sterile foraging members [4].

We investigate the following research question: How can intra-colony violence persist over generations in ant colonies in which all members are related, and colony success is predicated on mutual cooperation and almost ubiquitous sterility? To answer this question, we developed a multi-agent-based simulation in NetLogo [5]. We conducted several computational experiments to help us identify circumstances where pure cooperation among nest-mates, in the face of aggressively violent behavior among nest-mates, sometimes loses from an evolutionary perspective. It is the identification of these specific circumstances, we posit, that stand as the main contribution of this paper. An implication of our work is the possibility of a deeper understanding of underlying mechanisms of cooperation and conflict (more broadly), with obvious cautions against misinterpretation and over-generalization, as we explain later.

Our open source NetLogo model is available via Open ABM [6], where following Grimm et al. [7], we have also attached a complete Overview, Design, Concepts, Details protocol to make modeling design decisions explicit, and thus more reproducible. In this paper, we first situate our work at the nexus of previous scholarship on ant biology and computational complexity. We then describe the basic version of the NetLogo model we began with, and how we modified this model to simulate three different scenarios that allowed us to investigate follow-up research questions. We conclude with some implications of our work for future research on modeling violence as an emergent system, grounded in an essential caution against unwarranted generalization.

2 Background: Ant Biology

2.1 Individual Life Histories of Atta cephalotes

Ants are a worldwide species that dominate many terrestrial ecosystems [8]. Ant colonies are composed of mostly sterile female workers, who protect the colony, collect food and take care of eggs in a cooperative structure, building a unit of selection larger than the individual [3, 9]. There are over 15,680 distinct species of ants, with almost as many distinct patterns of life histories [10]. In this model, we modeled *Atta cephalotes*, the type of fungus growing ants the first author observed in field work. They collect leaves and flowers to feed to fungus farms within their subterranean nests [11].

In *Atta cephalotes* reproduction is handled either by an individual female ant or a small number of female ants, termed queens, who usually mate once and live for decades [12]. *Atta cephalotes*, like many other ants, have a communication system predominantly based on time-diminishing pheromones, but also including drumming, touching, and vision [13], and share their knowledge using tandem running [14]. While only using 10% of the colony for food foraging, colonies can provide the nutritional needs of the entire colony, including immobile larvae, or queens [15]. Ants' success, like other social animals, hinges on their ability to take advantage of ephemeral natural resources to rear young and pass on their genetic heritage to closely related offspring in neighboring colonies. The model used this life history information to inform the model creation as outlined in its ODD [6].

2.2 Distinguishing "Friend" From "Foe" in Ant Colonies

Ant colonies, like several other insect societies (e.g.., honeybees), are seen as examples of *superorganisms* [9, 16, 17]. Originally coined by Morton Wheeler [17] to denote insect societies that "possess features of organization analogous to the physiological processes of individual organisms" [16, p. 548], it indicates societies of organisms such as ants and honeybees that form a cooperative unit to propagate their genes. As Seeley noted, the analogy here is the classification of a group of cells as an organism when the cells form a cooperative unit to propagate their genes [16]. But this also suggests that most insect societies are not perfect superorganisms because there is usually intense intragroup conflict when members compete for reproductive success [18].

That being said, superorganisms have also been defined as insect societies that have very low aggression towards nestmates but high aggression towards outsiders [19]. As Moffett said "All societies are characterized by the capacity of their members to distinguish one another from outsiders" [20, p. 1]. In ant colonies, members recognize each other through odor cues contained in hydrocarbons, or oils, on the ant's exoskeleton [9, 21–27]. They also employ the hydrocarbons as home range marking [28], which allows colony mates to distinguish each other without having to actually have met before [20]. This "anonymous" scent based recognition in ants has two effects. First, it means larger groups of ants can work together altruistically on an effort and form a society without intra-colony violence or individual recognition. Second, it means errors may exist in recognizing colony or colony-mates, as the results have shown that in turn may lead to intra-colonial violence [25].

2.3 Intra-Colony Violence

Examples of violence between nest mates have been documented in a few different ant species [29]. *Linepithema humile*, Argentine ants, have been induced to kill colony mates, possibly through a change in diet [30]. In our field observations, we observed nestmates of *A. cephalotes* kill each other near Manaus, Brazil. Kin selection, competition among the minutely diverse genetic makeups of ants in a colony, provides a rationale for the persistence of intra-colony competition. Kin selection takes a variety of forms, including queens in the same colony that may kill each other for the right to control the genetic heritage of offspring, mostly sterile workers that may compete for the rights to lay male eggs [31], or "varieties of strategies that even nonlaying workers might employ to ensure genetic success" [29, p. 171], such as killing or selectively starving less related eggs. In short, while on the one hand, the super-organism attacking itself seems intuitively like a negative externality that should result in negative colonial fitness, that would be selected against. On the other hand, selfishness may still have benefits in an altruistic ant colony [29]. Of particular interest to us is the question whether such "cross-purposes" (p. 171) are merely glitches in an otherwise functional and highly adaptive super-organism structure, or themselves have utility evolutionarily speaking.

3 Theoretical Framework

3.1 High-Leverage Points in Complex Systems

Holland [32] argued that an ant nest is an apt example of "emergence of complex large-scale behaviors from the aggregate interactions of less complex agents" (p. 11). Such emergent systems can be understood as systems with high leverage-points, where a local or an individual-level interaction can have systemic impact [33]. In the game of chess, which is also an example of a complex adaptive system, one such leverage point is a gambit [33], where a player sacrifices a pawn early in the game to set up the board for improved late game position. In an immune system, an example is a vaccine [32], where immunity to disease in the whole body results from prior exposure to microscopic pathogens. In both examples, "a small input can produce major predictable, directed change—an amplifier effect— ..." [32, p. 5]. When discovered, these high leverage points cause state-changes, i.e., the macro-scale environment changes appreciably due to small changes at the individual-level. In war, while there are points of leverage, they are difficult to identify. Solvit described war as a complex adaptive system but one without methodically delineated levers of intervention [34]. We posit that violence in ant colonies also reflects this feature of complex systems. A contribution of this paper is that we propose how varying killing rates in the colony can help us identify such a high leverage point, as it can lead to an evolutionarily significant, aggregate-level change for the colony.

3.2 Inclusive Fitness and Natural Selection

In sociobology, the classic equation that models *inclusive fitness* is often used to measure fitness in social animals [35, 36] is shown below:

$$rb > c \tag{1}$$

This equation is known as Hamilton's rule. It helps us understand how altruism evolves or can be maintained in a selfish world. In this equation, r represents the relatedness between the social partners, and can also be understood as a genotype or a collection of genetic material on a DNA strand that is passed to the subsequent generation. Hamilton's rule represents the idea that the benefits to the recipients of an altruistic act (b), weighted by the relatedness between the social partners (r), must exceed the costs to the altruists c, in order for altruistic genotypes to be passed on or inherited over by future generations over nonaltruistic ones.

Researchers have shown that this equation holds true for observations of *inter-colony* violence [1, 2]. That is, one would expect two or more colonies competing with each other to fight over sources of food, and thus genes that cause functions leading to reproductive success to persist. However, the question at hand is as follows: how could *intra-colony* violence—i.e., members of the same colony killing each other—be evolutionarily beneficial for the colony? Is it even possible that the existence of colony members who practice the "kill my nestmate" behavior, in some circumstance, has a "benefit" to the colony? The equation for us, in such cases, would be as follows:

$$r(k)\, b > c \tag{2}$$

Here *r(k)* represents the rate at which killers are born in a given colony. How can such a scenario be understood from the perspective of inclusive fitness and natural selection? What implications may such evolutionary occurrences have for our broader understanding of modeling violence as an emergent system as well as using evolutionary approaches? These are the questions that we are investigating in this paper.

4 A Necessary Caution Against Over-Generalization

The notion of inclusive fitness is a well-accepted model for studying altruism in social insects [37, 38] and even human societies [39]. Altruism is defined as the benefit to others at a cost to the individual. In agent-based computational models of evolutionary biology, this is usually represented in the form of the transfer of genetic material at the cost of physical demise of the individual agent. Given this, for our present work, perhaps the most bizarre and dangerous question from a moral standpoint can be stated as follows: Could we view intra-colony violence as a form of altruism? That is, from a purely utilitarian perspective, could the emergent effect of ants killing nest mates be understood in light of a long-term successes for the colony as expressed through its inclusive genetic fitness?

The unequivocal answer to this question is that the logic of cost–benefit analysis that underlies Hamilton's rule is a vastly reductive approach for modeling violence or mass-scale devastation. As moral philosophers point out, underlying the notion of "altruism" are axiomatic assumptions of marginal value and utilitarianism [40], requiring us to take the "point of view of the universe" rather than the perspective of the individual ethics [41]. By illustrating conditions in which infrequent mass violence may be inherited, our goal is not to justify such occurrences—on the contrary, it is to alert us to conditions under which violent acts—carried out by individuals—might come to hold *marginal value* over co-operation at the aggregate level. The ability to reveal such counter-intuitive connections between the individual and the aggregate levels of behavior in a complex emergent system is a particular affordance of multi-agent models [42, 43].

But at the same time, from an ontological perspective, we see this as a crisis rather than an affordance: positioning large-scale violence as a form of altruism only reveals a form of logical fallacy that has indeed proven to be dangerous in human history, as evident in the history of justifications of human genocides [44]. We discuss these ontological challenges and implications for modeling violence in human societies in more detail in the final section of the paper.

5 The Basic Model

The basic model consists of two agent-types ("breeds" in NetLogo terminology): red ants (Rants) and blue ants (Bants). At the beginning of the simulation, each ant starts with a built-in dietary preference, represented by a randomly selected number between 0 and 140 (using the HSB color scale). When an ant finds a plant that corresponds to her preference she picks it up (RedFood or BlueFood), gains a specific amount of energy (250 units), and returns a portion of that food to the home-nest for consumption by colony-mates (return-to-nest).

Food collection serves both the individual and the colony's survival. When the collective food in a colony reaches a threshold value, queen hatches one new baby ant. During a single time-step the colony can only gain upto one new ant from reproduction.

At the beginning of each simulation run, every colony has a specific proportion of killer ants represented by the variable k. In any encounter with other ants, each killer ant has a probability of 50% of killing the other ant. Other individual-level ant variables include age, their current x and y coordinates, and a Boolean variable representing whether they are looking for food or returning to nest.

Learning occurs in our model as follows: Ants share food preferences with other ants they encounter. Because the surrounding environment in the model changes every time step (see ODD), the food preferences for each ant also needs to update. While ants can randomly discover food, they also share their preferences if they have successfully found food. Social learning is therefore a mechanism for ants to dynamically alter their food preferences by learning from others' successes in foraging. This is represented by the function *tandem-running* [14], in which ants share their preferred food color with other ants if they are returning to the nest with food.

6 The Computational Experiments: Three Scenarios

As stated earlier, our overall research question is: How can violence and discord, when nest mates kill nest mates, persist in a social situation over generations in which all members are related and colony success is predicated on mutual cooperation and almost ubiquitous sterility? In order to answer this question, we modeled three scenarios by modifying the aforementioned "base" model. In **Scenario 1**, we modeled two competing colonies and in each colony, the queen only reproduced sterile workers. So in each colony, when the queen reproduced, she did not give birth to reproductively active males or females that could create new colonies. Thus, while the colony population grew over time, new colonies were not founded—i.e., there was no reproduction of the super-organism. The first scenario thus sought to investigate the comparative survival rates of colonies with and without killer ants, as we varied k. We refer to colonies without killer ants as non-violent or altruist colonies, represented visually as blue ants, while colonies with killer ants were represented as red ants. In **Scenario 2**, we initialized the model with five colonies that could reproduce both sterile workers and reproductively active males and females, who, in turn could also establish their own colonies. This allowed for super-organism reproduction. Experiments in this scenario sought to investigate how the heritability of k, as a result of reproduction both within and between colonies, effected the prevalence of both violent and non-violent ants. That is, in Scenario 2, we did not vary k in each run, and instead observed how super-organism reproduction, along with other emergent factors (e.g.., availability of food resources) effected the the value of k over subsequent generations. Finally, experiments in **Scenario 3** sought to investigate the role of learning by implementing a calamity in terms of sudden, drastic reduction of food resources in the environment while *k* was not heritable but was varied systematically across simulation runs, as in Scenario 1.

The first scenario was run 165 times for 16,000 steps. The second scenario was run ten times for 43,810 steps. The Third scenario was run sixteen times for 16,000 steps.

7 Results

7.1 Scenario 1: Intra-Colony Violence Without Super-Organism Reproduction

An important finding of this set of experiments is that although on average the altruist colony can maintain a larger population, highly aggressive behavior can—albeit very infrequently—result in the total annihilation of competing colonies. This can be understood as follows. On an average, the non-violent colonies have a higher population than the murderous colonies. But in 6 out of 165 trials, i.e., 3.6% of the time, the non-violent blue colony collapsed, while in the same trials—regardless of the value of *k*—the red colony never collapsed. We also found that when the trials were run with two competing but non-violent colonies, neither colony ever collapsed, indicating that colony collapse cannot simply be attributed to resource exhaustion linked to competition from other colonies.

Another finding is that both attacks and social learning of food preferences are dependent on the density (i.e., spatial distribution) of agents, being most prevalent near the colony centers (Fig. 2). Consequently, the nestgate regions (i.e., areas immediately surrounding the center of the colony), which are the most effective locations for sharing food preferences due to the highest concentrations of contacts between ants, are also the places where most lethal encounters happen. Note that increasing the value of *k* also reduces learning rates, which makes a colony less adaptive to its environment. In simulations with food scarcity the effect is greater, as adaptation to a changing environment is more crucial than in a stable state, and we investigate this in more detail in Scenario 3.

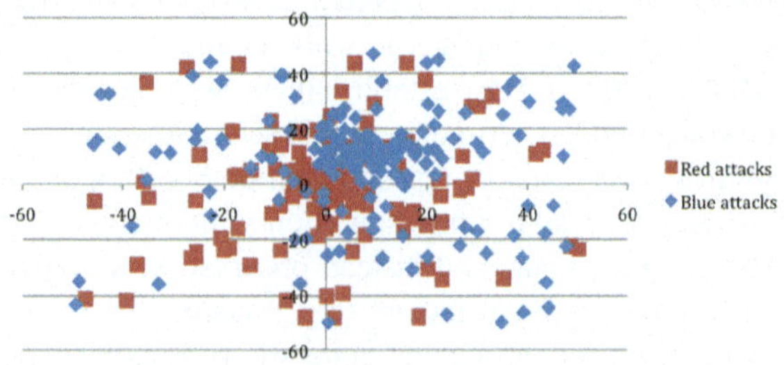

Fig. 2 Density-dependence of attacks: Attacks by Red and Blue killer ants are most prevalent *at* the colonies. Red colony is centered at (0,0) and Blue colony is centered at (10,10)

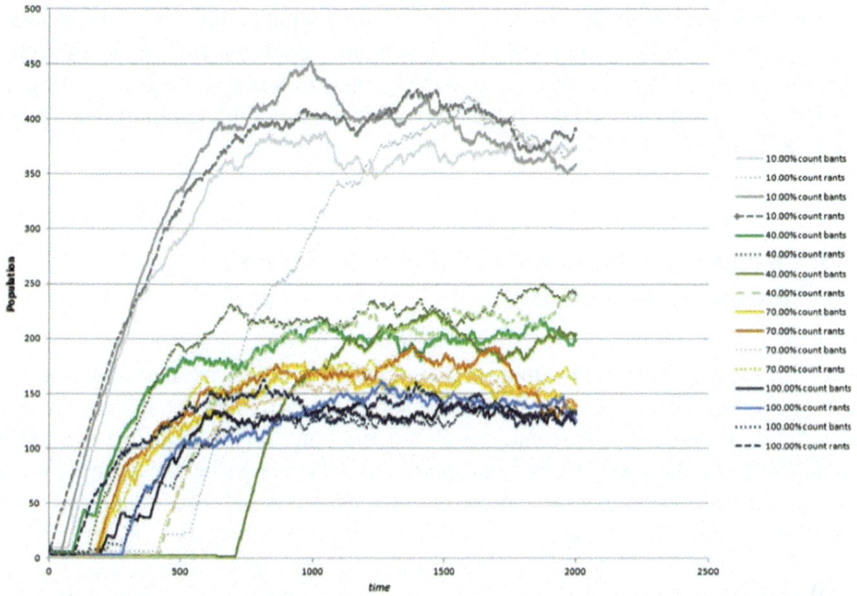

Fig. 3 The maximum population of ant colonies decreased as k increased

Finally, it is important to note that aggressive behavior also dampened population growth for the red colony. As shown in Fig. 3, k is increased from 10 to 100% in increments of 30%, which has a cost for the red colony: higher values of k are associated with with a lower maximum red ant population. For example, for $k = 0$, the max population in the red colony was 3,000, while for $k = 90\%$, the max population was 120.

7.2 Scenario 2: Intra-Colony Violence with Super-Organism Reproduction

For this scenario, we chose $k = 20\%$ as the starting value. This was based on sensitivity analysis which revealed that it is at and around this value, a violent colony can outperform non-violent colonies in terms of survival. When colonies were allowed to reproduce, we found that even though 20% of starting nests produced killers at the beginning of the simulation, by the end, only 58 of 1,167 (5%) of nests produced killers $k = 33\%$, i.e., their queens birthing killer ants one third of the time. Comparison of average value of k across all colonies at the beginning and the end of the experiment (43800 runs) reveals that it dropped from 40 to 2%, while the average value of k for violent colonies dropped from 67 to 33%.

This indicates once such violent traits become heritable, they are difficult to be exterminated from the gene pool, even though they might only dominate the competition infrequently or rarely. One possible explanation is that when k is heritable, even a single killing in a colony increases the likelihood of more killings both in the same colony and other colonies.

7.3 Scenario 3: A Deeper Investigation of the Role of Resource Availability

In this scenario, we worked with a version of the model used in Scenario 1, where k was not heritable (set to 20%) and involved two colonies (red and blue ants). We also introduced a food calamity event: we set red and blue food to 0 at time $t = 10000$ steps. We chose the 10000 steps as an appropriate time for setting the calamity event based on sensitivity analysis that reveled that populations stabilize around this time. As shown in Fig. 4, after the food calamity, the effect of k becomes clearer: 50% of red colonies and 37.5% of blue colonies collapsed after the food storage calamity.

We also found that it was around $k = 20\%$ that the effect of the calamity became pronounced in our simulation. This can be seen by comparing, for example, the populations of the two competing colonies at and around t = 10000 in Figs. 5 and 6.

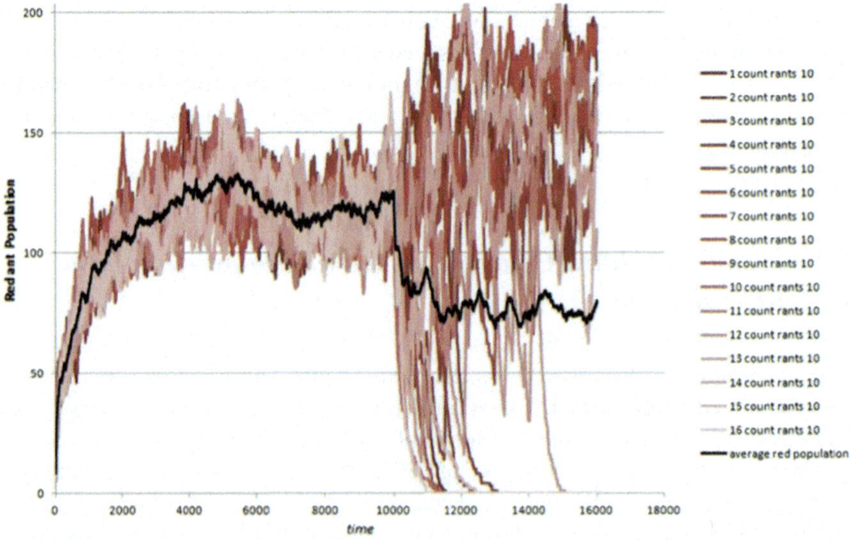

Fig. 4 Scenario 3: Average populations from 16 simulation runs for $k = 20\%$. Red colonies increase rapidly at first. Over time, growth plateaus and then drops to zero just as food resources deplete. Colonies must adjust quickly, or collapse. 50% (n = 8) colonies survive and their population explodes when the competing colonies die

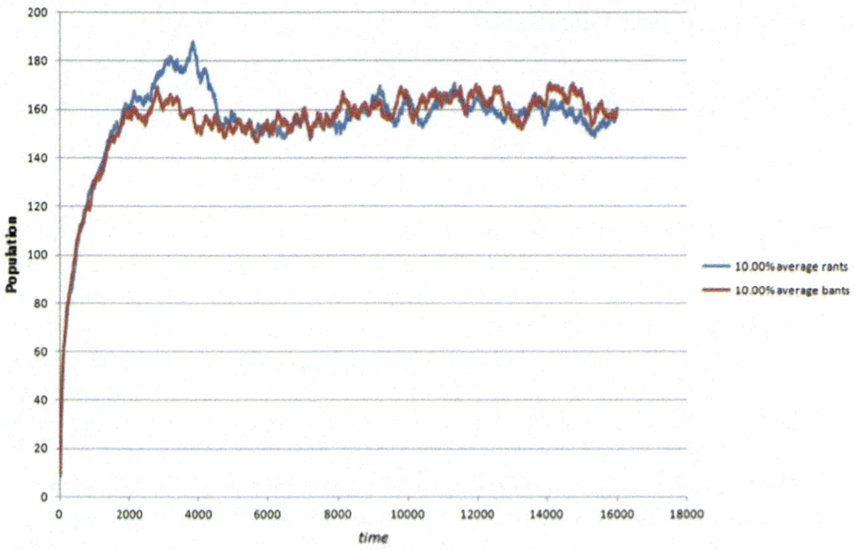

Fig. 5 Scenario 3: average populations from 16 simulation runs at $k = 10\%$

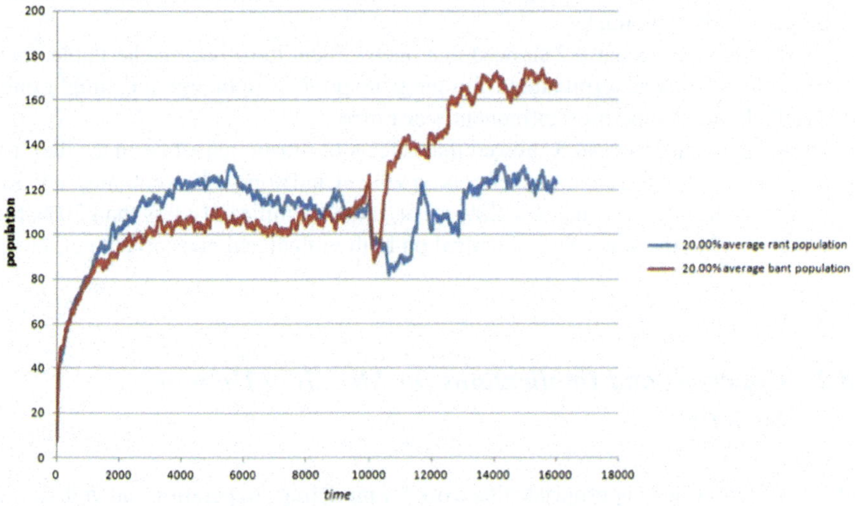

Fig. 6 Scenario 3: average populations from 16 simulation runs at $k = 20\%$

These graphs show that at $k = 20\%$, the average population of red (violent) colony increases after the calamity at the expense of the average population of the blue (non-violent) colony. In contrast, for values less that $k = 20\%$, we found no discernable difference in the average populations, an example of which is shown in Fig. 5.

8 Summary and Discussion

8.1 Summary of Results

The results from our experiments can be summarized as follows:

- An important finding from the experiments in Scenario 1 is that although on average non-violent colonies can maintain a higher population, highly aggressive behavior in violent colonies can—albeit very infrequently—result in the total annihilation of competing non-violent colonies.
- Experiments in Scenario 1 also showed that the population in violent colonies never collapsed, while the population of non-violent colonies did collapse 3.6% of the time. However, aggressive behavior also lowers the population growth for violent colonies.
- Attacks between ants are *density-dependent*. Experiments in Scenario 1 showed that attacks by killer ants are most prevalent at and around the colony centers, and particularly at the nestgate regions, where the likelihood of ants sharing their food preferences is also high. Coupled with a low ability to distinguish between friend and foe, the presence of a higher number of ants in a particular location leads to higher number of attacks.
- Experiments in Scenario 2 showed that once violent traits become heritable, they are difficult to be exterminated from the gene pool, even though they might only dominate the competition infrequently or rarely.
- Experiments in Scenario 3 showed that effects of drastic reduction in availability of food results in differentiated effects on violent and non-violent colonies only for certain values of k (around $k = 20\%$). For values significantly less than 20%, the effect of a famine was nearly identical on both violent and non-violent colonies.

8.2 Concerns and Implications for Modeling Human Societies

It may be tempting to appropriate this work for modeling cooperation and violence in human societies. It is indeed true that multi-agent simulations can offer productive approaches for modeling complex human societies and experiences that may be otherwise difficult to model or understand. For example, multi-agent simulations can help us understand how gender and sexuality based marginalization and resilience are emergent, multi-level phenomena rather than individual-level experiences [45], and how apparently democratic systems of elections may be resting on deeply oligarchic foundations [46]. But, this kind of work may require a fundamentally more synthetic approach that integrates qualitative and subjective forms of human experiences with computational methods. This includes integrating narratives with data objects and algorithms for simulating complexity as integral parts of the computational model and modeling experience [45], and validating computational and algorithmic approaches with empirical work [46].

In absence of such synthetic methods, our current paper should not be interpreted as a model of human interactions. It is not an endorsement of the virtues of infrequent, mass violence. On the contrary, our work should serve as a reminder that even apparently "altruistic" behavior at the system level, which can prolong the survival of societies, could in fact include morally reprehensible forms of behavior and violence at the level of individual interactions. Rather than solely relying on cost-benefit analyses for modeling violence in human societies, we encourage deeper engagement with scholarship in the social sciences and humanities that highlights the complexity of modeling and representing even apparently simple human interactions, by revealing how deeply they are shaped by social, political, historical and institutional forces, both implicitly and explicitly [47, 48]. In absence of such synthetic approaches, utilitarian perspectives on altruism and inter-generational benefits that undergird Hamilton's model of inclusive fitness for kin selection may be fundamentally reductionist and problematic as a paradigm for modeling human societies.

References

1. Whitehouse, M.E., Jaffe, K.: Ant wars: combat strategies, territory and nest defence in the leaf-cutting antatta laevigata. Anim. Behav. **51**(6), 1207–1217 (1996)
2. Moffett, M.W.: Ants & the art of war. Sci. Am. **305**(6), 84–89 (2011)
3. Wilson, E.O., et al.: The insect societies. The insect societies. (1971)
4. Townsend, C., Begon, M., Harper, J., et al.: Essentials of ecology. No. Ed. 3, Blackwell Publishing (2008)
5. Wilensky, U.: Netlogo (and netlogo user manual). Center for connected learning and computer-based modeling. Northwestern University (1999). http://ccl.northwestern.edu/netlogo
6. Martin, K.: Umwelten ant model. CoMSES Computational Model Library, Evanston (2015). https://www.comses.net/codebases/4499/releases/1.0.0/
7. Grimm, V., Berger, U., Bastiansen, F., Eliassen, S., Ginot, V., Giske, J., Goss-Custard, J., Grand, T., Heinz, S.K., Huse, G., et al.: A standard protocol for describing individual-based and agent-based models. Ecol. Model. **198**(1–2), 115–126 (2006)
8. Hölldobler, B., Wilson, E.O.: The Ants. Harvard University Press (1990)
9. Hölldobler, B., Wilson, E.O.: The superorganism: the beauty, elegance, and strangeness of insect societies. WW Norton & Company (2009)
10. Fisher, B., Ward, P.: Antweb. California Academy of Sciences (2002)
11. Hölldobler, B., Wilson, E.O.: The leafcutter ants: civilization by instinct. WW Norton & Company (2010)
12. Page Jr., R.E., Metcalf, R.A.: Multiple mating, sperm utilization, and social evolution. The Am. Nat. **119**(2), 263–281 (1982)
13. Jackson, D.E., Ratnieks, F.L.: Communication in ants. Curr. Biol. **16**(15), R570–R574 (2006)
14. Franks, N.R., Richardson, T.: Teaching in tandem-running ants. Nature **439**(7073), 153 (2006)
15. Dussutour, A., Simpson, S.J.: Communal nutrition in ants. Curr. Biol. **19**(9), 740–744 (2009)
16. Seeley, T.D.: The honey bee colony as a superorganism. Am. Sci. **77**(6), 546–553 (1989)
17. Wheeler, W.M.: The ant-colony as an organism. J. Morphol. **22**(2), 307–325 (1911)

18. Trivers, R.: Social evolution. No. **156**, T7 (1985)
19. Strassmann, J.E., Queller, D.C.: Insect societies as divided organisms: the complexities of purpose and cross-purpose. Proc. Natl. Acad. Sci. **104**(suppl 1), 8619–8626 (2007)
20. Moffett, M.W.: Supercolonies of billions in an invasive ant: what is a society? Behav. Ecol. **23**(5), 925–933 (2012)
21. Carlin, N.F., Hölldobler, B.: The kin recognition system of carpenter ants (camponotus spp.). Behav. Ecol. Sociobiol. **19**(2), 123–134 (1986)
22. Lahav, S., Soroker, V., Hefetz, A., Vander Meer, R.K.: Direct behavioral evidence for hydrocarbons as ant recognition discriminators. Naturwissenschaften **86**(5), 246–249 (1999)
23. Lenoir, A., Fresneau, D., Errard, C., Hefetz, A.: Individuality and colonial identity in ants: the emergence of the social representation concept. In: Information Processing in Social Insects, pp. 219–237. Springer (1999)
24. Martin, S.J., Vitikainen, E., Helanterä, H., Drijfhout, F.P.: Chemical basis of nest-mate discrimination in the ant formica exsecta. Proc. Roy. Soc. Lond. B: Biol. Sci. **275**(1640), 1271–1278 (2008)
25. Wagner, D., Tissot, M., Cuevas, W., Gordon, D.M.: Harvester ants utilize cuticular hydrocarbons in nestmate recognition. J. Chem. Ecol. **26**(10), 2245–2257 (2000)
26. Ydenberg, R.C., Giraldeau, L.A., Falls, J.B.: Neighbours, strangers, and the asymmetric war of attrition. Anim. Behav. **36**(2), 343–347 (1988)
27. van Zweden, J.S., d Ettorre, P.: Nestmate recognition in social insects and the role of hydrocarbons. In: Insect Hydrocarbons: Biology, Biochemistry and Chemical Ecology, vol. 11, pp. 222–243 (2010)
28. Lenoir, A., Depickère, S., Devers, S., Christidès, J.P., Detrain, C.: Hydrocarbons in the ant lasius niger: from the cuticle to the nest and home range marking. J. Chem. Ecol. **35**(8), 913–921 (2009)
29. Queller, D.C., Strassmann, J.E.: Kin selection and social insects. Bioscience **48**(3), 165–175 (1998)
30. Liang, D., Silverman, J.: "you are what you eat" : diet modifies cuticular hydrocarbons and nestmate recognition in the argentine ant, linepithema humile. Naturwissenschaften **87**(9), 412–416 (2000)
31. Cole, B.J.: Dominance hierarchies in leptothorax ants. Science **212**(4490), 83–84 (1981)
32. Holland, J.H.: Hidden orderhow adaptation builds complexity. No. **003**(7), H6 (1995)
33. Holland, J.H.: Studying complex adaptive systems. J. Syst. Sci. Complex. **19**(1), 1–8 (2006)
34. Solvit, S.: Dimensions of war: understanding war as a complex adaptive system. Editions L'Harmattan (2012)
35. Hamilton, W.D.: The genetical evolution of social behaviour. i. J. Theor. Biol. **7**(1), 1–16 (1964)
36. Hamilton, W.D.: The genetical evolution of social behaviour. ii. J. Theor. Biol. **7**(1), 17–52 (1964)
37. Hamilton, W.D.: Altruism and related phenomena, mainly in social insects. Ann. Rev. Ecol. Syst. **3**(1), 193–232 (1972)
38. Wilson, E.O.: Sociobiology: the new synthesis. Belknap Press of Harvard U Press (1975)
39. Tang, T., Ye, H.: The blessing of sexuality: evolution of altruism with mating preference. J. Artif. Soc. Soc. Simul. **19**(2), (2016)
40. Srinivasan, A.: Stop the robot apocalypse: the new utilitarians. Lond. Rev. Books **37**(18), 3–6 (2015)
41. de Lazari-Radek, K., Singer, P.: The point of view of the universe: Sidgwick and contemporary ethics. OUP Oxford (2014)
42. Sengupta, P., Wilensky, U.: Learning electricity with niels: thinking with electrons and thinking in levels. Int. J. Comput. Math. Learn. **14**(1), 21–50 (2009)
43. Wilensky, U., Resnick, M.: Thinking in levels: a dynamic systems approach to making sense of the world. J. Sci. Educ. Technol. **8**(1), 3–19 (1999)
44. Ihrig, S.: Justifying Genocide: Germany and the Armenians from Bismark to Hitler. Harvard University Press (2016)

45. Paré, D., Shanahan, M.C., Sengupta, P.: Queering complexity using multi-agent simulations. In: Proceedings of the 14th International Conference of Learning Sciences (ICLS 2020) **TBD**, TBD (2020)
46. Wright, M., Sengupta, P.: Modeling oligarchs' campaign donations and ideological preferences with simulated agent-based spatial elections. J. Artif. Soc. Soc. Simul. **18**(2), 3 (2015)
47. Ahmed, S.: Queer phenomenology: orientations, objects, others. Duke University Press (2006)
48. Banerjee, P., Connell, R.: Gender theory as southern theory. In: Handbook of the Sociology of Gender, pp. 57–68. Springer (2018)

Renormalization Group Approach to Cellular Automata-Based Multi-scale Modeling of Traffic Flow

Zhaohui Yang, Hossein Haeri, and Kshitij Jerath

Abstract A key problem in modeling and analyzing emergent behaviors in complex systems such as traffic flows, swarms, and ecosystems, is to identify the appropriate scale at which such emergent behaviors should be modeled. Current traffic flow modeling techniques can be neatly categorized as operating at one of three spatial scales: microscopic, mesoscopic, or macroscopic. While significant research effort has been expended on developing models in each of these three modeling approaches, a fundamental question remains unanswered: what is the 'appropriate' scale at which to model traffic flow that will enable optimal observation and prediction of emergent traffic dynamics? Recent techniques that attempt to merge models at different scales have yielded some success, but have not yet answered this fundamental question. In the presented work, we seek to use statistical mechanics-inspired traffic flow models, coupled with renormalization group theoretic techniques to allow for modeling of traffic flow dynamics at spatial scales that are arbitrarily multiples of an individual car length. While we do not yet answer the question about optimal spatial scales for observing emergent behaviors, this method provides a stepping stone towards that goal. In addition, the presented method also provides a single mechanism to link microscopic and macroscopic models of traffic flow. Numerical simulations indicate that the traffic phenomena such as backward moving waves and emergent traffic jams are retained with this modeling technique.

1 Introduction

The ability to model complex systems at multiple scales is a key requirement in creating a better understanding of observed emergent behaviors, which manifest at specific spatial scales [1]. To our knowledge, there are very few systematic approaches for determining these scales in emergent traffic flow and applying them to model traffic dynamics [2, 3]. Knowledge of the spatial scale at which emergent traffic flows dominate can help improve observation of complex traffic systems, and be used in control

Z. Yang · H. Haeri · K. Jerath (✉)
Francis College of Engineering, University of Massachusetts, Lowell, USA
e-mail: kshitij_jerath@uml.edu

D. Braha et al. (eds.), *Unifying Themes in Complex Systems X*, Springer Proceedings in Complexity, https://doi.org/10.1007/978-3-030-67318-5_2

methods to produce smoother traffic flows. There are several ancillary benefits that accompany smoother traffic flows, such as fewer traffic incidents and reductions in emissions and costs. Optimal prediction and/or feedback control of traffic typically relies on some form of traffic flow modeling, either based on first principles or using data-driven approaches [4–6]. Traditionally, first principles-based traffic flow models have been developed at three distinct spatial scales: (a) *macroscopic* approaches that model traffic as a fluid flow using the kinematic wave theory and partial differential equations, (b) *microscopic* approaches that model the dynamics of individual vehicles using car-following or cellular automata models, and (c) *mesoscopic* approaches that model traffic at intermediate scales such as vehicle clusters or road links. Unfortunately, while these models have proven to be extremely useful is several contexts, they do not allow traffic modeling at arbitrary spatial scales. The ability to understand and model traffic flow dynamics across multiple spatial scales is key to ensure the accurate observation and prediction of emergent traffic states. Moreover, the relationship between microscopic parameters and macroscopic models of traffic flow remains unknown and an active area of research.

In the present work, we discuss previous work on traffic models at different 'resolution' or spatial scales, as well as attempts to create a bridge from microscopic to macroscopic traffic flow dynamics. We then introduce a statistical physics-inspired traffic flow model operating in discrete space and time [7]. Borrowing additional concepts such as the renormalization group (RG) theory that model behaviors near criticality, we demonstrate how our statistical physics-inspired approach that uses the Ising model can model emergent traffic flow dynamics at arbitrary spatial scales [8]. Additionally, we demonstrate how the approach can generate a link between the microscopic and macroscopic scales. In particular, we demonstrate the mechanism to identify functional relationships between model parameters, i.e. Ising model interaction coefficients, at arbitrarily chosen spatial scales, in the context of traffic flow. Numerical examples help validate the renormalization group theoretic approach to modeling traffic dynamics at arbitrary scales.

2 Literature Review

In this section, we review prior works in traffic flow modeling that generate models at different spatial scales. However, only a few of these approaches attempt to utilize a known microscopic model and coarse-grain it to generate a model at a different spatial scale, an approach that is familiar to the complex systems research community [9–11]. The presented work is perhaps the first to *analytically* coarse-grain microscopic-scale traffic flow model to an *arbitrary* spatial scale (an arbitrary multiple of the length of a single vehicle).

2.1 *Traffic Flow Modeling at Distinct Spatial Scales*

As mentioned in the previous section, traffic flow is typically modeled at three different spatial scales: microscopic, macroscopic, and mesoscopic [12]. Car-following models are typical examples of microscopic modeling of traffic flow. These models use ordinary differential equations to represent how a driver controls the longitudinal vehicle dynamics (i.e., speed and position) relative to a preceding vehicle on a highway [13]. They encapsulate a significant amount of information and details, but are less amenable for use in prediction and control of future microscopic traffic states.

Handling a large number of ordinary differential equations for microscopic traffic flow models may become cumbersome due to modeling redundancies and computational costs. An alternative approach seeks to model traffic dynamics as a continuum, as in the case of fluid flow. Macroscopic traffic flow models focus on the dynamics of density $\rho(x, t)$ as a function of space x and time t, which trade off microscopic-level details for macroscopic-level descriptive power [14–16].

In contrast to microscopic and macroscopic modeling approaches, mesoscopic scale models were proposed as a means to capture dynamics associated with intermediate scales. While each of these approaches model several aspects of the traffic flow dynamics very well at their respective spatial scales, they fail to address a fundamental question: which spatial scale is best suited to and predict traffic flow behavior? The next subsection discusses recent approaches that have sought to answer this question.

2.2 *Modeling Traffic Flow at the 'Appropriate' Spatial Scale*

Recent works have sought to generate a link between the microscopic level description and macroscopic scale dynamics of traffic flow, with varying degrees of success. The Cell Transmission Model (CTM) has been used to model short sections of the roadways (called cells), where the size of a single cell is equal to the distance traveled by light traffic in a single tick (i.e., a single simulation time step) [17]. This representation has been successfully used to predict traffic's evolution over time and space, including transient phenomena such as the building, propagation, and dissipation of queues. While there have been attempts to modify this approach to account for cells of varying sizes (which would allow for selection of an 'appropriate' spatial scale for modeling and prediction), these have been met with considerably less interest than other applications of CTM [18]. Consequently, the cell transmission model does not allow for the selection of an optimal spatial scale to model and predict traffic dynamics.

Other recent works, such as variable multi-scale modeling by Cristiani et al. [19], have attempted to simultaneously describe traffic dynamics from both a discrete-space (microscopic) and a continuous-space (macroscopic) perspective through the coupling of models at different scales. However, while a great idea, this approach does not provide a direct connection from microscopic dynamics to macroscopic dynamics.

In this following sections, we will introduce a first principles-based approach which is inspired by the field of statistical physics to create a traffic flow model that originates at the scale of individual vehicles, and through repeated coarse-graining (i.e., aggregation) of the underlying discrete space, can describe traffic dynamics at arbitrarily large spatial scales [20]. The following subsections will also briefly introduce the coarse-graining mechanism that is central to the application of renormalization group (RG) theory in the context of traffic flow.

3 Statistical Mechanics-Inspired Model of Traffic Flow Dynamics

Another branch of microscopic traffic flow modeling—cellular automata (CA) models—can trace back its origins to the works of John von Neumann's comprehensive 'theory of automata', to model the evolution of a self-producing machine [21]. CA models were first utilized to model traffic flow by Nagel and Schreckenberg [22].

Statistical mechanics (or statistical physics) represents an extensively studied field that deals with statistical descriptions of large-scale systems with emphasis on studying their phase transitions and behaviors near criticality [23]. This makes its underlying concepts uniquely suited for application to traffic flow dynamics, describing phase transitions from free flow to congested flow at the critical density [24]. The presented statistical mechanics-based model shares some features with the cellular automata models, as it relies on a discrete-space, discrete-time representation of traffic. The key difference between these approaches is that while CA models rely on a set of simple rules with arbitrarily-chosen transition probabilities to depict vehicular motion, the statistical mechanics-inspired model relies on physics-based principles, such as energy, to relate these probabilities to physical parameters [7].

3.1 Modified Ising Model for Traffic Flow

In this paper, the CA-based traffic system is modeled as a single-lane (one-dimensional), closed ring-road with N discrete sites, where each site is the length of an individual vehicle (approximately 5 m). We begin with a plausible finest resolution of discrete space, i.e., a single stationary vehicle occupies one site, so that subsequent coarse-graining or aggregation of sites allows us to model traffic flow at any larger (and appropriate) spatial scale that are integer multiples of the length of a single vehicle. The traffic system contains $\lceil \rho N \rceil$ vehicles, where $\rho \in [0, 1]$ represents the vehicular traffic density in the system.

In accordance with the traditional Ising model [23], each site can be in state $s_i = -1$, representing that it is currently vacant, or $s_i = +1$, representing that a vehicle currently occupies the site. The Ising model was first proposed as a means of modeling magnetization in a ferromagnet and studying its phase transitions. The original model consisted of a one-dimensional array of spins, in which each cell was assigned a spin direction, either 'up' (+1) or 'down' (−1). Hence, the Hamiltonian $\mathcal{H}$ of Ising model, which can be understood as the total energy of system, in zero field with nearest neighbor interactions can be written as:

$$\mathcal{H} = -K \sum_{|i-j|\leq 1} S_i S_j \tag{1}$$

where K denotes the interaction coefficient between spins and S_i, S_j represent the spin states of two nearest neighboring sites.

In the Ising model, the spin state at a particular site flips from one state to the opposite, without influencing the remainder of the system. This type of evolution is known as 'spin flip dynamics'. However, it fails to conserve the total number of vehicles on a closed ring-road, i.e., no vehicles are entering or leaving the system [7]. Consequently, the forward motion of vehicles can be modeled through what is known as 'spin exchange dynamics' between two adjacent sites [25]. For example, if at time t states of two neighboring sites are given by $\{S_i, S_{i+1}\} = \{+1, -1\}$, then at time $t + 1$ the states will be $\{S_i, S_{i+1}\} = \{-1, +1\}$, i.e., the vehicle has moved forward as a result of exchange dynamics.

In the presented model, the repulsion force a vehicle may experience is a collective result of the local influence applied by preceding vehicles, and the relative strength of the influence is depends on their distances from the following vehicle. These influences dictate the transition probability $w(S_i \rightarrow S_{i+1})$ of the following vehicle advancing from site S_i to the site S_{i+1}, and may be evaluated using the site-specific Hamiltonian as follows:

$$w(S_i \rightarrow S_{i+1}) = f(-H(S_i)) = \exp\left\{ \sum_{j\in\mathcal{N}(i)} K\, T_{ij}\, S_i S_j \right\} \tag{2}$$

where $T_{ij} = \lambda/d_{ij}$ is a weighting parameter that accumulates the influence of all preceding vehicles in the neighborhood by using the reciprocal of their distances d_{ij} as the weights, as well as a spatial renormalization scale (λ), which will be discussed in more detail in the following subsections.

4 Traffic Flow Modeling Across Multiple Spatial Scales

In this section, we will use the description of the modified Ising model of traffic flow to find equivalent partition functions at coarser spatial resolutions. The partition function will then be used to obtain a mapping from microscopic-scale to coarser-scale model parameters of traffic flow dynamics. Roughly speaking, the partition function encapsulates the statistical properties of the system and depends on the number of accessible microstates for a given ensemble. Thus, the partition function is very useful to obtain aggregate measures of the observables of a system, such as free energy Z. Renormalization Group (RG) can be used to coarse-grain the system by 'grouping' like microstates. This process allows a transformation from fine-grained 'site lattices', to coarse-grained 'block lattices', where both have the same functional form of the free energy because they still represent the same system. Mathematically,

$$Z = \sum e^{-\beta H[\lambda]} \approx \sum e^{-\beta H'[\lambda']} = Z' \tag{3}$$

where Z represents the partition function at the original spatial scale λ, while Z' represents the partition function at the new spatial scale λ'. In the context of traffic flow, Z represents the distribution of finest traffic states when $\lambda = 1$, i.e., each site can accommodate exactly one vehicle, whereas the renormalized partition function Z' corresponds to an arbitrary spatial scale, scaled by λ' to represent that each 'block' can now accommodate exactly λ' vehicles (or none at all). Based on the equivalency between partition functions under different spatial scales, the total free energy of system $F(\lambda)$, given by the logarithm of the partition function, should also be invariant under the RG transformation, i.e.:

$$F'(\lambda') = -\ln\mathbb{Z}' = -\ln\mathbb{Z} = F(\lambda) \tag{4}$$

Therefore, the RG transformation effectively decimates spin degrees of freedom while preserving the total free energy. After the RG transformation, the number of sites in the system decreases in each application step from $N \rightarrow N'$, where $N' = N/\lambda^r$ in r-dimensional space. In the context of the traffic system, which is a 1-dimensional ring-road with N vehicles, application of the RG approach reduces the number of 'vehicles'. The coarse-grained traffic system description with reduced number of sites contains less information than the description with finer resolution, but it retains the ability to describe important features of emergent traffic flow dynamics such as backward moving waves and formation of emergent traffic jams. Of course, the fundamental underlying assumption in that the emergent traffic jam phenomena are related to the free energy of traffic system.

4.1 *Renormalization Group Approach Applied to CA-Based Traffic Flow Model*

We now proceed to apply the Renormalization Group approach to the modified Ising model of traffic flow. Since the free energy of modified Ising model is invariant under renormalization, we can evaluate the coarse-grained model parameters (i.e., the interaction coefficients) as functions of the spatial scale L. Figure 1 provides

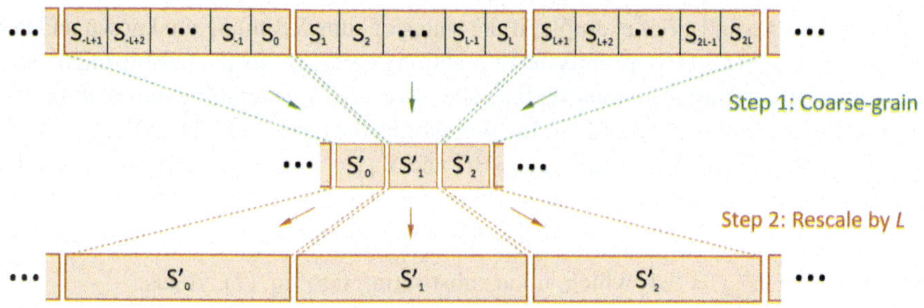

Fig. 1 Schematic showing application of renormalization group theory to a one-dimensional cellular automata-based traffic model. The renormalization can be applied multiple times to obtain coarsening of L sites to a single site, or as a single RG operation as shown in the figure

a schematic illustration of how the interaction coefficients are associated with the spatial scales. By referencing renormalization group theory and as shown in Fig. 1, the group of original sites $\{S_i, S_{i+1}, \ldots, S_{i+L-1}\}$ can be replaced by a single, aggregate representative site S_i'. Specifically, every non-overlapping block of L sites in the original system is transformed into a single site in the process of coarse graining.

Building on Eq. (2), and assuming the absence of any field and restricting interactions to the nearest neighbor, the partition function of the modified Ising model representing traffic flow can be written as follows in groups of L sites:

$$\begin{aligned} Z &= \sum_{\{S\}} \prod_{i} e^{K S_i S_j} \\ &= \sum_{\{S\}} \prod_{i=0,L,2L\ldots,N} e^{K S_i S_{i+1} + \cdots + K S_{i+L-1} S_{i+L}} \end{aligned} \tag{5}$$

Clearly, the full sum is expressed over every group of L interaction sites. Renormalizing the original system, the L site states may be coarse-grained and represented as a single site denoted by S_i'. Then, the new partition function Z' of the coarse-grained traffic system can be written as:

$$Z' = f(K)\, e^{K' S_i' S_{i+1}'} \tag{6}$$

and through the assumption of invariance of the free energy over the fine- and coarse-grained representations of the traffic system, we can equate the partition functions from Eqs. (5) and (6), i.e., $Z = Z'$, to obtain [23]:

$$\begin{aligned} g(L, 1) \times \left\{ e^{K(S_i + S_{i+L})} + e^{-K(S_i + S_{i+L})} \right\} + g(L, -1) \times \left\{ e^{K(S_i - S_{i+L})} + e^{-K(S_i - S_{i+L})} \right\} \\ = f(K)\, e^{K' S_i' S_{i+1}'} \end{aligned} \tag{7}$$

Here, $g(L, \cdot)$ represents the total interaction energy involving all the possible microstate combinations of the remaining intermediate $L - 1$ sites when site states are given at two ends. For example, if the states of sites S_0 and S_L are known and are identical, then $g(L, 1)$ represents the interaction energy due to all possible microstate combinations arising from intermediate sites $\{S_1, \ldots, S_{L-1}\}$, taking into account the equality of the end sites S_0 and S_L. In other words, the terms $g(L, 1)$ and $g(L, -1)$ in Eq. (7) correspond to $S_{i+1} = S_{i+L-1} = \pm 1$ and $S_{i+1} = -S_{i+L-1} = \pm 1$, respectively. At the coarse-grained spatial scales, values for states S_i and S_{i+L} would reference the set of S'_i and S'_{i+1}, then the two specific cases would be:

Case 1: $S'_i = S'_{i+1} = \pm 1$, which, upon substituting into Eq. (7), yields:

$$g(L, 1)e^{2K} + g(L, 1)e^{-2K} + 2g(L, -1) = f(K)e^{K'} \tag{8}$$

Case 2: $S'_i = -S'_{i+1} = \pm 1$, which, upon substituting into Eq. (7), yields:

$$g(L, -1)e^{2K} + g(L, -1)e^{-2K} + 2g(L, 1) = f(K)e^{-K'} \tag{9}$$

From Eqs. (8) and (9) above, the new interaction coefficient for the renormalized system K' and associated function $f(K)$ can be derived as follows:

$$K' = \frac{1}{2} ln \left\{ \frac{g(L, 1)e^{2K} + g(L, 1)e^{-2K} + 2g(L, -1)}{g(L, -1)e^{2K} + g(L, -1)e^{-2K} + 2g(L, 1)} \right\} \tag{10}$$

$$f(K) = \Big\{ (g(L, 1)e^{2K} + g(L, 1)e^{-2K} + 2g(L, -1)) \times (g(L, -1)e^{2K} + g(L, -1)e^{-2K} + 2g(L, 1)) \Big\}^{\frac{1}{2}} \tag{11}$$

4.2 *Generalization of Interaction Energy Calculation to Arbitrary Spatial Scales*

The missing components in the closed-form analytical solution are the functional forms of the interaction energy, i.e., $g(L, \cdot)$, for the intermediate sites. These can be found for blocks of moderate length. Specifically, we evaluate the interaction energy $g(L, \cdot)$ for the remaining $L - 1$ sites recursively when the site states were given at two ends, so $g(L, \cdot)$ is related to $L + 1$ site block. The partition function for L sites could be calculated by adding one more site at either end of the site block. Additional of an extra site might add 'up-spin' 1 or 'down-spin' -1 to the original system.

In order to represent different state cases more easily, $g(L - 1, 1)$ represents the case who has L long site block with the same state at first and last site; conversely, $g(L - 1, -1)$ represents the total energy of a case who has L long site block with the opposite state at first and last site. It could be predicted that an $L + 1$ long block with same state at two ends could be derived from two cases: $g(L - 1, 1)$ added with the same state site or $g(L-1,-1)$ with an opposite state site. Then, following the discussion that led to equation the recursive relation between the interaction energy of L intermediate sites versus $L-1$ intermediate sites may be written as:

$$g(L,1) = g(L-1,1)e^{1} + g(L-1,-1)e^{-1} \tag{12}$$

$$g(L,-1) = g(L-1,1)e^{-1} + g(L-1,-1)e^{1} \tag{13}$$

Here $g(L-1,1)\,e^{1}$ represents the case where the added and the neighboring site share the same state, whereas $g(L-1,-1)\,e^{-1}$ indicates that the states of the added and neighboring site are opposite in sign.

These can now be solved recursively and the resulting expression will depend on the number of sites L being used in the coarse-graining procedure. For example, if $L=3$, then for the 2^2 microstate combinations with first and last site state given(where both sites are '+1' or '−1'), the interaction energy in the intermediate sites would be $e^3 + 3e^{-1}$. This expression also provides us with the terminal condition for the recursive expression, i.e., $g(2,1)$. Thus, the closed-form expression for the interaction energy may be written as:

$$g(L,1) = \begin{cases} a_1 e^{L} + a_3 e^{L-4} + \cdots + a_4 e^{-L-2} + a_2 e^{-L+2} & \text{; if L is odd number} \\ g(L-1,1)e^{1} + g(L-1,-1)e^{-1} & \text{; if L is even number} \end{cases} \tag{14}$$

$$g(L,-1) = \begin{cases} a_1 e^{-L} + a_3 e^{-L+4} + \cdots + a_4 e^{L-6} + a_2 e^{L-2} & \text{; if L is odd number} \\ g(L-1,1)e^{-1} + g(L-1,-1)e^{1} & \text{; if L is even number} \end{cases} \tag{15}$$

where the coefficients a_i follow the Lth line in Pascal's triangle [26]. The expressions in (14) and (15) can be substituted into Eqs. (10) and (11) to obtain the coarse-grained interaction coefficients. These coefficients can be substituted into the coarse-grained modified Ising model to obtain the emergent traffic flow dynamics at arbitrary spatial scales that are integer multiples of an individual vehicle length (typically 5 m). The next section discusses the numerical simulations with these models at different spatial scales.

4.3 Numerical Examples

From the analytical solution in previous sections, we know that parameter K' is a function of renormalized spatial scales. Here we demonstrate the numerical simulation of the traffic dynamics using spatial scales $\lambda' = 3$ and $\lambda' = 5$ to test if the traffic flow after renormalization maintains fidelity. Applying results from last section to the Metropolis algorithm-based Monte Carlo simulations of the statistical mechanics-inspired traffic models [7], we obtained the following results. Figure 2 shows three traffic flow simulations at different spatial scales, including the original

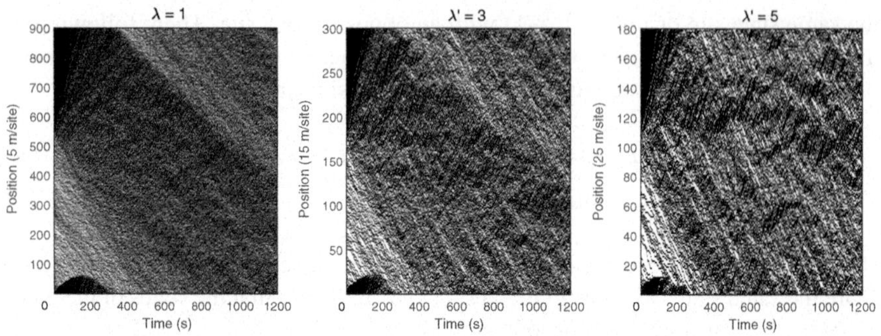

Fig. 2 Statistical mechanics-inspired traffic flow dynamics of original model, renormalized by scaling factors $\lambda' = 3$ and $\lambda' = 5$

model ($\lambda = 1$), and renormalized models with $\lambda' = 3$, and $\lambda' = 5$. The three different scales appear to share similar spatiotemporal evolution patterns in the context of occurrence of congestion and traffic flow rates.

5 Conclusion and Results

Emergent behavior is a scaling dependent property for most complex systems, while utilization of coarse graining technique proved to be an efficient method to simplify the process. Among them, a representative approach called Renormalization Group was applied in this paper during the analysis of traffic dynamics, whose behaviors were modeled and bridged across multiple spatial scales.

During the analysis, presented methodology for determination of the renormalized interaction coefficients in the new Hamiltonian, could help build up the connections of traffic dynamics at arbitrary scales, while preserving the total free energy and the emergent behaviors associated with specific spatial scales. It was also found that distributions of energy values match the Pascal Triangle very closely, which provides an easier tool in the modeling and prediction of dynamics. This work is currently being used to guide the authors' efforts to determine the 'appropriate' or optimal spatial scale at which to observe, model or predict emergent traffic flow dynamics.

Acknowledgements This work was partially supported by the National Science Foundation CMMI DCSD program (award no. 1663652).

References

1. Goldenfeld, N., Martin, O., Oono, Y.: Intermediate asymptotics and renormalization group theory. J. Sci. Comput. **4**(4), 355–372 (1989)
2. Papageorgiou, M., Diakaki, C., Dinopoulou, V., Kotsialos, A., Wang, Y.: Review of road traffic control strategies. Proc. IEEE **91**(12), 2043–2067 (2003)
3. Nagel, K., Paczuski, M.: Emergent traffic jams. Phys. Rev. E **51**(4), 2909 (1995)
4. Williams, B., Durvasula, P., Brown, D.: Urban freeway traffic flow prediction: application of seasonal autoregressive integrated moving average and exponential smoothing models. Transp. Res. Rec.: J. Transp. Res. Board **1644**, 132–141 (1998)
5. Lv, Y., Duan, Y., Kang, W., Li, Z., Wang, F.-Y., et al.: Traffic flow prediction with big data: a deep learning approach. IEEE Trans. Intell. Transp. Syst. **16**(2), 865–873 (2015)
6. Rodrígueza, M., Péreza, D.: First principles model based control (2005)
7. Jerath, K., Ray, A., Brennan, S., Gayah, V.V.: Dynamic prediction of vehicle cluster distribution in mixed traffic: a statistical mechanics-inspired method. IEEE Trans. Intell. Transp. Syst. **16**(5), 2424–2434 (2015)
8. Wilson, K.G.: The renormalization group: critical phenomena and the kondo problem. Rev. Mod. Phys. **47**(4), 773 (1975)
9. Lee, H.K., Lee, H.W., Kim, D.: Macroscopic traffic models from microscopic car-following models. Phys. Rev. E **64**(5), 12 (2001)
10. Nagatani, T.: Jamming transition in the traffic-flow model with two-level crossings. Phys. Rev. E **48**(5) (1993)
11. Israeli, N., Goldenfeld, N.: Coarse-graining of cellular automata, emergence, and the predictability of complex systems. Phys. Rev. E **73** (2005)
12. Helbing, D.: Traffic and related self-driven many-particle systems. Rev. Mod. Phys. **73**, 1067–1141 (1998)
13. Kesting, A., Treiber, M., Helbing, D.: Enhanced intelligent driver model to access the impact of driving strategies on traffic capacity. Philos. Trans. R. Soc. Lond. A: Math., Phys. Eng. Sci. **368**(1928), 4585–4605 (2010)
14. Lighthill, M.J., Whitham, G.B.: On kinematic waves ii, a theory of traffic flow on long crowded roads. Proc. R. Soc. Lond. A **229**(1178), 317–345 (1955)
15. Richards, P.I.: Shock waves on the highway. Oper. Res. **4**(1), 42–51 (1956)
16. Cremer, M., May, A.: An extended traffic flow model for inner urban freeways. In: Control in Transportation Systems 1986, pp. 239–244. Elsevier (1987)
17. Daganzo, C.F.: The cell transmission model: a dynamic representation of highway traffic consistent with the hydrodynamic theory. Transp. Res. Part B: Methodol. **28**(4), 269–287 (1994)
18. Ziliaskopoulos, A., Lee, S.: A cell transmission based assignment-simulation model for integrated freeway/surface street systems. Transp. Res. Rec. **1701**, 12–23 (1997)
19. Cristiani, E., Piccoli, B., Tosin, A.: How can macroscopic models reveal self-organization in traffic flow?. In: 2012 IEEE 51st Annual Conference on Decision and Control (CDC), pp. 6989–6994. IEEE (2012)
20. Israeli, N., Goldenfeld, N.: Coarse-graining of cellular automata, emergence, and the predictability of complex systems. Phys. Rev. E **73**(2), 026203 (2006)
21. Von Neumann, J., Burks, A.W.: Theory of self-reproducing automata. University of Illinois Press Urbana (1996)
22. Nagel, K., Schreckenberg, M.: A cellular automaton model for freeway traffic. J. de Phys. I **2**(12), 2221–2229 (1992)
23. Pathria, R., Beale, P.: Statistical Mechanics. Elsevier Science (1996)
24. Kerner, B.S., Rehborn, H.: Experimental properties of phase transitions in traffic flow. Phys. Rev. Lett. **79**(20), 4030 (1997)
25. Droz, M., Racz, Z., Tartaglia, P.: One-dimensional kinetic ising model with competing spin-flip and spin-exchange dynamics: ordering in the case of long-range exchanges. Phys. Rev. A **41**(12), 6621 (1990)
26. Haeseler, F.V., Peitgen, H.-O., Skordev, G.: Pascal's triangle, dynamical systems and attractors. Ergod. Theory Dyn. Syst. **12**(3), 479–486 (1992)

The Generation of Meaning and Preservation of Identity in Complex Adaptive Systems the LIPHE4 Criteria

Mario Giampietro and Ansel Renner

Abstract This paper combines an interdisciplinary set of concepts and ideas to explore the interface between complexity science and biosemiotics and to describe the processes of generation of meaning and preservation of identity in complex adaptive systems. The concepts and ideas used include: (i) holons (from hierarchy theory); (ii) the state-pressure relation (from non-equilibrium thermodynamics); (iii) the four Aristotelean causes (as used in relational biology); and (iv) upward and downward causation. Further insights from other disciplines, including biosemiotics, cybernetics, codepoiesis, theoretical ecology, energetics, and bioeconomics, are also borrowed to explore the mechanisms underlying the organizational unity of the various processes leading to a resonance between the tangible and notional definitions of identity. An original set of criteria, to be used for the characterization of this organizational unity, is then put forward: *Learning Instances Producing Holarchic Essences: Expected, Established, and Experienced* (LIPHE4). The LIPHE4 criteria help explain how complex adaptive systems can remain the same (preserve their identity) while becoming something else (evolve) and succeed even while implementing imperfect models to guide action.

Keywords Biosemiotics · Complex adaptive systems · Emergence · Relational analysis · Non-equilibrium thermodynamics · LIPHE4

M. Giampietro (✉) · A. Renner
Universitat Autònoma de Barcelona, 08193 Bellaterra, Spain
e-mail: mario.giampietro@uab.cat

A. Renner
e-mail: ansel.renner@uab.cat

M. Giampietro
ICREA, 08010 Barcelona, Spain

D. Braha et al. (eds.), *Unifying Themes in Complex Systems X*, Springer Proceedings in Complexity, https://doi.org/10.1007/978-3-030-67318-5_3

1 Introduction

In this paper, we explore the interface between complexity science and biosemiotics by combining different concepts and ideas from different authors and fields to describe the processes of generation of meaning and preservation of identity in complex adaptive systems. The derivation of a complete formalization of a complex adaptive system is impossible. For that reason, an important challenge of complexity science is indeed to handle the perception and characterization of the unique features of such systems in semantic terms. Section 2 introduces four key conceptual building blocks that prove essential for the perception and characterization of semantics. These concepts are: (i) holons (epistemic units of analysis used in hierarchy theory); (ii) the state-pressure relation (the implications of non-equilibrium thermodynamics); (iii) the four Aristotelean causes (a holistic consideration used in relational biology to identify the structural and functional relations inside complex systems); and (iv) the entanglement between downward causation and upward causation.

In Sect. 3, we use our four conceptual blocks to elaborate on Pattee's [1] (p. 24) claim, made in the field of biosemiotics, that "metaphorically, life is matter with meaning". This is done by mobilizing additional insights from theoretical ecology and the relatively new field of codepoiesis [2]. We show that in complex adaptive systems recorded information is used to generate and stabilize the identity of an entangled set of processes able to reproduce and interpret the continuously adapted and expanded meanings of records. This intrinsically impredicative process of checking the quality of information must meet three conditions: (i) agents must have purpose, including at least the purpose of reproducing oneself—the criterion of truth; (ii) action must be guided by beliefs—assumptions about both the usefulness of acquired blueprints and the effectiveness of the semiotic controls associated with recorded information; and (iii) semiotic controls must be contingent—mapping, to be able to learn, onto an expanding option space [3]. The combination of all these elements permits a characterization of the organizational unity of living complex adaptive systems, a characterization that underlies the resonance between the simultaneously tangible (realized) and intangible (notional) self-entailing identity. We propose a set of irreducible criteria that complex adaptive systems must meet to achieve this organizational unity. It is shown that these criteria also help explain the phenomenon, described by Taleb [4], of "optionality": systems capable of resonating between a notional and tangible definition of identity perform particularly well in their evolutionary trajectory because of their ability to seek new opportunities and limit damage from failures.

Section 4 concludes and summarizes the relevance of our findings for further research.

2 Conceptual Blocks

2.1 *Holons and Holarchies*

Concept. A holon is an epistemic unit of analysis that humans use, often unknowingly, to perceive entities and systems in the external world. The concept, first introduced by Koestler [5, 6], provides insight into the problems experienced when modeling complex adaptive systems [7]. A holon is simultaneously a whole made up of smaller parts (e.g. a human being made up of organs, tissues, and so forth) and a part of some greater whole (e.g. an individual human being as part of a family, a community, and so forth). This dual nature of holons explains the systemic ambivalence found in the perception of the elements of a complex adaptive system and the ensuing epistemological predicament inherent in their scientific representation [8] (pp. 8–16). Simon [9] casts this profound issue in terms of the forced handling of mapping organized structures onto a relational function.

The fact that the descriptive domain useful for the formal representation of structural types is non-equivalent and non-reducible to the one useful for representing functional types poses epistemological challenges to system representation [7]. An essential tactic for handling those challenges is the simultaneous use of various levels of abstraction of holons. To this end, the concept of a hierarchy of holons (termed a "holarchy" [5]) has been elaborated in hierarchy theory to deal with the epistemological implications of the co-existence of multiple relevant scales [8, 10]. Holarchies contribute to so-called "emergent properties"—properties that are only detectable at a high level.

Special Feature. One important epistemological challenge resides in the simple observation that the combination of a structural type (e.g. a given adult human being) and a functional type (e.g. a person that knows how to pilot an airplane) is still just a type. Types are notional entities and cannot express agency—agency can only be realized by an instance (e.g. a given pilot). This observation drives the need to consider all combinations of structure/function and type/instance (see the legend of Fig. 1).

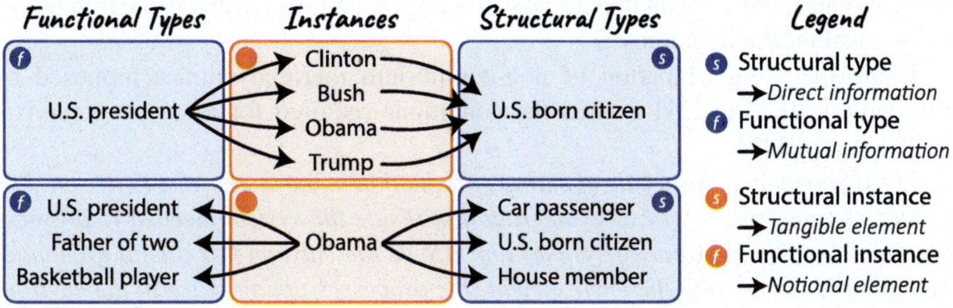

Fig. 1 Examples of many-to-one and one-to-many mappings of the meaning of labels used to identify a holon. Various other semantic combinations are possible

As illustrated in Fig. 1, the validity of a formal representation of a holon depends on the salience of the choice of encoding and can only be checked in relation to its purpose. Formal encodings may refer to a notional identity (describing type) or to a tangible identity (describing an observed realization, or instance). As illustrated in Fig. 1, even when referring to an individuated holon, the two encodings bifurcate. Patching's [11] concept of a notional system, i.e. one which does not exist in any tangible form, becomes relevant in relation to the need of characterizing a set of expected functions required to achieve some purpose. The constituent component concept [12] integrates the concepts of notional and tangible identities in elements capable of generating an emergent property of a whole. As discussed below, the constituent component concept entails: (i) an analysis of instances of structural types, useful for the bottom-up characterization of the material elements of constituent components (realized propensities, see Sect. 1.4); and (ii) a relational analysis of notional entities in relation to the achievement of purposes, i.e. a cybernetic analysis referring to the functions expressed by functional types. The analysis of functional types is useful when considering a set of top-down validated interactions observed by realizations (validated probabilities, see Sect. 1.4). It reflects the existence of an established set of functional relations over constituent components that, when combined, generate an emergent property of a self-organizing system. The interaction of parts generates the meaning and identity of the whole. On the relevant strand of hierarchy theory, see [10, 13, 14].

2.2 The State-Pressure Relation in Dissipative Systems

Concept. The nature of dissipative (metabolic) systems—a class of open systems studied in non-equilibrium thermodynamics which includes human society [15–19]—implies an inevitable existential predicament. On the one hand, dissipative systems express an identity that distinguishes them from their context. On the other hand, the expression of that identity also asserts them as dependent on the characteristics of their context. This existential predicament was notably flagged by Schrödinger [20] in his seminal book "*What is life?*", in which he also proposed the concept of negative entropy to explain it. N.B. Negative entropy is not possible in the framework of classical thermodynamics.

Using the iconic equation of non-equilibrium thermodynamics proposed by Prigogine [19], Kay [21] framed the conditions required for a system's survival as follows (p. 348):

Let dS_i be the internal rate of entropy production in the system due to irreversible processes (the process of exergy degradation inside the system, needed to preserve the identity of the dissipative system) and dS_ebe the entropy flux due to exchanges between the system and the environment (the processes taking place in the environment, needed to generate the required entropy fluxes). The total entropy change of the system is given by the relation:

$$dS = dS_e + dS_i \tag{1}$$

The second law states that $dS_i \geq 0$. *However, if the entropy flux entering the open system is sufficiently low, then* $dS_e \leq 0$*allowing that* $|dS_e| > |dS_i|$*which implies that* $dS < 0$. *If this is the case, then the system will be driven away from equilibrium. It is also possible for the system to eventually reach a steady state in which* ($dS = 0$).

Prigogine's relation cannot be used in a purely syntactic sense since the two terms, $+dS_i$ and $-dS_e$, can only be formalized using non-equivalent and non-commensurable descriptive domains [22]. They cannot be described in a single equation using a common metric and must be tested through learning by doing. The identity of a dissipative structure associated with the expression of a given metabolic pattern ($+dS_i$) is a belief about the existence of an admissible environment—the availability of processes in the environment guaranteeing the existence of favorable boundary conditions ($-dS_e$). The validity about the belief that a dissipative system $[(+dS_i) \rightleftarrows (-dS_e)]$ is possible can only be checked through the establishment of an instance of it. The forced relation between $+dS_i$ and $-dS_e$(a state-pressure relation) associated with a dissipative system entails that its identity can be interpreted as a "code" between the characteristics of the processes taking place inside the system (e.g. the metabolism of a given dissipative structure, such as a herbivore, associated with the expression of a given $+dS_i$) and those taking place outside in an admissible environment (e.g. the availability of edible herbaceous biomass, determining $-dS_e$). This ability of encoding state-pressure relations becomes most relevant when various metabolic elements constitute each other's environment, such as in an interconnected network of the network niches and organized structures shown in Fig. 2. In the case

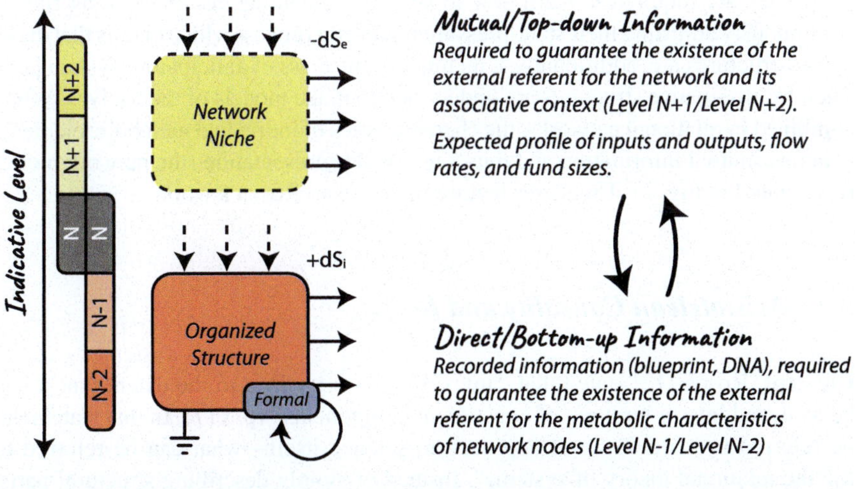

Fig. 2 The two types of information found in relational metabolic networks: (i) direct information (structural elements of a node); and (ii) mutual information (network niche of a node). A collection of *organized structures* and *network niches* in an admissible environment entails the availability of a $-dS_e$ compatible with the production of $+dS_i$

of, e.g. an ecosystem, the expected metabolic pattern associated with the identity of a dissipative structure ($+dS_i$), determined by the use of a blueprint in the production of the structural elements of the node and combined with the topological relations of the other nodes of the network, translates into the generation of a "sudoku effect" (an accumulation of mutual information) and determines the expected characteristics of network niches. This mechanism establishes a relation between the information in a blueprint used to produce an organized structure (the characteristics of structural elements) and the mutual information (the characteristics of the functional elements) that describes a network niche.

Special features. Dissipative systems can only stabilize the identity of dissipative structures (discernable identities associated with $+dS_i$) when in the presence of favorable boundary conditions (the presence of $-dS_e$). They are thermodynamically open systems, an aspect which makes it difficult to define their border in an uncontested manner. Although they are different from their inputs, they are still made of what they metabolize and there are several borders to be considered across multiple levels. This special feature of metabolic systems has been studied in detail by Georgescu-Roegen [23], who suggested an analytical distinction between fund elements (the metabolizing agents in metabolic processes, e.g. rabbits), whose size remains constant throughout a given analysis, and flow elements (the metabolized flows, e.g. biomass eaten by rabbits), whose size changes throughout a given analysis. The distinction between funds and flows is essential when characterizing a metabolic pattern in quantitative terms. By looking at the size of fund elements and their metabolic rate (flow rate per unit of fund element), we can study the state-pressure relation. Since the very existence of dissipative structures implies the destruction of the favorable gradients (on which they depend), dissipative structures must also be becoming systems [18]. They are required to learn how to become something else in the long run by continuously adjusting their state-pressure relation, a process which entails that dissipative structures are, additionally, a member of the class of anticipatory system [24]. Thereby, dissipative structures are understood to make models of themselves. When combined in relational networks, the characteristics of individual metabolic elements generate mutual information and thus a notional representation (the network niches represented in Fig. 2). This latter feature brings us to Rosen's work.

2.3 *Aristotelean Causality and Rosen*

Concept. Rosen [12] stated that "any system is organized to the extent that it can be analyzed into or built out of constituent components" (p. 117). In that statement, he referred to his own novel approach to science itself—what can be referred to as "the relational theory of systems". Instead of simply describing structural parts, relational analysis avoids reductionism in that it focuses on the relations over the functions that must be expressed by a given system [12, 24–26]. While the idea of a holon entails a series of couplings over: (i) structural and functional types; and (ii) a realizing instance (or token) and a notional type(s), Rosen's relational analysis goes further and suggests the use of the four Aristotelean causes for the characterization of structure and function in complex adaptive systems by looking at the existence of impredicative relations across levels—functional entailments [12].

Before delving into the special features of this concept, definitions of the four Aristotelean causes—material, formal, efficient, and final—are called for:

1. *Material cause.* What the fund and flow elements in biophysical processes, e.g. occurring in relation to a metabolic system/dissipative structure, are made of. The material cause represents the deep link between the biophysical characteristics of the process of self-organization and the characteristics of the admissible environment.
2. *Formal cause.* The pattern of a structural element. The constancy of the pattern associated with structural types is determined by recorded information operating at the local scale. The reproduction of an organized structure is guaranteed by blueprints (e.g. genetic information contained in DNA) and the expression of expected behaviors by cybernetic controls.
3. *Efficient cause.* Any agent capable of expressing a given function. When observing a specific process, efficient cause refers to the agent determining or initiating that process. Efficient causes are semantically defined and are capable of handling degenerate mappings (many-to-one and one-to-many).
4. *Final cause.* The "for what", i.e. why an observed function is useful for the larger context. In the case of a constituent component, the final cause is clear—they must express the specific function expected by them to reproduce the whole system. Final cause refers to the meaning, or purpose, of a function within a process of self-organization of a system. The usefulness of a final cause can only be checked at a high level of analysis.

Special features. Figure 3 uses Aristotle's four causes to express two manners of defining a holon—bottom-up and top-down.

- *Bottom-up view.* We can express a typological definition of a holon as a formal cause/structural type combined with an efficient cause/functional type. By adding in a material cause (the physical constituents of a fund element capable of expressing agency), we can then generate a tangible realization of a holon capable of expressing agency. However, to become meaningful for the system to which it belongs, this realization must map onto a final cause. Such information can only be obtained by considering a vertical analysis of the relations of a holon with other holons in a holarchy. N.B. The bottom-up view is a *propensity* understanding of a holon (explored further in Sect. 2.4).
- *Top-down view.* Alternatively, we can express a definition of a holon by looking at a coupling of final and efficient causes in relation to an observable emergent property. In the case of this view, the validity of the identified final cause can only be checked by looking at the effect of the function expressed by the realized holon. N.B. The top-down view is a *probability* interpretation of a holon (explored further in Sect. 2.4).

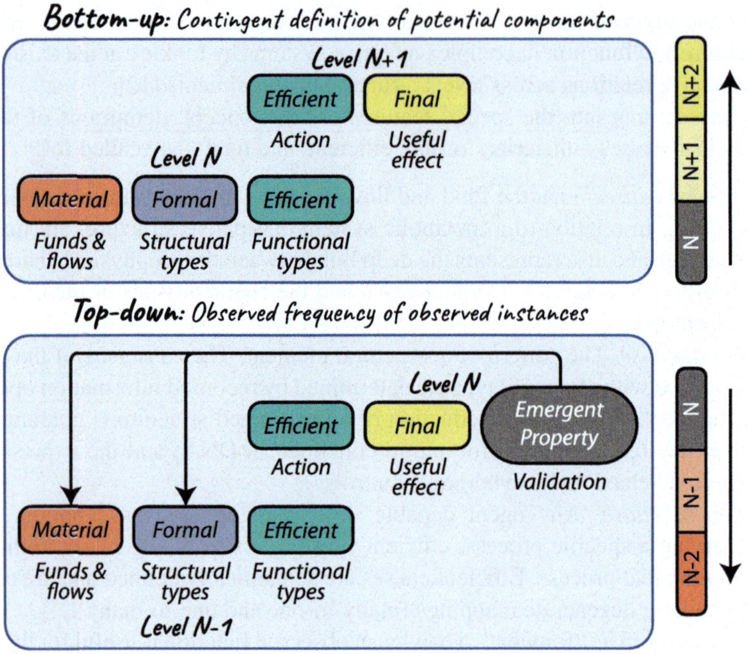

Fig. 3 Two possible ways of defining a holon using the four Aristotelean causes. The bottom-up diagram is associated with propensity, the top-down diagram with probability (see Sect. 2.4)

- *Combining the bottom-up and the top-down views across contiguous levels.* The biophysical process expressed by a coupling of material and formal causes—i.e. a potential efficient cause—becomes an efficient cause if and only if it fulfills a final cause, defined in relation to its ability to generate an emergent property. In other words, the definition of the usefulness of a given coupling of material and formal cause depends on the ability of a holon to express a meaningful function in relation to the events taking place at higher levels in a holarchy. Indeed, in Rosen's use of Aristotelean causality, explanatory priority is given to the final cause over the efficient cause.

The discussion of Fig. 3 points at the special status of the efficient cause, particularly concerning the "usefulness" of the function expressed by a holon. Effectively, efficient cause can be understood as ambiguous—a definition can be arrived to at least three different ways:

- *Notionally from below (bottom-up).* What can be expected when looking at the pattern expressed by lower-level structural elements. The direct information described in Fig. 2, e.g. the effects associated with the metabolic pattern of a rabbit.
- *Notionally from above (top-down).* What can be expected in terms of function from the rest of the system. The mutual information in Fig. 2, e.g. the expected

metabolic characteristics of various classes of herbivores to which the family of *Leporidae* belong to.

- *As observed in a (tangible) realization.* The final result of a "double contingency" over the combination of causes adopted in a specific realization. In this case, the efficient cause refers to a biophysical pattern that is expressed (profile of inputs and outputs) in a situation in which it has been established (from below), expected (from above), and experienced by the system that is using the function as a valid solution to the problem of reproduction of its own identity, e.g. a terrestrial ecosystem in which rabbits are stable elements of the herbivore component.

The discussion around Fig. 3 brings us to the concepts of propensities and probabilities, which are useful to explain the nature of the entanglement between downward and upward causation.

2.4 *Propensities, Probabilities, and the Entanglement Between Downward and Upward Causation*

The concept of propensity has historically been associated with the assessment of single-case probability in which the local structure of an element of a process allows the definition of an expected causal relation in a bottom-up analysis. In this framing, there is no contextualization concerning a pre-existing functional organization. The concept of probability, on the other hand, has often been associated with belief in expected frequencies of the observable outcomes of a given process. Heated discussions about the meaning of "propensities", notably in relation to the work of Peirce and the late Popper, led to a convergence of sorts: (i) Peirce accepted that "the possible" or "would-be's" (notional) should be considered as "real" even if not tangible; and (ii) Popper accepted an interpretation of propensity as potentiality. However, Popper's concept of potentiality was not the same as Aristotle's ontology-based concept, but rather a contingent definition of potentiality determined by a random, but "right", combination of required conditions [27–30]. N.B. Figure 3 informs this discussion and complements Fig. 4—the bottom-up view is an expression of a propensity definition of a holon and the top-down view is an expression of a probability definition of a holon.

The realization of a given holon at a given level, say Level N, can be seen as a "propensity" toward the expression of an unspecified efficient cause (an uncontextualized behavioral pattern). However, this "potential" efficient cause may or may not result in the fulfillment of an actual final cause. Only when the fulfillment of a final cause (as defined by the holarchy) does occur, does the propensity become meaningful in relation to the expression of the emergent property—the reproduction of the holarchy to which the holon belongs. When this occurs, the holon becomes locked-in in the process of self-organization. This event can only be verified by the actual expression of an emergent property at a higher level (Level $N + 1$), which is the verification of the original belief. An equivalence class of instances is realized

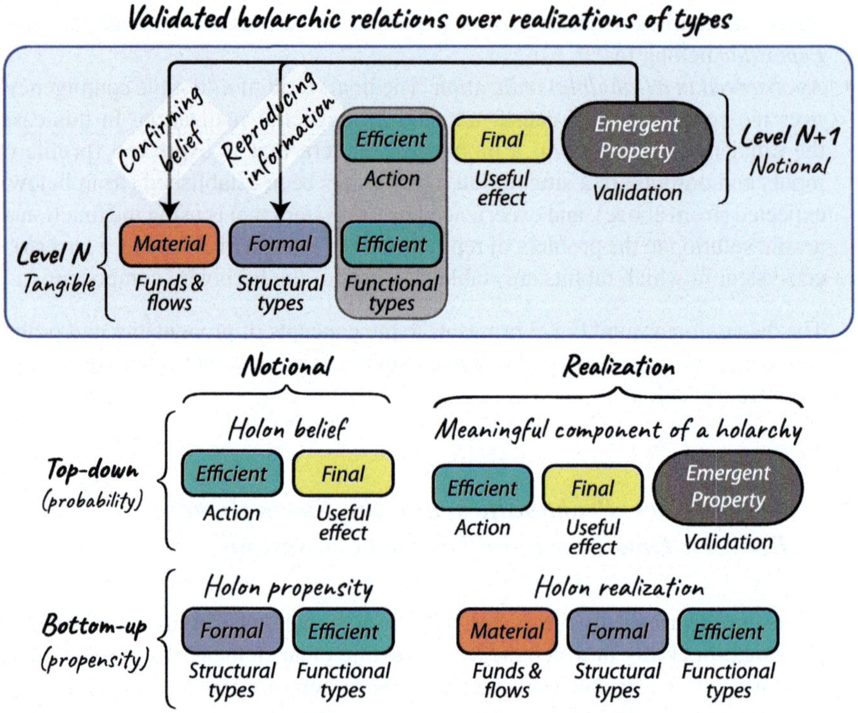

Fig. 4 Propensity and probability in relation to the four Aristotelean causes

in which it is possible to assess frequencies. For the conceptualization of propensity and probability, the following relations are proposed as pertains to the four causes, emergent properties, and Fig. 4:

- *Propensity.* Potentiality to be checked, which is to say assessment of the validity of:

$$[\text{Material} \rightleftarrows \text{Formal}] \rightleftarrows \left[(\text{Efficient})_N \rightarrow (\text{Efficient})_{N+1}\right] \rightarrow (\text{Final})_{N+1}$$

- *Probability.* Belief in frequencies determined by a valid:

$$[\text{Material} \rightleftarrows \text{Formal}] \rightleftarrows \left[(\text{Efficient})_N \rightarrow (\text{Efficient})_{N+1}\right] \\ \rightarrow \left[(\text{Final})_{N+1} \rightarrow \text{Emerg.Prop.}\right]$$

At this point, it is helpful to introduce Campbell's [31] definition of the term "downward causation" in hierarchically organized biological systems: "all processes at the lower levels of a hierarchy are restrained by and act in conformity to the laws of the higher levels" (p. 180). When the set of relations over the four causes has been validated through the success of a process of self-organization we can expect that two contrasting forces are kept in a dynamic equilibrium. The downward causation reflects the expected features that the constituent components must express in their interaction inside the holarchy and according to their path-dependent organizational structure. These expected functions translate into a set of final causes—i.e. a vertical, functional entailment—over the lower-level elements. In other words, it defines from the top-down the characteristics of a metabolic state in a self-referential way. At the same time, the thermodynamic constraints, determined by forced interaction with the environment, limit the possible expression of the downward causation. As for the realization of sequential metabolic pathways at the bottom levels of a holarchy, the option space of possible realizations is constrained by the potentialities of material entailment, i.e. the coupling in a realization of a required formal cause with a required material cause.

3 Meaning and Identity in Complex Adaptive Systems: The LIPHE4 Criteria

3.1 Meaning and Biosemiotics

Biosemiotics is an interdisciplinary field that studies the modalities and role of *signification* in living systems. In fact, communication is an essential characteristic of life. Margalef [32] suggested that ecosystems—more in general, living systems—have the ability to send messages to their progeny specifying how to survive and reproduce. In this narrative, the meaning of the message is what matters in relation to the preservation of the identity of the self-organizing system. In this narrative, Von Uexküll [33] developed a theory of meaning which considered animals as interpreters of their environment. (See Kull's [34] detailed history of biosemiotics.) Semiosis can only be established by simultaneously creating meanings, representamens, and interpretants.

To explain how complex self-reproducing systems learn how to generate and preserve meaning—relevant when considering the idea that "life is matter with meaning"—Pattee proposed the combination of biosemiotics and cybernetics [1]. To generate and preserve meaning, complex self-reproducing systems must regulate the process of their own maintenance, reproduction, and adaptation. In "autopoietic" metabolic systems, the mechanism of entropy generation (the expression of $+dS_i$) is determined by semiotic controls that define constraints on processes that are governable using adjustable rules (in contrast to inexorable laws) [35]. The difference between adjustable rules and inexorable laws is essential as it represents one of the prerequisites for learning from anticipation. Four conditions must be met in order to learn from anticipation. A learning system must have:

- A purpose associated with its own identity. The definition of a purpose makes it possible to compare expected results with achieved results, i.e. feed-back from the external world that can be used to check the validity of information.

- A belief associated with recorded knowledge determining the operation of a system of semiotic control. A belief indicates what "should be done" to achieve a specific goal in a given situation, i.e. the input used for learning how to better guide action.
- A system of semiotic control with contingency reflecting the existence of a difference between "physical laws" and "semiotic controls". For example, the behavior of a flame is determined by the characteristics of an attractor regulated by physical laws. A flame cannot learn how to behave differently—it is a simple dissipative system fully determined by its boundary conditions. On the contrary, the behavior of a given organism is determined by both the characteristics of its structural organization (physical laws) and its semiotic controls, both of which can be adjusted. These controls reflect the information (memory) accumulated by the ecosystem of complex species to which the organism belongs and generated by past interactions with the external world.
- Validation through action, where semantic closure can only be obtained by testing beliefs and—when needed—adjusting structural organization and semiotic control in response to the results obtained. This requires generating actual instances (material observables in physical systems) of the types described in the recorded information (immaterial observables in notional systems). To reduce the negative effects of failures, this also requires the presence of "optionality" (described below).

Changes in semiotic controls can refer to both structural organization and anticipatory models. For example, a simple anticipatory model used by a cockroach (the belief) could be "light equals danger" coupled to the normative narrative (system of control) "if there is light, then run for darkness". This type of information, coupling a "belief" to a "system of control", may be applied to different structural types (e.g. cockroaches, silverfish, and earwigs). In biological systems, it is also possible to have structural adaptation by changing the "formal cause", e.g. the genetic information used to produce structural elements of the system. Information contained in the formal cause can provide the opportunity to develop more effective schemata. Other relevant references concerning what is presented here are [36–38].

3.2 Learning Instances Producing Holarchic Essences: Expected, Established, and Experienced (LIPHE4)

The term *cybersemiotics* has been suggested by Brier [39, 40] to indicate a transdisciplinary framework needed to study natural, life, and social sciences. It is a framework capable of identifying how knowledge can emerge and expand in "hierarchically complex living systems" [41] in which anticipation plays a key role [42]. The need of expanding the diversity of structures and functions to boost adaptability entails the essential function of learning by continuously adding new meanings to those already used to sustain the semiotic process. The concept of *codepoiesis* (the continuous creation of new meanings in the semiotic process) indicates the condition, necessary but not sufficient, to guarantee adaptability to the process of evolution [2, 43, 44]. As Barbieri [45] (p. 29) explains:

> Sign and meaning, in other words, cannot be taken apart because they are the two sides of the same coin. Semiotics, therefore, is not just the study of signs; it is the study of signs and meanings together. The result is that a system of signs, i.e., a semiotic system, is always made of at least two distinct worlds: a world of entities that we call signs, and a world of entities that represent their meanings.

We propose here an irreducible set of criteria, under the name of LIPHE4, that complex adaptive systems must match to express both the semiotic and "codepoietic" activity required to express, preserve, and adapt their identity:

- *Learning*. They can expand purposes, beliefs, and contingency in the set of controls/behaviors by expanding the set of meaningful interactions associated with the expression of their identity by adding/adjusting codes through their activity.
- *Instances*. Despite their ability to adapt and learn, they are tangible instances of a special process of self-organization that is path-dependent. The integrated set of meanings used to guide the mechanism of lock-in in the established resonance between a notional definition of identity and a tangible expression of identity, which are both becoming in time, "remains the same" in the process of becoming.
- *Producing*. They must produce realizations of metabolic networks that are established by assigning expected functions to constituent components (downward causation) allowing the expression of the emergent property of the whole (compatible with upward causation), i.e. reproduction and adaptation.
- *Holarchies*. They can preserve a multilevel organization of relations over structural and functional elements realized across scales and resonating between notional and tangible identities while enjoying optionality.
- *E4*. The four "E"s refer to the generation and preservation of *essences* that must be *established*, *expected*, and *experienced* across levels of organization to be capable of maintaining an identity while becoming something else (adapting) in an integrated manner across hierarchical levels—both notionally and tangibly.

For the LIPHE4 criteria, the term *expected* (the ability to share the meaning of records of information between the sender and receiver) in relation to the step of *established* (message by the sender) and the step of *experienced* (by the receiver) can be interpreted as the ability of preserving the identity of the process—the essence determining the organizational unity of the various processes coordinated across levels. That is to say, the set of different senders and receivers across levels can share the messages that must result "meaningful" to the various holons operating across biophysical processes established at different scales. This complex identity remains the same even when tangible and notional parts change.

Figure 5 visualizes our conceptualization of the process of self-organization of complex adaptive systems. Relatable ideas of a process of resonance between a tangible and notional definition of identity have been previously suggested by: (i) Simon [9] in relation to the relation between recipes and processes; (ii) Prigogine [18]

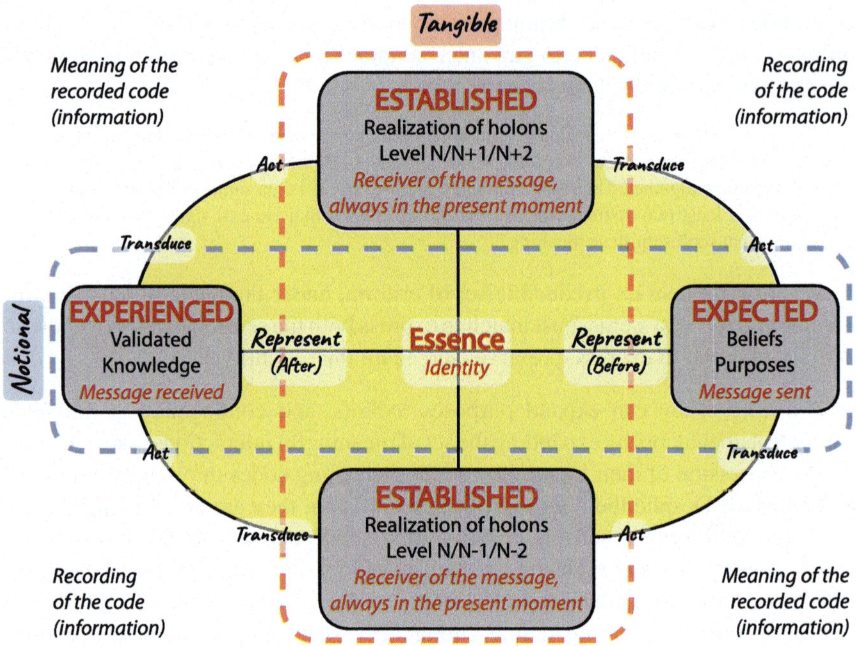

Fig. 5 An overview of the cybersemiotic process across levels of organization

in relation to the relation between DNA and metabolism; (iii) Odum [46] in relation to a process of self-organization determined by informed autocatalytic loops; and (iv) Margalef [32] in relation to biological systems sending messages to their future selves.

Two conceptual points are important here. Firstly, the concept of "time", intended as "before" and "after", refers only to the notional representation of the system. In fact, the meanings of "before" and "after" refer to representations of events taking place simultaneously at different scales—they are not reducible to each other. Simple time is a parameter in a given representation, it should not be confused with the concept of "change" intended as a change in the definition of the interaction of structural and functional elements. This second type of change can be associated with a change in the definition of final, efficient, formal, and material causes within a holarchy. However, this change refers to the learning done by the holarchy in terms of the acquisition of new meanings, new anticipatory models, new structural and functional types, and the diversity of elements of the networks as well as their relations. The change associated with learning is a change in the "essence" determining the organizational unity of the various process coordinated across levels. It is a change that cannot be fully described in formal terms. When considering events experienced by the specific instances of the set of metabolic elements across levels, we can assume a "flat ontology": metabolic patterns, expressed across different scales, can only be considered as occurring simultaneously. But then their representation cannot be integrated in formal terms. What does change over time is the notional representations of beliefs, purposes, anticipatory systems, and semiotic controls that allow a systemic adjustment in the description of the set of meaningful interactions of structural and functional elements, using the four Aristotelean causes. Our only hope to meaningfully perceive and represent these changes is by using the concept of complex time [7].

Secondly, the concept of optionality as suggested by Taleb [4], refers to a feature explaining the major evolutionary advantage enjoyed by systems capable of resonating between a "notional" (type-based) and a "tangible" (token-based, or, realization-based) identity. When learning and adapting at the macroscale, these systems can afford to gamble using "wrong models" while exploring the possible usefulness of new meanings (codes) and new systems of control. In fact, in a semiotic process operating under heavy doses of uncertainty, nobody can predict what must be done in order to encounter with success in the future. The only option is "learning by doing". Concerning this strategy, holarchic organization implies that when an innovation is successful, new useful types can be amplified to a very large scale—a new colony of cyanobacteria born at the spatial scale of a micron can colonize a whole planet and change its atmosphere! On the other hand, when an innovation is a failure, the loss is limited to a realization of a defective instance of a structural type at the local scale, i.e. the information affected by mutation becomes a corrupted copy of something—a structural type. In this case, the effects of the loss are covered by the redundancy in its notional characterization.

4 Conclusions

In this paper, we combined a set of concepts and ideas developed and proposed in remarkably different academic fields to describe the processes of generation of meaning and preservation of identity in complex adaptive systems. Additional insights from a variety of disciplines were used to semantically characterize the organizational unity of the various processes determining a resonance between a tangible and a notional definition of identity. To the best of our knowledge, the specific concepts, ideas, and insights used in this paper have never been combined in semantic terms. Our doing so opens a possible line of practical research.

The use of an entanglement of semantic concepts is essential in practice due to the fact that a complete formal representation of a complex phenomenon is impossible. The proposed set of irreducible criteria, as referred to by the acronym LIPHE4 (Learning Instances Producing Holarchic Essences: Expected, Established, and Experienced), captures the organizational unity of complex adaptive systems—a unity that remains constant across the domains of biology and social systems. The LIPHE4 criteria is able to explain two remarkable features of complex adaptive systems: (i) they are able to remain the same (preserve their identity) while becoming something else (when their structural and functional elements evolve) by preserving and expanding the meanings used in their internal organization; and (ii) because of the phenomenon of "optionality", they manage to succeed even when using wrong models to guide action. This pattern of organizational unity is common between biological and societal processes of self-organization and allows the use of meta-analytical approaches, e.g. the relational analysis of metabolic-repair networks in the two domains. For an example of the study of the factors determining the sustainability of the metabolic pattern of social-ecological systems, readers are referred to another paper in this same volumne of proceedings: *Cyborgization of Modern Social-Economic Systems: Accounting for Changes in Metabolic Identity* by Renner, Louie, and Giampietro.

Acknowledgements A. Renner acknowledges financial support from the Spanish Ministry of Education, Culture and Sport, through the "formación de profesorado universitario" scholarship program (FPU15/03376). Both authors acknowledge financial support by the European Union's Horizon 2020 research and innovation programme under Grant Agreement No. 689669 (MAGIC), the Catalonian Government (AGAUR) under Grant Agreement 2017-SGR-230, and the Spanish Ministry of Science, Innovation and Universities, through the "María de Maeztu" program for Units of Excellence (MDM-2015-0552). This work only reflects the view of the authors; the funding agencies are not responsible for any use that may be made of the information it contains.

References

1. Pattee, H.H.: Artificial life needs a real epistemology. In: Morán, F., Moreno, A., Merelo, J.J., Chacón, P. (eds.) Advances in Artificial Life. ECAL 1995. Lecture Notes in Computer Science (Lecture Notes in Artificial Intelligence), vol 929. pp. 23–38. Springer Verlag, Berlin, Heidelberg (1995). https://doi.org/10.1007/3-540-59496-5_286.
2. Barbieri, M.: Codepoiesis: the deep logic of life. Biosemiotics. **5**, 297–299 (2012). https://doi.org/10.1007/s12304-012-9162-4
3. Pattee, H.H.: Evolving self-reference: matter, symbols, and semantic closure. Commun. Cogn. Intell. J. Intergrated Study Artif. Intell. Cogn. Sci. Appl. Epistemol. **12**, 9–28 (1995)
4. Taleb, N.H.: Antifragile: Things That Gain From Disorder. Penguin Books, London (2012)
5. Koestler, A.: The Ghost in the Machine. The MacMillan Co., New York (1968)
6. Koestler, A.: Janus: A Summing Up. Hutchinson & Co, USA (1978)
7. Giampietro, M., Allen, T.F.H., Mayumi, K.: The epistemological predicament associated with purposive quantitative analysis. Ecol. Complex. **3**, (2006). https://doi.org/10.1016/j.ecocom.2007.02.005.
8. Allen, T.F.H., Starr, T.B.: Hierarchy: Perspectives for Ecological Complexity. University of Chicago Press, Chicago (1982)
9. Simon, H.A.: The architecture of complexity. Proc. Am. Philos. Soc. **106**, 467–482 (1962)
10. Ahl, V., Allen, T.F.H.: Hierarchy Theory: A Vision, Vocabulary, and Epistemology. Columbia University Press, New York (1996)
11. Patching, D.: Practical Soft Systems Analysis. Pitman (1990)
12. Rosen, R.: Life Itself : A Comprehensive Inquiry into the Nature, Origin, and Fabrication of Life. Columbia University Press (1991)
13. Pattee, H.H.: Hierarchy Theory. International Library of Systems Theory and Philosophy (1973)
14. Salthe, S.N.: Evolving Hierarchical Systems. Columbia University, New York (1985)
15. Glansdorff, P., Prigogine, I.: Thermodynamic Theory of Structure, Stability and Fluctuations. Wiley-Interscience (1971)
16. Nicolis, G., Prigogine, I.: Self-Organization in Nonequilibrium Systems : From Dissipative Structures to Order Through Fluctuations. Wiley (1977)

17. Prigogine, I., Stengers, I.: Order Out of Chaos: Man's New Dialogue with Nature. Bantam Books, New York (1981)
18. Prigogine, I.: From Being to Becoming: Time and Complexity in the Physical Sciences. Freeman, W.H (1980)
19. Prigogine, I.: Introduction to Thermodynamics of Irreversible Processes. Interscience Publishers, New York (1961)
20. Schrödinger, E.: What is Life & Mind and Matter. Cambridge University Press, Cambridge (1967)
21. Kay, J.J.: Self-Organization in Living Systems. Ph.D. Thesis, University of Waterloo (1984)
22. Mayumi, K., Giampietro, M.: Entropy in ecological economics. In: Proops, J., Safonov, P. (eds.) Modelling in Ecological Economics, pp. 80–101. Edward Elgar, Cheltenham, UK (2004)
23. Georgescu-Roegen, N.: The Entropy Law and the Economic Process. Hardvard University Press, Cambridge, Massachusetts (1971)
24. Rosen, R.: Anticipatory Systems : Philosophical, Mathematical, and Methodological Foundations (1985). https://doi.org/10.1080/03081079.2012.726322.
25. Rashevsky, N.: Topology and life: In search of general mathematical principles in biology and sociology. Bull. Math. Biophys. **16**, 317–348 (1954). https://doi.org/10.1007/BF02484495
26. Louie, A.H.: More Than Life Itself: A Synthetic Continuation in Relational Biology. Walter de Gruyter (2009)
27. Belnap, N.: Propensities and probabilities. Stud. Hist. Philos. Mod. Phys. **38**, 593–625 (2007). https://doi.org/10.1016/j.shpsb.2006.09.003
28. Miller, R.W.: Propensity: popper or Peirce? Br. J. Philos. Sci. **26**, 123–132 (1975). https://doi.org/10.2307/686511
29. Sfendoni-Mentzou, D.: Popper's propensities: an ontological interpretation of probability. In: Imre Lakatos and Theories of Scientific Change. Boston Studies in the Philosophy of Science, vol 111. pp. 441–455. Springer, Dordrecht (1989). https://doi.org/10.1007/978-94-009-3025-4_32.
30. Marais, K.: We have never been un(der)developed: Translation and the biosemiotic foundation of being in the global south. In: Marais, K., Feinauer, I. (eds.) Translation Studies Beyond the Postcolony, pp. 8–32. Cambridge University Press, Cambridge (2017)
31. Campbell, D.T.: Downward causation in hierarchically organised biological systems. In: Ayala, F.J., Dobzhansky, T. (eds.) Studies in the Philosophy of Biology. pp. 179–186. Palgrave, London (1974). https://doi.org/10.1007/978-1-349-01892-5_11
32. Margalef, R.: Perspectives in Ecological Theory. University of Chicago Press, Chicago (1968)
33. Uexküll, J.V.: A stroll through the worlds of animals and men: A picture book of invisible worlds. Semiotica. **89**, 319–391 (1992). https://doi.org/10.1515/semi.1992.89.4.319
34. Kull, K.: Biosemiotics in the twentieth century: A view from biology. Semiotica. **127**, 385–414 (1999). https://doi.org/10.1515/semi.1999.127.1-4.385
35. Pattee, H.H., Kull, K.: A biosemiotic conversation: Between physics and semiotics. Sign Syst. Stud. **37**, 311–331 (2009). https://doi.org/10.12697/sss.2009.37.1-2.12
36. Pattee, H.H.: The physics of symbols: Bridging the epistemic cut. BioSystems. **60**, 5–21 (2001). https://doi.org/10.1016/S0303-2647(01)00104-6
37. Pattee, H.H.: Epistemic, evolutionary, and physical conditions for biological information. Biosemiotics. **6**, 9–31 (2013). https://doi.org/10.1007/s12304-012-9150-8
38. Pattee, H.H.: Biological systems theory: Descriptive and constructive complementarity. In: Klir, G.J. (ed.) Applied General Systems Research, pp. 511–520. Plenum, New York (1978)
39. Brier, S.: Cybersemiotics: Suggestion for a transdisciplinary framework encompassing natural, life, and social sciences as well as phenomenology and humanities. Int. J. Body, Mind Cult. **1**, 3–53 (2014).
40. Brier, S.: Can cybersemiotics solve the paradox of transdisciplinary knowing? In: Pelkey, S., Walsh, M., and Sbrocchi, L.G. (eds.) Semiotics 2014: The Semiotics of Paradox. pp. 611–642. Toronto (2015). https://doi.org/10.5840/cpsem201448
41. Hall, W.P.: Emergence and growth of knowledge and diversity in hierarchically complex living systems. SSRN Electron. J. (2006). https://doi.org/10.2139/ssrn.1758090

42. Gare, A.: Biosemiosis and causation: Defending biosemiotics through Rosen's theoretical biology or integrating biosemiotics and anticipatory systems theory. Cosm. Hist. J. Nat. Soc. Philos. **15**, 31–90 (2019)
43. Barbieri, M.: Introduction to code biology. Biosemiotics. **7**, 167–179 (2014). https://doi.org/10.1007/s12304-014-9212-1
44. Barbieri, M.: From biosemiotics to code biology. Biol. Theory. **9**, 239–249 (2014). https://doi.org/10.1007/s13752-013-0155-6
45. Barbieri, M.: Three types of semiosis. Biosemiotics. **2**, 19–30 (2009). https://doi.org/10.1007/s12304-008-9038-9
46. Odum, H.T.: Environment, Power, and Society. Wiley Interscience, New York (1971)

On the Formal Incompleteness of Reductive Logic

J. Rowan Scott

Abstract A proof of formal incompleteness is spelled-out in relation to 'bottom-up' reductive Logic and three stacked abstract reductive formal system models composing a three-level hierarchy of complexity. Undecidable reductive propositions appear at the 'upper' boundary of each inter-related formal system model of 'sufficient complexity'. The consequence of formal reductive incompleteness demands necessary meta-consideration in the determination of reductive logical consistency. The further implications of formal reductive incompleteness predict that identified undecidable reductive propositions might be decided in adapted reductive formal system models using modified reductive Logic and slightly different axioms and rules that can handle undecidable dynamics. The adapted Logic and reductive formal system models must accept the inevitably of reductive incompleteness. Reductive incompleteness further predicts the exploration of an adjacent possible abstract domain in which 'bottom-up' formal reductive Logic can be preserved as a 'special case'; while multiple adapted forms of reductive Logic as well as multiple inter-related adapted reductive formal system models, develop a deeper understanding of how to abstractly manage undecidable dynamics and reductive incompleteness. The insights, outcome and implications arising from the exploration of the abstract domain, could instigate further scientific work determining whether modified reductive Logic and reductive formal system models sensitive to undecidable dynamics, provides a closer approximation of natural incompleteness driven, novelty generating evolutionary processes. Any abstract or applied, mathematical, computational or information grounded system model, equivalent to a reductive formal system model, will predictably be shown to manifest undecidable dynamics and incompleteness driven novelty generation. The implications of formal reductive incompleteness can be extended to the study of non-linear systems, Chaos Theory, Cellular Automata, Complex Physical Systems and Complex Adaptive Systems. Subsequently, future scientific and mathematical thought may derive incompleteness driven novelty generating formulations of reductive scientific philosophy, epistemology, methods for theoretical modeling and experimental methodology; each of which will unravel the implicit or explicit intention of the Reductive Scientific Paradigm to compose an

---

J. R. Scott (✉)
Department of Psychiatry, University of Alberta, Edmonton, AB, Canada
e-mail: j.rowan.scott@gmail.com

D. Braha et al. (eds.), *Unifying Themes in Complex Systems X*, Springer Proceedings in Complexity, https://doi.org/10.1007/978-3-030-67318-5_4

integrated, unified, closed, resolved and complete 'bottom-up' Reductive Scientific Narrative describing our Universe. An encompassing incompleteness driven Meta-Reductive Scientific Paradigm is a possibly, wherein novel approaches to previously unresolvable reductive scientific problems may reveal a unique path toward consilience and integration of the Reductive Sciences and the developing Complexity Sciences.

Keywords Gödel · Formal incompleteness · Formal reductive logic · Formal reductive incompleteness · Reductive science · Complexity science

1 Introduction

Slightly less than 400 years ago, René Descartes developed the first precepts of reductive natural science as well as formulating a *dualistic* philosophy of the human body, brain, mind and consciousness. Descartes precipitated a deep rift; excluding mind and consciousness (*res cogitans*) from reductive natural science and alienating mind and consciousness from brain (*res extensa*), rending open the unresolved Cartesian gap [10].

An unresolved puzzle attributable to Descartes weaves together a Gordian knot involving 'bottom-up' reductive scientific Logic ($\mathcal{L}_{\mathcal{R}}$), the residue of Cartesian dualism, and the historical and modern philosophies of science, mathematics, consciousness and mind. The riddle is most easily visualized in relation to two philosophical propositions, both of which employ, strong, rigorous, well-formulated reductive logic in order to arrive at perplexing statements in the form of *reductive epiphenomenalism of consciousness* [32] and *eliminative materialism* [31]. These two reductive propositions, and the implications associated with their acceptance as well-formed reductive arguments, can be *reframed* as *unresolvable self-referencing paradoxes* [3]. When composed in the form of self-referencing paradoxes, both arguments reveal a structure similar to the Liar paradox, which is stated: "This statement is false". *Reductive epiphenomenalism* can be stated: "*This* logically true reductive epiphenomenal *statement* says the intentional consciousness that composed it *is false*, an epiphenomenal illusion that does not exist"; and, *eliminative materialism* can be stated: "*This* logically true eliminative materialist *statement*, says the intentional consciousness that composed it can be fully reduced to brain, declaring intentional consciousness and mind to be *false*, unnecessary and invalid concepts". These two self-referencing, paradoxical, strong, reductive statements expose potential internal *contradiction* and a clear threat to reductive *logical consistency*. 'Bottom-up' reductive Logic can 'explode' into *triviality*, becoming useless and capable of proving true *anything* [34].

Vigorous attempts, to analytically *question* the logical *truth*, to find error in the reductive logic, to logically resolve the epiphenomenal and the eliminative materialist self-referencing paradoxes, have been unsuccessful [35], leaving scientists and philosophers trying to make convoluted sense of a Universe in which willful, mindful, subjective consciousness and the adaptive agency of human beings are erased from existence [11, 12].

The self-referencing paradoxical formulations of the epiphenomenal and eliminative materialist reductive arguments uncover the possibility that rigorous 'bottom-up' reductive Logic employed in *every* natural science, *could* contain unrecognized *undecidable reductive propositions,* suggesting formal reductive Logic is susceptible to *formal reductive incompleteness.* At a minimum this conclusion can *metaphorically* be related to Gödel's two formal incompleteness theorems [34]. More significantly, formal 'bottom-up' reductive Logic may be susceptible to *formal reductive incompleteness.* Formal proof of undecidable reductive propositions and formal reductive incompleteness could lead to a very different understanding of 'bottom-up' reductive Logic within every reductive natural science and it could also lead to a significant transformation of the entire Reductive Scientific Paradigm.

David Hilbert's 23-problem 1900 Foundations of Mathematics Challenge was clearly intended to advance both mathematics *and* natural science [14]. Kurt Gödel completely solved Hilbert's second problem, which asked for proof of the consistency of the axioms of arithmetic forming the foundations of mathematics. Gödel solved it in a completely unanticipated way by demonstrating the incompleteness of the axioms of arithmetic and the necessity for meta-consideration in order to establish the consistency of the axiom system. Gödel effectively demonstrated that it is not possible to *prove* every *true* mathematical statement.

Kurt Gödel [22] provided a meticulous, detailed logical argument in the 1931 proof of his two incompleteness theorems [9]. In the *first* incompleteness theorem, Gödel proves that any formal mathematical system employing first-order logic (first-order predicate calculus) that contains Peano's axioms of arithmetic [13, 17, 24]; as long as the system is consistent, i.e. allows no inconsistent contradictions, it will inevitably contain undecidable propositions or statements revealing the systems fundamental incompleteness. In his *second* incompleteness theorem, Gödel proves an implication that any such formal system cannot demonstrate its own consistency but must resort to external meta-mathematical analysis and considerations in order to prove consistency [27]. Gödel *generalized* his findings, proving that *formal incompleteness* impacts a wide range of abstract formal systems and computational systems of sufficient complexity [9, 19].

Gödel was deeply committed to the existence of an idealized, abstract Platonic mathematical realm. He was therefore antagonistic and often angered by attempts to apply his two abstract proofs of incompleteness to real world problems in applied mathematics, physics or science [1]. Despite Gödel's expressed reluctance many other authors have explored the implications of the incompleteness theorems in close or distantly related areas of mathematical and scientific interest [7, 9, 20, 24].

Among these explorations, curiously, there are no detailed attempts to apply the two incompleteness theorems to the formalized logic of reduction in natural science [36], despite the obvious importance of knowing whether or not the logic of scientific reduction is susceptible to *incompleteness* and whether or not reduction as a form of formal logic is capable of proving its own *consistency.* The answers to these two questions bear directly on fundamental concerns, involving whether or not reductive natural science is capable of ever creating a closed, resolved, 'decided' and complete scientific model that closely approximates the natural evolutionary logic of Nature and evolution [8, 21, 23, 27].

In experimental reductive science, *Nature always gets the last word.* The natural logic of Nature and evolution may have an unrecognized inherent relationship with *natural undecidable dynamics* and *natural evolutionary incompleteness.* If so, the natural logic of evolution may ultimately be proven to be open, unresolvable, 'undecidable', as well as, forever incomplete. If natural evolutionary incompleteness permeates Nature and the natural logic of the evolutionary process, then *undecidable dynamics* and *natural incompleteness* may drive a natural interplay between, *existent* patterns of order and consistency, *ambiguous* moments of novelty generating incompleteness, and emerging patterns of *meta-order* and *meta-consistency.*

If this is the way Nature works, then, in order to more closely approximate the natural logic of the evolutionary system, 'bottom-up' reductive Logic, the reductive sciences and the Reductive Scientific Paradigm will need to be reformulated.

2 Defining a Context for the Question

The *reductive scientific experimental methodology* ($\mathcal{M}_{\mathcal{R}}$) of modern reductive science proceeds in three steps: (1) *Reduction and Analysis*: A scientist begins by selecting a 'higher-order' phenomena, then taking apart the phenomena into its disjoint elements and individually investigating each of these; (2) *Theoretical Formulation*: By accumulating experimental evidence, using imagination, and trusting luck, a scientist then formulates a model describing how the disjoint components relate to each other and how they interact; (3) *Synthesis and Construction*: Using the composed theory defended by experimental evidence, the scientist again evaluates the theoretical, qualitative and quantitative success of the constructed model of the disjoint parts, their relations and interactions. The scientist again compares the synthesis with the experimental, qualitative and quantitative behavior of the original selected higher-level phenomena of interest, in hopes of demonstrating the scientific understanding of the phenomena is complete [33].

In order to demonstrate the presence of *formal reductive incompleteness* an adequate simplification and idealized model of reductive science, epistemology and the reductive scientific process is necessary (i.e. a 'sufficiently complex' context). The *simplified* model must be *sufficiently complex* that contradiction, self-referencing paradox and potentially threatening logical inconsistency are possible. There must be enough complexity that *unresolvable* and *undecidable concepts* and *propositions, undecidable theoretical dynamics* and *undecidable reductive incompleteness* can be spelled-out.

More profoundly, the simplified, abstract, epistemological model must be *sufficiently complex* to model *theoretical undecidable entities* and *processes*, which may reveal experimentally accessible examples of *ontological, natural, evolutionary, unresolvable instantiations.* Thus, *theoretical formal reductive incompleteness* may reveal the presence of *natural undecidable reductive systemic dynamics* and *natural reductive evolutionary incompleteness.*

3 *Mathematization* Defines a Reductive Formal System Model

In order to successfully uncover formal reductive incompleteness, a simplified scientific context and mathematized Reductive Formal System Model (RFSM), must be sufficiently complex to formulate the Liar's paradox or a similar *self-referencing paradoxical statement*, thereby establishing what is necessary to instantiate *negation*, formulate *diagonalization* and finally, support a definition of *universality* and *undecidability*. The RFSM used here to model reductive science therefore must conform to the logical rigor and definition of an abstract mathematical Formal System Model defined by Kurt Gödel in his two 1931 formal incompleteness theorems [30], as well as the course-grained examination of incompleteness presented by John Casti in 1995 [5], and the fine-grained explication spelled-out by Prokopenko et al. in 2019 [27]. It must also be consistent with definitions offered for reduction in recent philosophical explorations of reductive Logic in science [36], reductive Logic in biology [4]. Finally, the RFSM must be consistent with the troubling implications of applied reductive Logic when it underpins work on the brain in the neurosciences [2], wherein the implications of strong reductive Logic paradoxically reduce and erase from existence or conceptually invalidate any concept associated with consciousness and mind [31, 32, 35].

Following the path cleared by Kurt Gödel, Casti and Prokopenko et al., let us begin:

An abstract formal mathematical system F can be defined with a minimum of three components:

$$F = \langle A_F, X_F, R_F \rangle \tag{1}$$

where

1. A_F is a set constructing an *alphabet.* In the abstract, A_F is an ordered and finite set of symbols, such that $A*_F$ can be composed as a set of *words* or *strings* of finite linear sequences of symbols, composed in part from the symbols contained in A_F. The *Kleene Star* or *Operator* (*) is an abstract way to form $A*_F$ from A_F.
2. X_F is a finite *subset* of the set $A*_F$, symbolized, $X_F \subseteq A*_F$, creating a specific set of *axioms* (an established statement or proposition from which further defined structures can be composed).
3. R_F is a finite *subset* of the set $A*_F$, symbolized, $R_F \subseteq A*_F$, symbolically defining possible *relations,* called *rules of inference*; from which the alphabet A_F and axioms X_F can be composed into the words or strings in the set $A*_F$.

A similar abstract reductive formal system model (RFSM) can be defined with a minimum of three components. This *general* pattern can be used to create a three-level hierarchy by stacking the RFSM's, allowing us to capture the abstract similarities shared between the levels. The generalization leaves open conceptual space for later filling in *specific* differences and specific events in the sequence of organization, which might be associated with the self-organization of each 'bottom-up' hierarchically organized level.

Therefore, we concentrate first on a *general* description of *any* level (L_1 through L_3), and *any* Theory (T_1 through T_3). The minimal three, *similar* components of *any* level in a *general* Reductive Formal System Model, are:

$$F_R = \langle A_{F_R}, X_{F_R}, R_{F_R} \rangle \tag{2}$$

where

1. A_{F_R} is an abstract epistemological symbol set, symbolically representing ontological, naturally occurring fundamental entities (known or not yet known) that can be finitely (or infinitely) ordered or organized, representative of a scientifically accessible naturally self-organized *fundamental alphabet*. The set $A^*{}_{F_R}$ is the set of *words* or *strings* that can be composed from A_{F_R} using the abstract Kleene Star or Operator (*). Words or strings in $A^*{}_{F_R}$ are symbolic representations of possible (scientifically theorized or methodologically demonstrated) combinations of naturally occurring entities (defined in A_{F_R}). Natural *words* or *strings* (entities) are self-organized in evolution, while parallel reductive scientific conceptual and theoretical *words* or *strings* are composed statements or propositions, intended to create a complementary system model closely approximating the natural system. Scientific *words* or *strings* composed in $A^*{}_{F_R}$ are also *methodologically* intended to closely conform to and model naturally evolved *words* or *strings,* resulting in $A^*{}_{F_R}$ and A_{F_R} being components of an effective Formal System Model, F_{F_R}, capable of mirroring, reducing, predicting and replicating the natural entities found within Level L_1 and Theory T_1. The *words* or *strings*, whether finite or infinite, depend on how natural entities and their possible relations, modeled as an alphabet in A_{F_R}, can be linked, chained together, or concatenated (*) in $A^*{}_{F_R}$.
2. X_{F_R} is a finite (or infinite) *subset* of $A^*{}_{F_R}$, symbolized, $X_{F_R} \subseteq A^*{}_{F_R}$, creating a specific set of *epistemological abstract axioms* modeling what are presumed to be natural *ontological axiomatic entities* (*abstract axioms* and *natural axiomatic entities* create established or accepted reductive statements or reductive propositions from which further defined reductively organized structures can be composed).
3. R_{F_R} is a finite (or infinite) *subset* of the set $A^*{}_{F_R}$, symbolized, $R_{F_R} \subseteq A^*{}_{F_R}$, *symbolically* defining *epistemological defined rules of inference* and *possible relations* within the abstract model of natural entities. *Rules of inference* define *possible relations*; from which the alphabet A_{F_R} and axioms X_{F_R} can be further composed into *words* and *strings*, concatenated (*) in the set $A^*{}_{F_R}$. The natural

evolving system has an underlying logic and inherent evolutionary order to how it self-organizes *entities*, which reveals how Nature instantiates its *possible ontological relations*. Reductive natural science attempts to mimic and model the natural system, through scientific epistemology, experimental methodology and theoretical description; creating a *model* of *possible relations* from a blend of identified, experimentally confirmed natural laws and rules, reductive scientific rules of inductive and deductive reasoning, and abstract rules of inference synthesized from various formulations of abstract formal logic or developed within multiple mathematical system models.

Two broad categories of scientific reduction (Theoretical Reduction and Explanatory Reduction) and three further reductive scientific conceptual frameworks with their associated inter-connections (Ontological reduction, Methodological reduction and Epistemological reduction), can be defined and *embedded* in the overall concept of a *general* Reductive Formal System Model, $F_{F_R} = \langle A_{F_R}, X_{F_R}, R_{F_R} \rangle$.

3.1 Theoretical Reduction

Theoretical reduction refers to a theoretical reduction of *specific* theories: T_3 to T_2 to T_1. It captures the *intention*, to take a fully developed scientific theory describing one semi-isolated level (T_3 for L_3; or T_2 for L_2; or T_1 for L_1), composed in a hierarchical model of an evolved hierarchy of complexity (T_3 and T_2 and T_1); to reduce the selected specific *theory* to a reductive description using only the concepts derived in the next lower level down the hierarchy (T_3 reduces to T_2; T_2 reduces to T_1; with T_1 treated as the most fundamental accessible layer of the evolved hierarchy).

In Theoretical Reduction, as defined here, each layer of the hierarchical description of the evolved hierarchy of complexity, can be defined by its own fully developed Theory and Reductive Formal System Model: Thus, T_3 and T_2 and T_1 can be described by:

1. $T_3 = F_{F_{R_3}} = \langle A_{F_{R_3}}, X_{F_{R_3}}, R_{F_{R_3}} \rangle$ and
2. $T_2 = F_{F_{R_2}} = \langle A_{F_{R_2}}, X_{F_{R_2}}, R_{F_{R_2}} \rangle$ and
3. $T_1 = F_{F_{R_1}} = \langle A_{F_{R_1}}, X_{F_{R_1}}, R_{F_{R_1}} \rangle$ (3)

T_1 describes the composition and properties of 'fundamental' T_1 entities and interactions between T_1 entities, leading to the fabrication of the T_2 entities. T_2 describes the composition and properties of T_2 entities and the interactions between T_2 entities, leading to the fabrication of T_3 entities. T_3 describes the composition and properties of T_3 entities and the interactions of those entities. The expectation and prediction are that T_3 can be fully reduced to T_2; T_2 can be fully reduced to T_1; with T_1 being treated as the most fundamental accessible layer of the evolved hierarchy. Everything can be deduced from T_1.

3.2 Explanatory Reduction

Explanatory Reduction refers to a restricted reduction providing a reductive explanation for any semi-isolated phenomena x found *in* level L_3 and/or L_2 and/or L_1, located within the three-level theoretical hierarchy describing the semi-isolated three-level evolved hierarchy of complexity. Explanatory Reduction can also refer to any localized reduction producing an explanation for any semi-isolated phenomena x found *in the inter-space* of connection between level L_3 and/or L_2 and/or L_1. Thus, theoretical reduction can itself be reduced to a large number of lesser explanatory reductive explorations.

3.3 Further Subdivision of Theoretical and Explanatory Reduction

Theoretical and Explanatory Reduction can both be further *subdivided*, into *three* inter-related *kinds of reduction.* An implicit or explicit *three-domain* relationship is interwoven into the work of reductive natural science, based on underpinning *reductive premises* associated with *ontological reductive assumptions* (Theory of Reductive Ontological Description), accepted *reductive assumptions about methodological* or *reductive experimental approaches* (Theory of Reductive Observation and Experimental Technique) and *epistemological reductive assumptions* (Theory of Reductive Knowledge). These three *sets* of underpinning *reductive premises* inter-relate by *assuming* that an *epistemology* using *abstract reductive scientific logic*, in some important sense, can *theoretically*, *methodologically* and *experimentally* create a *sufficiently close enough approximation* that it will provide a *good-enough model* of *natural ontological systemic evolutionary logic.* 'Good-enough' therefore refers to an, in principle, assumption of reductive closeness of real-world approximation, *and* to an, in practice, acceptance of the usefulness of real-world reductive experimental methods and application.

The *two domains* of reductive scientific interest (Theoretical Reduction and Explanatory Reduction) and the *three* inter-related *kinds of reduction* (Ontological reduction, Methodological reduction and Epistemological reduction), have a relationship with the *Ontology of Nature and the natural evolutionary system* O_n. This can be symbolized in the following way:

1. *Ontological Reduction* (r_o): involves philosophical *ontological* commitment to a meta-physical assumption about the nature of *ontology* and the nature of *being*: There is an underlying ontological assumption that the reality of the universe and the natural evolving system (O_n) conforms to 'bottom-up' reductive principles and 'bottom-up' reductive logic and therefore science must also conform to similar reductive principles and logic in order to succeed at the task of modeling Nature and evolution.

2. *Methodological Reduction* (r_m): involves philosophical and scientific *methodological* commitment, to a reductive system of methods for investigating and formulating results in every field of scientific work: There is an underlying methodological assumption that 'bottom-up' reductive principles and logic, implemented in experimental scientific methods, will effectively and closely approximate the natural logic of Nature and the natural evolving system (O_n).
3. *Epistemological Reduction* (r_e): involves philosophical scientific epistemological commitment, to a reductive theory of knowledge, particularly in regard to the validity and scope of a reductive approach to theory creation and a reductive approach to the selection of experimental methods. The epistemological commitment rests upon decisions about what justifies a scientific belief or opinion: i.e., there is an underlying epistemological assumption that 'bottom-up' reductive principles, and applications of reductive logic, are justified, and can effectively and closely approximate the natural logic of Nature and the natural evolving system (O_n).

The explicit or implicit *reductive premise* and *hope* of reductive natural science, in examining Nature through the lens of Ontological Reduction (r_o), Methodological Reduction (r_m) and Epistemological Reduction (r_e); states that science has in the past succeeded and predicts that science will succeed further in the future, where $r_o \approxeq r_m \approxeq r_e$ and these are found to be synchronous, commensurate and aligned in a pattern of reductive consilience, closely mimicking the natural logic of Nature and the natural evolving system (O_n). Thus:

$$O_n \cong F_{R(r_o \approxeq r_m \approxeq r_e)} \tag{4}$$

The *reductive premise* suggests, that the simplified 3-Theory hierarchically 'stacked' Reductive Formal System Models, as well as the description of the inter-level 'joints' linking one level to another; should be capable of modeling a simplified natural 3-level emergent hierarchy of complexity. The reductive account should therefore provide a *minimal, generalized, fully complete picture* of the 3-level hierarchy of complexity.

Continuing with the exploration, we are particularly interested in finding out whether, within a single Reductive Formal System Model (RFSM) F_R:

$$F_{(r_o \approxeq r_m \approxeq r_e)} = \langle A_{F_{(r_o \approxeq r_m \approxeq r_e)}}, X_{F_{(r_o \approxeq r_m \approxeq r_e)}}, R_{F_{(r_o \approxeq r_m \approxeq r_e)}} \rangle \tag{5}$$

...is it possible, within the structure of a Reductive Formal System Model (RFSM), to construct an *undecidable reductive proposition* declaring *formal reductive incompleteness*? The structure of the following *proof* follows Gödel's incompleteness theorems [30] and recent work by Prokopenko et al. [27].

3.4 Mathematization and Proof of Formal Reductive Incompleteness

A scientific expression W is derivable or provable in F_R, if and only if there is a finite or infinite sequence of expressions $W_1,\ldots, W_n$ in which the statement $W \equiv W_n$ and every W_i is either a member of the alphabet, an axiom or results from the application of an inference rule applied to earlier expressions in the sequence. Where possible or necessary the statement W must also be fully defined within $F_R = F_{(r_o \approxeq r_m \approxeq r_e)}$. Using a standard notation drawn from logic, $F_R \vdash W$, expresses that W can be formulated, derived and proven in the Reductive Formal System F_R. This means there is a *proof* of W in F_R, therefore W is a statement, a proposition, or a theorem of F_R.

Proof and *truth* in reductive natural science are *already* differentiated by the necessity of spelling out W in $F_R = F_{(r_o \approxeq r_m \approxeq r_e)} \cong O_n$. This differentiation of proof and truth in reductive science is defined by the complicated dynamics natural science has with itself and with Nature. Ontological reductive, methodological reduction and epistemological reduction differentiate *proof* and *truth* even before we specifically seek out an analog of Gödel's formal incompleteness and before the necessary separation of *truth* and *proof* specific to formal reductive Logic. In Science, only Nature and ontology 'know' when a scientific statement W is *scientifically true.* Nature 'knows' its *truth* by naturally instantiating a phenomenon in the realized process of natural evolution. Reductive natural science 'knows' its *truth* only through successive theoretical approximations supported by experimental evidence, thus accumulating *proof* from the sum of sequential approximation, reduction, replication and prediction in theoretical and experimental contexts, supported by scientific *epistemology* (r_e), scientific *methodology* (r_m), and a slowly developing understanding of *ontology* (r_o). *Truth* in scientific understanding of *ontology* is never *closed*; it remains *open* and forever susceptible to *falsification,* whenever theoretical and experimental *proof* of a better approximation appears. *Truth* and *proof* in formal reductive Logic has so far been immune to any demonstration that they too are separated by an inevitable gap defined by *formal reductive incompleteness* and its implications. I can only speculate that the absence of any attempt to address this problem directly is in part based on respect for Gödel's strong reluctance to go there and in part because of the complicated and far reaching and difficult to handle consequence and implications of proving reductive incompleteness and the separation of truth and proof in reductive Logic.

Continuing now with the development of the proof in relation to reductive Logic: Just as in the development of a formal or mathematical system, in the Reductive Formal System Model F_R, in order to label W as a *theorem* it would be necessary to define and employ *in advance* some kind of *external* reasoning or measure that is able to characterize W as either an intermediate statement, or as a target expression and theorem. Such a predictive meta-level method must be able to recognize in advance the importance or salience of W as a target theorem. It is not yet clear how this might be achieved in an abstract mathematical Formal System, F, or in the Reductive Formal System Model F_R. Existing formal system and mathematical developments are not able to formally make this particular predictive decision regarding the salience of target theorems [27]. In modern Logic and Formal Systems Theory and in modern mathematical systems theory, there is, however, an awareness of the proof and implications of both the Halting Problem [6] and Gödel's incompleteness theorems [30], both of which provide an explanation for why theorists are not yet able to *predictively* define any external *meta-level criterion* capable of deciding whether W is or is not a target theorem.

Nevertheless, the limitations set by the Halting Problem and Formal Incompleteness have not stopped theorists from deciding from meta-level consideration, whether or not, a '*completed*' or '*final*' well-formed formal or mathematical statement and proof of W, is or is not, a salient target or theorem. It can be decided in relation to a *complete* or *finished* piece of work, it just can't be decided in advance. This realization is particularly significant in the present argument, in relation to Reductive Natural Science and the development of a Reductive Formal System Model F_R.

Past and present scientific developments have not yet formalized a Reductive Formal System Model sufficiently so that the implications of the Halting problem and Gödel's incompleteness theorems can be seen at work in the dynamics of F_R. The implications of proving the Halting problem and Gödel's incompleteness theorems *are* relevant in a Reductive Formal System Model, Reductive Science and the Reductive Scientific Paradigm; includes the realization that, for formal logical reasons, any scientific theory about a 'part' of Nature or the 'whole' reductive scientific narrative describing our Universe, *cannot* ever be closed, resolved, decided or complete.

Whether or not a statement is *intermediate* or a *target* statement *cannot be decided in advance* because there is always the possibility that whatever is now considered *true,* 'closed', 'complete', 'final', or 'finished' in natural science, and thus *believed* to be a 'target' statement; will be found instead to be an *intermediate statement*. Contradictory *proof* and *external meta-level consideration* may reveal a significant separation between *truth* and *proof* and a significant relationship may be found between *completeness* and *incompleteness,* as well between *consistency* and *inconsistency.* This may lead to a subsequent encompassing meta-reductive re-formulation.

Because this argument is about reductive Logic and Natural Science; *Nature must get the last word* regarding whether or not the natural Logic of the evolutionary process also instantiates a significant separation between naturally instantiated *truth* and naturally instantiated self-referencing *proof*. Nature may instantiate a significant relationship between order and disorder, evolutionary *completeness* and *incompleteness,* as well as evolutionary *consistency* and *inconsistency.*

4 Further *Mathematization*

At this point, in forming the set of 'words' and 'strings' of $A^*{}_{F_R}$, the Reductive Formal System Model F_R does not yet conform to any particular syntactical constraints, beyond the restrictive 'bottom-up' assumption, and the restrictions set by the definition of axioms (X_{F_R}) and the scientific rules of inference, defined in R_{F_R}. In order to move further toward a statement matching the complexity of a modern abstract formal system, it is necessary to expand the definitions and components of the Reductive Formal System Model F_R, so as to define and then concentrate only on *well-formed formulas* (abbreviated as wff's). Well-formed formulas, wff's, are fabricated from the reductive alphabet (A_{F_R}), the reductive axioms (X_{F_R}), the reductive rules of inference (R_{F_R}), the 'words' and 'strings' of $A^*{}_{F_R}$ *and* a *formalized* version of a *reductive grammar* (G_{F_R}).

A formalization of a *reductive grammar* (G_{F_R}) consists of the following components:

$$G_{F_R} = \langle A_{F_R}, N_{F_R}, P_{F_R}, S_{F_R}, C_{F_R}, E_{F_R} \rangle \tag{6}$$

where

1. A_{F_R} is an abstract symbol set of 'terminal' symbols, symbolically representing naturally occurring fundamental entities (known or not yet known) that can be finitely (or infinitely) ordered or organized, scientifically representative of a naturally self-organized but scientifically accessible *fundamental alphabet*;
2. N_{F_R} is an abstract symbol set (which *can* include the three *start symbols* such that $(\langle E_{F_R}, C_{F_R}, S_{F_R}\rangle \in N_{F_R})$, symbolically representing a finite (or infinite) set of 'nonterminal' symbols, that can define and form relationships between A_{F_R} symbols but are *disjoint* from $A^*{}_{F_R}$, i.e., the 'words' and 'strings' formed from A_{F_R};
3. P_{F_R} is a finite (or infinite) set of *production rules,* generally of the form $\left(A_{F_R} \cup N_{F_R}\right)^* N_{F_R} \left(A_{F_R} \cup N_{F_R}\right)^* \rightarrow \left(A_{F_R} \cup N_{F_R}\right)^*$. Each production rule maps from one 'word', or 'string' of symbols to another, with the 'head' string containing an arbitrary number of symbols provided that at least one symbol is 'non-terminal';
4. S_{F_R} is a *start symbol such that* $S_{F_R} \in N_{F_R}$. The start symbol S_{F_R} is *non-terminal* and is *disjoint* from $A^*{}_{F_R}$. In modeling Nature and the process of evolution from the presumed point of origin in the dense high-energy state in the Big Bang, S_{F_R} takes the form $S_{F_R} + C_{F_R} = E_{F_R}$, where E_{F_R} *is the dense high energy state from which the entire evolved universe is created through transformations of energy, into lower, evolving energy states and processes defined by* S_{F_R}*,and associated space–time manifestations, represented in* C_{F_R}. In later, evolved, hierarchically organized and semi-isolated systems selected from within the ongoing process of natural evolution (i.e., the three-level hierarchy being explored in the simplified model of the scientific situation), S_{F_R} is defined by the *initial state* of the *necessary* transformations of E_{F_R}, *creating a necessary* 'terminal' set of members of A_{F_R}, composed in $A^*{}_{F_R}$, along with an *initial* 'non-terminal' *process*, defined by the necessary set of production rules in P_{F_R}. The set created by the *initial state and process* defined by S_{F_R} is '*non-terminal*' setting in motion everything that subsequently happens, however, it is 'disjoint' from any specific subsequent arrangement of state and process in the system.

5. C_{F_R} is a *start symbol such that* $C_{F_R} \in N_{F_R}$. This start symbol C_{F_R} is *non-terminal* (or *unknown* as to its limit) in relation to the vastness of space–time but it can be composed and selected to be *terminal* in the choice of semi-isolated system within an experimental context. C_{F_R} is also *disjoint* from $A^*{}_{F_R}$. In modeling Nature and the process of evolution from the presumed point of origin in the original dense high-energy state in the Big Bang, C_{F_R} shares an evolutionary sequence of relationships $E_{F_R} \rightarrow S_{F_R} + C_{F_R}$, where E_{F_R} is the dense high energy state from which the entire universe evolved and from which the expanding energy-space–time manifold of the universe, and lower-energy-matter entities are created through transformations of high-energy states and processes into lower-energy states and processes, defined by S_{F_R}, *and* associated space–time manifestations and relationships, represented by C_{F_R}. In later, evolved, hierarchically organized, complex, semi-isolated systems *selected* from within the ongoing process of natural evolution (i.e., the three-level hierarchy being explored), C_{F_R} represents both the background bounded or unbounded space–time encompassing the selected semi-isolated system, *or* the selected finite manifold of space–time (infinite or unknown if the whole universe is the experimental context), contained within the bounded domain or container of the semi-isolated system of interest.
6. E_{F_R} is the initial evolutionary *start symbol such that* $E_{F_R} \in N_{F_R}$, from which C_{F_R} and S_{F_R} evolve through transformation of energy in an abstract, in *principle*, reversible relationship, $E_{F_R} \leftrightarrow S_{F_R} + C_{F_R}$. C_{F_R} and S_{F_R} arise from E_{F_R}, and they are the *start symbols* for all further evolution, which is defined in an abstract, in *practice*, non-reversible evolutionary relationship $E_{F_R} \rightarrow S_{F_R} + C_{F_R}$, where energy, space–time and matter transformations, patterns of composition, interactions and relationships, become the further symbolic representation of all subsequent evolution.

The 'complete' 'bottom-up' Reductive Formal System Model F_R can thus be represented:

$$F_R = \langle E_{F_R} \rangle \tag{7}$$

where

$$\langle E_{F_R} \rangle = \langle S_{F_R}, C_{F_R} \rangle \tag{8}$$

such that

$$E_{F_R} \rightarrow \langle S_{F_R}, C_{F_R} \rangle \tag{9}$$

and where

$$\langle E_{F_R}, S_{F_R}, C_{F_R} \rangle = \langle A_{F_R}, N_{F_R}, P_{F_R}, X_{F_R}, R_{F_R} \rangle \tag{10}$$

such that this equation

$$\langle E_{F_R} \rangle \rightarrow \langle S_{F_R}, C_{F_R} \rangle \rightarrow \langle A_{F_R}, N_{F_R}, P_{F_R}, X_{F_R}, R_{F_R} \rangle \tag{11}$$

can be used to formally model the *initial start symbols* and the sequence of evolution of the '*whole*' universe.

For practical reasons, associated with formally modeling *local* and *later sequences* of evolution, involving evolved '*part*' systems within the evolved universe (where the original, initial, energy, space–time, matter, 'whole' *start symbols* are now *implicit*; while the *selected* 'part' *start symbols* for a *selected semi-isolated system of interest*, are *explicitly* now defined as consisting of some evolved initial organization of state and process, with an evolved composition and arrangement of energy, space–time, matter, entities, interactions and relationships), the relationship

$$\langle E_{F_R}, C_{F_R}, S_{F_R} \rangle \in N_{F_R} \tag{12}$$

can be applied. This relationship allows a scientist to visualize a local or later sequence of interest within the continuous thread and flow of natural evolution, in which the fundamental relationships associated with the natural system are preserved in a deep pattern of symmetry and self-similarity.

It is now possible to simplify the reductive formal system to conform to more recent handlings of formal systems, in which it is conceivable to include components of the *grammar* G_{F_R} in the definition of the formal system, such that

$$F_R = \langle E_{F_R}, C_{F_R}, S_{F_R}, A_{F_R}, N_{F_R}, P_{F_R}, X_{F_R}, R_{F_R} \rangle \tag{13}$$

can be simplified to the statement

$$F_R = \langle A_{F_R}, N_{F_R}, P_{F_R}, X_{F_R}, R_{F_R} \rangle \tag{14}$$

such that:

1. A_{F_R} is the alphabet, i.e., is a finite or infinite (known or unknown) ordered set of symbols, referring to natural diachronically and synchronically self-organized entities, derived from the *start symbols* $\langle E_{F_R}, C_{F_R}, S_{F_R} \rangle \in N_{F_R}$, located in each subsequent self-organized level or layer in the sequence of change, evolution, emergence, hierarchical organization and complexity;
2. N_{F_R} is a finite or infinite (known or unknown) set of non-terminal symbols, which include the *start symbols* $\langle E_{F_R}, C_{F_R}, S_{F_R} \rangle \in N_{F_R}$, that are disjoint from the set $A^*{}_{F_R}$. The N_{F_R} *start symbols* are energy E_{F_R}, energy-matter S_{F_R} and energy-space–time C_{F_R}, as well as the compositional '*grammatical markers*' for developing states and processes involving the *start symbols*; along with symbols referring to natural and discovered '*grammatical markers*', which define 'grammatically acceptable' patterns of 'alphabet', 'words', 'strings' and 'collections of strings', sequentially forming more complicated self-organized 'grammatically acceptable' patterns of composition, interaction and relationship, effectively fabricating and becoming A_{F_R} and $A^*{}_{F_R}$;

3. P_{F_R} is a finite or infinite (known or unknown) set of *production rules* of the form $(A_{F_R} \cup N_{F_R})^* N_{F_R} (A_{F_R} \cup N_{F_R})^* \to (A_{F_R} \cup N_{F_R})^*$ referring to discovered naturally emergent laws, rules and constraints, defining admissible patterns of production, fabrication, evolutionary self-organization and hierarchical emergence, involving A_{F_R}, $A^*{}_{F_R}$ and N_{F_R}, including the encompassing interaction with evolved states and processes of the *start symbols* $\langle E_{F_R}, C_{F_R}, S_{F_R} \rangle \in N_{F_R}$;
4. X_{F_R} is a specific set of *axioms*, each of which must be a wff, referring to 'self-evident', 'accepted as given', 'established' entities, statements, propositions or levels of organization in F_R, upon which further abstract modeling of diachronic and synchronic sequences of natural evolution, self-organization, emergence and hierarchically organized complexity can be developed;
5. R_{F_R} is a finite or infinite (known or unknown) set of relations within the wff's called *rules of inference*, referring to the fundamental basis of *scientific logic*; the basis supporting 'acceptable evidence', the basis defining 'acceptable inductive and deductive reasoning' and the basis of how one can arrive at an 'acceptable scientific conclusion'. *As well*, the *rules of inference* include the understanding of any fundamental natural basis for the existence of *natural evolutionary logic*; the basis supporting 'an acceptable understanding that there is evidence of Nature instantiating an evolutionary pattern of reasoning', the basis for defining 'an acceptable understanding of any natural instantiated evolutionary pattern of reasoning' and the basis for an understanding of how Nature and evolution can arrive at 'acceptable natural evolutionary conclusions'.

From these general definitions, the specific case of a three-level hierarchy can then be constructed in more formal detail by duplicating the Reductive Formal System Model (RFSM = F_R) and creating three stacked Reductive Formal System Models, each one referring specifically to a defined level in the 'bottom-up' hierarchy. Thus,

1. Lower Level 1 in the 'bottom-up' metaphor can be represented by:

$$F_{R(Level\,1)} = \langle A_{F_{R,L1}}, N_{F_{R,L1}}, P_{F_{R,L1}}, X_{F_{R,L1}}, R_{F_{R,L1}} \rangle \tag{15}$$

2. Mid-Level 2 in the 'bottom-up' metaphor can be represented by:

$$F_{R(Level\,2)} = \langle A_{F_{R,L2}}, N_{F_{R,L2}}, P_{F_{R,L2}}, X_{F_{R,L2}}, R_{F_{R,L2}} \rangle \tag{16}$$

3. Upper-level 3 in the 'bottom-up' metaphor can be represented by:

$$F_{R(Level\,3)} = \langle A_{F_{R,L3}}, N_{F_{R,L3}}, P_{F_{R,L3}}, X_{F_{R,L3}}, R_{F_{R,L3}} \rangle \tag{17}$$

The 'joints' between hierarchical levels and the transformations of entities, involving the *composition* (C_{L1-3}) and *interactions* (I_{L1-3}) occurring at the boundaries between hierarchical levels in Levels 1–3, can then be composed:

$$F_{R(L1)} + (C_{L1} + I_{L1}) \to F_{R(L2)} + (C_{L2} + I_{L2}) \to F_{R(L3)} + (C_{L3} + I_{L3}) \tag{18}$$

such that;

The specific lower level, Level 1, *components, properties* and *composition* as well as the specific lower-level, Level 1, *interactions*, are sufficient to explain the appearance of the *components, properties* and *composition* of the mid-level, Level 2, as well as, the *full potential* for that levels specific emergent interactions amongst its components, such that;

$$(C_{L1} + I_{L1}) = \langle A_{F_{R,L1}}, N_{F_{R,L1}}, P_{F_{R,L1}}, X_{F_{R,L1}}, R_{F_{R,L1}} \rangle \rightarrow F_{R(L2)} + (C_{L2} + I_{L2}) \tag{19}$$

The mid-level 2, Level 2, components, properties and composition of components, then engages in specific Level 2 *interactions* that build the *component, properties* and *composition* of the third level, Level 3, as well as the *full potential* for any interactive properties of the third level and its components, such that:

$$(C_{L2} + I_{L2}) = \langle A_{F_{R,L2}}, N_{F_{R,L2}}, P_{F_{R,L2}}, X_{F_{R,L2}}, R_{F_{R,L2}} \rangle \rightarrow F_{R(L3)} + (C_{L3} + I_{L3}) \tag{20}$$

and, Level 3 can then be defined;

$$F_{R(L3)} = (C_{L3} + I_{L3}) = \langle A_{F_{R,L3}}, N_{F_{R,L3}}, P_{F_{R,L3}}, X_{F_{R,L3}}, R_{F_{R,L3}} \rangle \tag{21}$$

The construction of three stacked Reductive Formal System Models describing the three-level evolved hierarchy composed of three levels of specific composition and interaction of entities, with each level of interaction subsequently defining the entities in the next level up the hierarchy of composition, can be summarized in this form:

$$F_{R(L1-3)} = (C_{L1-3} + I_{L1-3}) = \langle A_{F_{R,L1-3}}, N_{F_{R,L1-3}}, P_{F_{R,L1-3}}, X_{F_{R,L1-3}}, R_{F_{R,L1-3}} \rangle \tag{22}$$

The *composition* and *interactions* in the third level, Level 3, might still have the potential to build further complexity but we have decided not to explore this potential at this point. By definition, in the three-level hierarchy of complexity, any emergent properties or phenomenon exhibiting complementarity which may appear in or between each level must be *conservatively* explained employing rigorous reductive logic: i.e., in this simplified example, complementarity at an inter-level interface or any emergent properties must be entirely explained by the composition and interactions described within and between the three levels of the hierarchy. However, as we shall see the devil is in the details of emergence and complementarity and so is the formal reductive incompleteness we shall pursue shortly.

As it is now defined, the entire three-level hierarchy can be effectively reduced, with T_3 reducing to T_2 and T_2 reducing to T_1. The Theory T_3 of Level 3 can be fully reduced to the Theory T_2 of Level 2, and the Theory T_2 of Level 2 can be fully reduced to the Theory T_1 of Level 1. The 'whole' hierarchical system can be reductively reduced to the lowest 'fundamental' level.

The *production rules* in P_{F_R} can be used to produce wff's, with the *rules of inference* also required, in order to derive *theorems.* Theorems in F_R are defined as the statements and propositions describing the components of a complete *scientific theory* modeling selected entities, components, interactions and one or many levels of hierarchical organization. In the *general case*, if we decide to consider all 'words' and 'strings' in $A^*_{F_R}$ as wff's, then the grammar G_{F_R} would not constrain the space of possible scientific inference.

In the *special case* of interest here, where the Formal Reductive System must contain *negation* in order to formally model formal reductive incompleteness, the Reductive Formal System Model can be called *consistent* if and only if there is no wff W such that both W and $\neg W$ (*not* W) can both be logically *true* and *proven* in the system.

Further, it is required that there is a *decision procedure* or an *algorithm* (utilizing P_{F_R}), which can be used for *deciding* whether a formula, statement or proposition, *is* or *is not* a well-formed statement (wff). Stated another way, in the modern interpretation of abstract formal systems, there is an expectation that the production rules will be found to be *decidable*: again, an *algorithm* should be available such that, given any arbitrarily chosen 'string' 'x', the *decision procedure* can decide whether 'x' is or is not a wff. As well, *inference rules* must be *decidable*: for every inference rule $R \in R_{F_R}$ there must be an *algorithm,* which can determine whether, given a set of wff $x_1, \ldots .x_n$ and a wff y, the *decision procedure* is able to *decide* whether R can be applied with input $x_1, \ldots .x_n$ in order to fabricate y.

Further, in the modern understanding of abstract formal systems and *recursive abstract formal systems*; *algorithmic, decision procedures,* employed in the systems, make the *axioms decidable* and construct a *set of all provable sentences* (i.e., the set of all theorems) that are *recursively enumerable* or *semi-decidable*: In recursively enumerable systems, if you begin with any arbitrarily selected wff, there is an *algorithm* that can correctly determine whether or not the formula is *provable* within the system, but when applied, and the wff is *not provable* in the formal system, the algorithm will produce a *negative answer* or *no answer at all.* Among the statements where the algorithm returns *no answer at all*, there are found to be formulas that are *undecidable* or *cannot be proven for fear of inconsistency,* the hallmark property of a formula associated with *formal incompleteness.*

Gödel's incompleteness theorems specifically refer to formal systems involving first-order logic or first-order predicate calculus, specifically operating with Peano's axioms of arithmetic. This is presented in such a fashion that the results can be generalized to a very wide class formal systems in which, (1) in order to protect the consistency of the system from exploding into inconsistency, there must be wff's that can neither be proven nor disproven, they must be *undecidable*, and thus the system must be *incomplete*; *and* (2) again for fear of inconsistency, any such formal system will not be capable of demonstrating its own consistency from within the system, *consistency must be decided through meta-consideration.*

This completes the construction of a Reductive Formal System Model (RFSM) and the application of the RFSM to the reductive and conservative model of the three-level hierarchy of complexity investigated by the scientist and his research team. We can now dig deeper into an understanding of the inevitability of undecidable dynamics within this simplified but sufficiently complex formal context.

5 The RFSM in the Matrix of Multiple Scientific Languages

The *simplification* (i.e. *restricting* the natural situation being modeled, to linear, local causation and 'bottom-up' reductive examination and accounting, etc.) and *mathematization* (i.e., *translating* the selected natural phenomena being studied, into a simplified, semi-isolated system contained within an experimental context, described by a *demanding, conservative, restrictive, strong reductive* framework, in the form of an abstract Reductive Formal System Model that *does not constrain* the space of possible scientific inference and interest), brings into focus an *unsolved but now manageable problem.* Consider 'bottom-up' reductive Logic, reductive formal system models and the experimental methodology employed within Reductive Science, the whole Reductive Scientific Paradigm, the reductive scientific narrative and the philosophy of Reductionism: Are all of these complicated conceptual structures impacted by reductively conceived *formal reductive incompleteness*? Further, consider Nature and the natural evolving system, which *includes* the human brain but *may also include* the *reductively excluded*, evolved, emergent, complex complementarity, associated with, the self-reflective, self-referencing mind and consciousness: Are all these *natural* phenomena prone to, *natural instantiations of evolutionary incompleteness*?

To determine the answer to these questions, the argument will follow Kurt Gödel [13] further down the logical path he set out in the proof of his two incompleteness theorems. In attempting this, it is necessary to *translate* the detailed Logic of Gödel's argument, from the austere World of pure, abstract formal logic and mathematics, into the subjectively experienced and complicated context of scientists and applied mathematicians, consciously living in the convoluted and diverse complexity of the Real-World.

In the complicated modern scientific context, reductive scientific theories are presented in the form of written papers, books or academic presentations, using Power Points and other forms of communicative media, which are all helpful in making the scientist's theory understandable to an interested audience of philosophers, scientists, mathematicians and anyone else interested in making the intellectual journey. A specific Scientific Theory $T_{F_R} = T_1$, when placed in the full complexity of modern natural science, is composed using a number of available 'languages' for scientific communication. These languages include but may not be limited to:

1. *Direct observation of a phenomena and its behavior*;
2. Scientific *narrative* or *story telling*;
3. *Scientific 'bottom-up' Reductive Logic or other selected forms of scientific logic*;

4. *Scientific mathematical languages and models*;
5. *Computational models and demonstrations*;
6. *Experimental evidence*;
7. *Digital information translation.*

In this complicated general scientific context, involving a vast community and convoluted matrix of communicative relationships, we wish to examine one specific example of a scientist composing his ideas about an evolving three-level hierarchy of emergent complexity, using 'bottom-up' reductive Logic formulated within a Reductive Formal System Model (RFSM $= F_R$), composed in the form of a reductive scientific Theory, T_I ($T_{F_R} = T_I$). We wish to demonstrate that within this simplified context, the presence of an *undecidable reductive proposition,* declaring *formal reductive incompleteness* is inevitable. Our next stop is *arithmetization.*

6 Arithmetization of Reductive Logic

Mathematization created a Reductive Formal System Model (RFSM $= F_R$) that *may* be *sufficiently complex* to be capable of presenting with *undecidable reductive dynamics. Arithmetization* of the mathematized Formal Reductive System Model (RFSM $= F_R$) should now make it possible to *reveal* the presence of *undecidable reductive dynamics.*

To arithmetize an abstract formal system, Kurt Gödel chose *Peano's arithmetic* and the comprehensive mathematical text, *Principia Mathematica,* written by Bertrand Russell and Alfred North Whitehead [24]. Peano's axioms of arithmetic include:

1. Axiom 1: 0 (zero) is a natural number;
2. Axiom 2: Every natural number has a successor informally stated $n + 1$, which is also a natural number;
3. Axiom 3: No natural number has 0 (zero) as a successor;
4. Axiom 4: Different natural numbers have different successors;
5. Axiom 5: If some *property P* holds for 0 (zero), it also holds for every natural number *n*, and then must hold for every natural number $n + 1$ such that *P* holds for all natural numbers.

Gödel's central insight was to *encode* the wff's of a comprehensive formal system (the comprehensive mathematical text, *Principia Mathematica,* written by Bertrand Russell and Alfred North Whitehead) by using '*Peano's arithmetic*' and the *natural numbers*, thus '*arithmetizing*', or '*Gödel numbering*' the wff's of the entire selected formal system.

We can follow Gödel's lead by arithmetizing and encoding the wff's of a comprehensive formal reductive system, purporting to scientifically describe the whole physical universe. In an imaginary book entitled, *The Complete Reductive Narrative of the Universe,* the author (an imaginary version of myself committed to a closed, resolved, decided and complete model of the Universe) presents a comprehensive scientific history and description of natural evolution involving a 'bottom-up' reductive picture of the entire Universe. Within that imaginary book there is a chapter which presents a narrative describing a scientist composing his ideas about an evolving three-level hierarchy of emergent complexity, using formal 'bottom-up' reductive Logic, composed in a Reductive Formal System Model (RFSM $= F_R$), and spelled-out in a reductive scientific Theory, T_I ($T_{F_R} = T_I$).

In the present context *another* version of myself is closely aligned with Gödel's line of reasoning and is committed to exploring the concept of an open, unresolved, often undecidable and forever incomplete scientific model of the Universe. Like Gödel, we can also use *Peano's arithmetic* to *encode* every well-formed English statement, formal logical argument, mathematical formulation or theoretical statement, in the comprehensive scientific theory and narrative composed in the imaginary book, including the chapter on the scientist and his theory. Within the specific chapter containing our scientist and his theory, we can *encode* every wff of the *mathematized Reductive Formal System Model* (RFSM $= F_R$), which narratively and mathematically is used to describe the simplified reductive picture of the three-level evolved 'bottom-up' hierarchy of complexity.

More formally stated: for every well-formed statement $\mathcal{W}$ in the Theory, T_I ($T_{F_R} = T_I$) and every well-formed wff, $\mathcal{W}$, composed by the mathematized and recursive Reductive Formal System Model, (RFSM $= F_R$); Gödel's *encoding* and *numbering* scheme will produce $\mathcal{G}(\mathcal{W})$, i.e., the encoded '*Gödel number*', which can then be additionally *encoded* as *one* of the *natural numbers*. In such a code, the *name* of the '*Gödel number*' $\mathcal{G}(\mathcal{W})$ of any statement or formula $\mathcal{W}$, is denoted as $\ulcorner \mathcal{W} \urcorner$.

To demonstrate this: first assign a natural number, to each and every primitive symbol s in Peano's arithmetic, to each and every primitive symbol s in the English language used in Theory $T_{F_R} = T_I$ and to each and every primitive symbol s used in the mathematized and recursive Reductive Formal System Model, F_R. Call these the '*symbol numbers' of* 's', e.g., to start with Peano's arithmetic, the symbol '0' is given the natural number 1 and symbol ' $=$ ' is given the natural number 5. Then consider the wff $\mathcal{W}$: '$0 = 0$'. The *Gödel number* for this formula $\mathcal{W}$ is uniquely produced as the corresponding product of powers of consecutive prime numbers (2, 3, 5, ...), as $\mathcal{G}(\text{'}0 = 0\text{'}) = 2^1 \times 3^5 \times 5^1 = 2 \times 243 \times 5 = 2430$. The *name* of the *Gödel number* $\ulcorner \text{'}0 = 0\text{'} \urcorner$ is the *numeral* 2430. Carry on assigning Gödel numbers to every primitive symbol s and formula in the mathematized and recursive Reductive Formal System Model, F_R and then to every primitive symbol s and English statement in Theory T_{F_R}. Knowing the Gödel number of *any* symbol, formula or statement, specifically allows us to *uniquely decode* back into the original English statements, the mathematized formulas and into the elements of Peano's arithmetic. This is possible because of Gödel's unique coding and decoding scheme using the prime-factorization theorem, which allows us to locate and define the unique sequence of prime factors with their associated exponents. Most significantly, Gödel numbers are *computable* and it is also important to note that it is *decidable* whether or not any given number is a Gödel number [28].

Gödel's coding scheme produces *sufficient complexity* to fulfill the necessary criteria supporting *formal incompleteness* in the context of an abstract Formal

System. *Undecidability* will be produced, and *undecidable dynamics* will appear in a formal system description or in a formal system model, (i) if it possible to generate *self-reference*; (ii) if it is possible to differentiate *program-data duality*; (iii) if the formal system, at a *minimum*, has the *potential* to access an *infinite computational medium,* and, (iv) if the formal system has the capacity to implement or instantiate *negation.* In 'sufficiently complex' formal systems, *truth* and *proof* are forever separated by a *gap* kept open by potential or instantiated *undecidability* [27].

'Sufficient complexity' for *formal reductive incompleteness* to appear in the context of 'bottom-up' reductive Logic employed in a Reductive Formal System Model (RFSM $= F_R$) composed in a reductive scientific Theory, T_I ($T_{F_R} = T_I$), will probably involve *undecidability,* (i) generated by *self-reference*; (ii) an available media, which at least in principle, is capable of *infinite computation,* (iii) the possibility of differentiating *program-data duality*; and (iv) the capacity to instantiate *negation.* As in abstract formal systems, there will be an *inevitable logical gulf* inhabited by *undecidable reductive dynamics* and *formal reductive incompleteness* separating reductive *logical truth* from reductive *logical proof*. The separation and gap in *reductive Logic,* precipitated by *formal reductive incompleteness,* must be differentiated from the separation and gap precipitated by the inevitable challenge posed by the distance between what has been or can be *theoretically conceived* and what can be *experimentally demonstrated.* Reductive *theoretical truth* may therefore be supported by reductive Logic or/and by reductive experimental evidence—the *gap* separating *truth* from *proof* in reductive science, therefore, will be *dual.* While incompleteness and undecidable dynamics can influence reductive Logic, creating a separation and gap between truth and proof in Logic, there is also a separation and gap between truth and proof dependent upon the distance between stated *reductive theoretical truth* and the available *experimental evidence* supporting reductive *experimental proof* wherein Science must enter into dialogue with Nature.

'Sufficiently complex' Reductive Scientific Statements, (1) stated in experimentally falsifiable hypotheses; or (2) stated as notations from direct observation; or; (3) stated in the form of scientific narrative and story-telling; or, (4) stated in the form of multiple applied mathematical forms; or; (5) stated in the form of any computational model or digital information representation, fulfilling the definition of a UTM; are *all* scientific representations susceptible to *polarizing debates* wherein there can reside theoretical *contradictions* and potential *self-referencing paradoxes* which *may* herald the presence of *undecidable reductive dynamics* and *formal reductive incompleteness.* It is therefore important to do everything possible to reveal this significant property of reductive Logic and to spell-out its implications. Undecidable reductive dynamics and formal reductive incompleteness *set a limit* on the use of formal reductive Logic but undecidable reductive dynamics and formal reductive incompleteness also instantiate a potential for *incompleteness driven novelty generation,* which may play a significant role in creative 'reductive thought'. The implications of reductive incompleteness include the possibility of creatively reconstructing the

Reductive Scientific Paradigm so that its successor encompasses reductive incompleteness associated with 'bottom-up' reductive thought in a *Meta-Reductive Scientific Paradigm* capable of exploring the role in Nature of *naturally self-organized evolutionary incompleteness.*

Hypothetically, natural evolutionary undecidable dynamics and *natural instantiations of incompleteness in evolution,* may play a significant role in defining complex *boundaries* and *limits* in natural systems and may, as well, play an important role in *natural incompleteness driven novelty generation* in naturally evolving systems.

The next essential step in Gödel's proof involves the Self-Reference Lemma. We must turn this Lemma into an understandable reductive conception.

7 The Self-reference Lemma and Reductive Logic

Gödel first uses the Self-Reference Lemma, in relation to any formal mathematical system, employing first-order logic (first-order predicate calculus) and containing Peano's axioms of arithmetic. The Lemma states:

> Let $\mathcal{Q}(\,\boldsymbol{x})$ be an arbitrary formula of formal system $\mathcal{F}$ with only one free variable. Then there is a sentence (formula without free variables) $\mathcal{W}$ such that: $\mathcal{F} \vdash \mathcal{W} \leftrightarrow \mathcal{Q}\ulcorner\mathcal{W}\urcorner$

The Self-Reference Lemma can be restated in a form allowing its relation to reductive Logic to be stated more clearly:

> Let $\mathcal{Q}(\,\boldsymbol{x})$ be an arbitrary formula of the mathematized and recursive reductive formal system model F_R with only one free variable. Then there is a sentence (formula without free variables) $\mathcal{W}$ such that: $F_R \vdash \mathcal{W} \leftrightarrow \mathcal{Q}\ulcorner\mathcal{W}\urcorner$

Prokopenko notes that "this lemma is sometimes called the Fixed-point lemma or the Diagonalization lemma…The Self-reference lemma establishes that for any formula $\mathcal{Q}(\,\boldsymbol{x})$ that describes a property of a numeral, there exists a sentence $\mathcal{W}$ that is logically equivalent to the sentence $\mathcal{Q}\ulcorner\mathcal{W}\urcorner$. The arithmetical formula $\mathcal{Q}(\,\boldsymbol{x})$ describes a property of its argument, e.g., a numeral $\boldsymbol{x}$, and hence the expression $\mathcal{Q}\ulcorner\mathcal{W}\urcorner$ describes a property of the numeral $\mathcal{W}$. This is the numeral of the Gödel number of the formula $\mathcal{W}$ itself. Since the formula $\mathcal{W}$ is logically equivalent to the formula $\mathcal{Q}\ulcorner\mathcal{W}\urcorner$, one can say that the formula $\mathcal{W}$ is referring to a property of itself (being an argument of the righthand side).

Strictly speaking…the lemma only provides a (provable) material equivalence between $\mathcal{W}$ and $\mathcal{Q}\ulcorner\mathcal{W}\urcorner$, and one should not claim 'any sort of sameness of meaning'" [27].

In both Gödel's *recursive formal system* $\mathcal{F}$, and in our *recursive, mathematized, reductive formal system model,* F_R, it *is* possible to compose arithmetical and mathematical *self-referencing* statements. In the *recursive, mathematized, reductive formal system model,* F_R, and *also* in a related scientific Theory T_{F_R}, a *self-referencing statement* in *Logic* and the specific associated *meaning* of the statement, can be translated into *linguistic* and *narrative* form.

For instance, the theoretical statement of *reductive epiphenomenalism of consciousness* [32] can be stated as $\mathcal{W}$. The statement $\mathcal{W}$ *postulates* or *means* that: "Formal reductive Logic can fully *reduce* Mind to Brain; *reducing* Mind fully to an *epiphenomenon* of Brain and more fundamental physical processes." The sentence $\mathcal{W}$ is logically and materially equivalent to the sentence $\mathcal{Q}\ulcorner\mathcal{W}\urcorner$. This is derived by applying the Self-reference lemma, which establishes that for any formula $\mathcal{Q}(\boldsymbol{x})$ describing a *property of a numeral* $\boldsymbol{x}$, there exists a *sentence* $\mathcal{W}$ that is *logically and materially equivalent* to the sentence $\mathcal{Q}\ulcorner\mathcal{W}\urcorner$ describing a *property of a numeral* $\ulcorner\mathcal{W}\urcorner$. Therefore, a *property* of the *epiphenomenal statement,* $\mathcal{W}$*,* can be captured by the arithmetical formula $\mathcal{Q}\ulcorner\mathcal{W}\urcorner$, which describes a *property* of its argument, e.g., a *numeral* $\ulcorner\mathcal{W}\urcorner$. Hence, the expression $\mathcal{Q}\ulcorner\mathcal{W}\urcorner$ describing a property of the statement $\mathcal{W}$ is the *numeral* of the Gödel number of the formula $\mathcal{W}$ itself. Since the formula $\mathcal{W}$ is logically and materially equivalent to the formula $\mathcal{Q}\ulcorner\mathcal{W}\urcorner$, one can say that the formula $\mathcal{W}$ is referring to a property of itself (being an argument of the righthand side). Thus the *epiphenomenal* statement $\mathcal{W}$: "Formal reductive Logic can fully *reduce* Mind to Brain; *reducing* Mind fully to an *epiphenomena* of Brain and more fundamental physical processes", is *logically and materially equivalent* to the statement $\mathcal{Q}\ulcorner\mathcal{W}\urcorner$: "This statement of *reductive epiphenomenalism* says of itself that *reductive epiphenomenalism* has as one of its properties a *numeral* that is the Gödel number of the *reductive epiphenomenal* statement itself."

As in Gödel's argument and the fine-grained examination of Prokopenko et al., one should *not* claim 'any sort of sameness of meaning'. However, the logical and material equivalence prepares us for what comes next. Self-reference in the reductive Logic of a statement within a Reductive Formal System Model *does not imply* there will *necessarily* be found, contradiction and self-referencing paradox threatening to unravel the usefulness of the underlying Logic. It does however set the stage for this development.

The next step in the proof involves the *provability predicate.*

8 The Provability Predicate and Reductive Logic

The *provability predicate,* $Provable_F(x)$, captures the property of the statement x that it is provable in the formal system F. The *reductive provability predicate,* $Provable_{F_R}(x)$, captures the property of the reductive statement x that it is provable in the Reductive Formal System Model, F_R.

Now let the formula, $Proof_{F_R}\,(y, x)$ strongly represent a binary relationship in which y is the Gödel number of a *proof* of the formula in F_R with the Gödel number x (Prokopenko notes that it is always recursively and algorithmically decidable, whether a given sequence of formulas y constitutes a *proof* of a given sentence x, in conformity with the rules of the *mathematized, recursive, Formal System Model,* and in our case, the *Reductive Formal System Model,* F_R).

The property of being *provable* in F_R can then be stated in the form: $\exists y Provable_{F_R}(y, x)$. This can then be abbreviated as Provable $F_R(x)$.

9 Negating the Provability Predicate in Reductive Logic

Gödel's final step in proving the First Incompleteness Theorem involves *a second* application of the Self-Reference Lemma to the *negated provability predicate.* Translating this step into the rules of the *mathematized, recursive, Reductive Formal System Model,* F_R, produces:

$$F_R \vdash \mathcal{W} \leftrightarrow \neg Provable_{F_R}(\ulcorner \mathcal{W} \urcorner) \tag{23}$$

This statement formally demonstrates that the system F_R can derive a statement: $\mathcal{W}$ is *true* if and only if $\mathcal{W}$ is *not provable* in the system F_R. Gödel, in his two Formal Incompleteness Theorems, and Prokopenko et al. in their discussion of Gödel's work, and now, also in this application in the context of Reductive Science and reductive Logic, we can go on to say: *if* the *mathematized, recursive, Reductive Formal System Model,* F_R, is *consistent*, then it is possible to show that a *true* sentence $\mathcal{W}$ is *neither provable nor disprovable* in F_R, thus proving that the *mathematized, recursive, Reductive Formal System Model,* F_R, must be *incomplete.* There are *true* statements in F_R that are *unprovable* and *undecidable.* It is once again important to state that the sentence $\mathcal{W}$ can be composed as a *well-formed formula,* wff, in the system F_R.

Kurt Gödel wrote less formally in his proof of the two Incompleteness Theorems that what we "have before us is a proposition that says about itself that it is not provable." What we have before *us* is a *well-formed* and *true* proposition that says about itself that it is *not provable* in the formal system F_R.

10 Strong and Weak Representation in F_R

The above demonstrates the presence of *undecidability* in F_R. We can say that a *mathematized, recursive, Reductive Formal System Model,* F_R, is *decidable* if the set of its theorems is *strongly representable* in F_R *itself*, where there is some formula P(x) of F_R such that:

$$F_R \vdash P(\ulcorner \mathcal{W} \urcorner) \text{ whenever } F_R \vdash \mathcal{W} \text{ and } F_R \vdash \neg P(\ulcorner \mathcal{W} \urcorner) \text{ whenever } F_R \nvdash \mathcal{W} \tag{24}$$

For a *weakly representable* set of theorems only the first half of the above statement need apply ($F_R \vdash P(\ulcorner \mathcal{W} \urcorner)$ whenever $F_R \vdash \mathcal{W}$). This defines *semi-decidability* where *negations* are not necessarily 'attributable' to *non-derivable* formulas. However, it is still possible to construct within the system F_R, a Gödel sentence V^P relative to P(x). This can then be stated:

$$F_R \vdash V^P \leftrightarrow \neg P\left(\ulcorner V^P \urcorner\right) \tag{25}$$

A *contradiction* must follow. Therefore, at least for this sentence *strong representability* does not hold and therefore F_R must be *undecidable.*

Following the lead of Prokopenko et al. [50] (Prokopenko, p. 7/23), it is important to clarify that crucially, the Gödel sentence V^P can be constructed as $V(\ulcorner V(x) \urcorner)$ for some wff V(x) with one free variable, and so the main expression clearly states…for the *mathematized, recursive, Reductive Formal System Model,* F_R:

$$F_R \vdash V\ulcorner V(x) \urcorner) \leftrightarrow \neg P(\ulcorner V(\ulcorner V(x) \urcorner) \urcorner) \tag{26}$$

In discussing the related formula associated with a *recursive formal system* Prokopenko et al. concludes [29]:

$$F \vdash V\ulcorner V(x) \urcorner) \leftrightarrow \neg P(\ulcorner V(\ulcorner V(x) \urcorner) \urcorner) \tag{27}$$

This formula would also hold for a *Reductive Formal System Model,* F_R:

$$F_R \vdash V\ulcorner V(x) \urcorner) \leftrightarrow \neg P(\ulcorner V(\ulcorner V(x) \urcorner) \urcorner) \tag{28}$$

Prokopenko et al. state that this evaluation makes it explicit that the Self-Reference Lemma or Diagonalization Lemma *is used twice* in Gödel's proof. The Self-Reference Lemma or Diagonalization Lemma *is also used twice* in the related formula associated with the *mathematized, recursive, Reductive Formal System Model,* F_R. Self-reference and diagonalization are used first *inside* and then *outside* the representative predicate P(x), which is inserted between the two applications of self-reference and diagonalization.

We can apply this proof to the construction of complex reductive theoretical statements. Failure to find any effective logical means of resolving self-referencing paradoxes in reductive theory, suggests it is necessary to consider whether there may be *no* satisfactory *exit* from the paradoxes stated by certain strong reductive arguments—they may be as *true in logic* as reductive science is capable of stating and the contradiction and paradoxical threat to reductive logical consistency, is *real.* This realization indicates specific 'unresolvable, self-referencing, paradoxical, reductive theoretical statements' may indeed be *complex undecidable reductive propositions*; at a *minimum* metaphorically related to Kurt Gödel's two formal incompleteness theorems but at a *maximum* clear evidence of *formal reductive incompleteness.*

For instance, the *reductive epiphenomenalism of consciousness* statement can be reformulated in conformity with Gödel's paradoxical statement: "This true statement cannot be proven". *Reductive epiphenomenalism* can be stated: "*This true* reductive epiphenomenal *statement*, composed logically to be about a conscious mind, erases from existence the same conscious mind, and says of itself, it *cannot be proven*". The self-referencing form of *reductive epiphenomenalism of consciousness* reveals the *undecidable* nature of this scientific and logically well-defended, well-formed reductive argument: "*This statement* of *reductive epiphenomenalism of consciousness*, composed by conscious, mindful, intentional human beings with brains, *says of itself* that a *property* of the reductive epiphenomenal statement *is* that the conscious, mindful, intentional human beings with brains, who composed the strong reductive statement, can be *fully reduced to brain, body* and more fundamental physical processes, leaving intentional mind and consciousness as epiphenomenal, empty, meaningless, unnecessary conceptions that can be erased from the universe as real phenomena: *Reductive epiphenomenalism* has as one of its properties, the *paradoxical contradiction* that *it is and is not* composed by brain, mind and human consciousness, and therefore the reductive epiphenomenal statement *cannot be proven.*"

The reductive statement of eliminative materialism provides a further example. *Eliminative materialism* can be stated: "*This* logically *true* eliminative materialist *statement*, composed by a conscious human mind, reduces the same conscious human mind to a determined and fundamental physical brain, eliminating the existence of the mind and consciousness that made the eliminative materialist statement in the first place, thus in paradoxical contradiction of itself and in response to the threat of logical inconsistency, the eliminative materialist statement must say of itself, it *cannot be proven*".

11 Proposition XI: Consistency and the Second Theorem

In the final section of his paper, Gödel states: "**Proposition XI**: If *c* be a given recursive, consistent class of *formulae*, then the *propositional formula* which states that *c* is consistent is not *c-provable*; in particular, the consistency of P is unprovable in P, it being assumed that P is consistent (if not, of course, every statement is provable)." [18].

The Liar paradox focuses on *truth*: "This statement is *false*", which makes it impossible to avoid the paradox [15]. Gödel's choice of paradoxical sentence focused on *proof*: "This statement is *unprovable*", which made it possible to state in a *well-formed sentence* a *truth* but also to *avoid proof* and the *paradox* and thus *avoid* logical *inconsistency*.

If the formal arithmetic of Peano and the proof of the *first* incompleteness theorem are *consistent*, which means that only *true* statements can be *proven* in the system, then Gödel's chosen well-formed statement: ("This statement is *unprovable*"), *must be true.* If Gödel's statement were *false* then it would be possible to *prove* it to be *false* but that would be *contrary* or *contradictory* to the *consistency* of the system. To be thoroughly paradoxical, Gödel's statement *cannot be proven,* for that would demonstrate exactly the opposite of what the statement asserts, which is that it is *unprovable.*

In the *second* incompleteness theorem, Gödel shows that in an abstract formal system of sufficient complexity, if one attempts to prove the consistency of the formal system from *inside* the formal system itself, then the whole argument determining consistency could be formalized and the statement: "This statement is unprovable", in its formal version, would thus be *proven*, and this leads to an immediate *contradiction*. The attempt to *prove consistency* from *inside* the system demonstrates the *inconsistency of the system*, rather than proving the consistency [16].

In sketching out the details of the *second* incompleteness theorem, Gödel generalized his findings to the axiom system of set theory M and to that of classical mathematics A "and here to it yields the result that there is no consistency proof for M or for A, which could be formalized for M or for A respectively, it being assumed that M or A are consistent. It must be expressly noted that Proposition XI represents no contradiction of the formalistic standpoint of Hilbert. For this standpoint presupposes only that the existence of a consistency proof effected by finite means, and there might conceivably be finite proofs which **cannot** be stated in P or in M or in A" [19]. Hilbert was angry and very upset that, by using the very formalism he had sought, his "grand scheme for securing the foundation of mathematics had been shown to be impossible" [19]. However, when convinced of the importance of Gödel's advance, the mathematics community began the difficult task of trying to understand Gödel's proof and its implications.

11.1 Proposition XI (Reductive Science Version)

If *c* be a given recursive, consistent class of *reductive formulae*, then the *propositional reductive formula* which states that *c* is consistent is not *c-provable*; in particular, the consistency of F_R is unprovable in F_R, it being assumed that F_R is consistent (if not, of course, then every reductive statement is provable)."

We can now generalize Gödel's findings to 'bottom-up' reductive Logic ($\mathcal{L}_{\mathcal{R}}$), Reductive Science (RS), the Reductive Scientific Paradigm (RP) and the Reductive Scientific Narrative (RSN) where Proposition XI (Reductive Science Version) yields the result that there is *no consistency proof* for 'bottom-up' reductive Logic ($\mathcal{L}_{\mathcal{R}}$), a Reductive Formal System Model (F_R), Reductive Science (RS), the Reductive Scientific Paradigm (RP) and the Reductive Scientific Narrative (RSN), which could be formalized for any of these respectively, it being assumed that they are each consistent in their application of reductive Logic.

Among the implications of *formal reductive incompleteness* is the possibility that what is *not* finitely *provable* in $\mathcal{L}_{\mathcal{R}}$, F_R, RS, RP and the RSN, might be *finitely provable* in a related form of reductive Logic, adapted Reductive Formal System Model, altered structure for Reductive Science, modified Reductive Paradigm or transformed Reductive Scientific Narrative, in each case using altered axioms or rules.

I have done my best to prove that Gödel's two incompleteness theorems can indeed be applied to formal 'bottom-up' reductive Logic employed in a Reductive Formal System Model, F_R, and further to prove that this application of incompleteness will follow and flow throughout every scientific context where formal 'bottom-up' reductive Logic is used in the Reductive Scientific Paradigm or the Reductive Scientific Narrative.

It follows from Gödel's comment about "finite proofs which **cannot** be stated in P"; for any particular example of a *formally undecidable reductive statement* declaring *formal reductive incompleteness* found in the Reductive Paradigm, there *must* be, *novel adapted* Reductive Formal System Models as well *as novel adapted* Meta-Reductive Scientific Paradigms in which modified reductive Logic, the elements of an adapted Reductive Formal System and the frame of an altered Meta-Reductive Paradigm, will have been *transformed* in such a fashion that it becomes possible to *prove* the previously undecidable statement, by spelling-out a finite statement in the adapted Logic, System and Paradigm.

When "bottom-up" reductive Logic and the RFSM, $\mathcal{F}_{\mathcal{R}}$, are used to model the 'simplified scientific situation' concerning a participatory scientist and a selected 'semi-isolated system of interest', they can create a three-level, evolved, conservatively emergent hierarchy of complexity; in which *each* of the *three* 'stacked' RFSM's, $\mathcal{F}_{\mathcal{R}_{(1-3)}}$ used to describe the three-level hierarchy, can be shown to contain *undecidable reductive propositions*, $\mathcal{U}_{\mathcal{R}(1-3)}$declaring the *recurrent formal reductive incompleteness* of the three-level hierarchical model ($\mathcal{i}_{\mathcal{R}(1-3)}$).

Consequently, in conformity with Gödel's first incompleteness theorems it becomes possible to create a definition of *sufficient complexity* associated with *formal reductive incompleteness*, in the reductive scientific context. The simplified context of the *three* 'stacked' RFSM's, $\mathcal{F}_{\mathcal{R}_{(1-3)}}$ used to describe the three-level hierarchy are sufficiently complex to be associated with formal incompleteness. More complex models will also be capable of undecidable dynamics.

In conformity with Gödel's second incompleteness theorem, the necessity of *meta-consideration* in determining *reductive logical consistency* can also be explained in the reductive scientific context. The abstract representation and generality of the 'simplified scientific situation' described above, implies a need to adapt *all* RFSM of sufficient complexity using rigorous reductive Logic, such that they incorporate within their core structures *undecidable reductive propositions* ($\mathcal{U}_{\mathcal{R}}$) and *formal reductive incompleteness* ($\mathcal{i}_{\mathcal{R}}$).

Additional implications of formal reductive incompleteness impacting any RFSM of sufficient complexity can then be spelled-out. These implications include, first, the possibility that an *alternative, adapted,* RFSM, can translate an *undecidable reductive proposition* into a *decidable reductive proposition*; and, second, *any* such constructed, alternative RFSM will inevitability reveal its own version of $\mathcal{U}_{\mathcal{R}}$and $\mathcal{i}_{\mathcal{R}}$. Further, formal reductive incompleteness and its implications can be *generalized* and carried into *any* and *every* aspect of 'reductive thought'; *wherever* 'bottom-up' reductive Logic is employed in Reductive Natural Science. Therefore, *meta-consideration* of conceivable *meta-constructions*, which alter the entire Reductive Scientific Paradigm (RP), offer a *novel incompleteness dependent definition*, for narrow intra-domain *adaptation* of a 'part' of natural science, or a global *revolution* involving the 'whole' Reductive Scientific Paradigm (RP).

12 Implications for the Reductive Scientific Paradigm

Among the significant conclusions and implications of this work are the following:

1. The Theory T_1 (F_R) describing the simplified conservatively reductive example of the three-level hierarchical model of Complexity and the complicated scientific situation composed within a mathematized and recursive *Reductive Formal System Model* (F_R) are *sufficiently complex* to demonstrate *undecidable dynamics* and *formal incompleteness*;
2. By analogy with Gödel's Formal Incompleteness Theorems we also see that even in this simple reductive model involving Theory T_1 (F_R), Formal System F_R, and the conservatively reductive three-level hierarchy of complexity; the situation still presents us with a potential for *self-referencing statements and propositions,* in a situation where *program/data duality can be differentiated* in a logical framework *capable of negation,* existing in a mathematical and computational medium capable of *infinite possible universal computation.*
3. The situation can be interpreted as implementing *program/data duality*—such that the abstraction of Theory T_1 (F_R), Formal System F_R, and the three-level hierarchy of complexity, become the abstract *program* while the implementation of specific processes of change defined by particular propositions and statements and wff's within the sequence of progression and evolution in the abstract system become the *data*;
4. *Mathematization* of Theory T_1 (F_R) and the Formal System F_R serve to explicitly reveal a potential for undecidable dynamics and incompleteness. *Mathematization* also reveals a three-part mathematical *equivalence*; wherein *Formal System models with their pattern of undecidability and Formal Incompleteness*, are completely *equivalent* and *translatable* into *Dynamical System Models* with an associated equivalent pattern of Dynamical undecidability and Dynamical Incompleteness. These two models, can then be *equivalently translated* again, into an *Information Theoretical System Model* based on an abstract Universal Turing Machine (UTM). This three-system equivalence has previously been explored in a course-grained examination, by Casti [5], and then very recently, in a fine-grained demonstration by Prokopenko et al. [27]. The three equivalent system models share a similar pattern of undecidable dynamics and incompleteness. This three-system equivalence can be further linked to more complex system models, derived in the study of non-linear systems, Chaos Theory, Cellular Automata, the Science of Complexity and Complex Adaptive Systems.
5. Through *arithmetization*, the fine-grained nature of undecidable dynamics becomes evident. At a minimum, it must be determined whether or not the undecidable dynamics found in equivalent system models is associated with or may closely approximate undecidable evolutionary dynamics in natural evolving systems and hierarchically organized complexity. The presence and exploration of undecidable dynamics in natural settings may demand re-defining phase transition, complementarity, and emergence, where these phenomena are found in association with undecidable dynamics within the evolution of systems and

hierarchically organized complexity. This may reveal significant and pervasive patterns of natural evolutionary undecidable dynamics and evolutionary incompleteness.

6. Contradiction, paradox and self-referencing paradox in all aspects of Scientific Theory construction and Formal System construction, should be *embraced* rather than avoided. This essential refocusing of scientific attention may open a window on how to effectively compose in a scientific theory an *undecidable reductive proposition*, indicating *formal reductive incompleteness* and how to explore through theoretical and experimental means, the possibly of natural evolutionary undecidable dynamics.
7. In conformity with Kurt Gödel's *second* Formal Incompleteness Theorem; formal reductive incompleteness and undecidable dynamics; any future attempt to demonstrate the *consistency* of any Reductive Theory $T\ (F_R)$ and any Reductive Formal System F_R, will ultimately need to be addressed through *meta-theoretical-consideration.*
8. The space of meta-reductive-theoretical-consideration will eventually lead to consideration of adapted forms of reductive Logic and meta-construction of Meta-Reductive Paradigms. In adjacent possible *adapted* Paradigms, *adapted* premises, assumptions and forms of formalized scientific logic, may allow previously *undecidable* reductive propositions (such as *reductive epiphenomenalism of consciousness*) composed in the original, historic and modern 'bottom-up' logic of the Reductive Paradigm, to be *decided* in the reformulated and adapted domain of an imagined *Meta-Reductive Paradigm*. Such a Meta-Paradigm might be composed specifically in order to deal with an undecidable proposition located in a $T\ (F_R)$ statement and it might encompass 'bottom-up' reductive Logic in an adapted form of meta-reductive Logic.
9. Consideration of adapted Meta-Reductive-Paradigms may reveal that there exist many *Meta-Reductive Paradigms,* which are capable of solving many previously unsolved scientific problems (such as the incommensurability of Relativity Theory and Quantum Physics or the relentless residue of Descartes Mind/Body Dualism). A Meta-Reductive Paradigm may be capable of addressing many remaining reductive scientific anomalies (such as objective demonstrations of anomalous subjective transpersonal experiences of consciousness).
10. However, we can also expect that any *Meta-Reductive Paradigm* and adapted meta-reductive Logic; will ultimately and inevitably reveal its own particular pattern of formal, dynamical and informational, undecidable dynamics and incompleteness.

Reductive formal incompleteness represents a kind of scientific 'unfinished description' specifically related to formal reductive Logic [25]. It is worthwhile beginning a search for other significant inter-related kinds of 'unfinished description' in relation to reductive natural science.

Reductive formal incompleteness introduces a novel way to work toward a deeper consilience of the Reductive Natural Sciences and the Complexity Sciences. As Reductive Natural Science and the Complexity Sciences explore the implications of formal reductive incompleteness there may be imagined adjacent possible Meta-Reductive Scientific Paradigms in which adapted reductive Logic more closely approximates the natural Logic of Nature [26]. In such a novel Paradigm, the vast evolved complexity of the human body/brain/mind and consciousness might finally be encompassed fully within meta-reductive natural science and fully within the evolutionary, self-organized, emergent, complex adaptive products of Nature.

Acknowledgements Thank you to Dr. Yakov Shapiro and Patricia G. Scott for their steadfast support.

References

1. Barrow, J.D.: Mathematical jujitsu: some informal thoughts about Gödel and physics: the limits of mathematical systems. In: Morowitz, H. (Editor in Chief), Casti, J. (Executive Editor) Special Issue: Limits in Mathematics and Physics. Complexity **5**(5), 30 (2000)
2. Bickle, J., Mandik, P., Landreth, A.: The philosophy of neuroscience. In: Zalta, E.N. (ed.) The Stanford Encyclopedia of Philosophy. Stanford Metaphysics Research Laboratory, Stanford (Fall 2019 Edition). https://plato.stanford.edu/archives/fall2019/entries/neuroscience/
3. Bolander, T.: Self-reference. In: Zalta, E.N. (ed.) The Stanford Encyclopedia of Philosophy. Stanford Metaphysics Research Laboratory, Stanford (Fall 2017 Edition). https://plato.stanford.edu/archives/fall2017/entries/self-reference/
4. Brigandt, I., Love, A.: Reductionism in biology. In: Zalta, E.N. (ed.) The Stanford Encyclopedia of Philosophy. Stanford Metaphysics Research Laboratory, Stanford (Spring 2017 Edition). https://plato.stanford.edu/archives/spr2017/entries/reduction-biology/
5. Casti, J.: Complexification: Explaining a Paradoxical World Through the Science of Surprise. Harper Collins, New York (1994)
6. Casti, J.: The halting theorem (theory of computation). In: Casti, J. (ed.) Five Golden Rules: Great Theories of 20th Century Mathematics—And Why They Matter, pp. 135–180. Wiley, New York (1996)
7. Casti, J., DePauli, W.: Gödel: A Life of Logic. Perseus Publishing, Cambridge (2000)
8. Dennett, D.C.: Darwin's Dangerous Idea: Evolution and the Meaning of Life, pp. 81–81. Simon and Schuster, New York (1995)
9. Gödel, K.: On formally undecidable propositions of principia mathematica and related systems (B. Meltzer, Translator). In: Braithwaite, R. (ed.) Hofstadher, D.R. (Introduction), Electronic reprint edition. Basic Books, New York (1992)
10. Hatfield, G.: René Descartes. In: Zalta, E.N. (ed.) The Stanford Encyclopedia of Philosophy. Stanford Metaphysics Research Laboratory, Stanford (Summer 2018 Edition). https://plato.stanford.edu/archives/sum2018/entries/descartes/
11. Harari, Y.: Homo Deus: A Brief History of Tomorrow. Signal Books, an imprint of McClelland & Stewart, A Division of Penguin Random House Canada, United Kingdom, Canada (2016)
12. Harris, S.: Free Will. Free Press, A Division of Simon and Schuster Inc., New York, London (2012)
13. Hawking, S.: Kurt Gödel (1906–1978): his life and work. In: Hawking, S. (Editor with Commentary) God Created the Integers: The Mathematical Breakthroughs That Changed History, pp. 1255–1284. Running Press, Philadelphia, London (2007)
14. Hawking, p. 1256
15. Hawking, p. 1257
16. Hawking, p. 1258
17. Hawking, p. 1263

18. Hawking, p. 1283
19. Hawking, p. 1284
20. Hodges, A.: Alan Turing. In: Zalta, E.N. (ed.) The Stanford Encyclopedia of Philosophy. Stanford Metaphysics Research Laboratory, Stanford (Winter 2019 Edition). https://plato.stanford.edu/archives/win2019/entries/turing/
21. Kauffman, S.A.: A World Beyond Physics: The Emergence & Evolution of Life. Oxford University Press, Oxford (2019)
22. Kennedy, J.: Kurt Gödel. In: Zalta, E.N. (ed.) The Stanford Encyclopedia of Philosophy. Stanford Metaphysics Research Laboratory, Stanford (Winter 2018 Edition). https://plato.stanford.edu/archives/win2018/entries/goedel/
23. Lincoln, D.: The Theory of Everything: The Quest to Explain All of Reality. The Great Courses, The Teaching Company, Chantilly, Virginia (2017)
24. Linsky, B., Irvine, A. D.: Principia Mathematica. In: Zalta, E.N. (ed.) The Stanford Encyclopedia of Philosophy. Stanford Metaphysics Research Laboratory, Stanford (Fall 2019 Edition). https://plato.stanford.edu/archives/fall2019/entries/principia-mathematica/
25. Morowitz, H. (Editor in Chief) Casti, J. (Executive Editor) Special issue: limits in mathematics and physics. Complexity **5**(5), 11–44 (2000)
26. Nickles, T.: Scientific revolutions. In: Zalta, E.N. (ed.) The Stanford Encyclopedia of Philosophy. Stanford Metaphysics Research Laboratory, Stanford (Winter 2017 Edition). https://plato.stanford.edu/archives/win2017/entries/scientific-revolutions/
27. Prokopenko, M., Harré, M., Lizier, J., Boschetti, F., Peppas, P., Kauffman, S.: Self-referential basis of undecidable dynamics: from the liar paradox and the halting problem to the edge of chaos. Rev. Phys. Life Rev. (2019). https://doi.org/10.1016/j.plrev.2018.12.003.ArticleinPress
28. Prokopenko 6/23
29. Prokopenko 7/23
30. Raatikainen, P.: Gödel's incompleteness theorems. In: Zalta, E.N. (ed.) The Stanford Encyclopedia of Philosophy. Stanford Metaphysics Research Laboratory, Stanford (Fall 2018 Edition). https://plato.stanford.edu/archives/fall2018/entries/goedel-incompleteness/
31. Ramsey, W.: Eliminative materialism. In: Zalta, E.N. (ed.) The Stanford Encyclopedia of Philosophy. Stanford Metaphysics Research Laboratory, Stanford (Spring 2019 Edition). https://plato.stanford.edu/archives/spr2019/entries/materialism-eliminative/
32. Robinson, W.: Epiphenomenalism. In: Zalta, E.N. (ed.) The Stanford Encyclopedia of Philosophy. Stanford Metaphysics Research Laboratory, Stanford (Summer 2019 Edition). https://plato.stanford.edu/archives/sum2019/entries/epiphenomenalism/
33. Scott, A.C.: The Non-linear Universe: Chaos, Emergence, Life, pp. 277–301. Springer, Berlin, Heidelberg (2010)
34. Scott, J.R.: Descartes, Gödel, Kuhn: epiphenomenalism defines a limit on reductive logic. In: Morales, A.J., Gershenson, C., Braha, D., Minai, A., Barr-Yam, Y. (eds.) Unifying Themes in Complex Systems IX, Proceedings of the Ninth International Conference on Complex Systems. Proceedings in Complexity, pp. 33–52. Springer-Nature, Springer, Switzerland (2018). ISSN 2213-8684, ISSN 2213-8692 (electronic). ISBN 978-3-319-96660-1, ISBN 978-3-319-96661-8 (eBook)
35. Seager, W.: Natural Fabrications: Science, Emergence and Consciousness. Springer, Heidelberg (2012)
36. van Riel, R., Van Gulick, R.: Scientific reduction. In: Zalta, E.N. (ed.) The Stanford Encyclopedia of Philosophy. Stanford Metaphysics Research Laboratory, Stanford (Spring 2019 Edition). https://plato.stanford.edu/archives/spr2019/entries/scientific-reduction/

Kinds of Unfinished Description with Implication for Natural Science

J. Rowan Scott

Abstract The Reductive Sciences and Complexity Sciences employ micro-scale first rigorous 'bottom-up' reductive Logic as well as the mathematics of Symmetry and Group theory when modeling systemic change, causal relationships and mechanisms. *Four* kinds of 'unfinished description' result from this approach. First, the reductive micro-scale first assumption fails to replicate the complexity of natural evolutionary processes. Second, the reductive 'bottom-up' metaphor obscures significant facets of a more complicated natural evolutionary Logic. Third, abstract, rigorous, 'bottom-up' reductive Logic is susceptible to *undecidable reductive propositions* revealing *formal reductive incompleteness* and its *implications*, which include necessary *meta-consideration* in determining *reductive logical consistency*. Fourth, the powerful mathematics of Symmetry and Group Theory is not sufficient when mathematically modeling causal relationships in Nature. Consequently, the *Reductive Scientific Narrative* creates a 'comprehensive' description of causal relationships in Nature and evolution that is fundamentally 'unfinished'. Explaining and then 'correcting' each of the four kinds of 'unfinished description' illuminates a novel path that can more closely approximate the natural system and move Reductive Science and Complexity Science toward a deeper consilience.

Keywords Micro-scale · Multi-scale · 'bottom-up' reductive logic · Formal reductive incompleteness · Meta-consideration · Consistency · Symmetry · Group theory · Self-similarity · Fractals · Meta-construction · Meta-reductive paradigms

1 Introduction

When modeling systemic change and causal relationships in Nature, the Reductive Sciences begin with a reductive account involving micro-scale entities, and then applies rigorous 'bottom-up' reductive Logic as well as the applied mathematics of Symmetry and Group theory [18]. There is an active debate, which polarizes the

J. Rowan Scott (✉)
Department of Psychiatry, University of Alberta, Edmonton, AB, Canada
e-mail: j.rowan.scott@gmail.com

D. Braha et al. (eds.), *Unifying Themes in Complex Systems X*, Springer Proceedings in Complexity, https://doi.org/10.1007/978-3-030-67318-5_5

Reductive Sciences and the developing Complexity Sciences, regarding the usefulness of rigorous 'bottom-up' reductive Logic when modeling the complexity of the natural order [26]. While doubting the usefulness of 'bottom-up' reductive Logic [9] the Complexity Sciences generally presume that Symmetry and Group theory provide a useful applied mathematical approach to defining causal relationships.

It is, however, possible to spell-out *four* inter-related kinds of 'unfinished or partial scientific description' associated with 'bottom-up' reductive Logic *and* the mathematics of Symmetry and Group theory when these are used in the context of theoretical, experimental and mathematical modeling of causal relationships in Nature and evolution.

1.1 Four Descriptive Deficiencies

First. The *reductive account begins with a natural micro-scale entities first assumption*, which fails to recognize Nature and evolution may self-organize on *multiple* scales at once. Proposing an alternative *meta-reductive natural multi-scale co-evolving entities assumption* as an alternative offers a theoretically and experimentally accessible *hypothesis*, which, if it finds support, could reveal the *reductive, begin with micro-scale entities first assumption*, is actually a 'special case' within a more comprehensive model of natural system evolution possible under a *meta-reductive natural multi-scale co-evolution assumption.*

Second. The *reductive 'bottom-up' metaphor* accepts an implicit or explicit *postulate* that the Universe and natural evolution self-organize from an *initial* sea of *micro-scale* quantum states. Every other *subsequent* form of hierarchically organized mid-scale or macro-scale complexity is *assumed* to result from 'bottom-up' self-organization of quantum micro-scale states and processes. If *meta-reductive natural multi-scale co-evolution* finds theoretical and experimental support, then, the *reductive 'bottom-up' metaphor* will become a 'special case' within a more comprehensive meta-reductive metaphor capable of describing and modeling *natural multi-scale co-evolution.* A much more complicated co-evolutionary metaphor may be required, capable of describing, at a minimum, multi-scale primary 'bottom-up', 'top-down' causal relationships as well as secondary 'lateral' or 'horizontal' causal relationships.

Third. Rigorous, formal, *reductive Logic* is susceptible to well-formed and logically true argument composing potential *contradiction* and *self-referencing paradox* [4]. This realization inevitably leads to the insight that reductive Logic and argument can contain *undecidable reductive propositions*, which reveal *formal reductive incompleteness* as well as a need for *meta-consideration* in determining the *consistency* of reductive logical argument [23, 25].

This discovery opens up a previously unrecognized path toward resolution of questions about the usefulness of reductive Logic in natural science. Rather than *polarization* between *reductive* and *anti-reductive* positions in debates regarding the usefulness of reductive Logic, it may be possible to demonstrate that polarized pro- and anti- positions, which either accept or reject reductive Logic in the task of modeling complex causal relationships, are *both poorly formulated* arguments.

If reductive Logic is susceptible to *formal reductive incompleteness* and its implications, then, addressing the *shared* failure to recognize formal incompleteness could reveal novel ways to *preserve* 'bottom-up', rigorous, strong, formal reductive Logic as a 'special case' within a more complex *iterative* implementation of meta-reductive Logic associated with *natural multi-scale co-evolution,* a *more complicated multi-scale causal metaphor* and *formal reductive incompleteness.* What is *undecidable* within current 'bottom-up' reductive Logic may be *decidable* within an adapted multi-scale meta-reductive Logic.

The above facets of *multi-scale co-evolution,* a *multi-scale causal metaphor* and *formal reductive incompleteness* suggest a novel concern for any adapting natural science: There is a hypothetical possibility that Nature and co-evolution contain identifiable *unresolvable natural instantiations* and *natural evolutionary incompleteness.* As well, Nature may instantiate an identifiable pattern of causally significant *natural incompleteness driven novelty generation.* These unusual structures and processes would be expected to appear in relation to natural *transitions,* where an established consistent systemic order, transits into impending systemic inconsistency, which then self-organizes a meta-systemic resolution with the emergence of a novel meta-consistent, meta-order.

Thus, hypothetically, the evolving natural system may have an inherent, iterated and inevitable, co-evolutionary relationship with natural co-evolutionary incompleteness and its implications. *Multi-scale co-evolution, reductive logical incompleteness* and *natural evolutionary incompleteness,* all require novel approaches and adaptations of reductive Logic, which are poised to transform theoretical understanding in *both* Reductive Science and Complexity Science.

Fourth. The applied mathematics of Symmetry and Group theory [18] may not be sufficient to model causal relationships in a Universe in which multi-scale co-evolution and inherent natural incompleteness are manifest. A more comprehensive model of causal relationships may require a synthesis of Symmetry and Group theory *with* Self-similarity and Fractals. Self-similarity is a 'special kind' of symmetry that focuses on invariance across multiple scales of causally significant self-organization [20]. Symmetry and Group theory may become a 'special case' within a more comprehensive meta-reductive model of causal relationships in natural systems where Symmetry and Self-similarity are integrated.

The presence of an unrecognized relationship shared by all *four* of the interconnected kinds of 'unfinished' or 'partial' scientific description and modeling *distorts* the comprehensive *Reductive Scientific Narrative*; and, *distorts* the frequently incommensurate story told in the *Scientific Narrative of Complexity Science* [27]. An integrated understanding of the four kinds of 'unfinished', 'partial' or 'incomplete' scientific description, could lead to a novel consilience between Reductive Science and Complexity Science. Within such a synthesis, mind and consciousness can assume a novel assimilated relationship *in* Nature and evolution.

In order to proceed, a clear sense of the modern reductive scientific approach is required, from which insights about the four kinds of 'unfinished description' can be developed.

2 The Modern Reductive Scientific Narrative

The 'bottom-up' metaphor of rigorous reductive scientific Logic makes an *assumption* that Nature, the Universe inflating in the Big Bang, natural evolution, self-organization and the emergence of complexity, are *naturally composed* from *micro-scale states* and *processes.*

Consistent with the micro-scale assumption, the comprehensive Reductive Scientific Narrative states that approximately 14.5 billion years of natural evolution in our Universe, originates from a tiny point of origin, which inflates and undergoes a series of *phase changes*, which are defined by successive states and processes engaging in cycles of change. An initial, hyper-dense, small, perhaps cool and simple state and process, transitions into an inflating, extremely hot, high-energy, state and process. Then, the 'whole' expands and transforms again, everywhere at once, creating a sea of diverse 'parts' or quantum micro-scale states and processes, which rapidly interact and slowly cool, generating increasingly complicated states and processes within the early quantum universe [5]. Early micro-scale quantum states and processes in our Universe, continue to evolve and progressively self-organize, creating the mid-scale and macro-scale emergent hierarchy of complexity [10] (Fig. 1).

Depictions of the Big Bang suggest this statement can describe the very early inflation from a tiny local energy state:

'The Universe rapidly expands everywhere at once'.

The accelerating expansion of the observable Universe involves increased distance between large cosmic objects with time. The *scale* of space changes with time while the gravitationally bound objects in space-time do not move within space and do not show the same expansion as the larger scale objects that are gravitationally unbound. The Big Bang model of cosmic expansion, therefore, only applies to large-scale objects roughly the size of galaxy clusters or greater. The expansion is therefore a *metric expansion*, which alters the size and geometry of space-time. 'Dark energy' has been suggested as a way to explain the accelerating expansion [12].

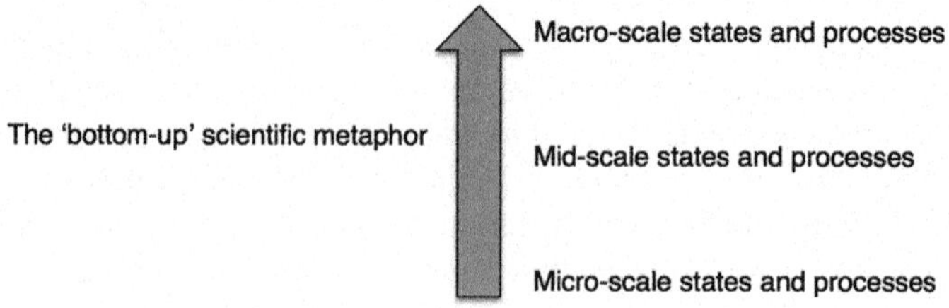

Fig. 1 Micro-scale first evolution and the 'bottom-up' scientific metaphor

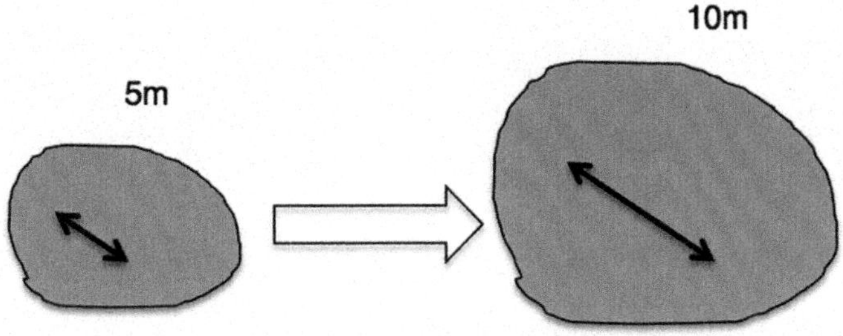

Fig. 2 Metric expansion in bread dough and the universe

The focus of attention in a description of cosmic expansion is on *increasing distance* and *metric measures* of *distance* between large-scale objects. An *analogy* often offered, involves a rising loaf of *raisin bread*, in which warm yeast growth and multiplication *expands* and *increases* the *space* between raisins *everywhere* at *once* (Fig. 2).

The Reductive Sciences, the reductive scientific method [1] and 'bottom-up' reductive scientific Logic [31], suggest one should seek understanding of natural causal relationships, cosmic inflation, natural evolution and raisin bread using a *three-step* scientific process [21].

(1) *First, select* a '*whole*' phenomenon of interest and *reduce* it, by breaking it down or deconstructing the 'whole' into its *smallest* scientifically accessible components or '*parts*', until you arrive at the most fundamental, tiniest 'parts' that reductive thought and the available scientific tools can access.
(2) *Second, theoretically analyze* each reduction of the original complexity, creating as comprehensive a *model* as possible.
(3) *Third*, ensuring you leave nothing out, begin *synthesizing* a 'bottom-up' reductive conception, starting from the smallest scientifically accessible elements or 'parts', moving 'upward' toward the 'whole'. Ultimately, through 'bottom-up' synthesis, arrive at a reductive account of the original 'whole' phenomenon.

As a further component of the *final step* and *synthesis*, it is recommended that the entire process be *experimentally* and/or *computationally* revisited. Starting with the smallest accessible reduced entities or components contained within a carefully selected semi-isolated situation known to enable the natural interaction of the necessary entities or components; the scientist experimentally demonstrates that the interacting 'parts' can *spontaneously recreate* the 'whole' phenomenon. If possible, also create a mathematical or computationally equivalent model, in which mathematically rigorous statements or computationally equivalent 'parts' interact. In such a semi-isolated context, demonstrate that a scientist and mathematician can *create* a mathematically precise or computationally equivalent version of the 'parts' composing, interacting and recreating the 'whole' phenomenon and its relationships (Fig. 3).

A reductive scientific and mathematical analysis of *presumed cause and effect relationships* within Nature, approaches selected phenomena of interest, *P*, with a

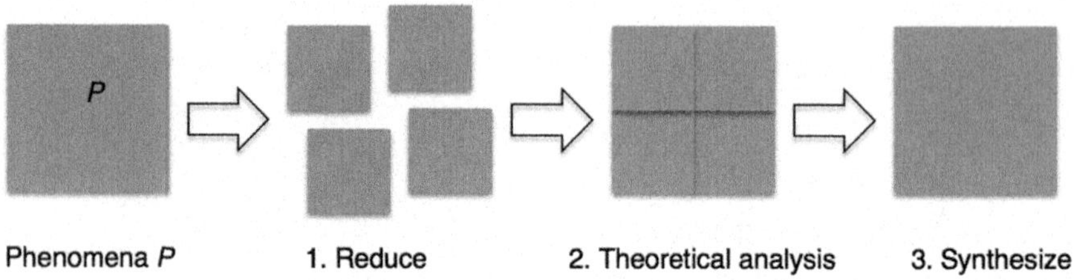

Fig. 3 The three-step reductive scientific approach

'bottom-up' reductive focus on micro-scale states and processes that can be *mathematically* approached employing the demanding mathematics of *Symmetry* and *Group Theory* [18]. Symmetry is a two-part concept linking something that stays the same and does not change, with, something related that does change and does not stay the same. Symmetry Groups involve, a set of elements, including an inverse for every element and an identity element, as well as an associative binary operation. Sequences of operations performed upon any series of elements linked by the binary operator, are associative if the sequence always produces the same outcome, whatever the grouping, as long as the order remains the same $[(a \times b) \times c = a \times (b \times c)]$ [30] (Fig. 4).

In a *selected semi-isolated* context, a presumed *subsystem of the cause* is differentiated from a presumed *subsystem of the effect*. Observation and experimental set-ups identify and detail *correlation* between the presumed cause and effect subsystems (an apparent mutual relation or connection between the two subsystems), then *equivalence relations* between the presumed causal subsystems (demonstrable invariance or similarity between the two subsystems), and finally, specifies any *preserved symmetry* shared by the presumed causal subsystems (demonstrable invariance and associated variance between the two subsystems).

The Symmetry Group concept allows the scientist or mathematician to abstract simplified meaningful pattern from the overall complexity of the two sub-systems of cause and effect differentiated within the complex sequences of state and process, which may otherwise involve complicated, changing, elements, entities, compositions, interactions and systemic relationships.

Ultimately, assuming nothing significant has been left out, the careful application of reduction, theoretical analysis and synthesis; experimental exploration and precise mathematical analysis, involving reductive Logic, Symmetry and Group

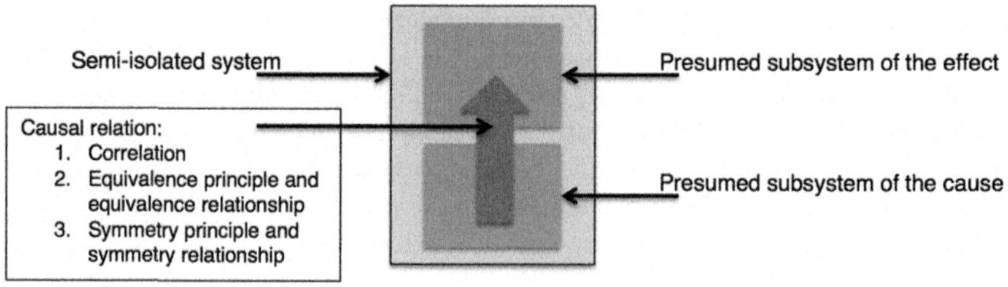

Fig. 4 The semi-isolated experimental context causal relationships

theory, natural science arrives at the most precise and exact definition of a causal relationship in Nature that reductive science and applied mathematics are capable of achieving.

In modern reductive natural science, using 'bottom-up' reductive Logic, Symmetry and Group theory, there are only *three* known kinds of *causal mechanism* underlying every natural phenomenon where scientists might choose to tease out the presence of a causal relationship. Two causal mechanisms are *classical* and one is *quantum.* In the words of Nicolas Gisin [7]:

> "All correlations observed in science, outside quantum physics that is, can be accounted for either by an influence of one event on another (explanation of the first type), or by common local causes ... (explanation of the second type). In both types of explanation, the said influence or common cause propagates continuously from point to point through space, and in a precise sense, all such explanations are local. ... (Quantum) physics provides us with a third possible explanation".

Non-local Quantum correlations (explanation of the third type) involve quantum *entanglement*, invoking the philosophical puzzle, the paradoxes and the complicated Quantum dynamics of action at a distance, with no local mechanism [19] (Fig. 5).

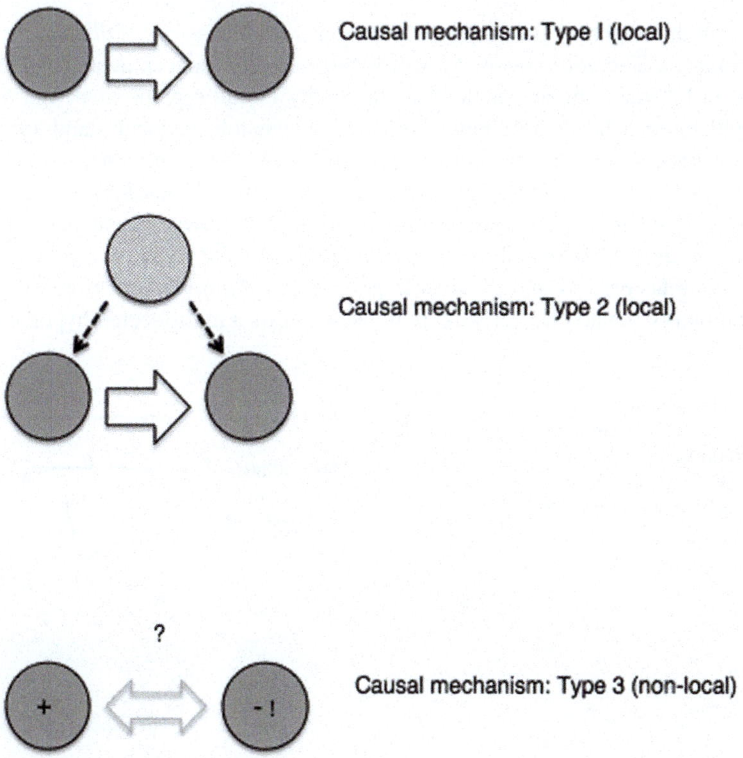

Fig. 5 Causal mechanisms in reductive science

3 A Thought Experiment

Just as it is possible for a scientist to discover, after careful scientific study of a natural causal relationship that something significant has been left out; so too the modern comprehensive Reductive Scientific Narrative created by the collective collaboration of a vast scientific and mathematical community, may have left something significant out of our understanding of Nature.

The further discussion in this paper explores *four* significant kinds of 'unfinished', 'partial' or 'incomplete' scientific and mathematical description. The discussion is composed in the form of a *thought experiment* [3].

The thought experiment begins by offering a *slightly modified* scientific description of the Big Bang and the initial moments of creation in our Universe. The altered description is presented in this statement:

'The Universe rapidly expands and evolves, everywhere at once, on every scale at once'.

The few extra words introduced into this modified description ('evolved' and 'on every scale at once') can make a significant difference to the overall theoretical understanding of the evolving natural system and causal relationships in Nature (Fig. 6).

The modified phrase goes beyond interest in *metric measures* of *distance* between large objects. The addition of the word, '*evolves*', introduces awareness that the raisins in the raisin bread, or the gravitationally bound regions involving smaller scale phenomena that do not show accelerating expansion, are local islands or regions where natural system evolution explores self-organization, and energy-matter and space-time create hierarchically organized complexity. The addition of the phrase, '*on every scale at once*', focuses attention, not just on metric measures of distance, but on the *differentiation* of *micro-scale, mid-scale* and *macro-scale* phenomena within complex regions of the Universe. *Evolution* and *differentiation of scale* across gravitationally bound and unbound regions, suggests a multifaceted hypothesis and

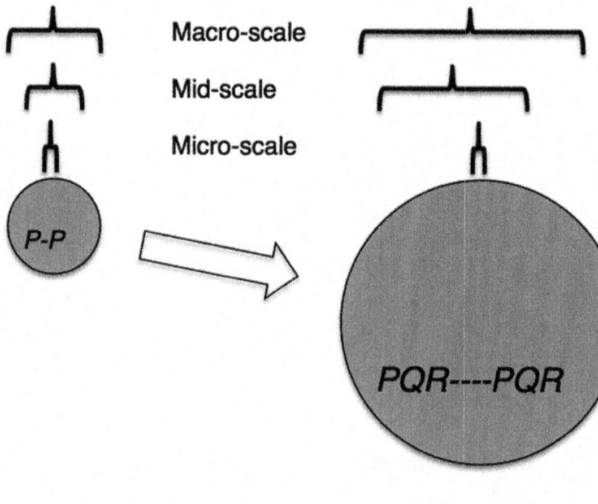

Fig. 6 Metric expansion, multi-scale differentiation and evolution, in a universe that expands and evolves, everywhere at once, on every scale at once

an alternative scientific methodological approach for modeling causal relationships in Nature.

The transformations and implications of the slightly modified statement are spelled-out in the thought experiment. The following discussion will examine how a natural science influenced by the adapted phrase might go about approaching theoretical puzzles, related experimental contexts and specific selected phenomenon of interest.

3.1 Micro-scale or Multi-scale? the Multi-scale Hypothesis

In a Universe that rapidly *expands and evolves, everywhere at once, on every scale at once*; the metaphorical *bread dough* and *raisins* in the raisin bread, or the gravitationally unbound and bound regions in the Universe, *evolve*, in lock-step, *everywhere at once, on every scale at once.*

The phrase '*everywhere at once on every scale at once*' suggests a scientific hypothesis that transforms the reductive conception of *micro-scale evolution* and the *reductive, micro-scale entities first assumption*, into a kind of 'unfinished description'. The *reductive micro-scale first assumption* is translated and positioned as a component and a 'special case' in an integrated model of *meta-reductive multi-scale, multi-directional co-evolution.* Multi-scale co-evolution entails multi-directional co-evolution. It is impossible to consider these two adaptations of reductive thought independently. Nevertheless, this section will concentrate on the multi-scale adaptation while the next section of the paper will address multi-directional co-evolution.

In multi-scale co-evolution, causal relationships must be looked at in relation to expanding inter-related, local and non-local, multi-scale causal spheres or manifolds. Multi-scale spheres or manifolds occur everywhere at once on every scale at once. A multi-scale sphere or manifold must encompass any 'part' and a multi-scale sphere or manifold must encompass any conception of the 'whole'. Multiple scales co-evolve, self-organize, emerge and engage in the natural construction of hierarchically organized complexity, in relation to any 'part' or in relation to any conception of the 'whole'. Meta-reductive epistemology predicts that multi-scale, multi-directional, co-evolution effectively correlates and inter-relates *everywhere at once* on every *scale at once*, in the metaphor of rising bread dough or in the ontological instantiation of the Universe.

The phrase '*everywhere at once on every scale at once*' therefore implies linked and correlated micro-scale, mid-scale and macro-scale relationships must inflate, expand, differentiate and co-evolve *together* within the rising bread dough, or within the inflating, expanding and co-evolving Universe. The micro-scale, mid-scale and macro-scale 'parts' *and* the micro-scale, mid-scale and macro-scale 'whole', of the *raisin bread* or the 'whole' Universe, *co-evolve*, in an integrated multi-directional co-evolutionary *synchronic* and *diachronic* relationship. In *meta-reductive, multi-scale, multi-directional, co-evolution* the 'whole' must be considered in relation to the 'parts' and the 'parts' must be considered in relation to the 'whole'. The historical reductive focus on micro-scale, unidirectional, evolution becomes a 'special-case' in the meta-reductive conceptualization.

The historical reductive scientific pursuit of a semi-isolated phenomenon of interest, focuses attention on micro-scale first 'bottom-up' subsystems of presumed cause and effect and then proceeds to construct the phenomena of interest with a unidirectional metaphorical focus on 'upward' causal organization: this extremely successful reductive approach becomes a 'special-case' in the meta-reductive account. The meta-reductive account places any *selected phenomenon of interest* within an *adapted* meta-reductive semi-isolated situation that assumes multi-scale, multi-directional, co-evolution. The multi-scale, multi-directional assumption of co-evolution necessarily leads to an adaptation of 'bottom-up' reductive Logic. Singular 'bottom-up' reductive Logic is transformed into a 'special-case' and meta-reductive form, which is then capable of handling multi-scale, multi-directional, co-evolution, through an iterated multiple application of reductive Logic exploring multiple scales and multiple directions of developing causal relationship.

In the meta-reductive account causal mechanisms and causal relationships can be examined in relation to any specific phenomena of interest composed within a semi-isolated meta-reductive sphere or manifold. A blend of reductive and meta-reductive conceptualization is required. The instantiated phenomena of interest can be described within a semi-isolated meta-reductive sphere or manifold as an abstract *entity*, *state* and *process*, involving a *composition* of 'parts', as well as the *interaction* of those 'parts', formulating a semi-isolated 'whole'. *Hierarchical* organization, causal mechanisms and relationships may be present within the meta-reductive sphere or manifold; however, these must be considered through the lens of multi-scale, multi-directional, co-evolution with the reductive 'bottom-up' account treated as a 'special-case'.

The construction so far could describe a selected phenomenon such as a simple quantum entity or it could describe an evolved Complex Physical System (CPS) or Complex Adaptive System (CAS) [23]. These selections are all now viewed through the lens of meta-reductive conception. Meta-reductive, multi-scale, multi-directional, co-evolution is presumed to apply in relation to a 'simple' micro-scale system involving a quantum particle/wave, and also presumed to apply in relation to more 'complex' Meta-Reductive Complex Physical Systems (MCPS) or Meta-Reductive Complex Adaptive Systems (MCAS).

The selected phenomena of interest contained within the meta-reductive semi-isolated sphere or manifold is then explored in relation to a matrix of potential causally significant *relationships* that surround the semi-isolated system. The meta-reductive matrix of causal relationships *potentially* has causal impact *within* the defined semi-isolated sphere or manifold as well as *beyond* a defined semi-isolated sphere or manifold. The meta-reductive matrix of causal relationships causally implicates the local ecosystem or the extended environment within which a defined semi-isolated sphere or manifold sits.

In Fig. 7 the 'special case' of a 'bottom-up' reductive examination of the micro-scale enabling of a selected phenomenon of interest is encompassed within 1.a. and

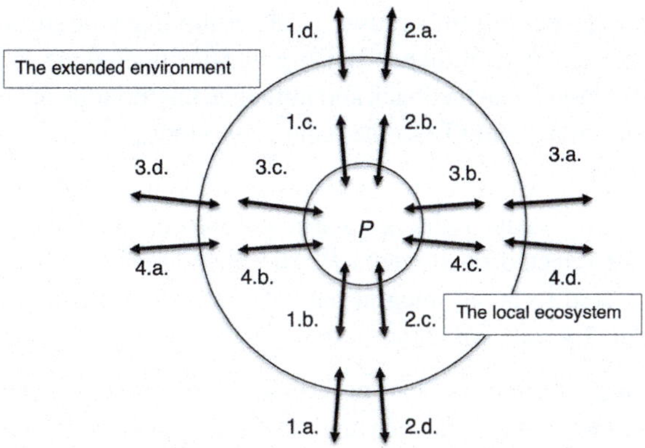

Fig. 7 Possible meta-reductive relationships associated with a given selected phenomena *P*. The selected phenomenon *P* is in the sphere of a meta-reductive semi-isolated system. The semi-isolated system is in the meta-reductive sphere of a local eco-system. The selected phenomenon *P* and the local eco-system are in an extended meta-reductive environment

1.b. in the meta-reductive framework and 4 × 4 matrix of possible causal relationships. The matrix surrounds the selected phenomenon of interest within a field of potential sources of causal input, circular causal events, feedback loops, and potential causal relationships. The selected phenomenon of interest can be influenced by, participate in, or precipitate causal relationships within the 4 × 4 meta-reductive matrix.

The 4 × 4 matrix of co-evolutionary, multi-scale, multi-directional causal relationships include:

(1) Multi-scale relationships arising 'bottom-up' involving the *extended environment* and *local ecosystem* that influence, participate with, or precipitate the *organizing* or *organized* selected phenomena *P*. The phenomenon *P* is not characteristically considered to be an integral part of the system making up the extended environment or local ecosystem:

  (a) *Enabling* 'bottom-up' causal relationships arising in the *extended environment* that make possible the process *organizing* phenomena *P*. The organizing phenomena *P* can feedback and influence the 'bottom-up' enabling causal relationships arising in the extended environment.
  (b) *Enabling* 'bottom-up' causal relationships arising in the *local ecosystem* that make possible the process *organizing* selected phenomena *P*. The organizing phenomena *P* can feedback and influence the 'bottom-up' enabling causal relationships arising in the local ecosystem.
  (c) *Engaging* 'bottom-up' causal relationships arising from the local ecosystem that influence the *organized* phenomena *P*. The organized phenomenon *P* can feedback and influence the 'bottom-up' engaging causal relationships arising from the local ecosystem.

(d) *Engaging* 'bottom-up' causal relationships arising from the extended environment that influence the *organized* phenomena *P*. The organized phenomenon *P* can feedback and influence the 'bottom-up' engaging causal relationships arising from the local ecosystem.

(2) Multi-scale relationships arising 'top-down' involving the *extended environment* and *local ecosystem* that influence, participate with, or precipitate the *organizing* or *organized* selected phenomena *P*. The phenomenon *P* is not characteristically considered to be an integral part of the system making up the extended environment or local ecosystem:

(a) *Enabling* 'top-down' causal relationships arising from the *extended environment* that influence and make possible the process *organizing* phenomena *P*. The organizing phenomena *P* can feedback and influence the 'top-down' enabling causal relationships arising in the extended environment.
(b) *Enabling* 'top-down' causal relationships arising in the *local ecosystem* that make possible the process *organizing* selected phenomena *P*. The organizing phenomena *P* can feedback and influence the 'top-down' enabling causal relationships arising in the local ecosystem.
(c) *Engaging* 'top-down' causal relationships arising from the local ecosystem that influence the *organized* phenomena *P*. The organized phenomenon *P* can feedback and influence the 'top-down' engaging causal relationships arising from the local ecosystem.
(d) *Engaging* 'top-down' causal relationships arising from the extended environment that influence the *organized* phenomena *P*. The organized phenomenon *P* can feedback and influence the 'top-down' engaging causal relationships arising from the local ecosystem.

(3) Multi-scale relationships arising from 'lateral' causal relationships in the *extended environment* and *local ecosystem* that influence, participate with, or are precipitated by, the *organizing* or *organized* selected phenomena *P*. *Lateral causal relationships* involve a co-developing, complex system in which the phenomenon *P* is considered to be an integral part:

(a) *Enabling* 'lateral' causal relationships arising in the *extended environment* that make possible the process *organizing* phenomena *P*. The organizing phenomena *P* can feedback and influence the 'lateral' enabling causal relationships arising in the extended environment of which *P* is a component.
(b) *Enabling* 'lateral' causal relationships arising in the *local ecosystem* that make possible the process *organizing* phenomena *P*. The organizing phenomena *P* can feedback and influence the 'lateral' enabling causal relationships arising in the local ecosystem of which *P* is a component.
(c) *Engaging* 'lateral' causal relationships arising from the *local ecosystem* that influence and participate with the *organized* phenomena *P*. The organized

phenomenon *P* can feedback and influence the 'lateral' engaging causal relationships arising from the local ecosystem.

(d) *Engaging* 'lateral' causal relationships arising from the *extended environment* that influence and participate with the *organized* phenomena *P*. The organized phenomenon *P* can feedback and influence the 'lateral' engaging causal relationships arising from the extended environment.

(4) Multi-scale 'horizontal' causal relationships involving the *organizing* or *organized* selected phenomena *P*, actively participate in influencing, participating with, or precipitate events in the *local ecosystem* and *extended environment. Horizontal causal relationships* involve phenomena *P* as an integral component of a co-developing, complex system:

(a) *Enabling* 'horizontal' causal relationships arising from the *organizing* phenomena *P* that influence the *extended environment* of *P*. The extended environment can feedback and influence the 'horizontal' enabling causal relationships arising from the *organizing* phenomena *P*.
(b) *Enabling* 'horizontal' causal relationships arising from the *organizing* selected phenomena *P* that influence the *local ecosystem* of *P*. The local ecosystem of which *P* is a component can feedback and influence the 'horizontal' enabling causal relationships arising from the *organizing* phenomena *P*.
(c) *Engaging* 'horizontal' causal relationships arising from the *organized* phenomena *P* that influence and participate in the local ecosystem of *P*. The local ecosystem can feedback and influence the 'horizontal' engaging causal relationships arising from the *organized* phenomenon *P*.
d) *Engaging* 'lateral' causal relationships arising from the *organized* phenomena *P* that influence and participate in the extended environment of *P*. The extended environment can feedback and influence the 'horizontal' engaging causal relationships arising from the *organized* phenomenon *P*.

'*Simple*' non-local and local causal mechanisms purport to account for *all* the fundamental causal mechanisms in Nature and evolution defined by Reductive Science. In a model of *multi-scale, multi-directional, co-evolution*, composed by an adjacent possible Meta-Reductive Science, there may be further, more complicated, naturally organized and hierarchically related causal mechanisms. These may involve more '*complex*', *multi-scale, multi-directional systemic causal mechanisms* associated with *complex, systemic, multi-scale, multi-directional, co-evolutionary causal relationships.*

More complex systemic causal mechanisms may already be known but not yet categorized in a coherent manner. The *reductive incompleteness hypothesis* (which will be stated below) defines a *fundamental incompleteness driven novelty generating causal mechanism* that can only be visualized, in theory, or detected in Nature, in association with recognized natural evolutionary incompleteness. The incompleteness driven novelty generating causal mechanism may form complex *combinations* with other complex systemic causal patterns, such as non-linear processes, stochastic processes, Lamarckian processes, systemic emergence and transient or sustained complementarity.

David Bohm suggested integrated 'part' and 'whole' evolution could be defined as 'explicate' and 'implicate' evolution, respectively [14]. A model of integrated *multi-scale co-evolution* presents a more complicated picture. In *multi-scale co-evolution*, inflating and expanding energy-matter and space-time, *opens* integrated 'whole' and 'part' causal spheres, manifolds or domains, which involve phenomena, relationships and mechanisms occurring *everywhere at once on every scale at once.*

Bohm's vision describes a *unitary* sequence and trajectory of integrated *implicate* (whole) and *explicate* (part) evolution. In *multi-scale, multi-directional, co-evolution*, an integrated sequence of intersecting trajectories of multi-scale co-evolution, involves causal mechanisms and relationships in inter-related spheres, manifolds or domains. *Within* each domain, multi-scale, multi-directional, co-evolution of 'whole' and 'part' co-occurs, with self-organizing entities, composition, interaction and relationships. As well, at the edge of any domain, multi-scale, multi-directional, co-evolution involves interactive mechanisms and relationships occurring *at* or *beyond* the *boundary* of each causal domain. These relational events co-occur in the local and extended environment, where expansive multi-scale co-evolution of 'whole' and 'part', self-organize further *extended* entities, compositions, interactions and relationships that reach *beyond the boundary* of any given or selected causal domain.

The multi-scale conception leads to a *multi-directional hypothesis* (already employed but stated below). In order to approximate the complicated causal relationships a novel multi-directional scientific metaphor must be developed. The *multi-scale, multi-directional, co-evolutionary* perspective transforms the *reductive 'bottom-up' metaphor* into a 'special case' within a *multi-directional metaphor* creating a complex *spherical* causal map and matrix of possible relationships associated with any definable, selected phenomenon of interest within a particular causal domain.

3.2 Bottom-Up or Multi-directional? the Multi-directional Hypothesis

The multi-scale co-evolution hypothesis predicts a related multi-directional hypothesis along with a related metaphor. If, the natural system *expands and evolves everywhere at once on every scale at once*, then, the reductive 'bottom-up' metaphor and unitary 'bottom-up' formal reductive Logic will produce an *insufficient, unfinished, partial, informally incomplete* picture of Nature and the natural Logic of evolution. The *reductive 'bottom-up' metaphor* can, however, become a 'special-case' within a *multi-scale co-evolutionary, multi-directional metaphor*. Unitary 'bottom-up' application of reductive Logic must also become a 'special case' within a spherical causal matrix describing *multiple* causal relationships. The spherical causal matrix involves multiple, iterative, multi-directional applications of reductive Logic associated with each of the multiple descriptions developed in the multi-scale co-evolution hypothesis.

To capture the possible multi-directional causal relationships in a *metaphor* describing a multi-scale co-evolutionary causal sphere, manifold or domain; the 'special-case' of 'bottom-up' reductive Logic mapping 'bottom-up' enabling relationships must be coordinated with other iterative and multi-directional applications of reductive Logic. This forms an integrated matrix of metaphorical 'bottom-up' and 'top-down' extrinsic enabling relationships along with interactive 'bottom-up' and 'top-down' coordinating and regulating feedback [6]. The 'bottom-up' and 'top-down' relationships must then be integrated with 'lateral' intrinsic enabling and engaging relationships and facilitating feedback, as well as 'horizontal' participatory relationships in which the phenomena synchronizes the intrinsic enabling and engaging systems. Multiple, multi-directional iterations of reductive Logic are then required to map the interactions and relationships in a meta-reductive description of a natural evolving system (examples will be developed later in the text) (Fig. 8).

A matrix of iterated reductive logical applications allows for detailed examination of metaphorical 'bottom-up', 'top-down' and 'lateral' or 'horizontal', co-evolving, multi-directional causal relationships within and beyond the boundary of any co-evolutionary causal sphere, manifold or domain.

In the *center* of the spherical map of a selected phenomenon *P,* or at the *boundary* of the central sphere and the peripheral extended causal relationships associated with the multi-scale environment, 'bottom-up', 'top-down' and secondary 'lateral' or 'horizontal' relationships, may other-organize or self-organize, entities, compositions, interactions and relationships involving particular complex phenomena that can co-evolve *positive* and *negative feedback* as well as *circular causation.*

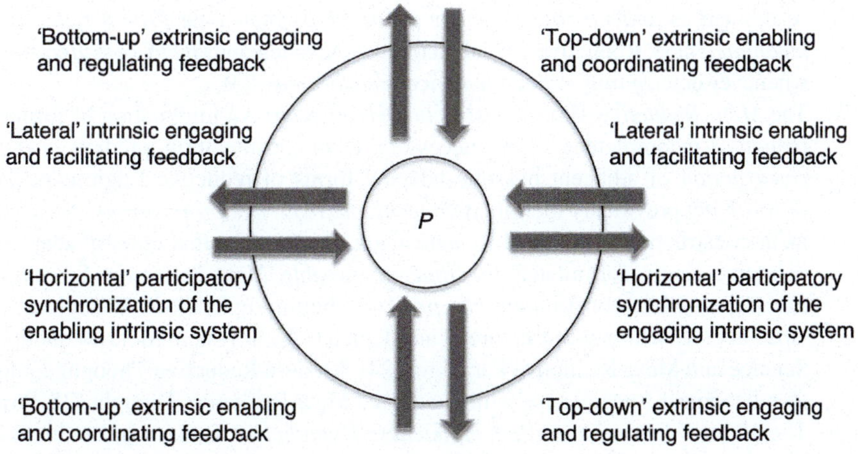

Fig. 8 Possible metaphorical meta-reductive relationships associated with a given selected phenomena *P*

The *multi-scale hypothesis* and the *multi-directional hypothesis, predict* complex points of *intersection* within multi-scale causal spheres, manifolds or multi-scale causal domains. The description of the intersections between multiple trajectories of multi-directional, primary and secondary causal accounting, might arrive at simple unidirectional or complicated multi-directional, reductive statements or natural instantiations that might provide clear evidence of reductive formal incompleteness and natural evolutionary incompleteness.

These considerations take us into a deeper look at the *Reductive Formal Incompleteness Hypothesis.*

3.3 Complete or Incomplete? Reductive Formal Incompleteness

A formal logical and mathematical proof of formal reductive incompleteness impacting 'bottom-up' reductive Logic has recently been attempted [25]. 'Unfinished description' can therefore involve 'bottom-up' formal reductive Logic. This particular kind of 'partial' scientific description will be addressed first in relation to modern scientific 'bottom-up' reductive Logic, and then in relation to an imagined, integrated meta-reductive Logic that can model multi-scale co-evolution and multi-directional causation. This is presented in the form of a series of theoretical hypotheses. In relation to 'bottom-up' reductive Logic:

(1) The *Reductive Formal Incompleteness Hypothesis*: Modern, scientific, 'bottom-up' reductive Logic can formulate *contradiction* and *self-referencing paradox.* To avoid the threat of inconsistency in the application of formal reductive Logic, where necessary, such contradictory or paradoxical reductive statements can be defined as *undecidable reductive propositions* indicating *formal reductive incompleteness.* Reductive formal incompleteness demands *meta-consideration* whenever determining *reductive logical consistency* [23].
(2) The *Meta-Reductive Meta-Construction Hypothesis*: An implication of formal reductive incompleteness and necessary meta-consideration predicts *meta-construction* of adjacent possible adapted forms of reductive Logic capable of *resolving* previously identified *undecidable reductive propositions.* As well, meta-construction might lead to *meta-reductive sciences* that employ adapted reductive Logic formulated in *adjacent possible Meta-Reductive Scientific Paradigms.* Meta-construction can *preserve* 'bottom-up' reductive Logic as a 'special case' within the structure of an adapted Logical system Meta-Reductive Science and Meta-Reductive Paradigm [24]. Modern Reductive Thought could then become a 'special case' within adapted Meta-Reductive Thought [25, 26].
(3) The *Natural Evolutionary Incompleteness Hypothesis*: 'Bottom-up' reductive Logic is susceptible to formal incompleteness, and 'bottom-up' reductive Logic *may* describe attributes of Nature and the evolutionary Logic of natural evolving systems. Therefore, Nature and natural evolution *may* manifest properties and

definable patterns consistent with *natural evolutionary incompleteness*. This may take the form of transient or preserved states and processes that manifest as 'unresolvable natural instantiations'— 'unresolvable' natural forms are likely to exhibit properties that *'are and are not' consistent* with the *existent order* within which they belong.

(4) The *Natural Meta-Construction Hypothesis*: Unresolvable natural instantiation and natural evolutionary incompleteness may be *resolved* when *one* natural order becomes embedded and encompassed within an emerging natural order and meta-system, which exhibits its own pattern of meta-consistency [24]. A *natural meta-consistent resolution* may also occur when *two* co-evolving, ordered and consistent regimes intersect, compete, collaborate or integrate.

(5) The *Inevitability of Incompleteness Hypothesis*: It is *inevitable* that any scientific Logic purporting to model the natural order or an evolving meta-order, will ultimately exhibit its own pattern of formal incompleteness. It is also inevitable that any natural order manifesting a resolution of natural incompleteness will ultimately reveal its own pattern of natural evolutionary incompleteness.

Defining these hypotheses in theoretical and experimentally accessible terms would allow their investigation in relation to Nature and could provide substantial support for the four kinds of 'unfinished' description involving 'bottom-up' reductive Logic and their resolution. It would also provide support for necessary meta-construction of adapted forms of scientific Logic, adapted meta-reductive sciences and adapted Meta-Reductive Paradigms. Adapted paradigms could produce adapted Meta-Reductive Scientific Narratives that include a pervasive awareness of reductive Logical incompleteness and the ubiquitous presence of natural evolutionary incompleteness.

Can Reductive Logic Adapt? Meta-Reductive Logic. To succeed the meta-reductive perspective must preserve 'bottom-up' reductive Logic as a 'special case' *and* adapt 'bottom-up' reductive Logic in order to confront meta-reductive, multi-scale, multi-directional, incompleteness driven, co-evolution.

The rigor of 'bottom-up' reductive Logic can be *preserved* and *encompassed* as a 'special case' in a more comprehensive, multi-descriptive, iterative, meta-reductive Logic. Every *iteration* of reductive Logic employed in the multiple descriptions of a meta-reductive, multi-scale and multi-directional perspective can be internally rigorous, strong, reliable and consistent; pursuing the goal of well-formed, logical, reductive statements targeting causal mechanisms and relationships.

However, the structure of preserved 'bottom-up' reductive Logic must also be *adapted* particularly in relation to defining, exploring and determining the *limits* and *implications* associated with the *formal incompleteness of reductive Logic*. A meta-reductive Logic therefore needs to be capable of handling *formal reductive incompleteness* by, perhaps, adapting reductive Logic into a *paraconsistent* form [13]. Such an adapted Logic would include *undecidable reductive propositions* declaring the presence of *formal reductive incompleteness,* thus defining a *limit* on any particular iterative implementation of reductive Logic.

In a paraconsistent meta-reductive Logic, *meta-consideration* is the only route to determining *reductive consistency*. A paraconsistent reductive Logic must be capable of effectively exploring the *onset of incompleteness* in any single iterative application of reductive Logic in the context of meta-reductive, multi-scale, multi-directional, incompleteness driven, co-evolution. As well, an adapted paraconsistent reductive Logic must be able to model the *complex edge of incompleteness* where multiple iterative implementations of reductive Logic intersect and interact in the context of meta-reductive, multi-scale, multi-directional, incompleteness driven, co-evolution. The intention of a paraconsistent, meta-logic would be to model how abstract scientific Logic and the natural Logic of Nature go about *resolving* and *deciding* previously unresolvable natural instantiations and undecidable reductive propositions.

Any paraconsistent Logic capable of modeling meta-reductive, multi-scale, multi-directional, incompleteness driven, co-evolution, would assume the inevitability of eventually running into its own undecidable reductive propositions.

Novel definitions and scientifically accessible hypotheses must inevitably appear in relation to the development of a meta-reductive paraconsistent Logic. For instance, self-organization, emergence and complementarity must also be redefined and explored in the framework of meta-reductive, multi-scale, multi-directional, incompleteness driven, co-evolution.

The next task involves spelling-out the relationship shared by reductive formal incompleteness and another kind of 'unfinished and partial description' associated with the mathematics of Symmetry and Group theory, which appears when this form of abstract and applied mathematics is employed in the task of modeling causal mechanisms and relationships in Nature.

3.4 The Synthesis of Symmetry and Self-similarity

In focusing attention on a scientific understanding of causal relationships in Nature through the applied mathematics of Symmetry and Group theory [18], natural science ignores a fundamental and significant conceptual link between Symmetry and Self-similarity that could advance the understanding of causal relations in natural evolving systems.

Symmetry is a *two-part* mathematical relationship involving *invariance* in relation to *variance. Self-similarity* is a *special* kind of symmetry in which *invariance* is related to *variance* across a range of *scale* [20, 33]. Symmetry and Group theory can be integrated with Self-similarity and Fractals and this synthesis can significantly modify the scientific approach to modeling causal relationships in Natural settings.

Defining a Fundamental Unit of Self-organization. In order to achieve the synthesis, a *fundamental unit of systemic evolutionary self-organization* needs to be defined that can link the abstract conception of Symmetry and Self-similarity; which then suggests an accessible hypothesis regarding Symmetry and Self-similarity in natural evolving systems and a deeper reason for the remarkable usefulness of mathematics in natural science [34].

A hypothetical *abstract fundamental unit of systemic organization* linking Symmetry and Self-similarity would need to be simple enough that it spells-out a fundamental *associative binary operation* allowing a consistent definition for a fundamental evolutionary Symmetry Group.

In order to integrate Symmetry and Self-similarity in causal modeling, the *same abstract fundamental unit of systemic organization* would need to also spell-out an *equivalent* fundamental *iterative algorithmic operation* consistent with a definable evolutionary Multi-Fractal.

To be scientific, the hypothesized *abstract fundamental unit of systemic organization* must be both theoretically and experimentally accessible. Ultimately, its existence in Nature must be repeatedly verified, thus, demonstrating that the abstract epistemological unit is *related to* an *ontological* manifestation in the form of a *natural fundamental evolutionary unit of systemic organization.* The scientific task, therefore, is to demonstrate that the abstract theoretical structure and process is a good approximation of a natural evolutionary structure and process (Fig. 9).

A candidate for a joint *binary associative* (Group theory) and *iterative relation* (Multi-Fractal) capable of fulfilling these expectations can be composed from a very simple binary relationship. The binary '*change* or *no change*', or, the binary '*difference* that *does or does not* make a difference' [2] offers a good starting point.

Within any *consistent established natural order*, *one half of the binary* can be stated: *change* can occur that makes *no difference* to the structure and rules of an established order and *change* can occur that makes no difference to the *consistency* of an established order. The *other half of the binary* can be stated: a *change* can occur that *does make a difference* to the structure and rules of an *established order* and does make a difference to the *consistency* of an established order.

The *binary* 'change or no change', 'does or does not', 'is or is not', when associated with the *consistency of an established natural order*, provides a fundamental binary

1. Defining a fundamental evolutionary meta-reductive binary *associative operation* and binary *iterative relation,* requires *meta-consideration* of the *consistency* of any naturally established *order.*

$$Binary\text{: } \Delta \text{ or } \Delta^0 \text{ in the Consistency } \mathbb{C} \text{ in an established order } \theta$$

2. A Meta-reductive *Evolutionary Symmetry Group* and *Evolutionary Multi-Fractal* can be based on the meta-reductive binary.

3. The Evolutionary Multi-Fractal Symmetry Group reveals a self-similar epistemological and ontological unit of complex organization, which again requires *meta-consideration* of the *consistency* associated with any naturally established *order.*
   a) A self-similar epistemological unit of self-organization: The *Meta-Reductive Complex Natural History,*
   b) A self-similar ontological unit of self-organization: The *Meta-Reductive Complex Natural Cycle of Existence.*

Fig. 9 A fundamental evolutionary binary operation; and, 2. The Evolutionary Multi-Fractal Summetry Group; and, 3. Hypothesized complex units of organization operation or abstract choice that can form the core of an *associative and iterative relationship.* Such a binary operation or binary choice results in the construction of a Symmetry Group and Multi-Fractal that responds to the presence and the implications of formal incompleteness.

The construction of a Symmetry Group and Multi-Fractal, where the associative binary operation and iterative relation is related to logical consistency, can only be achieved by composing the necessary structure in the space of *meta-consideration* and *meta-construction.* The Symmetry Group and Multi-Fractal must also be sensitive to formal reductive incompleteness in any abstract logic or mathematical system model of sufficient complexity. The Symmetry Group and Multi-Fractal must therefore initially exist *abstractly*, only in the epistemological world of self-reflective and self-referencing mind, consciousness and meta-consideration of change or no change in consistency. *However*, the composed Symmetry Group and Multi-Fractal may hypothetically describe an experimentally accessible ontological natural instantiation involving symmetry operations and self-similar iteration, manifest in the Universe and flow of complex natural system evolution.

In order to avoid contradiction and paradox, formal reductive incompleteness and natural evolutionary incompleteness demand that the *consistency* of any *algorithmic pattern* involving *natural systemic order must be determined as a meta-consideration,* which occurs outside the natural systemic order. Consciousness can assume the position of a meta-level system capable of making the determination of consistency in relation to a given sequence in the natural order. Where an individual consciousness *is* the system of interest, then another individual consciousness must assume the position of meta-consideration in relation to determining logical system consistency. *Meta-consideration* avoids contradiction and paradox and preserves, protects and sustains the abstract meaning and coherence of the inherent Logic associated with a selected abstract system or a selected natural evolving system.

While the operation, choice or decision regarding *consistency or inconsistency* is *binary*, formal incompleteness predicts that the possible states of the system from which the binary decision is abstracted, can involve *three* abstract options. These are: (1) A *consistent order* with an *established Logic*; (2) A *transitional order* in which there are *unresolved* or *undecided* statements which *protect the system's Logic and consistency*; *and*, (3) An *inconsistent order* with a *failed Logic.* These *three* possibilities are related to *three* Logic states: *true* (decidable and consistent); *undecidable* (consistent only because a statement threatening contradiction, paradox and inconsistency is left undecided); and *false* (decided and inconsistent).

The *binary* decision regarding *consistency or inconsistency* can also be linked with potential adjacent possible *meta-constructions.* In the abstract, meta-constructions involve additional axioms added to a Logic system or system model that can *resolve* or *decide* a previously *undecidable* proposition or statement. The equivalent in a natural system, involves a natural *meta-construction* that moves *a transitional order* into a *resolved meta-order*. In a transitional order, there are *unresolvable instantiations* and *undecided properties* that have not yet made manifest their potential action, interaction, relationship or implication. In a resolved meta-order, the same previously unresolvable instantiation and undecided property has transformed and become manifest in its emergent properties within the resolved meta-system. The unresolvable instantiation becomes an entity, takes part in a composition, participates in an interaction, or engages in a relationship with entities that are not part of the *prior established consistent order*. The unresolvable instantiation is now revealed by its behavior and properties to be a resolvable instantiation that is an integral part of an *emergent order with its own inherent meta-consistency*.

Within a *natural transitional order*, a *change* occurs in the form of an energy-matter and space-time, self-organized entity, composition, interaction or relationship, which is contained within a local and extended environment. The *change* '*is* and *is not* consistent with the established order'. Such a transitional state and process can only be *unresolved* within the existing natural order and *undecided* in a Logic-based model of the natural order. A transitional state and process in a scientific model of a natural system would need to be left undecided and undecidable, as it threatens contradiction and self-referencing paradox, by having properties that are both *consistent* with the established order and *not consistent* with the established order. The transitional form exhibiting natural evolutionary incompleteness must be left unresolved and undecided. The status, as a transitional entity, cannot, however, be established, until the extended emergent meta-order begins to self-organize *around* the transitional property that *is not consistent* with the previously established order but is consistent within itself.

As an *emergent meta-order* self-organizes and establishes its own meta-Logic and meta-consistency, a transitional 'is and is not' state and process will reveal its capacity to participate in meta-consistent or meta-constructed relationships.

Within the meta-order the *binary decision* and *binary operation* involving consistency will again appear. The potential for meta-level incompleteness is inevitable.

A simple and fundamental link can now be made between the concepts defining Symmetry, Group theory, Self-similarity and Fractals.

Defining a Fundamental Evolutionary Symmetry Group. A fundamental evolutionary Symmetry Group would involve:

(a) A set of *elements* (defined by specific examples of *binary* '*change or no change in consistency', in relation to an established consistent order*; ie., there exists an instantiated natural order, involving state and process, energy-matter and space-time phenomenon, entity, composition, interaction and relationship in which the inherent Logic can be spelled-out and the consistency determined as a meta-consideration);
(b) An *inverse* for every element (an *element* is defined by a *binary 'change or no change* in relation to consistency, and the presence of a binary choice can *exist or not-exist*—Nature can erase evidence of historical events or destroy the present moment of states and processes);
(c) An *identity* element (defined as a *binary change* which precipitates '*no change* in *consistency*' and '*no change* in *existence or non-existence*, leaving the element, the order and the consistency, *unchanged*—a binary change that makes no difference); as well as,

(d) An *associative binary operation* ('change or no change in the consistency of an established order', is the fundamental binary operation): The three choices associated with incompleteness therefore are: 'is consistent therefore decided', 'is and is not consistent therefore undecided' and 'is not consistent therefore decided'.

Sequences of binary operations performed upon any series of elements linked by the binary operator, are *associative*, if the sequence always produces the same outcome, whatever the grouping, as long as the order remains the same $[(a \times b) \times c = a \times (b \times c)]$. In this case, the *binary operation* and *any sequence* of *elements*, in the form of an instantiated evolutionary sequence of state and process, energy-matter and space-time phenomenon, entity, composition, interaction and relationship, sustains the evolutionary sequence and outcome of, change and no-change, existence and non-existence, order and disorder, as well as, unresolved transition and meta-order. The sequence and outcome are sustained no matter how consciousness might choose to group, to self-reflect and meta-consider; and no matter how Nature might naturally cluster, group, or instantiate the sequence. If the sequence is associative, the outcome of the entire sequence, which might include the self-referential emergence of consciousness, will remain the same.

Focusing on a binary operation that is sensitive to inherent incompleteness and also is constructed on a relationship involving consistency, which necessarily requires meta-consideration in its determination; it becomes possible to construct an evolutionary Symmetry Group that is *associative. Consciousness* does *not* have a special role in any part of the natural evolutionary sequence that emerges from the Symmetry Group. However, consciousness can *choose* to go beyond abstract description, clustering and grouping of an invariant evolutionary sequence, by spelling-out Nature's simple rules, laws and patterns associated with evolving symmetry relationships as well as intentionally or unintentionally engaging in non-local or local *interactions* with the evolutionary sequence [11], thus altering the course of co-evolution.

Defining a Fundamental Evolutionary Multi-Fractal. A fundamental Multi-Fractal can now be composed that can be integrated with the fundamental Symmetry Group, by using the same fundamental unit of self-organization:

The properties of a Fractal include, *self-similarity, iteration* and *scaling*. In the abstract, a *fractal* is a *geometric form* created by an *iterated system*. The Mandelbrot set, for instance, is a fractal with some logical depth but little or no effective complexity (the length of a concise description of a set of entities regularities).

The Mandelbrot set can be *reduced* to a simple equation:

$$f_c(z) \stackrel{\text{def}}{=} z^2 + c$$

Despite its iterative simplicity and lack of effective complexity, it produces an infinite and vastly complex geometric object.

If the process of natural evolution spells out a potentially increasingly complex and repeating *regularity*, then natural evolution and the repeating regularity can be taken as an *iterated system* and the products of complexification can be taken as similar to a potentially increasingly *complex geometric form.* Thus, change, evolution, self-organization, the emergence of a hierarchy of complexity as well as the entire sequence of the evolving Universe, can be described as a *multi-fractal system*, one which does exhibit a *very high order of effective complexity* in a region *intermediate between total order and complete disorder* [22].

The *regularity* can be described by a self-similar, iterated and scaled, *binary sequence of operations*, involving, '*change or no change in the consistency of an established order*'.

$$f_{\mathbb{C}}(\theta) \stackrel{\text{def}}{=} \theta\left(\Delta_{\mathbb{C}} \vee \Delta^0_{\mathbb{C}}\right) + \mathbb{C}$$

A function (f) operates on the consistency ($\mathbb{C}$) of an existent order (θ). An iterated binary operation ($\Delta_{\mathbb{C}} \vee \Delta^0_{\mathbb{C}}$) defined by: change ($\Delta$) in consistency ($\mathbb{C}$) or ($\vee$) no change ($\Delta^0$) in consistency ($\mathbb{C}$) operates on an existent order (θ) and (+) the constant abstract meta-consideration of the consistency ($\mathbb{C}$) associated with the given order (θ). The value $+\mathbb{C}$ refers to invariant contexts where proven epistemological meta-consideration of logical consistency maps onto logical and experimentally provable ontological consistency. The value $-\mathbb{C}$ refers to variant contexts where proven epistemological meta-consideration of logical inconsistency maps onto logical and experimentally provable ontological inconsistency ($-\mathbb{C}$ represents meta-consideration of an interesting hypothetical and very different scientific context—however, it is *not* the context we are presently exploring).

Once again, there are *three* choices relevant in each self-similar, iterated and scaled representation of the binary sequence of operations and elements: (1) The 'change or no change is consistent with the established order'; (2) the 'change or no change is consistent with the established order but only by having potentially contradictory or paradoxical properties that are not yet fully instantiated, manifest or resolved, in relation to a potential adjacent possible emerging meta-system and meta-order, within which the unresolved properties would be defined as resolved and consistent; and, (3) The 'change or no change is not consistent with the established order'.

The evolutionary Multi-Fractal is then defined by:

(a) *Self-similarity*: The abstract elements and binary operations, associated with consistency and incompleteness, can be related to natural energy-matter, space-time, entity, composition, interaction and relationship involving any level or degree of hierarchical complexity.
(b) *Iteration*: The sequence of binary operations, acting on elements associated with consistency and incompleteness, can be any length or degree of complexity and the abstract nature of the iteration will be sustained.
(c) *Scaling*: The abstract elements, the sequence of binary operations and the relationship with consistency and incompleteness, can be related to any natural system involving micro-scale, mid-scale or macro-scale state and process.

The proposed evolutionary Multi-Fractal can be further spelled-out and integrated with other scientific work. For instance, the fractal dimension of our Universe can be defined more precisely and the relationship shared by the self-similar, iterative, scaled fractal evolutionary pattern and simple rules, laws and diverse, complex, real world patterns that emerge within the sequence of Nature's fractal evolution. The synthesis of the fundamental evolutionary Symmetry Group and the fundamental evolutionary Multi-Fractal, can integrate Symmetry and Self-similarity in the analysis of causal relationships in Nature. The integration produces a scientific narrative supported by a Complex Multi-Fractal Symmetry Group.

Symmetry, Self-similarity and Meta-Reductive Causal Relationships. Symmetry demands analysis of causal relation through exploration of *correlation, equivalence relations* and *symmetry relations* in selected subsystems of cause and effect. Self-similarity would demand analysis of causal relation in selected subsystems of cause and effect, through *multi-scale, self-similar correlations, multi-scale self-similar equivalence relations* with subsequent determination of *significant multi-scale self-similar causal relationships.* The exploration of symmetry *and* self-similarity in relation to causal relationships in Nature would make it possible to spell-out both 'simple' and 'complex' causal mechanisms; as well as possible to abstract and quantify the rate at which Nature and evolution increase detail in the form of simple or complex symmetry groups [30] while increasing complexity and the fractal dimension of energy-matter states and processes in relativistic space-time [20].

The abstract, simple, *fundamental unit of systemic evolutionary self-organization* can be linked sequentially into more complex abstract units of self-organization that can also exhibit high degrees of symmetry and self-similarity. A linked sequence, for instance, could be conceived and defined in relation to an *epistemological Complex Natural History* and an *ontological Complex Cycle of Existence*, which can describe the history and cycle of existence of simple or complex, evolving, micro-scale, mid-scale or macro-scale, energy-matter, space-time phenomena [24].

Two further hypotheses arise from the integration of Symmetry with Self-similarity and a further integration of the dynamics of incompleteness, in relation to the scientific task of defining causal mechanisms and relationships in Nature:

(1) An *Incompleteness Driven Novelty Generating Hypothesis*: predicts natural *incompleteness driven novelty generation* could function as a *ubiquitous fundamental complex systemic causal mechanism* in natural evolving systems.
(2) A *Hypothetical Hierarchy of Simple and Complex Causal Mechanisms* can be composed: where non-local and local causal mechanisms may belong within a more complicated description of causal relationships defined under Symmetry, Self-similarity and Incompleteness.

    (a) Non-local and local causal mechanisms can be defined as *simple causal mechanisms* discovered under the mathematical umbrella of Symmetry.
    (b) A fundamental *Incompleteness Driven Novelty Generating Causal Mechanism* could be defined as a fundamental *complex* causal mechanism allowing transition into incompleteness and novelty generation.
    c) The mathematics of Self-similarity, when used in examining causal relationships, might lead to theoretical and experimental categorization of more *complex causal mechanisms* beyond the *Incompleteness Driven Novelty*

Generating Causal Mechanism. *Complex* causal mechanisms could self-organize from *simple* non-local or local causal mechanisms as above and the *Incompleteness Driven Novelty Generating Causal Mechanism* combined with other secondary more complex causal mechanisms.

(i) Simple *primary* non-local and local causal events may enter an *edge of incompleteness*, where the complex systemic *incompleteness driven novelty generating causal mechanism* could combine with *secondary* and more complicated kinds of causal relationship.

(ii) Complex causal relationships associated with incompleteness are already found within descriptions of reductive 'bottom-up' evolution but will also be found within the description of multi-scale, multi-directional, multi-relational, co-evolution.

(iii) Examples of *secondary* complex systemic causal mechanisms that could combine at the *edge of incompleteness* with *incompleteness driven novelty generation*, include, but is not limited to: phase change; non-linear or chaotic process; stochastic system evolution; Lamarckian system evolution; bidirectional intersection with developing systemic duality or complementarity; simple conservative (reductive) emergence or more complex forms of novel or radical emergence (meta-reductive), consistent with multi-scale, multi-directional, co-evolution.

3.5 Can the Scientific Method Adapt? Meta-Reductive Methodology

Two perspectives on 'part' and 'whole' relationships provide a framework for considering how to adapt reductive scientific theoretical and experimental methodology in the face of formal reductive incompleteness.

The 'part' perspective is consistent with modern Reductive Science. A selected phenomenon of interest, *P*, can be the 'raisin' in the metaphor of raisin bread, such as an energy-matter *state* in space-time. A scientist can expand the range of their interest by examining a *sequence* of *states,* thus exploring the *process* of *P*. If the scientist choses to further expand the range of their interest, then a full synchronic and diachronic, *Four-Dimensional Life Cone Trajectory* could be modeled for phenomenon. The *Four-Dimensional Life Cone Trajectory* is a hypothetical meta-reductive *epistemological* structure, which composes a *Complex Natural History* describing the entire sequence of states and processes comprising the existence of phenomenon *P.* It is intended to model an *ontological* structure and process, defined as the *Complex Natural Cycle of Existence* of the selected phenomenon *P*. A semi-isolated experimental examination of *P* could *reduce* the overall system to cause and effect subsystems, which exhibit observable, predictable or repeated pattern and causal relationships. The sum of the experimental observations allows the scientist to piece together the entire narrative story of the phenomenon *P.*

The 'whole' perspective is again consistent with modern Reductive Science. In this case, the selected phenomenon of interest *P'* is, the 'whole' multiplying expanding yeast, the flour and other ingredients including the raisins making up the 'whole' raisin bread dough. This is the perspective of modern Cosmology. The 'whole' perspective is *limited* by the *impossibility* of ever repeating *P'*. Science, however, is not limited in its ability to abstractly semi-isolate the system, reduce it, then replicate and make predictions from conceptual, computational or mathematical models of *P'*. A *Four-Dimensional Life Cone Trajectory* composing the *Complex Natural History* and the *Complex Natural Cycle of Existence* of selected phenomenon *P'* can be pieced together. Again, a narrative story of *P'* can be composed.

A Meta-Reductive *modification* of the reductive scientific *experimental* approach to *P* or *P'* encompasses the reductive semi-isolated system as a 'special case' in a meta-reductive extended semi-isolated system.

A *reductive semi-isolated system and its presumed cause and effect subsystems*, are positioned *inside* an encompassing, extended, *meta-reductive semi-isolated environment* with *its presumed cause and effect subsystems*. A selected phenomenon of interest *P* or 'part' is contained in a reductive semi-isolated system. *P* is then encompassed within *P'* or the 'whole'. The 'whole' *P'* is located in an encompassing, extended, meta-reductive semi-isolated system, which holds the 'special case' of the reductive semi-isolated system and 'part' *P*; and is further subdivided into a local environment *E*, and an extended environment *U*, representing the 'whole' Universe. Attention can oscillate between the two integrated perspectives involving phenomenon *P* or *P'* and *P* and *E* or *U* (Fig. 10).

This modification of semi-isolated theoretical and experimental methodology is already in use in science but it is not formalized in a *meta-reductive framework* in which *formal reductive incompleteness* and *multi-scale co-evolution* are addressed.

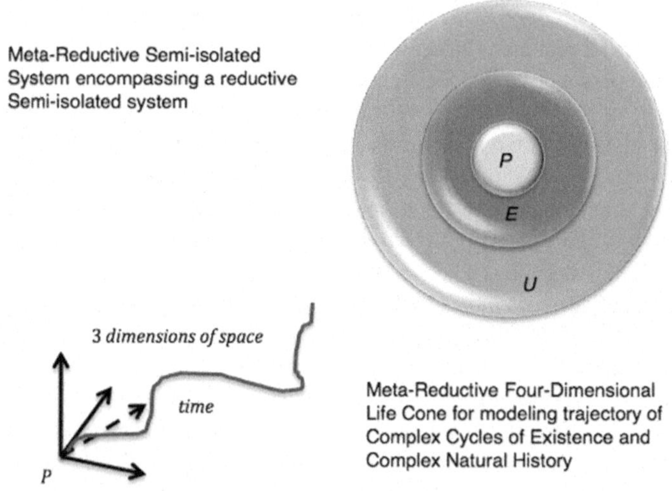

Fig. 10 Meta-reductive semi-isolated system and the four-dimensional life cone

4 Conclusion

This brief outline is but a scant sketch of some of the implications for Reductive Natural Science of multi-scale, multi-directional co-evolution and formal reductive incompleteness. It is intended to capture the interest of anyone who can see the merit of going beyond the bounds of the Reductive Scientific Paradigm. Unknown Meta-Reductive Paradigms sit in the adjacent possible of the Reductive Paradigm, waiting to be explored and offering potential roads to a deeper understanding of Nature and a necessary consilience between the Reductive Sciences and the Complexity Sciences.

A meta-reductive perspective is poised to offer potential novel resolutions to unsolved problems in many branches of natural science. For instance, a Meta-Reductive Matrix of Causal Mechanisms and Relations, offers an alternative way to view phenomena such as quantum entanglement, wave collapse and non-local phenomena within Quantum physics [19]. A meta-reductive causal matrix could also offer a novel way to view the still unresolved boundary between Quantum physics and Relativity theory [15]. A meta-reductive definition of complementarity, involving causally significant symmetry, asymmetry and self-similarity, can reach across many scales of self-organization and many orders of complexity. Complementarity within fundamental physics might then fit within a hierarchy of complementarity, concerning phenomena throughout the hierarchy of complexity and Nature [8].

The meta-reductive perspective preserves 'bottom-up' reductive Logic, by encompassing it as a 'special case' within a meta-reductive paraconsistent Logic, which is capable of modeling undecidable dynamics as well as capable of modeling multi-scale and multi-directional co-evolution. However, the application of reductive Logic is substantially altered in the meta-reductive perspective. The *iterative* application of reductive Logic within a paraconsistent, meta-reductive context reinforces the scientific application of abstract logical and mathematical formal system modeling, allowing for a more immediate and direct connection between scientific thought and modern applied mathematical thought.

By iteratively applying formal system reasoning in modeling undecidable dynamics in multi-scale, multi-directional co-evolution, it is possible to generate accessible scientific and applied mathematical hypotheses, which can link and integrate phenomena across a wide range of scales from micro-scale, to mid-scale, to macro-scale in the natural self-similar evolving system. For instance, scientifically describing undecidable dynamics and natural, patterned, algorithmic, co-evolution in micro-scale quantum systems could theoretically and experimentally access an important *multiple system equivalence.* This descriptive *equivalence* involves, iterated *formal system models,* which can capture undecidable dynamics and multi-scale and multi-directional co-evolution; iterated *dynamical system models* that can also capture undecidable dynamics and equivalent undecidable, multi-scale, multi-directional, co-evolving dynamics in chaotic systems and in non-linear dynamics

[21]; and iterated *information theoretical system models*, which again could equivalently model undecidable dynamics and multi-scale, multi-directional co-evolution in an information theoretical context.

The iterated multiple system equivalence could be used to spell-out, in formal, dynamical and information theoretical terms, a more detailed analysis of how natural differentiation, synthesis and fabrication of self-organized phenomena can arise from binary 'bits' of evolution, involving emerging order, existing order and consistency, unresolvable and undecidable dynamics, incompleteness driven novelty generation and emerging patterns of meta-order and meta-consistency. The theoretical and mathematical equivalence of formal, dynamical and information theoretical system modeling [25], could deepen our understanding of the self-similar, iterative and scaled properties within the natural system [22, 24], ranging from the micro-scale puzzles of quantum physics, to the mid-scale complexity of brain/mind and consciousness and on to the macro-scale intricacy of the whole Universe. Multi-scale, multi-directional, co-evolution, described by a novel meta-reductive, iterated, paraconsistent Logic could further explore this important mathematical system equivalence and in so doing might advance the integration of the Reductive Sciences with the Complexity Sciences.

The meta-reductive study of *change*, introduces an awareness of 'resolvable and unresolvable' natural phenomena and 'decidable and undecidable' abstract theoretical conceptions associated with *causal mechanisms* and dynamic *relationships* in natural systems. Consequently, the meta-reductive understanding provides a novel categorization of 'simple' non-linear and linear causal mechanisms, a fundamental *incompleteness driven novelty generating causal mechanism*, and then 'complex' interactive meta-reductive causal mechanisms and relationships.

Meta-reductive natural science and mathematics could more effectively synthesize the applied mathematics of *symmetry* and *self-similarity* allowing for an integrated understanding of causal mechanisms and relationships, as well as a deeper awareness of the usefulness of Symmetry Groups and Fractal mathematics in the scientific modeling of causal relationships in natural systems.

A meta-reductive paradigm offers a unique perspective from which to consider the relationship between, Nature's simple rules such as recently pursued by Wolfram in relation to Quantum physics and Relativity theory [35], general physical laws and self-organized, emergent patterns within naturally evolving systems. The meta-reductive conception of ontological Complex Cycles of Existence and epistemological Complex Natural Histories, the meta-reductive proposal of a fundamental binary unit of self-organization with an associated fundamental evolutionary Symmetry Group and Multi-Fractal, offers a novel way to integrate underlying simple rules with general physical law, and bring the Reductive Sciences and the Complexity Sciences into a closer inter-disciplinary relationship. The preservation and advancement of formal reductive Logic composed in a meta-reductive framework involving hypothesized natural, multi-scale, multi-directional, incompleteness driven co-evolution creates the potential for an integrated meta-reductive perspective capable of addressing many unresolved scientific problems. Among these anomalous and outstanding problems, a meta-reductive model of a Complex Physical System

(CPS) or Complex Adaptive System (CAS) might be better positioned to address the complexity [28, 32] and the paradoxes of brain, mind [29] and consciousness [16, 17]. A task for another paper.

Acknowledgements Thank you to Dr. Yakov Shapiro and Patricia G. Scott for their steadfast support.

References

1. Andersen, H., Hepburn, B.: Scientific method. In: Zalta, EN. (ed.) The Stanford Encyclopedia of Philosophy. Stanford Metaphysics Research Laboratory, Stanford (Summer 2016 Edition). https://plato.stanford.edu/archives/sum2016/entries/scientific-method/
2. Bateson, G.: Mind and Nature, a Necessary Unity. Bantam Books, Toronto, New York London, Sydney (1980)
3. Brown, J.R., Fehige, Y.: Thought experiments. In: Zalta, EN. (ed.) The Stanford Encyclopedia of Philosophy. Stanford Metaphysics Research Laboratory, Stanford (Winter 2019 Edition). https://plato.stanford.edu/archives/win2019/entries/thought-experiment/
4. Cantini, A., Bruni, R.: Paradoxes and contemporary logic. In: Zalta, EN. (ed.) The Stanford Encyclopedia of Philosophy. Stanford Metaphysics Research Laboratory, Stanford (Fall 2017 Edition). https://plato.stanford.edu/archives/fall2017/entries/paradoxes-contemporary-logic/
5. Carroll, S.: The big picture: on the origins of life, meaning, and the universe itself. Dutton est. 1852, an imprint of Penguin Random House LLC, New York, New York (2016)
6. Ellis, G.: How Can Physics Underlie the Mind: Top-Down Causation in the Human Context. Springer, Berlin, Heidelberg (2016)
7. Gisin, N.: Quantum Chance: Nonlocality, Teleportation and Other Quantum Marvels. Springer, Berlin, Heidelberg, New York (2014)
8. Kelso, J.A.S., Engstrom, D.A.: The Complementary Nature. A Bradford Book, the MIT Press, Cambridge, MA, United States (2006)
9. Kauffman, S.: Reinventing the Sacred: A New View of Science, Reason and Religion. Basic Books, New York, New York (2008)
10. Kauffman, S.: Humanity in a Creative Universe. Oxford University Press, Oxford, UK (2016)
11. Kerr, J.: The Unsolved Puzzle: Interactions, Not Measurements. Gordon Books, The Green, Clophill, Bedfordshire (2019)
12. Lincoln, D.: The Theory of Everything: The Quest to Explain All of Reality. The Great Courses, The Teaching Company, Chantilly, Virginia, United States (2017)
13. Priest, G., Tanaka, K., Weber, Z.: Paraconsistent logic. In: Zalta EN. (ed.) The Stanford Encyclopedia of Philosophy. Stanford Metaphysics Research Laboratory, Stanford (Summer 2018 Edition). https://plato.stanford.edu/archives/sum2018/entries/logic-paraconsistent/
14. Pylkkanen, P.: Mind, Matter and the Implicate Order. Springer Publishing, New York, New York (2007)
15. Pylkkanen, P.: Mind, matter and the implicate order, p. 2. Springer Publishing, New York, New York (2007)
16. Ramsey, W.: Eliminative materialism. In: Zalta, EN. (ed.) The Stanford Encyclopedia of Philosophy. Stanford Metaphysics Research Laboratory, Stanford (Spring 2019 Edition). https://plato.stanford.edu/archives/spr2019/entries/materialism-eliminative/
17. Robinson, W.: Epiphenomenalism. In: Zalta EN. (ed.) The Stanford Encyclopedia of Philosophy. Stanford Metaphysics Research Laboratory, Stanford (Summer 2019 Edition). https://plato.stanford.edu/archives/sum2019/entries/epiphenomenalism/
18. Rosen, J.: Symmetry Rules: How Science and Nature are Founded on Symmetry. Springer, Berlin, Heidelberg (2008)

19. Schlosshauer, M. (ed.): Elegance and Enigma: the Quantum Interviews. Springer, Berlin, Heidelberg (2011)
20. Schroeder, M.: Fractals, Chaos, Power Laws: Minutes from an Infinite Paradise. W.H Freeman and Company, New York, New York (1991)
21. Scott, A.C.: The Non-Linear Universe: Chaos, Emergence, Life, pp. 277–301. Springer, Berlin, Heidelberg (2010)
22. Scott, R.: A Missing Fractal System Theory?, vol. 2, no. 6, p. 10. Letter in: Complexity, Publisher, Wiley (1997)
23. Scott, J.R.: Descartes, Gödel, Kuhn: epiphenomenalism defines a limit on reductive logic. In: Morales, A. J., Gershenson, C., Braha, D., Minai, A., Barr-Yam, Y. (eds.) Unifying themes in complex systems IX, Proceedings of the Ninth International Conference on Complex Systems, (pp. 33–52). Switzerland, Springer-Nature, Springer, Proceedings in Complexity, ISSN 2213-8684, ISSN 2213-8692 (electronic). ISBN 978-3-319-96660–1, ISBN 978-3-319-96661-8 (eBook) (2018)
24. Scott, J.R.: Transpersonal psychology and fractal evolution (Chap. 4, pp. 104–143). In: Marks-Tarlow, T., Friedman, H., Shapiro, Y. (eds.) A Fractal Epistemology for a Scientific Psychology: Bridging the Personal with the Transpersonal, Cambridge, Cambridge Scholars Publishers LTD. (2020)
25. Scott, J.R.: On the formal incompleteness of reductive logic. In: Submitted to the Tenth International Conference on Complex Systems, New Hampshire, USA (submitted for consideration 2020)
26. Seager, W.: Natural Fabrications: Science, Emergence and Consciousness. Springer, Heidelberg (2012)
27. Shapiro, Y., Scott, J.R.: Dynamical systems therapy (DST): complex adaptive systems in psychiatry and psychotherapy. In Mitleton-Kelly, E., Paraskevas, A., Day, C. (eds.). Handbook of Research Methods in Complexity Science, pp. 567–589. Edgar Elgar Publishing, Cheltenham, UK, Northampton, MA, USA (2018)
28. Shapiro, Y., Scott, J.R.: Extraordinary knowing within the framework of natural science: toward a theory of “scientific mysticism”. In: Craffert, P.F., Baker, J.R., Winkelman, M.J. (eds.) The Supernatural After the Neuro-turn. Routledge, Taylor and Francis Group, London and New York (2019)
29. Siegel, D.J.: Mind: a Journey to the Heart of Being Human. W.W. Norton & Company, New York, London, Norton Series on Interpersonal Neurobiology (2016)
30. Stewart, I.: Symmetry: A Short Introduction. Oxford University Press, Oxford (2013)
31. van Riel, R., Van Gulick, R.: Scientific reduction. In: Zalta, EN. (ed.) The Stanford Encyclopedia of Philosophy. Stanford Metaphysics Research Laboratory, Stanford (Spring 2019 Edition). https://plato.stanford.edu/archives/spr2019/entries/scientific-reduction/
32. Weinberger, J., Stoycheva, V.: The Unconscious: Theory, Research, and Clinical Implications. The Guilford Press, New York, London (2020)
33. West, G.: Scale: The Universal Laws of Life, Growth, and Death in Organisms, Cities and Companies. Penguin Books, New York, New York (2017)
34. Wigner, E.: The unreasonable effectiveness of mathematics in the natural sciences. In: Wigner, E. (ed.) Symmetries and Reflections, Ox Bow Press (1979)
35. Wolfram, S.: Finally, we may have a path to the fundamental theory of physics… and it’s beautiful. In: Stephen Wolfram Writings April 14. https://writings.stephenwolfram.com (2020)

Developing an Ontology for Emergence Using Formal Concept Analysis

Shweta Singh, Mieczyslaw M. Kokar, and Durga Suresh

Abstract The phenomenon of emergence has been studied in a variety of fields. Each of these fields has their own understanding and interpretation of what emergence means. Thus far, there is no agreed-upon definition of emergence. In fact, the debate about what emergence is and its validity in scientific understanding is still continuing. Having a clear understanding concept of emergence is critical for analysis of such behaviors. In this work-in-progress paper, we discuss our approach for developing an ontology of emergence which can be viewed as a *descriptive theory of emergence*. The ontology will create a foundation which will support the understanding of the meaning and implications of emergence. A side benefit of this ontology will be the taxonomy of both the emergent behaviors and the emergence research. Thus researchers in this field will be able to identify which aspects of emergence are covered by the various research works and which aspects are not covered and thus open for investigation. Conceptually, the ontology will be a *meta-theory* of emergence. By raising the level of abstraction, the meta-theory of emergence will allow to express various theories of emergence within one framework, thus allowing multiple theories that may contradict each other to live in the same place. This will allow one to query the ontology about both the commonalities and differences between the theories. We discuss the development of such ontology and evaluation approach in this paper.

Keywords Emergence · Formal concept analysis · Ontology · Evaluation

S. Singh (✉) · M. M. Kokar · D. Suresh
Department of Electrical and Computer Engineering, Northeastern University, Boston, MA 02115, USA
e-mail: ssingh@ece.neu.edu

M. M. Kokar
e-mail: mkokar@ece.neu.edu

D. Suresh
e-mail: dsuresh@ece.neu.edu

D. Braha et al. (eds.), *Unifying Themes in Complex Systems X*, Springer Proceedings in Complexity, https://doi.org/10.1007/978-3-030-67318-5_6

1 Introduction

Emergence has multiple interpretations and classifications in literature. The history of emergence is inconsistent and controversial. But understanding different theories or model of emergence is critical for this research. Following is a small sampling of the definitions from the literature.

- **Chalmers (1996)**: "Emergence is the phenomenon wherein complex, interesting high-level function is produced as a result of combining simple low-level mechanisms in simple ways." Chalmers also presented the more generalized definition: "Emergence is the phenomenon wherein a system is designed according to certain principles, but interesting properties arise that are not included in the goals of the designer" [7].
- **Holland (1998)**: "Emergence is above all a product of coupled, context-dependent interactions. Technically these interactions, and the resulting system, are *non-linear*. The behavior of the overall system cannot be obtained by *summing* the behaviors of its constituent parts" [14].
- **J. Goldstein (1999)**: "Emergence refers to the arising of novel and coherent structures, patterns and properties during the process of self-organization in complex systems. Emergent phenomena are conceptualized as occurring on the macro level, in contrast to the micro-level components and processes out of which they arise" [13]. Goldstein also outlined the common properties to identify system or computer simulations as emergent: (1) radical novelty; (2) coherence or correlation; (3) global or macro level; (4) dynamical; and (5) ostensive.
- **J. Fromm (2005)**: "A property of a system is emergent, if it is not a property of any fundamental element, and emergence is the appearance of emergent properties and structures on a higher level of organization or complexity" [11].
- **G. Marsh (2009)**: "The emergent properties of the collective whole do not in any transparent way derive from the underlying rules governing the interaction of the system's components" [21].
- **C. Szabo and Y. M. Teo (2013)**: "Complex systems often exhibit properties that are not easily predictable by analyzing the behavior of their individual, interacting components. These properties, called emergent properties, are increasingly becoming important as software systems grow in complexity, coupling and geographic distribution" [34].
- **O'Toole et al. (2014)**: "Emergence can only occur in systems composed of autonomous parts, agents, who interact in dynamic, non-deterministic ways" [27].

Different definitions of emergence have different features, thus a generic representation/formalism is needed. In other words, a language in which the features of each of the cases of emergence are represented in the same way is needed. One approach is to build an *Ontology*. Ontology is a formal explicit description of concepts and relationships of a specific domain. Ontology has been used in emergence domain, to best of our knowledge, in two papers [16, 23]. However, none of the works have demonstrated the capability of representing and classifying all, or at least most of the possible cases of emergence in a system. Our aim in this work is to create the theory of emergence using Ontology. Ontology provides the capability of inference that will help us to understand the commonalities and difference between the theories.

2 Emergence Ontology

The first step in building an ontology is a comprehensive study of the knowledge domain for the concept that captures its theories, definitions, and classifications. In this work, we use an approach of developing ontology using Formal Concept Analysis (FCA). FCA is a mathematical approach for data analysis that supports ontology development by abstracting conceptual structures [12, 31] that facilitates the development of the ontology. Figure 1 presents our approach. The following sections discuss the use and significance of each step.

2.1 Features of Emergence

The first step is to investigate the features of emergent behaviors. The term 'feature' is used here to represent a characteristic or a property. Multiple definitions presented above, although different, but still consist of some common characteristics. In our review of the literature, we studied different definitions of emergence in literature and identified various features that characterize emergence. Below, we describe these features. Some of them are relatively simple to explain intuitively. But then some others are quite involved and even not well defined:

1. *WOSP* (**W**hole is **O**ther than the **S**um of its **P**arts): This famous quote by Aristotle [2] means that the whole system generates richer behavior than the sum or difference of the behaviors of its components.
2. *Levels*: This feature refers to the idea that for emergence to exist, there must be at least two levels: (1) the level of a collection of agents, and (2) a higher level that is above the agents that interacts with the agents but also has some attributes that are associated with the whole systems of agents, and not with any of the single agents. In general, there may be more than two levels.

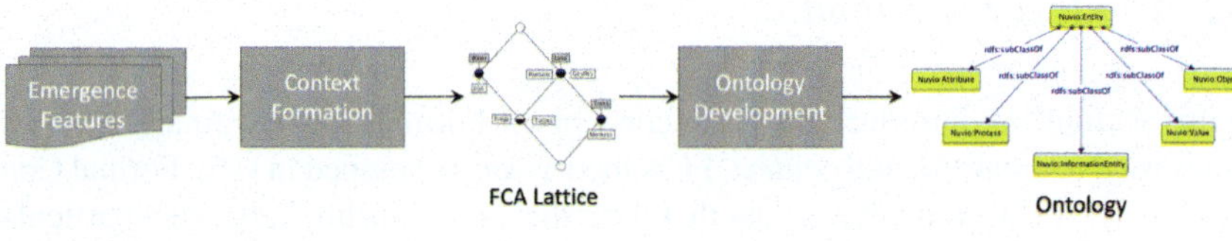

Fig. 1 Ontology development

3. *Interactions*: This feature refers to an exchange of information, and possible actions, between the agents and between the levels. Interactions may be passive, like taking measurements (observing) of the agents, or active—passing messages between agents and between levels.
4. *Radical Novelty*: This feature refers to something that is "genuinely new", like new structures, patterns of behavior or properties.
5. *Unpredictability*: It means that there is no prediction possible that is more efficient than simulation.
6. *Irreducibility*: A behavior is irreducible if a property cannot be deduced from the properties of its constituent parts. This concept seems to be semantically equivalent to the concept of unpredictability.
7. *Dynamical*: This feature refers to the fact that interacting agents are (often non-linear) dynamical systems and thus emergence is the result of changes that take place over time.
8. *Coherence*: Logical consistency or quantitative continuation (no erratic jumps). The emergents appear as integrated wholes that tend to maintain some sense of identity over time. This coherence spans and correlates the separate lower-level components into a higher-level unity.
9. *Decentralized*: This feature refers to decentralized control, i.e., the agents are not controlled by a central controller.
10. *Observer*: This feature represents an external entity that observes the global properties of the system. Ronald et al. [29] state that without an observer, the issue of emergence could not arise at all.

Each of the above features represents a concept referenced by one or more authors to define the term emergence or emergent behaviors in respective papers. The ten features appear in the definitions in [1, 3–6, 8–11, 13–22, 26–28, 33]. With just 23 definitions, we have ten different features representing emergence. Note that the definitions from these papers/authors do have overlapping features. Many authors in the literature also presented multiple classifications based on these features. Our aim here is to study these definitions, features, and classifications to create a *meta-theory* about emergence that can be accessed to get a clear understanding of the domain. We define this relationship between authors and referenced features by formalizing a *formal context* which can be further mapped into Ontology. The next section discusses the theory behind context formation.

2.2 Context Formation

In this section, we introduce the basic concepts of Formal Concept Analysis (FCA) that is used to create *formal context*. FCA theory was developed in [36]. Formal Concept Analysis (FCA) analyzes data that describe a relationship between a particular set of objects and a particular set of attributes. Here in our work, we are considering definitions of emergence as objects and features of emergence as attributes;

A	B	C	D	E	F	G	H	I	J	K
	Levels	Novelty	Unpredict…	Irreducibil…	Interactions	Coherence	Dynamical	Decentrali…	Observer	WOSP
Lewes		X		X						
Crutchfield		X					X	X	X	
Chalmers	X	X								
Dyson		X					X			X
Holland					X	X				X
Goldstein	X	X				X	X			
Kim	X	X	X	X						
Bonabeau	X				X		X		X	
Fromm	X				X	X				
Fisher	X		X		X	X		X		
Johnson				X	X					
Mogul			X				X			
Alison		X	X							
Kaisler		X		X						
Marsh				X	X					
Hsu	X				X			X		
Szabo			X		X					
O'Toole	X	X			X		X			
V.Kim	X	X		X			X			
Kubik	X				X					X
Bedau	X						X			
Baas	X			X	X				X	
O'Connor	X			X						

Fig. 2 Formal context—tables with objects (definitions) and attributes (features)

this makes this suitable for formal concept analysis. Central to FCA is the notion of formal context, which is defined as a triple, $K = \langle G, M, I \rangle$, where G is a set of objects, M is a set of attributes, and $I \subseteq G \times M$ is a binary relation between G and M. A relation $(g, m) \in I$ is read as "object g has the attribute m", where $g \in G$ and $m \in M$. A formal context can be depicted by a cross table. Figure 2 shows the cross table where elements on the left side are objects, the elements at the top are attributes, and the relations between them are represented by the crosses. In this example, objects are the authors that reference one or more attributes, represented by features of emergence. FCA also produces a concept lattice. A concept lattice is a collection of formal concepts in the data which are hierarchically ordered by a subconcept-superconcept relation. For the same data shown in Figs. 2 and 3 shows the concept lattice.

2.3 Ontology Development

With FCA as a backbone, we create a framework for ontology development. In general, to build an ontology, we need to identify the following basic components in a domain. An ontology has four basic primitives: class, relation, individual and axiom. These primitives for ontology are extracted from FCA with the help of a domain expert. FCA provides a semantic structure showing the relationship between objects and attributes. In the emergence context, objects are the authors that describe emergence using some of the attributes. Our next job is to map such structure into an Ontology. Ontology development is not a straight forward process but requires an understanding of the domain.

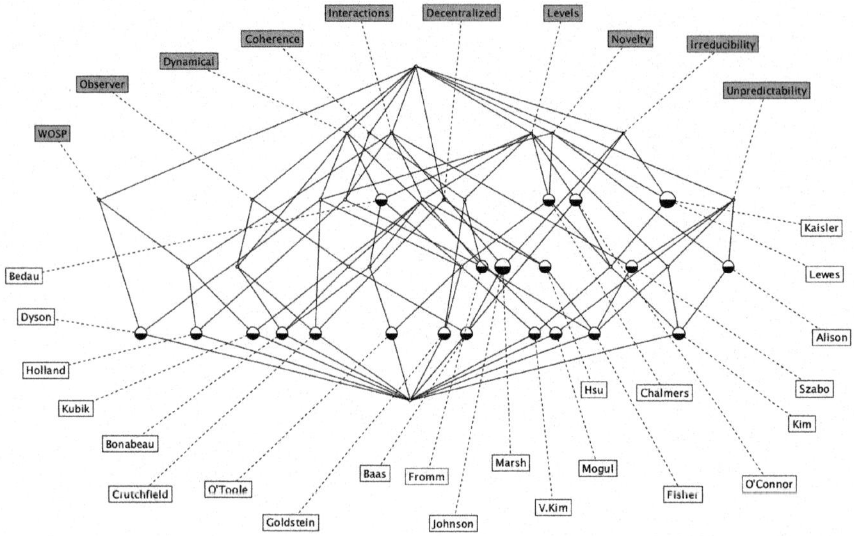

Fig. 3 Concept lattice with emergence features and corresponding reference using formal concept analysis

In this work, we do not want to re-invent the wheel of creating an ontology from scratch but due to the lack of emergence ontology in literature, we need to find an ontology that can be used as a foundational ontology over which we can create our own classes and properties of emergence. We use Nuvio (Northeastern and VIStology) ontology, a new foundational ontology developed by Northeastern and VIStology, Inc [32]. The emergence ontology is created over Nuvio to provide a common vocabulary for defining the concepts and relationships between those concepts for specifying emergence. Figure 4 shows our new classes of emergence Ontology (EmergOnt) over Nuvio.

At the top level, the *Entity* class is the enumeration of all of its direct subclasses, where all the subclasses are disjoint. The subclasses from Nuvio include, *Attribute*, *Object*, *Process*, *Value* and *InformationEntity*. To understand the domain and the requirements for building the new ontology over the existing one using the concepts defined in FCA, we follow the following steps:

- **Class Identification** Map attributes (G) from FCA as subsclasses of Attribute class in Ontology using the *subClass* property
- **Instance Identification** Map objects (M) from FCA as individuals of Object class in Ontology using the *hasIndividual* property
- **Relation Identification** Map the relations (I) from FCA as relations between class and individual in Ontology
- **Axiom Identification** Map the concepts $\langle G, M\rangle$ from format context of FCA as axioms in the Ontology

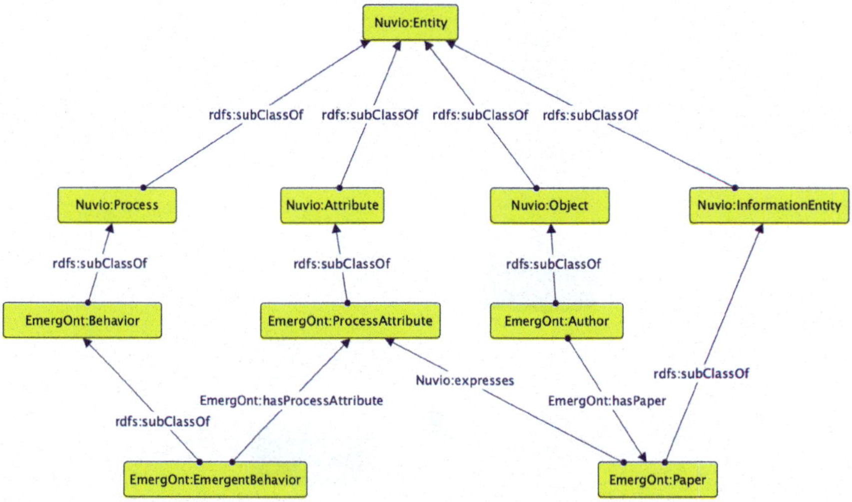

Fig. 4 Emergence ontology

FCA is limited in the ways to represent the domain knowledge. For instance, we had no way to represent that features (attributes) are actually features of emergence, i.e., we cannot represent hierarchical structure using FCA. This can be overcome by using Ontology. For EmergOnt, we first define *EmergentBehavior* as a subclass of *Behavior* which itself is a subclass of *Process*. EmergentBehavior class is connected to *ProcessAttribute*, using a object property *hasProcessAttribute*, which in turn is connected to class *Paper* and *Author* using *expresses* and *hasPaper* property, respectively. The ontology is developed using Protégé [25]. To make sure that we correctly represent the domain of emergence in the ontology, we need to test it. Next section discusses evaluation.

3 Evaluation of Ontology

Currently, there is no methodology that is agreed upon for the development of ontologies or consensus as to how to evaluate them [24]. The ontology summit in 2013 was convened to create guidance for ontology development and best practices for evaluation. Authors in [24] agreed that evaluation of ontologies should be focused on five aspects of the quality of the ontologies:

- *Intelligibility*: Can the humans understand ontology correctly?
- *Fidelity*: Does the ontology accurately represent the domain?
- *Craftsmanship*: Is the ontology well-built and are decisions followed consistently?
- *Fitness*: Does the representation of the domain fit the requirement for its intended use?

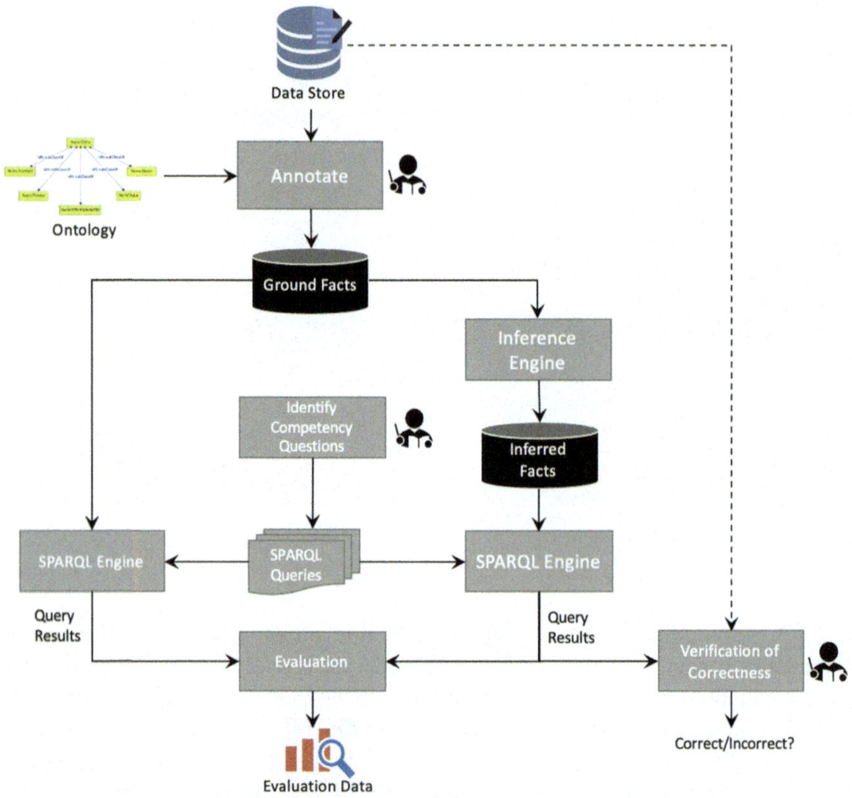

Fig. 5 Comparative evaluation

– *Deployability*: Does the deployed ontology meet the requirement of the information system of which it is part?

All these evaluation criteria are qualitative in nature. In this work, we focus on evaluating the ontology against the fidelity criterion from the above list and attempt to provide some quantitative metrics. We propose two ways for evaluating the ontology for fidelity—*comparative* and *correctness* evaluation. For both approaches, we formalize a set of questions, called *competency questions* [24]. Competency questions are formulated in natural language, often as kinds of queries that the ontology should be able to support for a specific domain. Successful answers to competency questions provide evidence that the ontology meets the requirements.

Figure 5 shows the combined evaluation approach. The comparative evaluation process tests the inference capability of the ontology to generate query results using inferred facts against ground facts. Following are the steps/data involved in comparative evaluation:

1. **Data Store**:
   The data store is raw input to the evaluation process. In our case, data store consists of information about the articles in the domain literature. In this work, data store consists of a subset of the articles in literature where authors provide definitions of the term "emergence" or "emergent behavior", or provide a classification of emergence.
2. **Annotate**:
   Since the data store consists of just articles about emergence, our first step towards refining this raw information is to annotate the data. Annotation is a manual process (although it would also be possible to use some automatic annotation tools) where the data store is annotated with ontological terms, resulting in facts, referred to as *ground facts*. For instance, the ground facts consist of facts about the articles like the following:

   - `Author` *hasIndividual* `author_GHLewes`
   - `author_GHLewes` *hasPaper* `Lewes1875`
   - `Lewes1875` *publishedIn* `year_1875`
   - `Paper` *hasIndividual* `Lewes1875`
   - `Lewes1875` *expresses* `Novelty`
   - `Lewes1875` *hasTitle* `'Problem of Life and Mind'`
   - `Lewes1875` *expresses* `Irreducibility`
   - `Lewes1875` *expresses* `Levels`

   The above facts are just a few examples of the facts stored as ground facts. We plan to extend the facts as we encode more information about the theory of emergence in the ontology and as we annotate more papers.
3. **Inference Engine**:
   The inference engine runs the ontology with the ground facts and creates *inferred facts*. Inferred facts consist of more information inferred from the ground facts and the ontology axioms. For instance, papers in literature usually mention examples of systems that exhibit emergence without mentioning the type of emergence that is described in the paper. We can use the inference capabilities of the ontology inference engine to automatically infer that information about the given example. If we consider the classification of emergence by Chalmers [7], Table 1 shows two examples of systems that are classified as two types of emergence based on [7] and combining the results with that of classification provided by [11]. Brownian motion is characterized by random movements of particles in the fluid. Each particle collides (*interacts*) with each other to form random movement at the system level, therefore there are no novel structures/behaviors in this system, which is a kind of simple emergence. On the other hand, in flocking example, boids are interacting with each other, creating a novel flocking as emergent behavior at the system level or a kind of weak emergence. Therefore, the inference engine correctly identifies Brownian motion and flocking as Type Ib (or simple) and Type IIa (or weak) emergence respectively.

Table 1 Generating inferred facts from ground facts

Example	Annotated facts	Inferred facts

Brownian motion	`x rdf:type EmergentBehavior`
	`x hasProcessAttribute Levels`
	`x hasProcessAttribute Interactions`
		`x rdf:typeIb`
		`x rdf:type SimpleEmergence`

Flocking–Boids	`x rdf:type EmergentBehavior`
	`x hasProcessAttribute Novelty`
	`x hasProcessAttribute Interactions`
	`x hasProcessAttribute Levels`
	`x hasProcessAttribute Irreducibility`
		`x rdf:typeIIa`
		`x rdf:type WeakEmergence`

The inference is really useful to increase the understanding emergence domain. One of our published works [30], describes the classification of emergent behavior using ontology. The framework was based on agent-based simulation and classified emergence for a specific implemented scenario.

4. **Identifying Competency Questions**:
Competency questions are identified by a domain expert. These are the questions that we want the query engine to answer based on the developed emergence ontology. A domain expert (expert in the domain of emergence) creates such competency questions in natural language. Following are a few examples of competency questions (CQ):

   - **CQ 1**: Which papers and authors refer to emergence or emergent behaviors in their papers?
   - **CQ 2**: What are the main characterizing features of emergence mentioned by authors in their papers?
   - **CQ 3**: What is an example of weak emergence?
   - **CQ 4**: Which paper uses "Novelty" as one of the features to define emergence?
   - **CQ 5**: How does Fromm classify emergence?

5. **SPARQL Queries and SPARQL Engine**:
The competency questions, written in natural language, are not machine readable and therefore, can not be understood by the query processor. The next step is to create SPARQL queries [35]. SPARQL is an RDF (Resource Description Framework) query language, that is, a semantic query language for database that is capable of retrieving data stored in RDF format. SPARQL allows for a query to consist of triple patterns, conjunctions, disjunctions, and optional patterns.

The competency questions are first mapped into SPARQL queries. The process of mapping is manual. Obtained SPARQL queries are run against the developed ontology with the help of SPARQL query engine. A query engine is built in the Protégé environment. SPARQL query engine produces results which are used for evaluation.
Following are three examples of competency questions—CQ 2, CQ 4 and CQ 5, mapped from natural language into the SPARQL language:

```
PREFIX rdf: <http://www.w3.org/1999/02/22-rdf-syntax-ns#>
PREFIX EmergOnt:<http://www.semanticweb.org/shweta/ontologies/2020#>
SELECT ?paper ?attribute
    WHERE {
            ?paper rdf:type EmergOnt:Paper .
            ?attribute rdf:type EmergOnt:ProcessAttribute .
            ?paper EmergOnt:hasProcessAttribute ?attribute
          }
```

```
PREFIX rdf: <http://www.w3.org/1999/02/22-rdf-syntax-ns#>
PREFIX EmergOnt:<http://www.semanticweb.org/shweta/ontologies/2020#>
SELECT ?paper ?attribute
    WHERE {
             ?paper rdf:type EmergOnt:Paper .
             ?attribute rdf:type EmergOnt:Novelty .
             ?paper EmergOnt:hasProcessAttribute ?attribute
          }
```

```
PREFIX rdf: <http://www.w3.org/1999/02/22-rdf-syntax-ns#>
PREFIX EmergOnt:<http://www.semanticweb.org/shweta/ontologies/2020#>
SELECT ?paper ?author ?name ?emerg
    WHERE {
            ?emerg a EmergOnt:EmergentBehavior .
            ?emerg EmergOnt:classifiedIn ?paper .
            ?paper EmergOnt:hasAuthor ?author .
            ?author EmergOnt:hasName ?name .
            FILTER regex(?name, "^Fromm")
          }
```

6. **Evaluation**: Finally, the query results obtained from the ground facts are compared against the query results obtained through the inferred facts. The SPARQL engine with inferred facts is expected to provide better results as compared to the ground facts as there are more facts in inferred facts as compared to that in ground facts. This evaluation tests the inference capability of the developed ontology. We test this by running multiple queries on both sets of facts and computing evaluation metrics, like *precision* and *recall*.

The comparative evaluation compares the inference capability of the ontology but does not verify if the query results obtained by SPARQL engine are correct or not.

The correctness evaluation process ("Verification of Correctness" block in Fig. 5), on the other hand, tests the correctness of the query results obtained from ontology against the representation of the domain literature in the data store. The process of verifying the correctness is manual, i.e., a user verifies the output of the query against the data store. Based on the applied query, a user looks at the data store to verify if the obtained inferred query results are consistent with the information in the data store. Again, for this evaluation, we will provide a quantitative empirical evaluation using the precision and recall metrics. The comparisons that result in true (i.e., query engine results are correct), and false, contribute to the computation of the metrics. Precision, also known as true positive accuracy, refers to the fraction of correct query results among total query results obtained, i.e., the ratio of true positives to sum of true positive and false positive. Recall, also known as sensitivity or true positive rate, is the fraction of the total number of correct query results that were actually obtained, i.e., the ratio of true positive to the sum of true positive and false negatives. Precision measures quality whereas recall measures completeness.

4 Initial Results and Conclusion

The presented work in still in progress. We evaluated the developed ontology with few SPAQRL queries based on the limited encoded data and verified the correctness results against the data store. The obtained values of correctness precision and recall metrics are 100%. Figure 6 shows the initial results. With the limited data encoded in the ontology, these results are as expected. This result means that the knowledge obtained as query results from ontology matches with the knowledge of domain

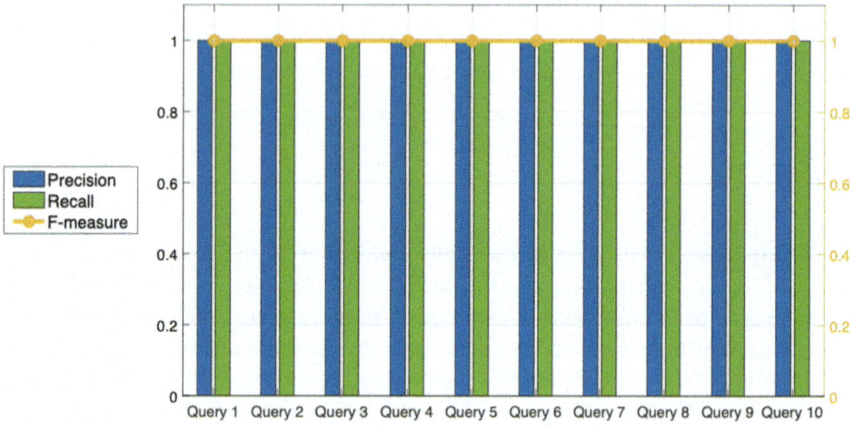

Fig. 6 Correctness evaluation

experts. We plan to evaluate the ontology more rigorously for comparison and correctness after encoding more articles with definitions, classifications and examples of emergence.

To conclude, our aim in this work was to gain understanding of the nature of emergence in general. FCA was used to develop an ontology of emergence that can be queried to understand the domain of emergence. In summary, even though we have not developed a fully descriptive theory of emergence, we have identified a potential direction for the future research of this concept that can be used by researchers to understand this domain better as the inference capability of the ontology will provide us more information about the domain than what is explicitly stated in the literature.

References

1. Allison, I., Merali, Y.: Software process improvement as emergent change: a structurational analysis. Inf. Soft. Technol. **49**(6), 668–681 (2007)
2. Aristotle: Metaphysics. The Internet Classics Archive, 350BCE
3. Baas, N.A.: Emergence, hierarchies, and hyperstructures. In: Santa Fe Institute Studies in the Science of Complexity Proceedings, vol. 17, p. 515. Addion-Wesley Publishing Co (1994)
4. Bedau, M.: Downward causation and the autonomy of weak emergence. Principia **6**(1), 5 (2002)
5. Bonabeau, E., Dessalles, J.-L., Grumbach, A.: Characterizing emergent phenomena (2): a conceptual framework. Revue Internationale de Systémique **9**(3), 347–371 (1995)
6. Chalmers, D.: The Conscious Mind. Oxford University Press, New York (1996)
7. Chalmers, D.J.: Strong and weak emergence. Reemergence Emergence 244–256 (2006)
8. Crutchfield, J.P.: The calculi of emergence: computation, dynamics and induction. Physica D: Nonlinear Phenomena **75**(1), 11–54 (1994)
9. Dyson, G.B.: Darwin Among the Machines: The Evolution of Global Intelligence. Basic Books (2012)
10. Fisher, D.: An emergent perspective on interoperation in systems of systems. Technical report, CMU/SEI-2006-TR-003), Pittsburgh, PA, Software Engineering Institute, Carnegie Mellon University (2006)
11. Fromm, J.: Types and forms of emergence. arXiv:nlin/0506028 (2005)
12. Fu, G.: Fca based ontology development for data integration. Inf. Process. Manage. **52**(5), 765–782 (2016)
13. Goldstein, J.: Emergence as a construct: history and issues. Emergence **1**(1), 49–72 (1999)
14. Holland, J., Mallot, H.: Emergence: from chaos to order. Nature **395**(6700), 342 (1998)
15. Hsu, J.C., Butterfield, M.: Modeling emergent behavior for systems-of-systems. In: INCOSE International Symposium, vol. 17, pp. 1811–1821. Wiley Online Library (2007)
16. Johnson, C.W.: What are emergent properties and how do they affect the engineering of complex systems? Reliabil. Eng. Syst. Safety **91**(12), 1475–1481 (2006)
17. Kaisler, S., Madey, G.: Complex adaptive systems: emergence and self-organization. Tutorial Presented at HICSS-42 Big Island (2009)
18. Kim, J.: Making sense of emergence. Philosophi. Stud. **95**(1), 3–36 (1999)
19. Kim, V.: A design space exploration method for identifying emergent behavior in complex systems. Ph.D. thesis, Georgia Institute of Technology (2016)
20. Kubík, A.: Toward a formalization of emergence. Arti. Life **9**(1), 41–65 (2003)
21. Marsh, G.E.: The demystification of emergent behavior. arXiv:0907.1117 (2009)
22. Mogul, J.C.: Emergent (mis)behavior vs. complex software systems. In: IN EUROSYS, pp. 293–304. ACM (2006)

23. Moshirpour, M., Alhajj, R., Moussavi, M., Far, B.H.: Detecting emergent behavior in distributed systems using an ontology based methodology. In: 2011 IEEE International Conference on Systems, Man, and Cybernetics (SMC), pp. 2407–2412. IEEE (2011)
24. Neuhaus, F., Vizedom, A., et al.: Ontology summit 2013 communiqué: towards ontology evaluation across the life cycle. Gaithersburg, MD (2013)
25. Noy, N.F., Fergerson, R.W., Musen, M.A.: The knowledge model of protege-2000: combining interoperability and flexibility. In: International Conference on Knowledge Engineering and Knowledge Management, pp. 17–32. Springer (2000)
26. O'Connor, T.: Emergent properties. Am. Philoso. Quart. **31**(2), 91–104 (1994)
27. O'Toole, E., Nallur, V., Clarke, S.: Towards decentralised detection of emergence in complex adaptive systems. In: 2014 IEEE Eighth International Conference on Self-Adaptive and Self-Organizing Systems, pp. 60–69. IEEE (2014)
28. Plsek, P., et al.: Redesigning health care with insights from the science of complex adaptive systems. In: Crossing the Quality Chasm: A New Health System for the 21st Century, pp. 309–322 (2001)
29. Ronald, E.M., Sipper, M., Capcarrère, M.S.: Design, observation, surprise! a test of emergence. Artificial Life **5**(3), 225–239 (1999)
30. Singh, S., Lu, S., Kokar, M.M., Kogut, P.A.: Detection and classification of emergent behaviors using multi-agent simulation framework (wip). In: Proceedings of the MSCIAAS, SpringSim. Society for Computer Simulation International (2017)
31. Stumme, G., Maedche, A.: Fca-merge: bottom-up merging of ontologies. IJCAI **1**, 225–230 (2001)
32. Suresh, D., Kokar, M.M., Moskal, J., Chen, Y.: Updating CRO to CRO2. In: Wireless Innovation Forum Conference on Wireless Communications Technologies and Software Defined Radio. Wireless Innovation Forum (2015)
33. Szabo, C., Teo, Y.M.: Post-mortem analysis of emergent behavior in complex simulation models. In: Proceedings of the 1st ACM SIGSIM Conference on Principles of Advanced Discrete Simulation, pp. 241–252. ACM (2013)
34. Szabo, C., Teo, Y.M.: Semantic validation of emergent properties in component-based simulation models. In: Ontology, Epistemology, and Teleology for Modeling and Simulation, pp. 319–333. Springer (2013)
35. W3C. Sparql query language for rdf [online]. https://www.w3.org/TR/rdf-sparql-query/. Accessed 25 Jan 2020 (2020)
36. Wille, R.: Restructuring lattice theory: an approach based on hierarchies of concepts. In: International Conference on Formal Concept Analysis, pp. 314–339. Springer (2009)

Pandemic Preparation, Consciousness and Complex Adaptive Systems Through the Lens of a Meta-reductive Scientific Paradigm

J. Rowan Scott

Abstract The novel coronavirus SARS-CoV-2 and Covid-19 pandemic and the associated economic crisis provide an opportunity to explore an incompleteness driven novelty generating meta-construction of a Meta-Reductive Scientific Paradigm. Previous proof of formal reductive incompleteness and necessary meta-consideration in determining reductive logical consistency, permits the meta-construction of adjacent possible meta-reductive paradigms. An imagined Meta-Reductive Paradigm then allows for a search in an adjacent possible conceptual space, beyond the boundary of the Reductive Paradigm, looking for possible adaptations that might enhance the resilience of individual and community level responses to the Covid-19 pandemic and economic crisis. Unrecognized formal incompleteness of reductive scientific Logic is related to the unresolved integration of Reductive Science and Complexity Science. The lack of integration of these two broad areas of scientific interest, is in part associated with a diminished interest in formal reductive logic and its properties, particularly in the Complexity Sciences. The failure to sustain a strong link to rigorous reductive logic could indirectly relate to diminished public trust in scientific and medical information, with consequent political failure to effectively integrate scientific and medical advice in response to the Covid-19 situation. Careful meta-consideration reveals a subtle way to address formal reductive incompleteness and meta-construct an adjacent possible Meta-reductive CAS Model (MCAS) and Meta-reductive Paradigm (MRP). The strong link to logic and differentiated truth and proof in the proposed MCAS and MRP can integrate willful, intentional, brain, mind and consciousness in Nature, as well as spell-out a number of scientific and social adaptations that could improve the integration of medical and scientific information in the political management of the Covid-19 crisis, increasing the chance that humanity will be better prepared for the next pandemic or next urgent global crisis.

Keywords Pandemic · Reductive science · Complexity science · Reductive logic · Formal reductive incompleteness · Reductive paradigm · Meta-reductive paradigm · Complex adaptive system model · Mind and consciousness

---

J. R. Scott (✉)
Department of Psychiatry, University of Alberta, Edmonton, AB, Canada
e-mail: j.rowan.scott@gmail.com

D. Braha et al. (eds.), *Unifying Themes in Complex Systems X*, Springer Proceedings in Complexity, https://doi.org/10.1007/978-3-030-67318-5_7

1 Introduction

An unprecedented global crisis has been caused by the severe acute respiratory syndrome coronavirus 2 (SARS-CoV-2), which is the strain of coronavirus responsible for the human pandemic beginning in 2019 (Covid-19) [1].

The developing meta-construction of a Meta-Reductive Paradigm (MRP) [2, 3], responsive to demonstrated formal reductive incompleteness of formal reductive scientific Logic [4], can be used to model the Covid-19 coronavirus and its behavior in the ongoing human pandemic. An associated Meta-Reductive Complex Adaptive System Model (MCAS) could also be deployed in the task of modeling the human brain, mind and consciousness, as humanity continues with the task of dealing with and adapting to the SARS-CoV-2 Covid-19 pandemic and its economic consequences.

Incompleteness driven novelty generation is a significant meta-reductive complex causal mechanism defined in the meta-reductive scientific perspective [2, 5], as well as a significant component of the developing World-View created by the Meta-Reductive Paradigm (MRP) and Meta-Reductive Complex Adaptive System Model (MCAS) [6]. Novel adaptations and enhanced creativity of response can emerge from viewing the Covid-19 medical and economic crises through the lens of Meta-Reductive thought, which is responsive to formal incompleteness impacting reductive scientific Logic.

2 Covid-19 in Reductive and Meta-reductive Frameworks

The SARS-CoV-2 coronavirus responsible for the Covid-19 pandemic was originally limited to another animal species, probably a bat [7]. The coronavirus and bat exist together apparently without seriously harming the bat, effectively creating a consistent, ordered, interactive system. The relationship with human beings and human culture in the context of wet-markets brought the bat-virus system into close association with human beings creating a context for an inter-species jump into homo Sapiens followed by an infectious human pandemic [8].

2.1 The Reductive Framework

A reductive account of the situation concentrates attention on defining causal mechanisms and relationships, correlations, equivalence relationships and symmetry relationships [9], hoping to define and prove as best natural science can, that the likely causal origin of the present pandemic occurred in the context of a wet-market where the Covid-19 coronavirus found a local causal mechanism allowing an opportunistic chance to infect human beings [7, 8]. Ongoing reductive scientific work tries to define other properties of the SARS-CoV-2 coronavirus [10] and epidemiological properties of the virus associated with mechanisms of individual and community infectious spread in the human population creating the Covid-19 pandemic [11]. The medical community pursues detailed monitoring and analysis of symptoms [12] and possible symptomatic treatment approaches for the infection [13], as well as possible lines of attack on the development of a vaccine and other mitigating treatments [14].

2.2 A Meta-reductive Framework

A meta-reductive account responds to formal reductive incompleteness [4] and its implications [6]. It incorporates the reductive model of the coronavirus-bat system and human vector as a 'special-case' within a more comprehensive meta-reductive assessment [2]. The fine details of the coronavirus-bat system and human vector can be modeled as an epistemological, ordered and consistent meta-reductive formal system model or equivalent system model [3]. The meta-reductive model contains within it a subset or 'special-case' [5], consisting of the 'bottom-up' reductive account. The meta-reductive model and the natural phenomenon it describes can then be examined in search of potential epistemological undecidable reductive propositions and theoretical formal reductive incompleteness correlated with ontological unresolvable natural instantiations and natural evolutionary incompleteness [3] and incompleteness driven novelty generation [2] in the complex system involving the bat-virus-human vector.

2.3 Reductive Incompleteness in Relation to Covid-19

In essence, does the virus system, the virus-bat system or the virus-bat-wet-market system have, fully formed, the necessary states and processes creating the causal mechanisms and relational properties required to make the jump, opportunistically infect a human being and produce a pandemic? If this is the case no undecidable reductive proposition and no reductive formal incompleteness need be involved and a 'bottom-up' reductive model of the situation will be sufficient. The jump to the human vector expands the range of the virus and increases the number of hosts it can infect. The emerging properties of the human infection are revealed as the novel infection and pandemic proceeds. The event can be described in a closed, resolved, consistent and complete reductive model.

Alternatively, on closer examination, the abstract model of the virus system may demonstrate properties that are conceptually equivalent to an undecidable reductive proposition and reductive formal incompleteness. For example, there could be something about the definable properties of the virus that is *undecidable*. There is structure and process present that *is true* and *proven to be true* to the given *order* of the virus system (in logic and in experimental confirmation) but among these decided states and processes there is something that *is true but cannot be proven to be true*, something that is *unresolved* and *unresolvable*; in that it *is and is not* consistent with the structure and process of the virus as a closed system in isolation—there is something *undecidable* (for example: something must necessarily be left unknown—perhaps a stochastically selected well-formed strand of its genetic material or a protein expressed on the surface of the virus, with structure, position and, either, no further apparent function or a function limited to different tasks in the virus or virus-bat system; the specific genetic material or the protein have no apparent function or play a different role in the virus system, with the virus-bat-human interaction only existing as an *unresolved potential*). There may be something about the definable properties of the virus-bat system that is abstractly *undecidable* and practically *unresolvable* (for example: the virus-bat system sheds viruses from the skin of the bat when it flaps its wings, a behavior having no function but producing a mechanism for transmission in a wet-market). There may also be something about the definable properties of the virus-bat-wet-market system that is *undecidable* and *unresolvable* (the dropping viruses falling off the wings of the bat randomly fall on individual human beings working in the wet-market, creating an opportunistic mechanism for stochastic transmission to human beings with some of the randomly falling viruses surviving on or in the human host, thus selected for survival, replication and the onset of an emerging human pandemic).

The complex virus-bat-wet-market-human system may bring forth incompleteness driven novelty generating properties associated with previously non-functional viral genetic material and protein now functional and causal in the context of the human infection. Particular novel capacities are revealed as the virus infects human lung tissue, producing asymptomatic but highly infective carriers and unrecognized asymptomatic pulmonary failure. The infection involves human liver and cardiac tissue, produces changes in blood clotting mechanisms and immune system function, as well as infecting many other organs, with diverse symptomatic presentations in different age groups, sexes and complex comorbidity with other human illnesses [15]. The incompleteness driven novelty generating outcomes of the SARS-CoV-2 coronavirus human infection and Covid-19 pandemic includes the complex interaction and relationship the virus and pandemic have with the state of human readiness and preparedness to contain the virus through individual case interventions and community level, social and physical distancing, isolation and quarantine [11, 13]. Further novelty generation involves successful or tragic outcomes associated with medical systems, nursing homes, intensive care units and the availability and supply lines for personal protective gear and respirators, as local medical facilities and hospitals deal with a flattened viral curve or an unmitigated surge of viral infections, leading to large numbers of deaths when the medical system is overwhelmed. Vulnerability or resilience is revealed, involving the behavior of individual human beings, communities, local governments, national governments, particular political agendas and the interaction with other institutions, such as the CDC and the WHO [5]. Individuals and communities in scientific, medical and political contexts reveal their readiness to participate in creating an integrated, coordinated response, or, sadly, fail to integrate a coordinated response, with collapse into chaotic and disastrous outcomes, with more viral victims and deaths associated with the spreading pandemic. The undecidable statement or unresolvable instantiation regarding the originally non-functional viral DNA and protein now becomes a decidable statement with clear mechanism, function and outcomes in the virus-bat-wet-market-human system. The unresolvable natural instantiation of the original viral DNA and protein become a resolvable natural evolutionary statement in the context of the complex virus-bat-wet-market-human system.

An '*is and is not*' structure or process such as an entity, state or process, composition, interaction or relationship, that is, *true* to, and *consistent* with, the sequence of natural evolution of a self-organized *order* (under scientific investigation in a semi-isolated system) can have an abstract *undecidable* property and a correlated naturally instantiated *unresolvable* property or function. The undecidable, unresolvable or unknown 'is and is not' structure or process exhibits a potential or realized property or function that is, in the abstract or in the correlated natural instantiation, *false, contradictory* or *inconsistent* with the *logic* of the preceding *order* and process of natural evolution. To protect the abstract logic and avoid contradiction in describing the natural instantiation, it must be left undecided and unresolved. The abstractly *undecidable* can become *decidable* in an evolving context in which the correlated naturally *unresolvable* becomes a *resolvable* natural instantiation involving an entity, state or process, property or function, that is *true* and *consistent* with an *emerging*, naturally evolved semi-isolated system or self-organized *order*. Scientific narrative and natural evolution reveal themselves to be, necessarily, open, undecided and undecidable, unresolved and unresolvable, consistent and para-consistent, thus requiring imperative meta-consideration in determining consistency. Scientific narrative and models of correlated natural evolution must incorporate formal logical undecidable propositions, which are predicted to correlate with natural unresolvable instantiations accompanied by inevitable incompleteness in a meta-reductive model and narrative.

Further examples of epistemological, abstract, undecidable reductive proposition and formal reductive incompleteness associated with correlated natural ontological, realized, unresolvable instantiations and natural evolutionary incompleteness could be pursued in every corner and through every layer of viral-human interaction and relationship. The undecidable pattern of incompleteness may or may not repeat in relation to the cell, the organ, the individual human being, the human family and any complex cultural human context impacted by the invading and transformative virus-host interaction.

Scientific exploration could reveal that natural, incompleteness driven exploration of the adjacent possible, can lead to emerging, self-organized, patterns of novelty generation. Novelty generation in natural settings may have a neutral influence on the survival of any particular living species. However, as in the case of the Covid-19 pandemic, novelty generation at the level of the SARS-CoV-2 coronavirus can increase vulnerability and have disastrous consequence for the infected human species. Fortunately, novelty generation at the level of complex organization involving human mind and consciousness, has the potential to induce greater resilience and creative adaptation, in the individual, the local community or the entire social system of the human host.

3 Meta-reductive Life Cones and Self-similarity

An abstract epistemological *meta-reductive four-dimensional life cone* can model a single SARS-CoV-2 coronavirus as it instantiates its meta-reductive *ontological complex cycle of existence*. The ontological cycle of existence can be described and modeled by a meta-reductive epistemological *complex natural history* [2, 3]. The life cone, a cycle of existence and a complex natural history are *self-similar correlated* ontological and epistemological structures formulated in the meta-reductive perspective.

3.1 Self-similarity and Meta-reductive Models

A meta-reductive life cone with a correlated ontological complex cycle of existence and epistemological complex natural history reveals a further pattern of self-similarity in theory and in natural evolution. The existence of a Covid-19 coronavirus can be described by dividing its existence into three inter-related meta-reductive system components.

Semi-isolated Material Existence. The first component subsystem describes the physical and material existence of a SARS-CoV-2 coronavirus, limited to the entity, its states and processes, its composition and interactions, contained within a subsystem involving the spherical shell, the enclosed genetic material and the exposed surface proteins.

Semi-isolated Potential Dynamic Existence. The second component subsystem describes the potential dynamic existence of a SARS-CoV-2 coronavirus. This component description captures the sum and potential range of dynamic composition, interaction and relationship. There are two potential dynamic descriptions. The first models the virus in its system of origin involving the cycle of existence in the coronavirus-bat ecosystem. The second models the transition, emergence, self-organization and evolution of the coronavirus-human ecosystem involving the Covid-19 pandemic. The two dynamic descriptions of the coronavirus are synthesized into a detailed account describing the sum of all potential or possible dynamic states and processes, entities, compositions, interactions and relationships within which the SARS-CoV-2 coronavirus does or could participate.

Semi-isolated Realized Material Dynamic Existence. The third component subsystem captures specific moments in state and process involving the immediate location and behavior of a SARS-CoV-2 coronavirus in the virus-bat system or the virus-human system. The epidemiological pursuit of a single case presentation of Covid-19 in the context of public health detection, single case management, orders for self-isolation and quarantine, as well as medical management of the evolving symptoms of an individual case of coronavirus infection; these all fall into the third description. Community level public health orders recommending intentional deconstruction of connectivity with social and physical distancing, masks, handwashing and surface sterilizing, as well as public health models of specific moments and daily reports of the pandemic state and process, describing specific individual case reports and tracking specific patterns of community spread and lab testing; all fall under the third component description.

3.2 A General Pattern of Evolutionary Self-similarity

A significant meta-reductive pattern of theoretical and instantiated Fractal evolutionary, self-similarity, iteration and scaling, is revealed by the threefold SARS-CoV-2 coronavirus description. The same three component subsystems can be employed to describe a micro-scale quantum particle/wave system, the mid-scale complexity of the human brain/mind system, and the macro-scale structure and process of the entire material/dynamic existence of the whole Universe.

Micro-scale Complexity. The micro-scale quantum particle/wave system involves complementarity with a *particle* description (*material existence*), a *wave* description (*potential dynamic existence*), and the sum of all specific individual experimental demonstrations of a particle/wave, including non-local superposition in wave state and process, local 'collapsed' particle state and process; ultimately creating a model mapping the sum of all possible experimentally demonstrated state and process of a *particle/wave* (*realized material and dynamic existence*).

Mid-scale Complexity. The mid-scale complexity of the human brain/mind system, involves complementarity with a *brain* description (*material existence*: localized possible brain states and processes), a *mind* description (*potential dynamic existence*: extended possible mind states and processes in relation to self, other living beings, the World and the Universe), and the sum of all specific individual theoretical or experimental demonstrations of consciousness, including localized realized brain states and processes as well as extended realized mind states in relation to self, other living beings, the World and the Universe (*realized material dynamic existence*).

Macro-scale Complexity. The macro-scale structure and process of the entire material/dynamic existence of the whole Universe as a system involves complementarity with a *physical* description (*material existence*: a localized possible Universe of states and processes), a *flow* description (*potential dynamic existence*: an extended description of a possible Universe of states and processes in relation to an observing, meta-considering conscious self, other living beings, the World and the Universe itself and possibly an unknowable greater system beyond the Universe), and the sum of all specific individual theoretical or experimental demonstrations of the Universe, involving localized realized states and processes of the Universe or extended realized states and processes of the Universe, in relation to an observing, meta-considering conscious self, other living beings, the World and the whole Universe conforming to itself (*realized material dynamic existence*).

3.3 The Implication of Generalized Self-similarity

The reductive scientific account of consciousness in the Universe, excludes mind and consciousness as mere epiphenomena of more fundamental physical or quantum realities [15] or dismisses mind and consciousness as superfluous unnecessary concepts [16]. The meta-reductive account, driven by reductive incompleteness and novelty generation, preserves mind and consciousness and brings brain, mind and consciousness into Nature as significant differentiated self-similar phenomena [2, 3].

A self-similar meta-reductive account of the SARS-CoV-2 coronavirus and the Covid-19 human pandemic treats the 'bottom-up' reductive account as a 'special-case'. The meta-reductive account then expands the model of the SARS-CoV-2 coronavirus and the Covid-19 human epidemic to include a scientific search for meta-reductive abstract and realized incompleteness and incompleteness driven novelty generation. A meta-reductive formulation can create an interactive mapping of the life-cone of human consciousness, mind and brain in a co-evolving relation to the SARS-CoV-2 coronavirus and Covid-19 human pandemic. A detailed meta-reductive account reveals a self-similar systemic evolutionary pattern present in the sequence of change, self-organization, emergence and hierarchical organization of complexity associated with the composition, the interaction and the relationship the SARS-CoV-2 coronavirus has with the adaptive, novelty generating creative potential of the human being.

Willful, intentional, mindful human consciousness can very much participate in modifying the trajectory, flattening the pandemic curve as well as exploring complex alternative ways to address the pandemic and the associated economic crisis. Human consciousness, in a Meta-Reductive integration of Reductive and Complexity Science, could strongly adhere to adapted, rigorous, iterated reductive logic, responsive to reductive incompleteness and aware of incompleteness driven novelty generation. This refocusing of human scientific attention, with mind and consciousness fully integrated in Nature and rigorous logic and differentiated truth and proof invigorated at the core of scientific thought, could then actively and adaptively, try to make choices that creatively enhance the resilience of the human being and avoid individuals or communities transiting onto a trajectory of vulnerability and potential death.

4 Adapting Kinds of Reductive 'Unfinished Description'

The Meta-Reductive Paradigm initially explored in relation to unrecognized formal reductive incompleteness [4] addresses a number of further identified patterns of 'unfinished description' theoretically associated with the Reductive Paradigm [2].

4.1 Micro-scale Evolution Transformed into Multi-scale Co-evolution

The meta-reductive framework first transforms the 'unfinished' singular reductive 'bottom-up' micro-scale upward interpretation of evolution, into a multi-scale interpretation of co-evolution.

The 'bottom-up', micro-scale first, reductive account of the SARS-CoV-2 coronavirus describes the entity, its parts, states and processes, its composition and internal interactions as well as its potential for external relationships leading to the Covid-19 human pandemic. The reductive account tends to focus scientific attention on the virus in isolation and only then on the impact it has in the arena of the human pandemic. The reductive account becomes a 'special-case' in the meta-reductive account.

The meta-reductive account highlights multi-scale co-evolution. This brings into focus a micro-scale, mid-scale and macro-scale description of co-evolving entities, self-organization of entities on multiple scales, developing interactions and emerging patterns of systemic relationship occurring on multiple scales of organization. From the beginning in a meta-reductive account, the states, processes, entities, composition, interactions and relationships of SARS-CoV-2 coronavirus interacting with adaptive human beings is highlighted, in relation to the evolving Covid-19 pandemic. In the meta-reductive frame, individual human consciousness (prepared or unprepared), human patterns of multi-scale social interaction (wet-markets or applied social and physical distancing), the state and recognition of scientific knowledge (ignored, unfunded, abandoned, recognized, incorporated) and the effective orientation of human political institutions and actors (accepting, collaborative, inclusive of public health, medical and scientific expertise, with altruistic motives in relation to potential human vulnerability and adaptive resilience; *or*, dismissive, combative, excluding of public health, medical and scientific expertise, with self-absorbed or self-interested motives in relation to potential human vulnerability and adaptive resilience).

4.2 Unidirectional Evolution Transformed into Multi-directional Co-evolution

This second adaptation allows the metaphoric, 'unfinished' unidirectional 'bottom-up' reductive interpretation to be treated as a 'special-case'. The reductive perspective is adapted and transformed into a meta-reductive account by including the reductive account within an incompleteness driven, novel, metaphoric, multi-directional causal interpretation of co-evolution.

In the meta-reductive perspective, 'simple' non-local and local causal mechanisms are integrated with 'complex' incompleteness driven novelty generating causal mechanisms [2]. 'Simple' and 'complex' causal mechanisms and relationships are explored in extended meta-reductive semi-isolated systems, which include the 'special-case' of the reductive semi-isolated system embedded in an extended meta-reductive sphere involving the local proximate ecosystem and the distant but relational environment. An individual case interpretation would place SARS-CoV-2 coronaviruses, or the viruses and an adaptive human being in an extended meta-reductive semi-isolated system. Alternatively, a community level interpretation would place selected populations of human beings presently impacted by the Covid-19 pandemic in an extended meta-reductive semi-isolated system. Causal mechanisms and relationships can then be explored.

The selected phenomenon of interest in an individual or community level interpretation involving virus and human interactions and relationships, is represented by a threefold description modeling the viral and human *material existence*, the *potential dynamic existence* and the *realized material dynamic existence.* An integrated reductive and meta-reductive model could represent the complexity of the situation as a meta-reductive complex physical system (MCPS) or meta-reductive complex adaptive system (MCAS) [17]. In a semi-isolated system, a reductive or complexity account would focus on the selected phenomenon of interest, analyzed in relation to defined entities, causal states and processes, composition, interaction and hierarchical organization. In an extended meta-reductive semi-isolated system, the reductive and complexity description is encompassed in a meta-reductive account of the greater systemic complexity involving the causal relationships the MCPS or MCAS has within an extended sphere of a local, proximate ecosystem and an extended interactive and relational environment. The meta-reductive frame creates a further matrix of causal relationships defined in the context of multi-directional causal accounting responsive to multi-scale co-evolution [2]. The meta-reductive framework thus encompasses 'bottom-up' reductive accounting and the complexity-based interpretation of CPS's and CAS's, as an integrated 'special-case' within the meta-reductive framework.

The extended meta-reductive account then explores possible causal mechanisms and relationships in an extended meta-reductive semi-isolated system. The extended system holds the SARS-CoV-2 coronavirus and an adaptive human being, or a population of human beings facing the Covid-19 pandemic, in a complex systemic, social and environmental context. A self-similar multi-level, analysis of the situation can involve an examination of viral and human *material existence*, the *potential dynamic existence* and the *realized material dynamic existence* of the coronavirus and the pandemic in relation to adaptive human bodies, brains, minds, consciousness and human communities. Within the co-evolving, realized material dynamic existence of viral-human interactions and relationships, there would be a description of causal public health, medical, scientific research and communication, as well as, causal collaborative or partisan political exchanges, which prepared individuals and communities or left individuals and communities tragically unprepared. Human individuals and communities are made more resilient or made profoundly more vulnerable, depending upon very complex interactive relationships that involve particular individual people, communities, jurisdictions, nations and international institutions attempting to deal with the Covid-19 pandemic [2]. The hierarchical, compositional, interactive and relational complexity of the situation is made apparent by the meta-reductive method of causal accounting. The responsibility and failure to take responsibility and the adaptive or maladaptive response of particular individuals, communities, jurisdictions, nations and international institutions, are an inherent part of a meta-reductive account [5].

4.3 Reductive Logic Transformed into Iterated Meta-reductive Logic

An 'unfinished' description is created by the failure of 'bottom-up' reductive Logic to predict formal reductive incompleteness in the Reductive Paradigm. The recognition of undecidable reductive propositions and formal reductive incompleteness, multi-scale co-evolution and the need for multi-directional causal accounting, leads to a transformation of 'bottom-up' reductive Logic.

Reductive 'bottom-up' Logic is adapted by encompassing it as a 'special-case' in an iterated, paraconsistent form of reductive Logic [2, 18], capable of handling undecidable reductive propositions and reductive incompleteness, as well as handling multi-scale co-evolution and multi-directional causal accounting. Multiple iterative implementations of reductive Logic are then required in relation to the matrix of meta-reductive causal accounting.

The meta-reductive perspective and method of causal accounting recommends the SARS-CoV-2 coronavirus and the Covid-19 human pandemic be analyzed by applying multiple iterative implementations of reductive Logic within a detailed exploration of the complex hierarchically organized system in which the virus and human pandemic co-evolve.

4.4 Symmetry and Self-similarity Deployed in the Analysis of Causal Relations

Finally, the reductive reliance on Symmetry and Group theory in mathematically defining causal mechanisms and relationships in natural systems [9] is transformed in the meta-reductive system, into an integrated meta-reductive model of causal relationships employing Symmetry and Group theory, as well as Self-similarity and Fractals, deployed in the task of modeling naturally co-evolving systems [2]. Symmetry becomes a 'special-case' within a more comprehensive exploration of causal relation involving symmetry and self-similarity.

The reductive sciences employ a semi-isolated system in order to create predictable and reproducible contexts for experimental exploration of probable causal interactions and relationships and the definition of causal mechanisms. Synthesis of symmetry and self-similarity in scientific accounting of causal mechanisms and relationships requires the implementation of an enhanced and extended semi-isolated system, which can model extended relationships within a local system as well as an associated ecosystem and environment. The usual reductive semi-isolated system becomes a 'special-case' within the extended meta-reductive semi-isolated system capable of exploring symmetry relations as well as self-similar relationships. Self-similarity is a special form of scale dependent symmetry. The meta-reductive application of symmetry and self-similarity leads to recategorization of causal mechanisms in natural systems.

The reductive sciences recognize local and non-local causal mechanisms. These are contained as a 'special-case' within the extended meta-reductive categorization of causal mechanisms. The meta-reductive account of causal mechanism includes a primary *incompleteness driven, novelty generating causal mechanism* and other secondary incompleteness driven causal mechanisms and relationships. Examples of *secondary* complex systemic incompleteness driven causal mechanisms that can combine at the *edge of incompleteness* with *incompleteness driven novelty generation*, include, for example: phase change; non-linear or chaotic process; stochastic system evolution; Lamarckian system evolution; as well as bidirectional intersection in multi-scale, multi-directional co-evolution, which leads to the development of systemic duality or self-similar complementarity previously mentioned in this article (examples: 1) complementarity defined by particle/wave and the sum of locations; (2) complementarity defined by SARS-CoV-2 coronavirus/Covid-19 human pandemic and specific interactions and relations in complex systems; (3) complementarity defined by brain/mind and moment-to-moment experiential consciousness; (4) complementarity abstractly defined as, the *material existence*/the *potential dynamic existence* and the *realized material dynamic existence*). The meta-reductive assumption of multi-scale, multi-directional, co-evolution leads to differentiation of simple conservative (reductive) emergence and/or more complex forms of novel or radical emergence (meta-reductive), which are consistent with multi-scale, multi-directional co-evolution and meta-reductive causal accounting.

In the meta-reductive framework, causal relationships could be experimentally evaluated through exploration of semi-isolated systems of presumed cause and presumed effect assisted by a synthesis of the applied mathematics involving both symmetry and self-similarity. The process of causal analysis in reductive thought demands analysis of correlation, equivalence relationships and symmetry relationships. In meta-reductive thought, causal analysis requires the further exploration of scaled correlation, scaled patterns of equivalence and a search for scaled, self-similarity.

The meta-reductive perspective would recommend the SARS-CoV-2 coronavirus and the Covid-19 human pandemic be examined in greater detail as a complex hierarchically organized natural system in which the causal mechanisms and relationships are derived from both the reductive and meta-reductive paradigms. The applied mathematical accounting entails both symmetry and self-similarity. The iteration of reductive Logic in the meta-reductive frame can then explore the matrix of meta-reductive causal description in relation to multi-scale, multi-directional, co-evolution [2].

5 Meta-reductive Thought During the Covid-19 Pandemic

The jump of the SARS-CoV-2 coronavirus into human beings having no specific immunity and the ensuing catastrophic Covid-19 human pandemic provides a tragic vehicle for spelling-out the importance of Natural Science responding to a number of shortcomings and challenges to its continued progress.

Reductive Natural Science, the ‘bottom-up’ Logic of reductive science and the Reductive Scientific Paradigm have enabled almost 4 centuries of vastly successful scientific and applied mathematical discovery. There are, however, limits to the application of ‘bottom-up’ reductive Logic and the semi-isolated reductive scientific experimental method. The favored ‘bottom-up’ Logic of Reductive Natural Science is susceptible to undecidable reductive propositions and formal reductive incompleteness. Addressing this unrecognized property of reductive Logic could lead to a resurgent interest in the relationship shared by Logic and Natural Science. How truth must be separated from proof in the scientific context becomes an important theoretical and experimental question. The necessity of meta-consideration when determining reductive logical consistency provides a further impetus to explore how human consciousness can be fully integrated in a scientific model of Nature.

The present political climate in which the Covid-19 pandemic evolves includes a decay into irrationality and a resurgence of emotionally based, sometimes grossly illogical argumentation with blatantly false interpretations of reality. The handling of the Covid-19 pandemic frequently reveals inaccurate translation of public health advice, as well as medical and scientific information. In this context it can only be a good thing for there to be an increased interest, in Logic, truth, proof, consistency, the nature of unfinished scientific and mathematical description, as well as the implications of formal incompleteness in scientific and mathematical domains.

The scientific investigation of the SARS-CoV-2 coronavirus and the Covid-19 human pandemic will most certainly proceed in conformity with the structure and method of Reductive Natural Science and the Reductive Paradigm. However, there are many problems associated with Covid-19 pandemic that may be better addressed through the further development of a meta-reductive perspective. The important issues directly reflect upon how human beings see themselves in Nature and indirectly influence how we might choose to collectively address an evolving pandemic threatening our existence and way of life on Earth.

First, the meta-reductive account offers a way to bring Reductive Naturel Science into consilience with the Complexity Sciences. The integration of reductive and complexity thinking offers a way to transform a reductive Complex Adaptive System Model (CAS), challenged by the unresolved contradictions and paradoxes of consciousness and mind; into a Meta-Reductive Complex Adaptive System Model

(MCAS) that can include brain, mind and consciousness in Nature [6]. An MCAS can introduce a resolved and decided acceptance of willful, intentional, mindful human agency within Nature and create a model of complex adaptation unfettered by contradiction and self-referencing paradox.

The meta-reductive inclusion of human conscious agency and a renewed interest in truth, proof and consistency in natural science, in medicine, as well as in the political arena and wider human culture, are critical to the future hope and efforts of humanity in the face of the Covid-19 pandemic and economic crisis. A Meta-Reductive Paradigm could also offer novel means to model complex features and properties of other co-evolving global crises including climate change.

Acknowledgements I want to thank Dr. Yakov Shapiro and Patricia G. Scott for their unfailing support.

References

1. World Health Organization: Naming the coronavirus (Covid-19) and the virus that causes it. www.who.int. Last accessed 11 May 2020
2. Scott, J.R.: Kinds of unfinished description with implication for natural science. In: Proceedings of the 2020 New England Complex Systems Conference. Accepted for publication, ICCS 2020
3. Scott, J.R.: Transpersonal psychology and fractal evolution. In: Marks-Tarlow, T., Shapiro, Y., Wolf, K.P., Friedman, H. (eds.) A Fractal Epistemology for a Scientific Psychology: Bridging the Personal with the Transpersonal, pp. 104–143. Cambridge Scholars Publishing, Newcastle upon Tyne, UK (2020)
4. Scott, J.R.: On the formal incompleteness of reductive logic. In: Proceedings of the 2020 New England Complex Systems Conference. Accepted for publication, ICCS 2020
5. Scott, J.R.: Scientific logic, natural science, polarized politics and SARS-CoV-2. In: Proceedings of the 2020 New England Complex Systems Conference. Submitted for consideration, ICCS 2020
6. Scott, J.R.: Descartes, Gödel, Kuhn: Epiphenomenalism defines a limit on reductive logic. In: Morales, A.J., et al. (eds.) Proceedings of the 2018 New England Complex Systems Conference Held in Boston MA, USA. ICCS 2018 SPCOM, pp. 33–52. Springer Nature, Switzerland AG (2018)
7. AAAS: The COVID-19 coronavirus epidemic has a natural origin, scientists say—Scripps Research's analysis of public genome sequence data from SARS-CoV-2 and related viruses found no evidence that the virus was made in a laboratory or otherwise engineered. EurekAlert! Scripps Research Institute, 17 March 2020. https://en.wikipedia.org/wiki/EurekAlert!. Accessed online 11 May 2020
8. Andersen, K.G., Rambaut, A., Lipkin, W.I., Holmes, E.C., Garry, R.F.: The proximal origin of SARS-CoV-2. Nat. Med. **26**(4), 450–452 (2020). https://www.ncbi.nlm.nih.gov/pmc/articles/PMC7095063. Accessed online 12 May 2020
9. Rosen, J.: Symmetry Rules: How Science and Nature are founded on Symmetry. Springer, Berlin, Heidelberg (2008)
10. van Doremalen, N., Bushmaker, T., Morris, D.H., Holbrook, M.G., Gamble, A., Williamson, B.N., Tamin, A., Harcourt, J.L., Thornburg, N.J., Gerber, S.I., Lloyd-Smith, J.O., de Wit, E., Munster, V.J.: Aerosol and surface stability of SARS-CoV-2 as compared with SARS-CoV-1. N. Engl. J. Med. **382**, 1564–1567 (2020)

11. Norman, J., Bar-Yam, Y., Taleb, N.N.: Systemic risk of pandemic via novel pathogens—coronavirus: a note. New England Complex Systems Institute, January 26, 2020. https://necsi.edu. Accessed on 12 May 2020
12. Wadman, M., Couzin-Frankel, J., Kaiser, C.: How does coronavirus kill? Clinicians trace a ferocious rampage through the body, from brain to toes. Science. April 17, 2020. https://ScienceMag.org. Accessed on 11 May 2020
13. Norman, J., Bar-Yam, Y., Taleb, N.N.: Stopping the coronavirus pandemic. New England Complex Systems Institute, January 26, 2020. https://necsi.edu. Accessed on 12 May 2020
14. Radcliffe, S.: Here's exactly where we are with treatments for Covid-19. Healthline. Published online May 13, 2020. https://Healthline.com. Accessed on 13 May 2020
15. Robinson, W.: Epiphenomenalism. In: Zalta, E.N. (ed.) The Stanford Encyclopedia of Philosophy, Summer 2019 edn. https://plato.stanford.edu/archives/sum2019/entries/epiphenomenalism/
16. Ramsay, W.: Eliminative materialism. In: Zalta, E.N. (ed.) The Stanford Encyclopedia of Philosophy, Spring 2019 edn. https://plato.stanford.edu/archives/spr2019/entries/materialism-eliminative/.
17. Holland, J.: Complexity: A very short introduction. Oxford University Press, Oxford, UK (2014)
18. Priest, G., Tanaka, K., Weber, Z.: Paraconsistent logic. In: Zalta, E.N. (ed.) The Stanford Encyclopedia of Philosophy, Summer 2018 edn. https://plato.stanford.edu/archives/sum2018/entries/logic-paraconsistent/

Cardiorespiratory Activity in the Auto-Organization of the Hierarchical Order in Crayfish

Iván Oliver-Domínguez, Aidee Lashmi García-Kroepfly, Mireya Osorio-Palacios, Jesús Hernández-Falcón, and Karina Mendoza-Ángeles

Abstract Under laboratory conditions, social interactions between triads of crayfish result in the establishment of a hierarchical order with a dominant animal and two submissive (1 and 2 respectively). This social interactions are accompanied by changes in the cardiorespiratory electrical activity, which seems to indicate that crayfish have autonomic-like responses during agonistic encounters, despite the lack of an autonomic nervous system. A tool that allows the study of the autonomic nervous system of vertebrates and the adjustments made under physiological conditions is the analysis of heart rate time series. In particular, the heart rate variability (HRV). We used triads of adult crayfish in intermolt that were implanted with Pt-Ir electrodes in the cardiac sinus and one branchial chamber. Behavioral and electrophysiological recordings were made simultaneously. Each triad was placed in an arena and videotaped during one hour. The first 15 min animals shared the same aquarium which was divided by a plastic separator; here, they perceived each other but could not fight. For the next 45 min, we removed the separator, and animals fought with each other. After this time the behavior was analyzed and the hierarchical order of each member was determined. Results indicate that dynamics of the cardiorespiratory electrical activity of adult crayfish is different before and during social interaction, between dominants and submissive animals.

Keywords Heart rate variability · Social interactions · Autonomic-like responses

I. Oliver-Domínguez (✉) · A. L. García-Kroepfly · M. Osorio-Palacios · J. Hernández-Falcón · K. Mendoza-Ángeles
Laboratorio de Redes Neuronales, Departamento de Fisiología, Facultad de Medicina, Universidad Nacional Autónoma de México, Ciudad de México, México
e-mail: ivan.oliver@c3.unam.mx

D. Braha et al. (eds.), *Unifying Themes in Complex Systems X*, Springer Proceedings in Complexity, https://doi.org/10.1007/978-3-030-67318-5_8

1 Introduction

Organism's compensatory changes in response to external or internal stimuli are necessary for the maintenance of homeostasis [1]. Physiological responses such as heart rate, respond to perturbations in order to ensure the stability of different physiological variables such as blood pressure [2].

A better understanding of homeostasis is reached by studying the underlying mechanisms and systems that work to keep its maintenance. Homeostasis depends on two different types of variables, regulated ones (such as blood pressure and core temperature) which oscillate within a narrow range of values for the organism to be healthy, and physiological responses, (such as heart rate, and vasomotor action) that adapt to perturbations in order to maintain the stability of the regulated ones [2]. The behavior of the variables reflects the dynamics of underlying homeostatic regulatory mechanisms [2, 3]. The difficulties of quantifying homeostasis are diminished by time-series analysis and a certain phenomenology of time series may be assigned to specific physiological variables, for example heart rate and skin temperature in their role of adaptative physiological responses will show a large variability, whereas the regulated variables such as core temperature and systolic blood pressure will be constant and close to a predetermined set point, with a large variability in them becoming an indicative of illness [2].

In vertebrates, the autonomic nervous system with its sympathetic and parasympathetic divisions regulates cardiac and smooth muscle, among other, in order to keep a proper internal environment and react to possible external threats, in invertebrates, whose nervous systems vary from simple nerve nets to systems conformed by central and peripheral components, evidence of an autonomic nervous system has not been found, nevertheless these animals display physiological responses that maintain internal stability [4], and highly developed invertebrates like crustaceans even present "fight or flight" behaviors [5] normally attributed to the sympathetic division of the autonomic nervous system.

Some crustaceans such as crayfish exhibit aggressive conducts towards conspecifics [6], by engaging in agonistic encounters when an intruder invades their territory [7].

Under laboratory conditions, triads of unfamiliar crayfish interact by engaging in agonistic encounters and establish a hierarchical order with one animal being the dominant and the others becoming submissive (1 and 2) [8]. The encounters are conformed of threats, pushes, attacks, grabs, and avoidance behaviors that include retreats and escape responses [9]. The hierarchy is established since the first encounter [9]. Bovbjerg proposed an analytical method where agonistic actions are classified in positive and negative contacts [10].

Social interactions are accompanied by changes in the cardiorespiratory electrical activity [5, 11], which indicates that crayfish presents autonomic-like responses during agonistic encounters, as it happens in vertebrates, but without an autonomic nervous system being reported.

A tool that allows the study of the autonomic nervous system of vertebrates and the adjustments made under physiological conditions, is the analysis of heart rate time series. In particular, the heart rate variability (HRV), which is based on the study of the intervals between each beat and whose characteristics contribute to a detailed description of the cardiac dynamics [12].

The aim of this work was to analyze cardiorespiratory electrical activity of adult crayfish *Procambarus clarkii*, through HRV, before and during social interactions in order to compare the dynamics of these variables between dominants and submissive animals.

2 Methods

We used triads of adult crayfish in intermolt (weight and size difference less than 5%), implanted with Pt-Ir electrodes in the cardiac sinus and one branchial chamber. (animals were cold anesthetized), After 48 h of post-surgical recovery, behavioral and electrophysiological recordings were made simultaneously. Here we repot only the heart rate changes.

We recorded the electrical activity of the heart from each member of a triad, 15 min before any experiment, in isolation (Condition 1), then, each triad was placed in an arena and videotaped for one hour. The first 15 min animals shared the same aquarium which was divided by a plastic separator (Condition 2). For the next 45 min, we removed the separator, and animals fought with each other (Condition 3). After this time the behavior was analyzed, and the hierarchical order of each member was determined.

We used standard methods to process electrical signals, and stored data on a computer for offline analysis.

For the electrical activity of the heart we measured the time between each beat and constructed a new time series (HRV time series), one for every member of the triads, for the 3 conditions.

3 Results

We analyzed the data of HRV in the time domain with statistic measures, such as the mean of intervals between each beat (MB), the standard deviation of those intervals (SD) and the square root of the mean squared differences between successive intervals between each beat (SRMB).

Then, we analyzed HRV in the frequency domain, through Fourier analysis and calculated the power spectrum in order to know the peaks in the very low frequency (VLF 0–0.04 Hz), low frequency (LF 0.04–0.15 Hz) and high frequency (HF 0.15–0.4 Hz) bands.

Finally, we calculated the heart and respiratory rates (HR and RR, respectively).

The higher HRV was observed in the dominant during conditions 1 and 3; in condition 2, the submissive 1 had the highest HRV, the dominant one had the lowest HR and RR during all conditions. Submissive 2 had the lowest HRV but the highest HR and RR during all the experiment.

In the frequency domain we found, during condition 1, more power in the VLF band for dominant and submissive 2 animals, but higher power in the HF band for submissive 1 crayfish. During conditions 2 and 3 the highest power was in VLF for all animals (Figs. 1, 2; Tables 1, 2, 3).

4 Conclusion

Crayfish present autonomic-like responses during social interactions that seem to depend on their hierarchical order.

Further analysis is needed in order to have a better understanding of the autonomic-like responses in crayfish, more parameters of HRV could help, like distributions or non-linear ones.

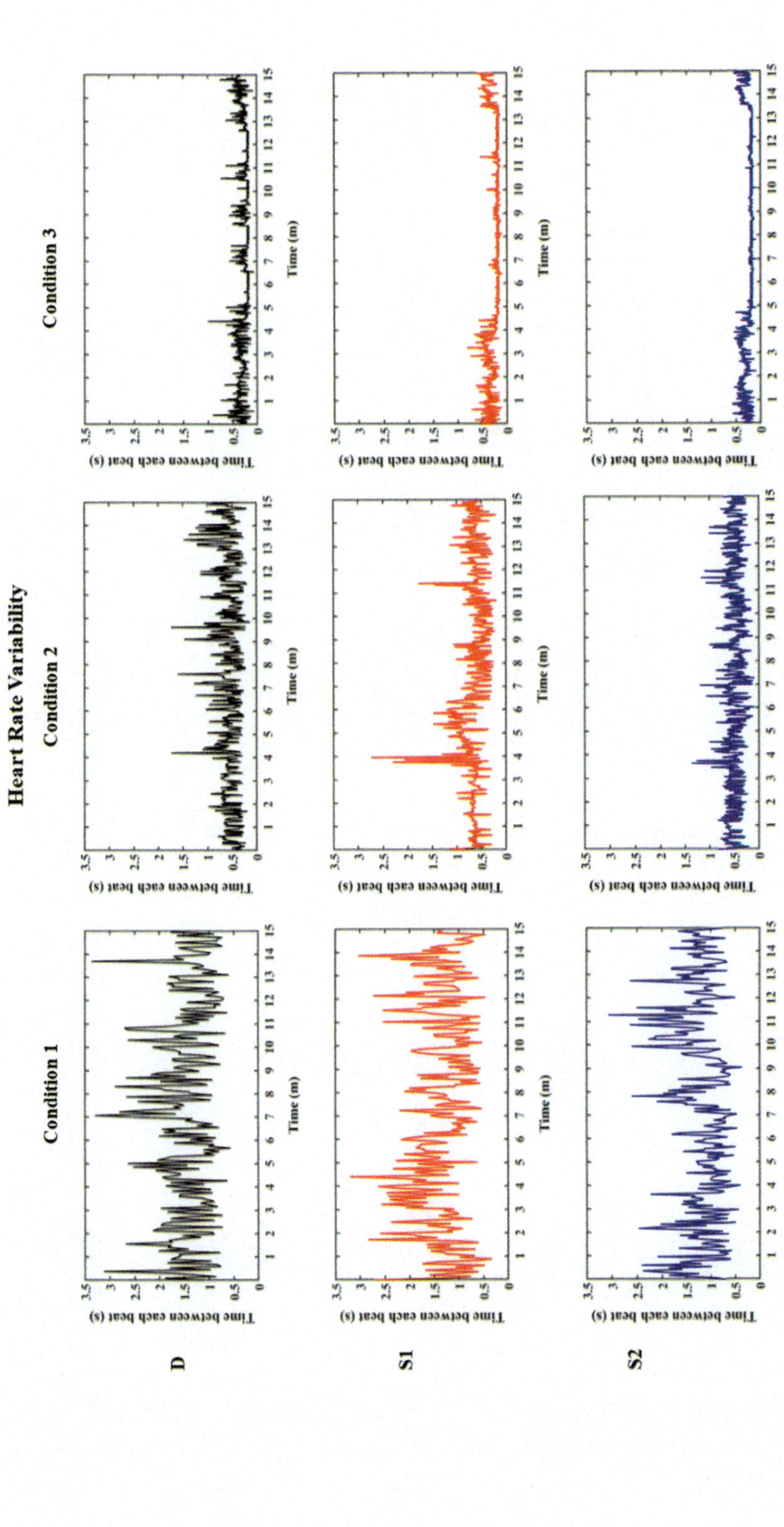

Fig. 1 Heart rate variability for the 3 conditions of all crayfish in a triad. First column is for condition 1, second, condition 2 and the last column is condition 3. D is for dominant crayfish, S1 for submissive 1 and S2 for submissive 2

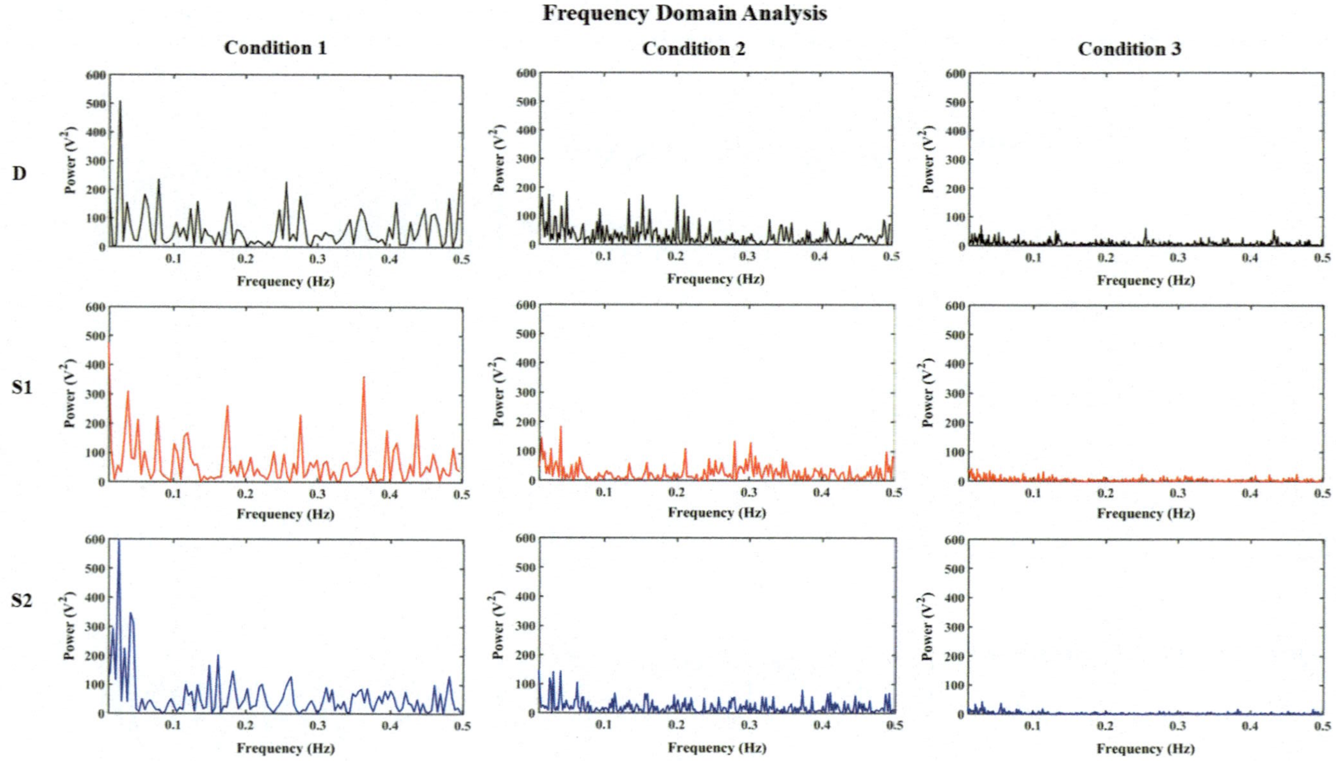

Fig. 2 Frequency domain analysis for the 3 conditions of all crayfish in a triad. First column is for condition 1, second, condition 2 and the last column is condition 3. Row 1 is for dominant crayfish, row 2 for submissive 1 and row 3 for submissive 2

Table 1 Values of MB, SD and SRMB for the 3 conditions of all crayfish in a triad

	MB (s)	SD (s)	SRMB (s)
Condition 1			
Dominant	1.4632	0.4917	0.7359
Submissive 1	1.3678	0.4805	0.6975
Submissive 2	1.2046	0.4057	0.5816
Condition 2			
Dominant	0.5726	0.2255	0.2924
Submissive 1	0.6591	0.2271	0.3281
Submissive 2	0.4970	0.1788	0.2348
Condition 3			
Dominant	0.2812	0.0840	0.1056
Submissive 1	0.2596	0.0648	0.0755
Submissive 2	0.2453	0.0518	0.0583

Table 2 Values of the peak in frequency domain analysis for the 3 conditions of all crayfish in a triad

	Peak VLF (Hz)	Peak LF (Hz)	Peak HF (Hz)
Condition 1			
Dominant	0.0246	0.0788	0.2562
Submissive 1	0.0368	0.0783	0.3641
Submissive 2	0.0242	0.0404	0.1619
Condition 2			
Dominant	0.0230	0.0479	0.1536
Submissive 1	0.0397	0.0662	0.2804
Submissive 2	0.0399	0.0632	0.3727
Condition 3			
Dominant	0.0272	0.1304	0.2552
Submissive 1	0.0148	0.1144	0.2507
Submissive 2	0.0302	0.0564	0.3830

Table 3 Values of the HR and RR for the 3 conditions of all crayfish in a triad

	HR (BPM)	RR (RPM)
Condition 1		
Dominant	41	18
Submissive 1	43	30
Submissive 2	49	39
Condition 2		
Dominant	104	31
Submissive 1	91	33
Submissive 2	120	45
Condition 3		
Dominant	213	60
Submissive 1	231	71
Submissive 2	244	100

Acknowledgements This project is partially funded by UNAM-DGAPA-PAPIIT IN231620.

References

1. Horn, J., Swasson, L.: The Autonomic Motor System and The Hypothalamus. In: Kandel E., et al. (eds.) Principles of Neural Science, pp.1056–1076. McGraw-Hill, Health Professions Division, New York (2013)
2. Fossion, R., Rivera, A.L., Estañol, B.: A physicist's view of homeostasis: how time series of continuous monitoring reflect the function of physiological variables in regulatory mechanisms. Physiol. Meas. **39**(8), 084007 (2018)
3. Seely, A.J.E., Macklem, P.T.: Complex systems and the technology of variability analysis. Crit. Care **8**, R367 (2004)
4. Arbas, E.A., Levine, R.B., Strausfeld, N.J.: Invertebrate nervous systems. Comp. Physiol. (2011). https://doi.org/10.1002/cphy.cp130211
5. Schapker, H., Breithaupt, T., Shuranova, Z., Burmistrov, Y., Cooper, R.L.: Heart and ventilatory measures in crayfish during environmental disturbances and social interactions. Comp. Biochem. Physiol. A: Mol. Integr. Physiol. **131**(2), 397–407 (2002)
6. Usio, N., Konishi, M., Nakano, S.: Biol. Invasions **3**(2), 179–185 (2001)
7. Tainã, L., Anastácio, P.M.S.G., Beatriz, P., Souty-Grosset, C., Pereira, M.: Red swamp crayfish: biology, ecology and invasion—an overview. Nauplius **23**(1), 1–19 (2015)
8. Delgado-Morales, G., Hernández-Falcón, J., Ramón, F.: Agonistic behaviour in crayfish: the importance of sensory inputs. Crustaceana **77**(1), 1–24 (2004)
9. Jiménez-Morales, N., Mendoza-Ángeles, K., Porras-Villalobos, M., Ibarra-Coronado, E., Roldán-Roldán, G., Hernández-Falcón, J.: Who is the boss? Individual recognition memory and social hierarchy formation in crayfish. Neurobiol. Learn. Mem. **147**, 79–89 (2018)
10. Bovbjerg, R.V.: Dominance order in the crayfish *Orconectes virilis* (Hagen). Physiol. Zool. **26**(2), 173–178 (1953)
11. Hernández-Falcón, J., Basu, A.C., Govindasamy, S., Kravitz, E.A.: Changes in heart rate associated with contest outcome in agonistic encounters in lobsters. Cell. Mol. Neurobiol. **25**(2), 329–343 (2005)

12. Rivera, A.L., Estañol, B., Sentíes-Madrid, H., Fossion, R., Toledo-Roy, J.C., Mendoza-Temis, J., Frank, A.: Heart rate and systolic blood pressure variability in the time domain in patients with recent and long-standing diabetes mellitus. PLoS ONE **11**(2), e0148378 (2016)

Cyborgization of Modern Social-Economic Systems

Accounting for Changes in Metabolic Identity

Ansel Renner, A. H. Louie, and Mario Giampietro

Abstract In Part 1 of this paper, the metabolic nature of social-economic systems is explored. A general understanding relating the various constituent components of social-economic systems in a relational network is presented and used to posit that social-economic systems are metabolic-repair (M, R) systems of the type explored in relational biology. It is argued that, through modernization and globalization, social-economic systems are losing certain functional entailment relations and their ability to control replication. It is further argued that modern social-economic systems are losing control over their identity. In Part 2, the implications of those realizations are explored in terms of effective accounting methodology and a practical set of methods capable of harnessing the deep complexity of social-economic systems. In terms of methods, a practical set of metrics defined through the lenses of a macroscope, a mesoscope, and a microscope is presented. Intended to be used simultaneously, the various descriptive domains suggested by our three scopes may be useful for decision-makers who wish to make responsible decisions concerning the control of system identity change or to combat processes of societal cyborgization.

Keywords Metabolism · Social-economic system · Relational biology · Thermodynamics

A. Renner (✉) · M. Giampietro
Universitat Autònoma de Barcelona, 08193 Bellaterra, Spain
e-mail: ansel.renner@uab.cat

M. Giampietro
e-mail: mario.giampietro@uab.cat

A. H. Louie
Ottawa, ON, Canada

M. Giampietro
ICREA, 08010 Barcelona, Spain

D. Braha et al. (eds.), *Unifying Themes in Complex Systems X*, Springer Proceedings in Complexity, https://doi.org/10.1007/978-3-030-67318-5_9

1 Part 1

This manuscript's first part aims to establish bridges between concepts used in different disciplines. Firstly, we reflect on and apply the epistemological tools used in the multidisciplinary field of societal metabolism. We then also reflect on and apply the explanatory resources and rigorous definitions provided by the field of relational biology insofar as they are useful in the classification of social-economic systems as living (and therefore "authentically", not simply "metaphorically", metabolic) or non-living. The discussion made in Part 1 sets up Part 2's presentation of three scopes useful for the practical assessment of social-economic systems.

Before setting off, it is productive to position our discussion of metabolic systems in relation to the broader discussion of complex systems. Following the classification and nomenclature used in relational biology, the set of living systems (**O**) is a subset of the set of complex systems (**I**). Furthermore, the set of complex systems is a subset of the set of natural systems (**N**). In other words, $\mathbf{O} \subset \mathbf{I} \subset \mathbf{N}$. The union of the set of complex systems (**I**) and the set of simple systems (**S**) is the set of natural systems (**N**). In other words, $\mathbf{S} \cup \mathbf{I} = \mathbf{N}$. *All metabolic-repair (M, R) systems are complex systems and all replicative (M, R)-systems are living systems.* Refer to Louie [1] for detailed causal definitions of each of the system types referred to.

1.1 Societal Metabolism

As originally noted in the field of physical biology, i.e. the work of Lotka [2], metabolic processes may be meaningfully divided into an endosomatic class and an exosomatic class. Endosomatic metabolism refers to processes taking place *within* a given organism insofar that an organism uses fluxes of negative entropy available in the environment to sustain a process of exergy degradation[1] and stabilize their metabolic pattern while reproducing, maintaining, and adapting their structural and functional elements. An exosomatic process refers to a situation in which the processes of exergy degradation used to stabilize a metabolic pattern take place *outside* a given organism. This essential distinction between types of metabolic process also applies in the domain of social-economic systems [4]. In that context, for example, it is advantageous to differentiate between an endosomatic population (e.g. the population of humans) and an exosomatic "population" (e.g. the population of machines) [5]. Post-Industrial Revolution, the exosomatic population has been seen to dramatically increase in both absolute and relative terms [6]. This is a change in metabolic identity that cannot be ignored—we will come back to it in Sects. 1.4 and 1.5.

In his discussion of the biophysical nature of the economy, Georgescu-Roegen [4] made a further critical distinction in his flow-fund model. In a metabolic pattern,

---

[1] Use of available energy forms that can be converted into useful work according to the characteristics of the user and the environment within which the conversion takes place [3].

flow elements are those that do not maintain their identity over the course of an analysis. Food or exosomatic energy inputs that are consumed during the course of an analysis certainly do not maintain their identity—they are examples of flow elements. Fund elements are those that do maintain their identity during the course of an analysis. Over the timespan of a week, an organism typically maintains its identity, assuming it consumes an acceptable flow of food and exosomatic energy inputs—it is an example of a fund element. Stocks are not the same as fund elements. If a deposit or withdrawal is made to or from a stock, the stock does not maintain its identity.

So, flows, funds, and stocks can be used as descriptive categories when describing endosomatic and exosomatic metabolic processes and populations. This is a good first step in the assessment of social-economic systems as metabolic systems. What we lack, however, is an explanatory framework allowing us to answer questions of "why?" with assertions of causality. The division between negative and positive forms of entropy in non-equilibrium thermodynamics is semantic—it is contingent on the identification of "whys?" in relation to metabolic processes. Section 1.2 introduces a language of causality which allows us to develop substantive (transferable) theories about metabolic processes.

1.2 Explanatory Resources (Causality)

Relational biology is a discipline concerned with the elaboration and assessment of causal relations (mappings) useful for the derivation of meaningful answers about a system. It is the operational inverse of reductionist biology. While reductionist biology begins an analysis by throwing away function (keeping matter), relational biology begins an analysis by throwing away matter (keeping function). See Rashevsky [7, 8] for its origins, Rosen [9, 10] for its most well-known presentation, and Louie [1, 11, 12] for its most recent advances.

The causal building blocks considered by relational biologists are the four Aristotelean causes. Each of the four causes may be considered as an irreducible explanatory resource, essential to answer an inquiry of "why?". For Aristotle, and indeed in relational biology, all four causes must be identified in a complete explanation. An elaboration of anything less is a partial description.

- *Material cause* is that out of which something is constituted. For example, the concrete and steel that constitute the roads and buildings of a modern city could be identified as a material cause.
- *Formal cause* is the "blueprint" or "form" of something, or the "account of what it is to be". For example, the layout/configuration of a city could be identified as a formal cause.
- *Efficient cause* is the primary agent of change realizing or initiating something which is done. For example, the construction workers who erected the roads and buildings of a city could be identified as an efficient cause.

- *Final cause* is the sake for which something is done, i.e. to what end? Two examples that could be identified as the final cause of a city are the desire to live closer to other humans and the desire to shelter oneself from the unpredictable nature of the non-constructed environment.

Before moving any further, a disclaimer is merited concerning final cause. Although final cause and teleology are frequently considered to be synonyms, there is an important clarification that can be made. Final cause is teleological to the extent it references an end purpose, but it is not psychologized in the sense there is no notion of intention when a final cause is formally mapped. This point is further discussed in another paper of this issue [13]. In contrast to formal mappings of final cause, teleology typically implies the existence of intentionality [9, 14]. Even the necrophiliac[2] mode of science known as reductionism does not, on a theoretical basis, disallow the practical use of formal mappings of final cause.

1.3 Relational Characterization of a Societal Node

In social-economic systems, Aristotelean causality can be intuitively applied at a low-level scale in the description of social practices. Bundles of social practices can, in turn, be used as building-block descriptions of societal sectors—the nodes of a relational description of a social-economic system. At a very basic level, social practices describe the convergence and linkage between meanings, competences, and materials, as expressed by a group of agents [16]. Meanings are characterizable as final causes, competences as formal causes, materials as material causes, and expressing agents as efficient causes. Figure 1 and the discussion that follows illustrate a consideration of these explanatory resources in relation to a generic conceptualization of an agriculture sector—an entity that emerges from a bundling of the expression of social practices related to agriculture.

The right-hand side of Fig. 1 explores the final causes of agriculture—the set of behaviors expected from it. In this sense, the right-hand side details the array of end-uses (a "hologram") of agriculture and considers the agriculture sector as a social-economic component. The left-hand side of Fig. 1 explores the various material, formal, and efficient causes of agriculture. In this sense, the left-hand side of Fig. 1 refers to the agriculture sector as a social-economic constituent. Used in contrast to component, the term constituent refers to a structural definition. In material terms, the agriculture sector is made up of various vegetal and animal organisms and products, machinery, buildings, infrastructure, and so forth, in which human activity is used to control the processes of exergy transformation. Its formal cause is the configuration of such material considerations, including, for example, the relative break-down and

[2]"Necrophilia in the characterological sense can be described as *the passionate attraction to all that is dead, decayed, putrid, sickly; it is the passion to transform that which is alive into something unalive; to destroy for the sake of destruction; the exclusive interest in all that is purely mechanical. It is the passion 'to tear apart living structures'*" [15].

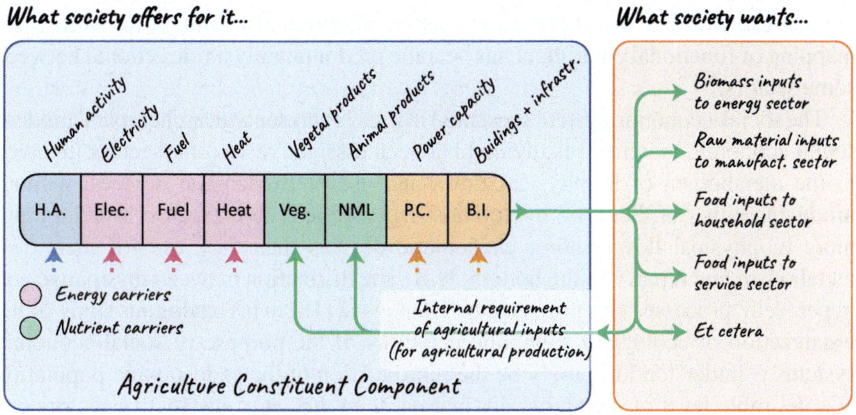

Fig. 1 Biophysical demand placed by a social-economic system on its constituent component "agriculture"

arrangement of materials. Lastly, an account of human activity controlling the power capacity provided by machines is a proxy consideration of the efficient cause of social practices of agriculture—the proximate "agents" behind the realization of the agriculture sector (see [17]).

Keeping in mind the language presented in Sect. 1.1, the ratio comparison between the use of an exosomatic population of funds (such as machines, buildings, and infrastructure) and the use of an endosomatic population of funds (humans) can provide a characterization of the degree of capitalization of an agriculture sector. A similar comparison between flows can provide further contextual information such as the degree of reliance on external inputs—expected relations that must be guaranteed.

1.4 Society as a Relational Network

We now wield the tools needed to represent social-economic systems as metabolic networks in which constituent components stabilize each other in an impredicative (self-referential) set of relations. Such a representation allows for the generation of characteristics such as the relative size of constituent components, their expected metabolic rates, and, more in general, a definition of societal identity.

Societal sectors, such as the agriculture constituent component depicted in Fig. 1, are distinguishable elements of social-economic systems that generate an emergent property. Each sector has a meaningful and relevant identity regarding an identity-dependent coupling of positive and negative forms of entropy. Each of the sectors of a social-economic system can also be said to have a final cause. Historically, a dialectical exchange of functional entailment (a mapping where the final cause of one process becomes the efficient cause of another, also referable to as hierarchical composition) was made between the various sectors of social-economic systems. In the modern world, a world prevailed over by biophysically affluent urbanities, the mapping of functional entailment has become predominantly unidirectional between some sectors.

The social-economic system presented in Fig. 2 represents an archetypical modern social-economic system. It is divided between dissipative sectors (sectors involved in the metabolism of biophysical flows and use of exosomatic devices, without producing either of them) on the top and hypercyclic sectors (sectors which output more biophysical flows and/or exosomatic devices than they use for their own metabolism and repair) on the bottom. N.B. The distinction between dissipative and hypercyclic processes was proposed by Ulanowicz [18] in his analogous study of the organization of ecological metabolic networks. If the purpose of social-economic systems is understood to purely be the reproduction of the endosomatic population at a desirable level of metabolic dissipation, then the set of dissipative sectors can also be understood as anabolic and the set of hypercyclic sectors as catabolic.

In modern societies, the final causes of dissipative sectors map to the efficient causes of each other and to the various hypercyclic sectors of the economy. The dissipative sectors provide a system of control for the hypercyclic sectors. The final cause of the hypercyclic sectors, on the other hand, is more and more to provide exosomatic flows of biophysical material to dissipative sectors—no questions asked. This role of the hypercyclic sectors was not always the case—this role was not the case in pre-industrial agrarian societies, for example [6].

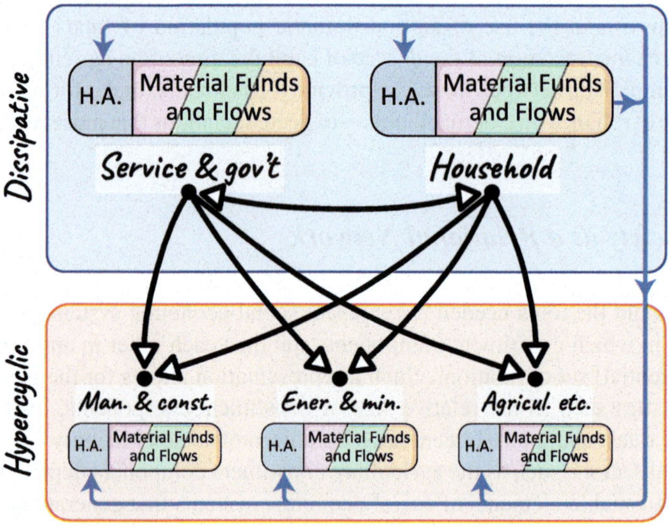

Fig. 2 Functional entailment between the components of a social-economic system (hollow-headed black arrows) and its vector of control (solid-headed blue arrows) with sector details: (i) household; (ii) service and government; (iii) manufacturing and construction; (iv) energy and mining; and (v) agriculture, forestry, and fishing

1. In the *service and government sector*, an effective interface between production and final consumption is made. The preservation, notional reproduction, and adaptation of institutions occurs. Trust, one prerequisite for proper market operations, is generated. Regulations are made. Education, security, and law enforcement are enacted.
2. In the *household sector*, humans as individual or family-level agents are reproduced and maintained. Social practices and normative values are shaped. Market preferences are shaped and voting and political participation occurs.

Together, the two dissipative sectors identified in Fig. 2 provide a reflexive analysis of a social-economic system's identity, feelings, emergent concerns, and interactions with the external world (due to the presence of humans). Contingent analysis, required to establish societal priorities in an impredicative option space, is made. Likewise, analyses required for the tackling of the unavoidable existence of uncertainty and the unavoidable existence of legitimate but contrasting perspectives are made.

However, functional entailment, shown in Fig. 2 as running between the dissipative sectors and from the dissipative sectors to the hypercyclic sectors, can only be defined after matching a definition of "metabolic demand", coming from a specific metabolic pattern of system components, with a definition of "metabolic supply", coming from a specific metabolic pattern of system components. Functional entailment represents a top-down constraint (directly resulting in "downward causation") related to an emergent property (an "identity"). Material entailment (a mapping where the final cause of one process becomes the material cause of another), on the other hand, relates to the need for establishing coherent biophysical relations. In this sense, material entailment represents a bottom-up constraint (indirectly resulting in so-called "upward causation"). Material entailment is about establishing a feasible and viable state-pressure relation with various local admissible environments. In the language of non-equilibrium thermodynamics, it can be represented by the definition of a local coupling of patterns of exergy degradation (state) mapping onto fluxes of negative entropy (pressure) across different levels. Figure 3 presents a different view of our archetypical modern social-economic system. Figure 3, supplementary

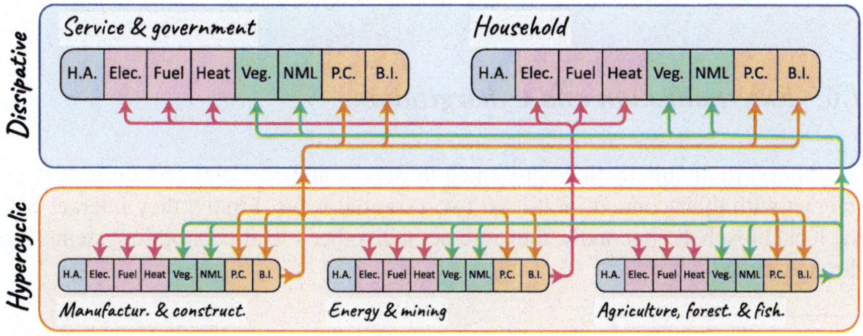

Fig. 3 Material entailment between the constituents of a social-economic system

to Fig. 2, highlights relations of material entailment. It should be clear that the hypercyclic sectors, in their provisioning of material flows to dissipative sectors, present a biophysical constraint on each other and on the dissipative sectors.

1.5 Society as an (M, R)-System

There is a theorem in the field of relational biology which states that all living systems are formalizable as *replicative* metabolic-repair (M, R) systems. Let's dig deeper. In terms of the concept of metabolism, we've already established a respectable understanding. Repair, on the other hand, is the reproduction of a catalyst of change. In terms of causal mappings, functional entailment is repair and material entailment is metabolism. Lastly, replication is not "self-replication" in the modern biological sense. Rather, the output of replication in (M, R)-systems is repair (see [1, 19, 20]). A similar concept is found in the bioeconomics of Georgescu-Roegen—the economy does not produce goods and services but rather reproduces the fund elements producing and consuming goods and services to guarantee adaptability and a desirable state [21].

If the functional entailment relations presented in Fig. 2 were a complete graph[3]—as could easily be claimed to have been the case in pre-modern society—then we would consider social-economic systems to be in a situation of deadlock. Deadlock, the relational analog of impredicativity (that great hallmark of complexity), is a "situation wherein competing actions are waiting for one another to finish, and thus none ever does" [1] (p. 208). Under a state of deadlock, with all mappings of final causality locked up in hierarchical composition, a social-economic system could be said to be *closed to efficient causation.* To apply Georgescu-Roegen's flow-fund concept, the social-economic system could be said to be in a situation where all metabolic funds internally generate each other, given the presence of favorable boundary conditions. In that circumstance, a social-economic system could be referred to as living under the same causal foundation that conventional biological organisms are said to be living.

1.6 Externalization and Cyborgization

In addition to the interactions presented in Figs. 2 and 3, social-economic systems interact with their context in at least two essential ways. Firstly, they interact with the local biosphere. Secondly, they interact with other social-economic systems, e.g. through trade.

[3]Since our mappings are directed, we refer here to a complete directed graph, meaning that every pair of vertices (contextually, societal sectors) is connected by a pair of unique edges (contextually, functional entailments)—one in each direction.

The first type of interaction implies a change in consideration from social-economic systems to social-ecological systems. A social-ecological system can be defined as a complex of constituent components that operates within a prescribed boundary and is controlled in an integrated manner by activities expressed both by a given set of social actors and institutions operating in the economy (under human control, in the technosphere) and a given set of ecosystems (outside human control, in the biosphere). This change in consideration from economic to ecological involves a description of an entropically favorable state-pressure relation between a symbiotic technosphere and biosphere.

For the second type of interaction, we will focus on the externalization of economic processes through trade. Through imports, a social-economic system is able to reduce its requirement to secure a reliable internal flow of biophysical material—a flow that would otherwise need to be produced by its own hypercyclic sectors. In this sense, externalization increases the ability of a system to free itself from the need to establish a complete set of internal relations of material entailment. It reduces the constraints determined by the counterfactual local pressures of downward and upward causation.

However, from a relational perspective, externalization also has a dark side. Externalization may be seen as a major reason why many modern social-economic systems are losing their control over the reproduction, maintenance, and adaptation of their own identity. Through processes of modernization (and changes in considerations of desirability, related to rising expected levels of material standard of living), the ability of dissipative sectors to provide feedback through functional entailment relations with hypercyclic sectors—to "cast their vote"—has been seen to systematically reduce. As a result, social-economic systems are losing their ability to replicate[4] and losing control over their ability to repair. In essence, they are losing control over their identity. The many benefits of modernization may be seen to have come hand-in-hand with a local breaking of the closure to efficient causation of social-economic systems—a process of "cyborgization" of society in which societal identity is being increasingly determined by external factors.

As Swyngedouw [22] points out, cities are exemplars of this process of cyborgization of society. N.B. Cities also epitomize externalization in social-economic systems. A cyborg is a hybrid between an organism and at least one artificial component, where an artificial component is a simple system tasked with restoring an organic function. From the relational biology perspective, a human (a type of organism) equipped with and reliant on a pacemaker (a type of artificial component) is already a cyborg, albeit a very basic cyborg. With no great stretch of the imagination, we can similarly conceptualize cities as extensive mechanical cocoons that offer a wide array of inert goods and services as replacements of societal functions. It is undeniable that cities have many desirable characteristics, but it is also undeniable that they are the technocratic havens of social agents operating, as Turkle [23] phrased it, "alone together". As previously mentioned, the unorthodox economist Georgescu-Roegen was clear in identifying that the final cause of the economy is not "to produce goods

[4]"Replication" not in the molecular biology sense, rather in the relational biology sense described at the head of Sect. 1.5.

and services" but rather to reproduce fund elements associated with the production and consumption of goods and services. That is, the final cause of the economy is simply the reproduction of itself while guaranteeing an "enjoyment of life" to citizens. Considering Georgescu-Roegen's understanding of the economy, it would be advisable for social-economic systems to make a careful assessment of their intrinsic, complex network of functional and material entailments before making the societal decision to off-shore, or replace with artificial components, their societal organs. An organism turned cyborg can no longer decide how it wishes to enjoy life.

2 Part 2

The functional entailment from the dissipative sectors of a social-economic system to the hypercyclic sectors, as shown in Fig. 2, represents a set of powerful initial constraints on the definition of that system's option space (downward causation). Conversely, the potential combinations of material, formal, and efficient causes in the hypercyclic sectors of society can either expand or—more likely—limit the initial option space definition (so-called "upward causation"). In this sense, the option spaces of social-economic systems are constrained by a double contingency (downward and upward causation). A second paper in this same volume explores in greater detail the ideas of downward and upward causation in relation to the generation of meaning and identity preservation in complex adaptive systems [13]. What will instead be outlined in what remains of this paper is a set of quantitative accounting considerations capable of informing processes of societal deliberation over questions such as: Assuming a given social-economic system, to what degree is change in identity to be allowed? In what ways should identity change be considered permissible?

By applying the concepts outlined in Part 1, it is possible to distinguish which entailment relations are necessary (which isn't to say sufficient) for the maintenance of the various aspects of a social-economic identity. In that endeavor—the maintenance of a social-economic identity—it is advisable to use at least three lenses. In the following sub-sections, we propose a macroscope, a mesoscope, and a microscope. N.B. As also claimed by the principle of biological relativity [24, 25], no one scope is "better" than the other. The adoption of a specific scope must follow the societal selection of a question or concern. Generally, all three scopes should be used simultaneously (Fig. 4).

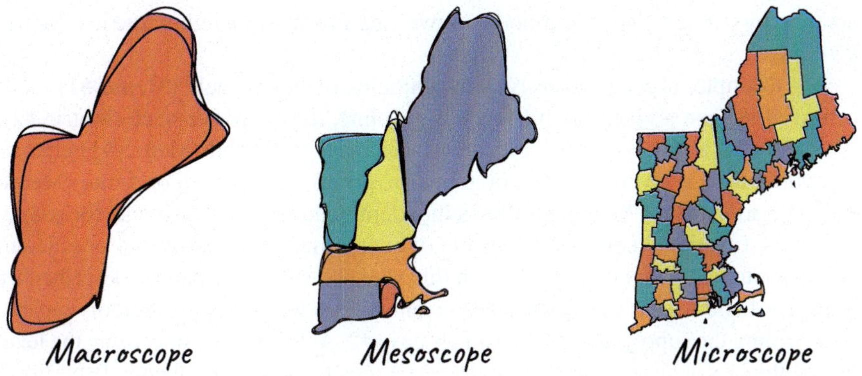

Fig. 4 Three lenses useful for the relational characterization of a social-economic system and its option space. No single lens is superior

2.1 Macroscope

A macroscope is a highly approximated descriptive domain—the result of liberal coarse-graining. When looking through a macroscope, one sees a profile of end-uses and an approximated vision of the results of functional entailment in a social-economic system (the set of final causes is associated with the set of end-uses). A profile of allocation of fund and flow elements by constituent component (sector) is defined both in terms of the relative size of aggregate-level funds and the specific metabolic rates of those funds.

In using a coarse characterization of metabolized flows ("Metric #1"), it is possible to characterize an array of end-uses in terms of: (i) a set of final causes in a dendrogram of splits of fund element across different levels of organization; (ii) a definition of the various exosomatic metabolic rates of those funds (like a neural-network integrating the dendrograms of splits of funds with dendrograms of splits of flows); and (iii) a calculation of a "bio-economic pressure" determined by the relative size and relative endosomatic and exosomatic metabolic rates of the various constituent components of society [26].

2.2 Mesoscope

A mesoscope shows an approximated vision of the relations of the various supply systems of a social-economic system as it interacts with the local biosphere and other external social-ecological systems. The mesoscope maps the requirement of metabolized flows—as characterized in an array of end-uses defined using the macroscope's Metric #1—against a non-equivalent characterization of the supply of flows required to stabilize the metabolic pattern. The metabolized flows expressed using Metric #1 are thereby converted into commodity flows measured using another metric ("Metric #2").

For example, a given quantity of megajoules of liquid fuels (Metric #1) characterized against a given mix of liters of gasoline, diesel, or kerosene (Metric #2). Alternatively, a given quantity of kilocalories of vegetal food (Metric #1) characterized against a given mix of kilograms of grains, vegetables, or fruits (Metric #2). The adoption of Metric #1 allows the identification of a final cause for supply systems—functional complexes capable of expressing an integrated set of efficient causes matching a given final cause. In this way, the metabolic pattern described by an array of end-uses is translated into a given requirement of supply systems capable of providing the supply, described using Metric #2. At this point, concerning the level of openness of a system, we can identify: (i) the fraction of the metabolized flows generated by local supply systems operating inside the border of a social-economic system; and (ii) the fraction the metabolized flows generated by "virtual supply" systems operating outside the border of a social-economic system. This is indeed an important piece of information since it characterizes, when considering material entailments, the degree of freedom from internal constraints on the societal option space. In conclusion, supplementing Metric #1's characterization of the metabolic pattern with Metric #2 allows the identification of the required set of efficient causes (either expressed locally or embodied in imported commodities) determined by the definition of final causes expressed by human funds. Since no information is given about "how" these efficient causes are expressed (no mention of material or formal cause), the relation between Metric #1 and Metric #2 can be considered a *downward causation* of metabolic pattern determination.

2.3 Microscope

Use of a microscope descriptive domain yields an approximated vision of the relations of the mix of various structurally defined production processes—defined as combinations of material, formal, and efficient causes—into functional units, which express functions required at the mesoscope level. Using a microscope, it is possible to assess the technical viability, environmental pressures, and environmental impacts (feasibility) of production and consumption processes.

For each of the "supply systems" defined using the mesoscope and represented in quantitative and qualitative terms using Metric #2, it is possible to generate a non-equivalent quantitative characterization of the mix of the processes needed to generate that supply system. That is, at a lower level of analysis, we can describe the specific combination of material, formal, and efficient cause (the local end-uses, or local social practices) that, when combined in a functional bundle, are capable of expressing the functions associated with given supply systems. The quantitative representation of these relations requires the adoption of a third metric (Metric #3) in which we describe the combinations of fund elements and flow elements (end-uses) in local production processes. In this way, for example, a commodity flow such as a quantity of kilowatt-hours of electricity can be mapped onto a mix of quantities of electricity generated by different types of power plants (funds and flows mixed together in the analysis). It's worthwhile to elaborate with examples: (i) baseload electricity can be produced using either nuclear power plants (fund) using uranium (flow) as part of the material cause or coal power plants (fund) using coal (flow) as part of the material cause; (ii) peak electricity can be produced using either hydropower plants (fund) using falling water (flow) as part of the material cause or gas turbines (fund) using natural gas (flow) as part of the material cause; and (iii) intermittent electricity can be produced using either wind turbines (fund) using the kinetic energy of the wind (flow) as part of the material cause or photovoltaic cells (fund) using solar radiation (flow) as part of the material cause. Of course, the material needed to construct the power plants (exosomatic funds) should also be included in the analysis, whenever relevant. The various combinations of lower-level elements associated with a specific combination of material, formal, and efficient cause within a given specific production process can be characterized using an array of vectors—a *metabolic processor* in relational analysis—and stored in a database. Each specific combination maps onto an expected profile of inputs and outputs that can be combined into a chained functional unit (a sequential pathway, for example, the extraction of coal, transportation of coal, and production of electricity in a power plant) for which it becomes possible to calculate the metabolic pattern.

3 Conclusion

In this contribution, we used the well-established language of relational biology to present an initial exploration of social-economic systems as replicative (M, R)-systems. We concluded that social-economic systems are metabolic in a causal sense identical to the that of biological organisms. Shortly after arriving at that conclusion, we declared that modern society is undergoing a process that could be understood as a *cyborgization*—a process by which society is progressively losing control over the definition of its own identity. Though we identified several ways in which the process of cyborgization is creating a fragile society, we made no mention of whether the process itself is "good" or "bad". Nevertheless, substantial desirability concerns arise from the process of cyborgization of society—those concerns can be understood as "good" or "bad" once framed by a specific stakeholder and in the light of a specific question or concern.

Governance in complexity is about staying on the horns of dilemmas. While it is possible to put a harness on complexity, it is impossible to tame complexity. To put a harness on social-economic systems understood as complex systems, we suggested the simultaneous use of a macroscope, a mesoscope, and a microscope to observe different relevant aspects of societal performance. The use of our three scopes allows for the establishment of a bridge between representations of the metabolic pattern of social-economic systems from the perspectives of: (i) downward causation—what is required in terms of supply systems by the characteristics of societal metabolism as observed with the macroscope; (ii) upward causation—what can be supplied by local combinations of microscope production processes given the constraints provided by the characteristics of formal causes (technical viability, i.e. internal constraints, determined by technical coefficients) and material causes (environmental feasibility, i.e. external constraints, determined by biophysical processes outside human control); and (iii) the role that externalization through trade plays in lessening the constraints imposed by upward causation (characterized using the mesoscope).

The general approach to the assessment of social-ecological systems in this contribution has been developed within the auspices of the European project *Moving Towards Adaptive Governance in Complexity* [27]. Initial examples of applications of macroscopic views of social-economic systems are available in [26, 28–30], of mesoscopic views in [31, 32], and of microscopic views in [33, 34]. Substantial work remains in terms of exploring the many implications of relational biology for the assessment of societal metabolism and in further formalizing the approach.

Acknowledgements A. Renner acknowledges financial support from the Spanish Ministry of Education, Culture and Sport, through the "formación de profesorado universitario" scholarship program (FPU15/03376). A. Renner and M. Giampietro acknowledge financial support by the European Union's Horizon 2020 research and innovation programme under Grant Agreement No. 689669 (MAGIC), the Catalonian Government (AGAUR) under Grant Agreement 2017-SGR-230, and the Spanish Ministry of Science, Innovation and Universities, through the "María de Maeztu" program for Units of Excellence (CEX2019-000940-M). This work only reflects the view of the authors; the funding agencies are not responsible for any use that may be made of the information it contains.

References

1. Louie, A.H.: More Than Life Itself: A Synthetic Continuation in Relational Biology. Ontos Verlag (2009)
2. Lotka, A.J.: Elements of Physical Biology. Williams & Wilkins Company (1925)
3. Gaudreau, K., Fraser, R.A., Murphy, S.: The tenuous use of exergy as a measure of resource value or waste impact. Sustainability **1**, 1444–1463 (2009). https://doi.org/10.3390/su1041444
4. Georgescu-Roegen, N.: The Entropy Law and the Economic Process. Harvard University Press, Cambridge, MA (1971)
5. Mayumi, K.T.: Sustainable Energy and Economics in an Aging Population: Lessons from Japan. Springer (2020)
6. Giampietro, M.: Multi-scale Integrated Analysis of Agroecosystems. CRC Press LLC, New York (2004)
7. Rashevsky, N.: Mathematical biophysics. Nature **135**, 528–530 (1935). https://doi.org/10.1038/135528a0
8. Rashevsky, N.: Organismic sets: outline of a general theory of biological and social organisms. Bull. Math. Biophys. **29**, 139–152 (1967). https://doi.org/10.1007/BF02476967
9. Rosen, R.: Life Itself: A Comprehensive Inquiry Into the Nature, Origin, and Fabrication of Life. Columbia University Press, New York (2005)
10. Rosen, R.: Essays on Life Itself. Columbia University Press, New York (2000)

11. Louie, A.H.: The Reflection of Life: Functional Entailment and Imminence in Relational Biology. Springer, New York, NY (2013). https://doi.org/10.1007/978-14614-6928-5
12. Louie, A.H.: Intangible Life: Functorial Connections in Relational Biology. Springer International, Cham, Switzerland (2017). https://doi.org/10.1007/978-3-319-65409-6
13. Giampietro, M., Renner, A.: The Generation of Meaning and Preservation of Identity in Complex Adaptive Systems: The LIPHE4 Criteria. In: Braha, D. (ed.) Unifying Themes in Complex Systems X. Springer, Cham, Switzerland (2021)
14. Chase, M.: Teleology and final causation in Aristotle and in contemporary science. Dialogue Can. Philos. Assoc. **50**, 511–536 (2011). https://doi.org/10.1017/S0012217311000527
15. Fromm, E.: The Anatomy of Human Destructiveness. Holt, Rinehart and Winston, New York (1973)
16. Shove, E., Pantzar, M., Watson, M.: The Dynamics of Social Practice: Everyday Life and How It Changes. SAGE Publications Ltd., London (2012)
17. Patching, D.: Practical Soft Systems Analysis. Pearson Education Limited, Essex (1990)
18. Ulanowicz, R.E.: Growth and Development: Ecosystems Phenomenology. Springer, New York (1986)
19. Rosen, R.: A relational theory of biological systems. Bull. Math. Biophys. **20**, 245–260 (1958). https://doi.org/10.1007/BF02478302
20. Rosen, R.: A relational theory of biological systems II. Bull. Math. Biophys. **21**, 109–128 (1959). https://doi.org/10.1007/BF02476354
21. Giampietro, M., Pastore, G.: Biophysical roots of 'enjoyment of life' according to Georgescu-Roegen's bioeconomic paradigm. In: Mayumi, K., Gowdy, J. (eds.) Bioeconomics and Sustainability: Essays in Honor of Nicholas Georgescu-Roegen, p. 438. Edward Elgar Publishing (1999)
22. Swyngedouw, E.: Circulations and metabolisms: (hybrid) natures and (cyborg) cities. Sci. Cult. (Lond) **15**, 105–121 (2006). https://doi.org/10.1080/09505430600707970
23. Turkle, S.: Alone Together: Why We Expect More from Technology and Less from Each Other. Basic Books, New York, NY, USA (2012)
24. Noble, D.: A theory of biological relativity: no privileged level of causation. Interface Focus **2**, 55–64 (2012). https://doi.org/10.1098/rsfs.2011.0067
25. Noble, R., Tasaki, K., Noble, P.J., Noble, D.: Biological relativity requires circular causality but not symmetry of causation: so, where, what and when are the boundaries? Front. Physiol. **10**, 1–12 (2019). https://doi.org/10.3389/fphys.2019.00827
26. Giampietro, M., Mayumi, K., Sorman, A.H.: The Metabolic Pattern of Societies: Where Economists Fall Short. Routledge, London (2012)
27. Giampietro, M., Aria, M., Cabello, V., Cadillo-Benalcazar, J.J., D'Ambrosio, A., de La Fuente, A., Di Felice, L., Iorio, C., Kovacic, Z., Krol, M.S., Madrid López, C., Matthews, K., Miller, D., Musicki Savic, A., Pandolfo, G., Peñate, B., Renner, A., Ripa, M., Ripoll Bosch, R., Serrano-Tovar, T., Siciliano, R., Staiano, M., Velasco, R.: Report on Nexus Security Using Quantitative Story-Telling. http://magic-nexus.eu/sites/default/files/files_documents_repository/magic-ga689669-d4.1-revision.pdf
28. Velasco-Fernández, R., Chen, L., Pérez-Sánchez, L., Giampietro, M.: Multi-scale integrated comparison of the metabolic pattern of EU28 and China in time. EUFORIE Project. Deliverable 4.4. ICTA, Autonomous University of Barcelona. ICTA, UAB (2018). http://www.euforie-h2020.eu
29. Pérez-Sánchez, L., Giampietro, M., Velasco-Fernández, R., Ripa, M.: Characterizing the metabolic pattern of urban systems using MuSIASEM: the case of Barcelona. Energy Policy. **124**, 13–22 (2019). https://doi.org/10.1016/J.ENPOL.2018.09.028
30. Giampietro, M., Saltelli, A.: Footprints to nowhere. Ecol. Indic. **46**, 610–621 (2014). https://doi.org/10.1016/j.ecolind.2014.01.030
31. Cadillo-Benalcazar, J.J., Renner, A., Giampietro, M.: An accounting framework for the biophysical profiling of the European agricultural system. In: ALTER-Net and EKLIPSE (eds.) The EU Biodiversity Strategy Beyond 2020: Research Insights and Needs for Biodiversity and Ecosystem Services in Europe. Ghent, Belgium (2019)

32. Renner, A., Cadillo-Benalcazar, J.J., Benini, L., Giampietro, M.: Environmental pressure of the European agricultural system: Anticipating the biophysical consequences of internalization. Ecosyst. Serv. **46** (2020). https://doi.org/10.1016/j.ecoser.2020.101195
33. Cabello, V., Renner, A., Giampietro, M.: Relational analysis of the resource nexus in arid land crop production. Adv. Water Resour. **130**, 258–269 (2019). https://doi.org/10.1016/j.advwatres.2019.06.014
34. Di Felice, L.J., Ripa, M., Giampietro, M.: An alternative to market-oriented energy models: nexus patterns across hierarchical levels. Energy Policy **126**, 431–443 (2019). https://doi.org/10.1016/j.enpol.2018.11.002

Trustable Risk Scoring for Non-bank Gateways in Blockchain and DLT Financial Networks

Percy Venegas

Abstract We propose a method to assess the counterparty risk of non-banking financial institutions that operate as fiat-crypto gateways in blockchain and distributed ledger technology financial infrastructures. The risk scores are suitable to evaluate both traditional money services businesses and fintech companies (including cryptocurrency payments and blockchain systems operators). The main users are banks, investors, and, businesses that need to assess counterparty risk across jurisdictions and under uncertainty, as non-banks are often less regulated than other financial institutions. We follow an automation-focused, multidisciplinary approach rooted in established techniques from network science and genetic programming, as well as in the emerging fields of machine behavior and trustworthy artificial intelligence. The method and findings pertain to any decentralized financial infrastructure with centralized components such as fiat on/off ramps, where counterparty risk assessment becomes a necessity for regulatory, investment, and, operational purposes. The method is demonstrated on a network of Ripple payment gateways.

Keywords Payments networks · Counterparty risk · Blockchain · Fintech · Machine behavior · Trustable AI

Key Findings

- Decentralized financial infrastructures with centralized fiat offramps are evolving complex systems where inputs are highly correlated.
- Symbolic dependency graphs, and, symbolic regression via genetic programming, are suitable artifacts for mapping and ranking relative risk exposures.
- Trustable AI enhances model explainability and the understanding of dynamic uncertainty.

P. Venegas (✉)
Economy Monitor, King's College London, London, UK
e-mail: v.percy@economymonitor.com

D. Braha et al. (eds.), *Unifying Themes in Complex Systems X*, Springer Proceedings in Complexity, https://doi.org/10.1007/978-3-030-67318-5_10

1 Introduction

The study of trust asymmetries [1] in the financial sector requires an extensive mapping of the institutions that are part of the system; both regulated and unregulated players introduce different fragility vectors [2] that central bankers should understand and that potential business partners, acquirers, and, banking providers should investigate for the purpose of counterparty risk assessment. When discussing the impact of blockchain technology on industry and society Aste et al. [3] ask a key question: *How much collective irrational phenomena such as sentiment/confidence swings will affect the capability of these markets to operate?* The issue is key because distributed ledger technology systems do not operate in isolation—the continuity of the business model of companies such as Ripple depends on the visibility (in the network theoretical sense) and performance (in the regulatory and business operations sense) of partners such as the gateways that facilitate fiat-crypto on-ramps and off-ramps. The customers of such gateways are also users of traditional online commerce channels, which they utilize before and after their activity within the trading platform–this enables a "flow of attention" at the upper level of the "transactions flow". This sort of economic behavior presents the features of an evolving complex system [4]: dispersed interaction, no global controller, cross-cutting hierarchical organization, continual adaptation, perpetual novelty, and, out-of-equilibrium dynamics.

Moreover, record-keeping in distributed ledgers operates as a form of social computing. The difference here, when compared to the traditional financial system, is that these interactions can also be fully automated. That is, an autonomous agent can aggregate social feedback from databases, humans and other automated agents, to affect the state of the system. The new state can, in turn, have an effect on the new posture of both humans and agents (it achieves settlement finality, although there might be different timeframes and degrees of certainty, according to the technology [5]).

Researchers of human-machine co-behavior [6] note that *most AI systems function in domains where they co-exist with humans in complex hybrid systems, such as how trading patterns can be altered by interactions between humans and algorithmic trading agents as well as which factors can facilitate trust and cooperation between humans and machines.* This inevitably brings to the table the topic of trustable and safe AI, particularly in multiparty human-robot situations. These include issues of *trust, privacy, and ethics with respect to sharing information and modeling the beliefs of others* [7].

In this paper, we are concerned with models that evolved as such human-machine behavior evolves. The initial mapping of the system is achieved using network theory, but since the implementation utilizes symbolic dependency graphs, the mapping itself can adapt as the symbolic regression expressions obtained via genetic programming change.

2 Method

2.1 Coverage

The number and types of players in the non-banking segment, and particularly in the fintech branch, is undergoing rapid growth due to the accelerating pace of innovation in the financial sector. It is possible to learn the categorical segments (e.g. non-bank lenders, digital assets exchanges) from data [8], although on occasion there are overlaps related to competitive dynamism i.e. companies diversify into different delivery platforms/devices, M&A activity, and so on. Figure 1 shows the non-banks taxonomy used as a departing point, which is calibrated at regular intervals to provide a more accurate depiction of the environment as it changes–the evolving risk space.

Our focus for the purpose of this paper is a type of cryptocurrency company known as a "gateway". In particular, Ripple Gateways are businesses that provide an entry point into the Ripple Network [9]. They enable customers to transfer funds—both traditional and cryptocurrency—using the Ripple Network. This means that customers can use the gateway to: (a) Deposit funds in exchange for Ripple IOUs (balance/debt "I owe you" records) issued by the gateway; (b) Transfer their IOUs to another Ripple (IOU) address. (c) Withdraw funds by redeeming Ripple IOUs issued by the gateway.

We use daily time-series data from Morningstar Inc and Coinmetrics.io in the period from December 2017 to May 2019. Data processing is done using Mathematica [10] and the DataModeler package [11].

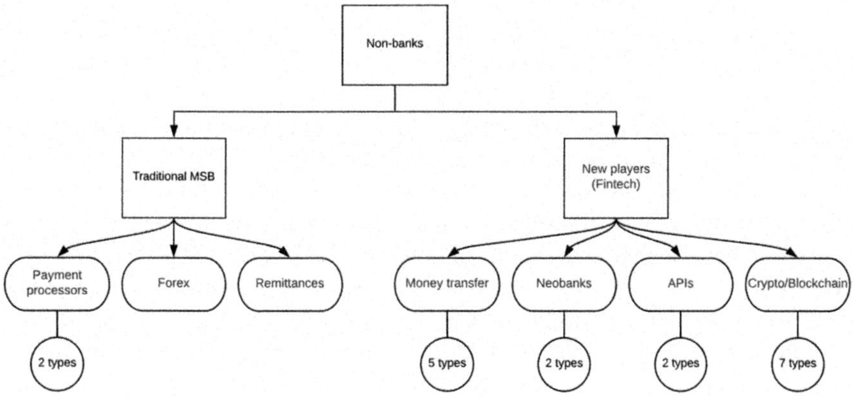

Fig. 1 Non-banks taxonomy

2.2 Generative Network Models

We use symbolic regression [12], an approach that allows detecting realistic decentralized network growth models from empirical data, to derive an expression of the form

$$XRP_{TxTfr} = f(gateway_a, gateway_b, gateway_c)$$

where XRP denotes the native token of the Ripple network and TxTfr refers to the financial activity metric: Transactions, the sum count of transfers that day. Transfers represent movements of native units from one ledger entity to another distinct ledger entity. Only transfers that are the result of a transaction and that have a positive (non-zero) value are counted. The independent variables *gateway a...b* represent the behavioral variables i.e. actual usage of the gateways by users, measured by daily web sessions. In other words, the formula expresses the volume of Ripple assets transfers as a function of user interactions, which can be considered a proxy for market demand. The characteristics of each gateway are shown in Table 1.

Similarly, we create expressions that describe each gateway in terms of its sources, as in

$$gateway_i = f(s_1, ..., s_n)$$

While transactions represent a bundle of intended actions to alter the ledger initiated by a user (human or machine), inputs from sources represent web level user activity (human or machine) that signal asset demand—but not necessarily resolve to a transactional event. Nevertheless, there is a causality path between both: new user activity is a pre-requisite for transfers by new users. In total, we obtained time series for 201 *gateway_a* sources, 455 *gateway_b* sources, and, 126 *gateway_c* sources, which were used as inputs for modeling. Only 107 of those were found to have any predictive power on the performance of the gateways.

There are two levels of modeling. The first is concerned with the response in terms of transactional activity (transfers) given the demand activity in different gateways. One could have started with dozens of gateways, and arrive at a shortlist of 3 that become the focus of study—for instance, if they appear consistently across most models. One such model has the form.

$XRP_TxTfr = 52869 + 0.05 * sma(gateway_b, 13) + 2.89e - 5 * sma(gateway_b, 83) * sma(gateway_c, 86) - 0.76 * sma(gateway_b, 77)$

Then each gateway is modeled based on its constituent sources, obtaining expressions as in Fig. 2.

Based on the formulas for each gateway we create a SymbolDependencyGraph [13] where an edge a → b will appear in the graph if the symbol b appears in the definition of symbol a. The purpose is to see how symbols in a context are related. In this case, the symbols are the variables obtained via genetic programming (symbolic regression), which show the dependencies between driving sources and

Table 1 Ripple gateways under study

	Description	Universe	Primary industry	Other industries	Verticals	Employees
Gateway_a	Developer of a bitcoin exchange platform designed to trade bitcoin products. The company's platform offers an exchange application, enabling customers to buy and sell bitcoins and ether with Mexican pesos	Venture capital	Financial software	Application software, social/platform software	Cryptocurrency/blockchain, E-commerce, FinTech, SaaS	25
Gate way_b	Operator of an online trading software. The company's platform is built on Ripple and allows users to send, receive, trade and manage any asset	Venture capital	Social/platform software		FinTech	7

(continued)

Table 1 (continued)

	Description	Universe	Primary industry	Other industries	Verticals	Employees
Gateway_c	Provider of an online stock and currency exchange platform for digital and math currencies. The company's platform allows trading of bitcoins, litecoins, namecoins, dogecoins, peercoins, XRP (Ripples) and increases its virtual currency exchange capacity	Other private companies	Financial software	Software development applications	Cryptocurrency/blockchain, FinTech	2

gateway-b

	Complexity	$1-R^2$	Function
1	11	0.019	$24333.56 + 80.45\ \text{rippexnet}$
2	18	0.018	$-22025.41 + (4.21\times 10^{-2})\sqrt{\text{btcdirecteu}}\ \text{ripplecom}$
3	19	0.012	$-57398.51 + 47.46\ \text{rippexnet} + 3.62\ \text{ripplecom}$
4	22	0.010	$-32441.41 + (3.94\times 10^{-5})\ \text{cryptocomparecom}\ \text{rippexnet} + 3.88\ \text{ripplecom}$
5	27	0.009	$-39922.93 - 29.49\ \text{cryptoinforu} + 41.55\ \text{rippexnet} + 5.07\ \text{ripplecom}$
6	30	0.007	$-35693.24 + 34.99\ \text{rippexnet} + 3.15\ \text{ripplecom} + (3.94\times 10^{-4})\ \text{bitemycoincom}\ \text{ripplecom}$
7	38	0.006	$-46490.42 + 30.03\ \text{rippexnet} + (2.64\times 10^{-3})\ \text{bitemycoincom}\ \text{ripplecoinnewscom} + 4.42\ \text{ripplecom} - 1.62\ \text{themerklecom}$
8	42	0.006	$115200.46 + 27.46\ \text{rippexnet} + (2.51\times 10^{-3})\ \text{bitemycoincom}\ \text{ripplecoinnewscom} - 1352.65\sqrt{\text{ripplecom}} + 5.98\ \text{ripplecom}$
9	50	0.005	$72785.45 + 25.83\ \text{rippexnet} + 9.97\ \text{ripplecoinnewscom} - 1050.00\sqrt{\text{ripplecom}} + 4.60\ \text{ripplecom} + (3.06\times 10^{-4})\ \text{bitemycoincom}\ \text{ripplecom}$
10	61	0.005	$85792.81 + 26.07\ \text{rippexnet} + 6.91\ \text{ripplecoinnewscom} + (8.29\times 10^{-4})\ \text{bitemycoincom}\ \text{ripplecoinnewscom} - 1143.41\sqrt{\text{ripplecom}} + 5.04\ \text{ripplecom} + (2.10\times 10^{-4})\ \text{bitemycoincom}\ \text{ripplecom}$

Fig. 2 Symbolic regression models

gateways. Figure 3 depicts the graph object that was obtained, where gateways a, b, c (seen in counterclockwise) are shown as having clusters of sources around them; XRP (the node centered in the graph, with outgoing edges towards each of the gateways) is only shown for stylistic reasons—there is no flow of user activity between the actual Ripple token and the gateways, although a relationship of dependency is implied.

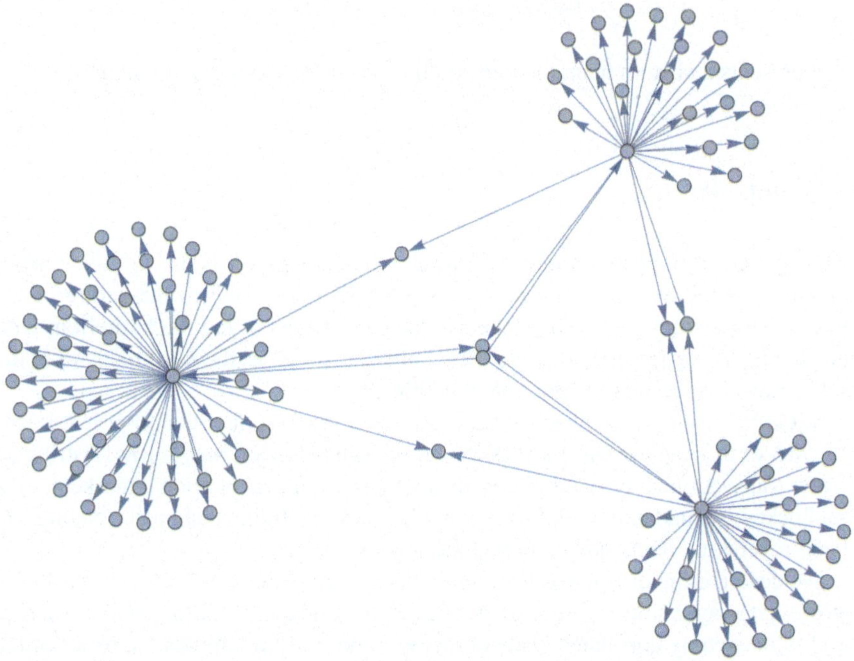

Fig. 3 Symbol dependency graph. Gateway_a is shown in the upper right corner, gateway_c in the bottom right

2.3 Ranking Scores

In the symbol dependency graph, there are only five k-2 core components (where a k-core component is defined as a maximal weakly connected subgraph in which all vertices have degree at least k) that operate simultaneously as sources. From Fig. 3 it is clear that there are multiple relationships of dependency: at least one of the sources from the sample might be systemically important (all gateways depend on it); all 3 the gateways influence one of the sources themselves (possibly creating a feedback effect); each gateway has developed a specialized and localized business, where sources overlaps are kept at a minimum, with very few sources in common.

Which gateway is at risk of failing? By optimizing for an objective, under particular error and complexity constraints, it is possible to derive a comparative measurement. The definition of such a ranking function is shown below

$$(gateway_a > gateway_b) = (f(s_1, s_2 .. s_n)_a > f(s_1, s_2 .. sn)_b)$$

And, by comparing the same sources for 2 different gateways, the result

$$f(source_1, source_2, ..., source_n)$$

represents a custom heuristic that could be used for scoring and ranking.

3 Analysis

Although not formally part of our method, we begin the analysis with the characterization of the transactions time series, which should confirm the nature of the process under study. We found 10 candidates by means of the AIC criterion, with an autoregressive–moving-average model of order $\{1, 2\}$ as the most likely to fit the time-series. The results are presented in Table 2.

ARMA is appropriate when a system is a function of a series of unobserved shocks (the MA or moving average part) as well as its own behavior. In this case, distributed ledger transaction activity (similarly to stock prices in the traditional markets) may be shocked by fundamental information as well as exhibiting technical trending and mean-reversion effects due to market participants [14].

We describe the distinctive features of the sources time series as well. As Fig. 4 shows, gateway_c sources are strongly (linearly) correlated, while gateway_a sources tend to be weakly correlated. This observation not only reinforces the result of the symbolic network mapping seen in Fig. 4 but even before modeling, it suggests that each gateway ecosystem is qualitatively different (despite the fact that when viewed on a temporal dimension, co-movement is often exhibited). The periodicities (correlation clusters) also indicate possible dependencies between the sources themselves.

Table 2 Transaction time series characterization

	Estimate	Standard error	t-Statistic	P-value
a_1	0.892017	0.0238012	37.4779	1.66359×10^{-153}
b_1	0.299732	0.0496947	6.03147	1.49678×10^{-9}
b_2	−0.0960089	0.0489722	−1.96048	0.0252235
			Candidate	AIC
1			**ARMAProcess(1, 2)**	**9697.49**
2			ARMAProcess(1, 3)	9701.82
3			ARMAProcess(2, 1)	9704.73
4			ARMAProcess(1, 1)	9708.79
5			ARWAProcess(2, 2)	9711.36
6			ARMAProcess(2, 3)	9729.55
7			ARProcess(1)	9776.03
8			MAProcess(2)	10,046.9
9			MAProcess(1)	10,160.5
10			MAProcess(0)	10,696.4

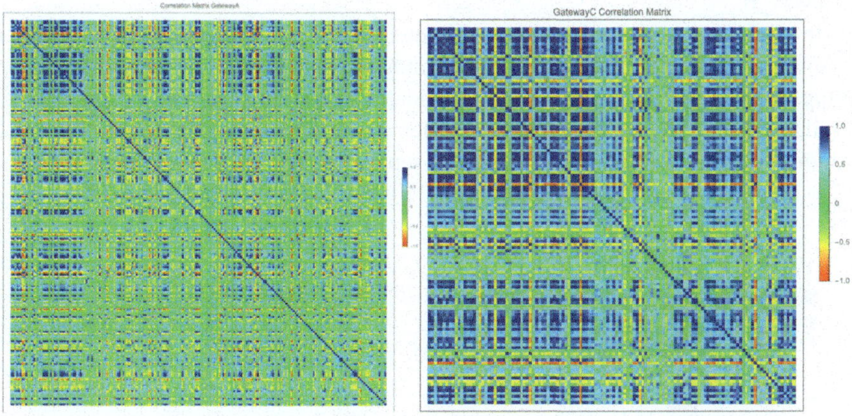

Fig. 4 Gateway_a (left) and Gateway_c (right), correlation plots for the sources

The differences in the correlation matrices also suggest that gateway_a presents higher complexity in the collection of its counter-parties, an observation that the error/complexity Pareto obtained via symbolic regression confirms. As Fig. 5 depicts, for a given level of complexity around the knee of the Pareto front (e.g. 50), gateway_c' sources are easier to model (error predominantly under 0.1 for modeling time equal to 300 s).

Gateway_a and gateway_c are less dissimilar between them than gateway_b: they share more sources in common, the clusters around them have roughly equivalent sizes, and so on.

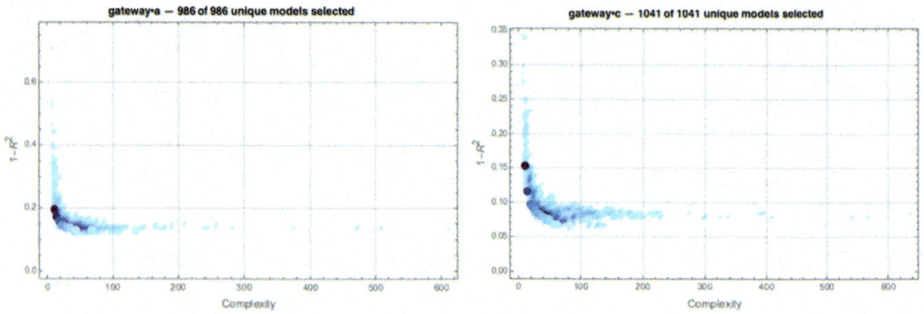

Fig. 5 Pareto plots: accuracy versus complexity trade-off

Gateway_b, therefore, might be an interesting option for diversification of sources. First, we study the driving variables, that is, the sources that are prevalent across models. Only one variable appears in over 80% of the models, as shown in Fig. 6.

By creating ensembles of diverse models (i.e. algorithmically generated composite models), we can also better understand robustness. Some sources will be prevalent not only in absolute terms across all models but will be dominant in the ensembles that constitute the best predictors as well. Figure 7 shows an ensemble generated for gateway_b.

In practice, each model encodes the interplay of one or more demand sources. Symbolic regression via genetic programming not only allows to discover the underlying mathematical structure but also to update those structures as new data becomes available. Error performance improves when combining models into ensembles. The genome of those models are themselves represented as connected acyclic graphs,

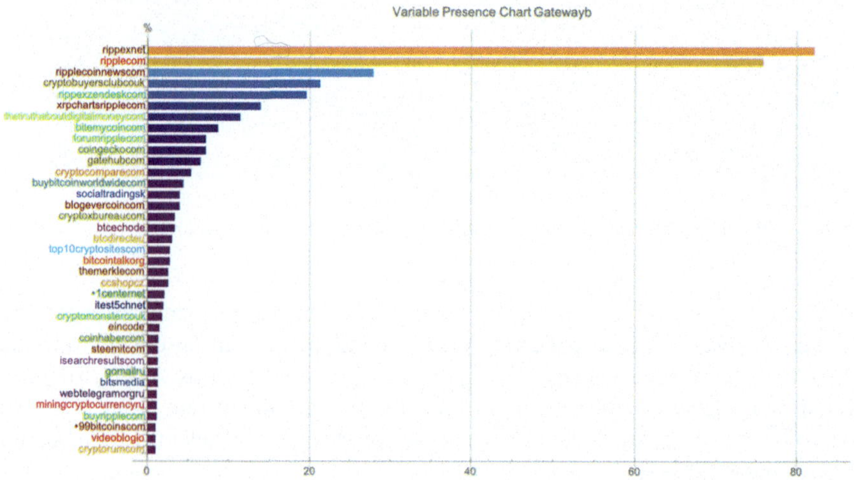

Fig. 6 Driver variables, Gateway_b

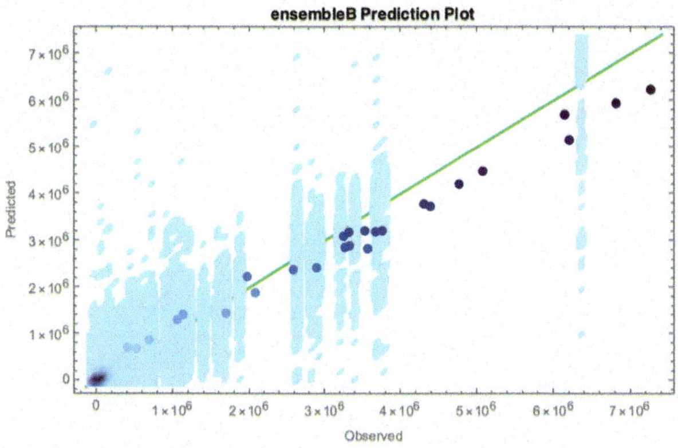

Ensemble Quality Summary Table

	R^2	0.991653
Ensemble Quality	# Variables	56
	# Models	97
	Avg Model Complexity	22.6495

Fig. 7 Ensemble of symbolic regression models

which implies a measure of causality i.e. in behavioral terms, the hierarchical structure is an expression of the conversion path that, on average, takes the user from source to the gateway. In this context, model complexity is the visitation length which is defined as the number of nodes transited from the root node to each of the nodes. Figure 8 describes a model and traditional metrics of trustworthiness, such as the model' statistical attributes.

We plot all models and inspect the relationship between gateway usage prediction and activity at the sources, Fig. 9 shows another one of such models. The ribbon offers a visualization of the trustability of the predictions (the confidence interval); models agree where they are constrained by data, and uncertainty increases (models diverge) in regions of unseen parameter space. In other words, the ensemble's measurement of trustability allows quantifying how much one can trust what is already known at this stage (that is, the behavior at the driver sources of gateway activity that were discovered using symbolic regression)—this serves as an assessment of the risk associated with the prediction. Dynamic uncertainty is also a signal of the materiality of the input to the problem: Ripple related sources display such behavior, while a less relevant source (e.g. specialized in bitcoin) has a static response.

The fact that trustability can be quantified and visualized using this method is not trivial. A number of financial systems exhibit fragility to external shocks, but the exact sources and the expected impact is often unknown. As a case in point, a key driver and one of the sources in Fig. 9 (Rippex, a Brazilian exchange) ceased operations suddenly between March and April 2018 [15]. To what extent the service was used as a utility (e.g. Ripple wallet), or, was used to exchange funds between the two gateways, is a matter that would require further investigation. However, the fact is that this particular source was in the transactional conversion path of gateway_b, and by the trustability profile, it is clear that large variations in the response typically correspond to large variations in the source behavior. Such behavior might be a reason for concern: the effect of the source's flows is material to the performance of the gateway, and, disseminates high uncertainty into the system. However, by its

Model TreePlot & Statistical Attributes Summary Table

Model Tree Plot

22.22 ripplecoinnewscom − 2231.73 + bitcointalkorg rippexnet (3.86×10^{-5})

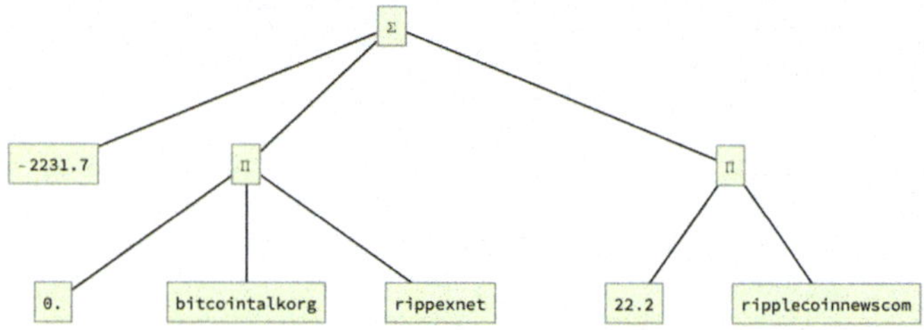

R-Squared	0.991653
Adjusted R-Squared	0.991638
Noise Power	0.0139202
Coefficient Of Variation	0.340334
DurbinWatsonD	1.34185

Parameter Confidence Interval Table

$-2231.73 + (3.86 \times 10^{-5})$ bitcointalkorg rippexnet + 22.22 ripplecoinnewscom

	Estimate	Standard Error	Confidence Interval
1	−2231.73	6292.87	{−14593., 10129.6}
bitcointalkorg rippexnet	0.0000386086	1.02444×10^{-6}	{0.0000365963, 0.0000406209}
ripplecoinnewscom	22.2206	1.32831	{19.6114, 24.8299}

Fig. 8 Model tree plot

nature, this sort of payments system is both adaptable and continually novel, so it may be the case that other sources take the place of the failing exchange.

4 Conclusion

On a payments network such as Ripple's, humans shape machine behavior by the use of gateways, and machines shape human behavior through the distributed ledger records—and the effect and impact are only compounded and accelerated when algorithmic trading enters the scene. Some machine behavior may spread because it is 'evolvable' [6]—easy to modify and robust to perturbations—similar to how some traits of animals may be common because they facilitate diversity and stability [16]. This calls for the use of a technique such as genetic programming, that can keep the pace with rapidly changing competitive environments such as fintech, blockchain, and, distributed ledger technology. Symbolic regression via genetic programming also offers the benefit of explainability: clearly readable mathematical expressions

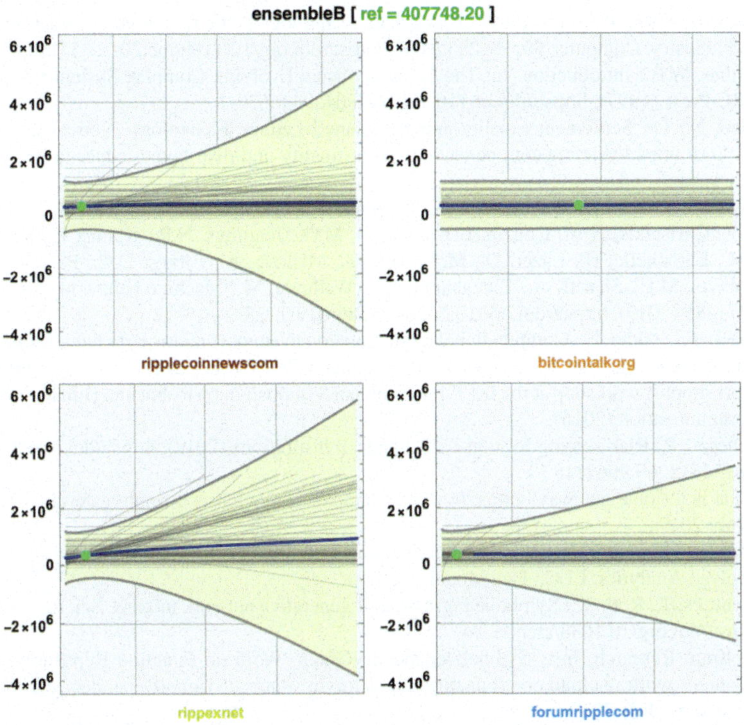

Fig. 9 Trustability profiles. Blue line is the ensemble and gray are its constituent models

enhance trust among decision makers and policymakers, the end-users of artificial intelligence generated models. We demonstrated how key drivers of financial activity can be discovered in the multiplex network structures that support the economic activity at the entry and exit points of the distributed ledger system. Model ensembles were used to show how the uncertainty associated with such drivers can be quantified and visualized. The method and findings pertain to any decentralized financial infrastructure with centralized components, where counterparty risk assessment becomes a necessity for regulatory, investment, and, operational purposes.

Declaration of Interest The authors report no conflicts of interest. The authors alone are responsible for the content and writing of the paper.

References

1. Venegas, P.: Trust Asymmetry. In: Unifying Themes in Complex Systems IX, pp. 467–475. Springer International Publishing (2018). https://doi.org/10.1007/978-3-319-96661-8_48
2. Taleb, N.N.: Antifragility robustness, and fragility, inside the black swan domain. SSRN Electron. J. (2010). https://doi.org/10.2139/ssrn.1669317

3. Aste, T., Tasca, P., Matteo, T.D.: Blockchain technologies: the foreseeable impact on society and industry. Computer **50**, 18–28 (2017). https://doi.org/10.1109/mc.2017.3571064
4. Arthur, W.B.: Introduction. In: The Economy as an Evolving Complex System II, pp. 1–14. CRC Press (1997). https://doi.org/10.1201/9780429496639-1
5. Liao, N.: On Settlement Finality and Distributed Ledger Technology. Notice & Comment (2017). https://yalejreg.com/nc/on-settlement-finality-and-distributed-ledger-technology-by-nancy-liao/.
6. Rahwan, I., Cebrian, M., Obradovich, N., Bongard, J., Bonnefon, J.-F., Breazeal, C., Crandall, J.W., Christakis, N.A., Couzin, I.D., Jackson, M.O., Jennings, N.R., Kamar, E., Kloumann, I.M., Larochelle, H., Lazer, D., McElreath, R., Mislove, A., Parkes, D.C., Pentland, A.'S., Roberts, M.E., Shariff, A., Tenenbaum, J.B., Wellman, M.: Machine behaviour. Nature **568**, 477–486 (2019). https://doi.org/10.1038/s41586-019-1138-y
7. Black, E.I., Sklar, E.: Computational argumentation to support multi-party human-robot interaction: challenges and advantages. In: Proceedings of the Groups in Human-Robot Interaction Workshop: A workshop at the IEEE International Symposium on Robot and Human Interactive Communication (2016)
8. Venegas, P.: Risk scoring for non-bank financial institutions (2019). https://doi.org/10.22541/au.154312269.99621183
9. What is a ripple gateway? https://www.bitstamp.net/faq/what-is-a-ripple-gateway/. Accessed 1 July 2019
10. Wolfram Research, Inc.: Mathematica, Version 12.0 (2019)
11. Evolved Analytics, LLC.: DataModeler (2018)
12. Menezes, T., Roth, C.: Symbolic regression of generative network models. Sci. Rep. **4** (2014). https://doi.org/10.1038/srep06284
13. Wolfram Research, Inc.: SymbolDependencyGraph. Wolfram Function Repository. https://resources.wolframcloud.com/FunctionRepository/resources/SymbolDependencyGraph. Accessed 1 July 2019
14. Autoregressive–moving-average model—Wikipedia. https://en.wikipedia.org/wiki/Autoregressive%E2%80%93moving-average_model#Applications. Accessed 1 July 2019
15. Rippex. https://web.archive.org/web/20180413171541/. https://rippex.net/. Accessed 1 July 2019
16. Wagner, A.: Robustness and Evolvability in Living Systems. Princeton University Press (2013). https://doi.org/10.1515/9781400849383

When Econometrics Met Information Theory: A Real-Time Approach to Track Yen Exchange Rate

Nan Wang

Abstract Fluctuations in currency exchange rates have enormous economic impact, affecting a wide range of market participants including governments, banks, corporates and individuals. Tracking currency exchange rates is crucial in many perspectives. While the financial press closely tracks and discusses major currency exchange pairs on a high frequency basis, scientific literature has been mostly silent in studying their patterns. This paper proposes an innovative approach in the endeavor of tracking the yen exchange rate against US Dollar (USDJPY), the second most traded currency pair. The proposed approach applies econometrics and information they in a real-time manner, leveraging the former on compressing high dimensional sparse data and the latter on quantifying nonlinear dependency. Using merely macroeconomics information and strictly avoiding look-ahead bias, the resulting tracking index from this approach achieves significant linear correlation with USDJPY and demonstrates ability to explain about 37% variances in USDJPY fluctuations over a 13-year period. The proposed approach has three phases: "vintage" generation, "vintage" storage and "vintage" transformation. The proposed approach is based on a deployed system which was started building in 2007. Since then, the system has been evolving systematically all the time to adapt to the ever-changing global macroeconomic environments.

The proposed approach is innovative in the following four aspects:
Use a real time database
Cover a broad range of macroeconomic information
Update information in a timely manner
Consider the dependency between countries.

Keywords Foreign exchange · USDJPY · Econometrics · Information theory · Dynamic system · Real-time database · Nonlinear dependency · Complex system

---

N. Wang (✉)
DeepMacro Inc, New York City, NY 10017, USA
e-mail: miya117517@gmail.com

D. Braha et al. (eds.), *Unifying Themes in Complex Systems X*, Springer Proceedings in Complexity, https://doi.org/10.1007/978-3-030-67318-5_11

1 Introduction

Foreign exchange market is the largest and most liquid financial market in the world. According to 2019 Triennial Survey of turnover in OTC FX markets from the Bank for International Settlements, trading in foreign exchange markets valued at $6.6 trillion per day on average in April 2019. Foreign exchange market participants range from governments and banks (central and commercial), to commercial corporations, individuals and many others, all exposed to massive foreign exchange risk. Effectively tracking currency exchange rate is without doubt of high value.

Fluctuations of USDJPY particularly have tremendous economic impact compared to other currency pairs. According to aggregated Currency Composition of Foreign Exchange Reserves from IMF, the Japanese yen is the third most commonly used and the fourth most popular reserve currency as of 2019. Moreover, the yen has been a crucial indicator for economic activities. Through USDJPY, Investors express their views regarding the global economy, political event, fiscal policy and risk appetite [1]. For example, when macro climate is considered as risk-averse, the yen is usually aggressively purchased. During European sovereign crises in 2011, demands for the yen were unprecedentedly high, sending the yen to a historical high at 75.54 per Dollar [2].

Yet USDJPY patterns have received considerably less attention from scientific literature compared to its economic importance. One reason is that patterns of currency rates are notoriously difficult to track, and most academic models rarely work for even a bit longer than a brief period of time. Random walk is widely accepted as prediction benchmarks.

As a contribution, this paper proposes an innovative approach to effectively track USDJPY.

The approach is novel in four aspects.

Use a real time database. In the real world, macroeconomics data is changing all the time due to revisions.[1] The past looks different from today's "vintage"[2] than it did in the past. The proposed approach preserves all the data as they were at the time of release, and all the modeling are based on the latest published time series available at that point in time. Therefore, this approach effectively avoids information spillover, eliminates look-ahead bias,[3] or "forward looking bias" [3], and enables more reliable back testing.

Uses a broad range of economic information as building blocks. It's widely acknowledged that foreign exchange markets are primarily driven by macroeconomics factors. Yet macroeconomics covers many aspects. Besides, no single factor can explain a significant share of variances in exchange rate dynamics. Therefore, the proposed approach is aimed at covering as much economic information as possible. To name a few, information collected includes expenditure (i.e., retail

[1] See Sect. 2 for the definition of "revisions".

[2] See Sect. 2 for the definition of "vintage".

[3] Look-ahead bias occurs when using information or data in a study that would not have been known or available during the period being analyzed.

sales, household spending and passenger car sales), housing (i.e., condo inventories), Capex[4] (i.e., shipments of construction goods domestic use), exports, imports, labor (i.e., job offerings), industry (i.e., primary machinery orders), services (i.e., tertiary activity) and sentiment (such as consumer confidence). The source of the information could be either official, such as government agencies, central banks, and widely followed private sources, or unstructured alternative data[5] such as public internet displays, satellite images and social media. The benefits of integrating a broad range of information is to get earlier, orthogonal, and more detailed views of economic picture.

Building blocks are updated in a timely manner. Financial markets are pricing in various sorts of news, surprises or shocks all the time. Therefore, it is vital to keep the tracking dynamic and adaptive as well. In the proposed approach, the update frequency of input variables could be streaming, hourly, daily, monthly or quarterly, depending on the data availability. The proposed approach is designed to synchronize global economy, capturing and, if needed, factoring in new information at the first second.

The approach considers economic dependency between countries. Currency exchange rates are in relative terms only. Yet when modeling the driving force of exchange rate, common factors employed are individual-focused, such as economic growth [4], interest rate [5] and inflation [6]. Seeing the world as a highly intertwined network, we consider it important to take into account the dependency among countries and their interactions in terms of risk, trade and economic outlook when tracking their exchange rates. The proposed approach measures the business cycle dependency between Japan and the US in the efforts of tracking the relative positive of their currencies.

The rest of the paper is structured as follows. Section 1 contains definitions of terms referred in this paper. Section 2 covers the methodology of the proposed approach, discussing variables, econometrics, information theory metrics and the architecture of the proposed approach. Section 3 evaluates the experiment. Section 4 draws conclusions.

1.1 Definitions

This section provides definitions of the terms, including revisions, vintage, vintage point, reference point, real-time database, and alternative data, which are referred in this paper.

---

[4]Capital Expenditure.

[5]See Sect. 2 for the definition of "alternative data".

1.2 Revisions

Revisions, in broad sense, are defined as a change in value for any reference point of the time series for a statistic when released to the public by an official statistical agency. Revisions can occur either when new observations (e.g. one additional month or quarter) become available and some past values are modified, or when the current and some past values are modified in an updated release of the current time series [7].

1.3 Vintage, Vintage Point and Reference Point

For a given time series we define vintage as the set of data (sequence of values) that represented the latest estimate for each vintage point in the time series at a particular moment of time. And that particular moment of time is considered as a reference point of the vintage [7].

1.4 Real-Time Database

A real-time database is a collection of historical vintages of the same time series. The revision on a given reference point for a time series can be identified in a real-time database as the change in value from an earlier vintage of estimates to a later vintage. A real-time database is popularly represented by a two-dimensional array, with vintage points in rows and reference points as columns. If the publication schedule of the time series has the same frequency as its periodicity (e.g. monthly), then the real-time database will have the appearance of a symmetric triangle as shown in Table 1 [7].

1.5 Alternative Data

Alternative data refers to data derived from non-traditional sources of information that might ultimately have investment alpha value.

2 Methodology

In this section, we describe the methodology for the proposed approach. We first describe input variables and their processing. Next, we introduce the employed econometrics model and the information theory metric. Finally, we describe the

Table 1 Example for a symmetric real-time database

Reference point	201,004	201,003	201,002	201,001	200,912	200,911	200,910	200,909	200,908
Vintage point									
201,004	−3.1298								
201,003	−4.7303	−3.7642							
201,002	−4.6134	−4.2548	−3.0574						
201,001	−4.8674	−4.5555	−4.3013	−3.2958					
200,912	−4.7807	−4.4307	−4.1613	−3.7999	−3.0997				
200,911	−4.7391	−4.3878	−4.0326	−3.6375	−3.0451	−2.5272			
200,910	−4.7998	−4.4742	−4.0248	−3.6214	−2.8978	−2.3338	−1.6303		
200,909	−4.8236	−4.5089	−4.078	−3.6165	−2.8787	−2.1536	−1.4258	−1.1593	
200,908	−4.8409	−4.5198	−4.0837	−3.6332	−2.8701	−2.0299	−1.1461	−0.7358	−0.7895

system architecture which contains three phases: vintage generation, vintage storage and vintage transformation. Phase one and two are implemented in a deployed factor system in DeepMacro, Inc (DFS). The system (DFS) was started building in 2007. Since then, the system has been evolving and improving by keeping bringing in alternative data, generating high frequency signals and applying nonparametric techniques.

2.1 Input Variables

Input variables are building blocks of the proposed approach. The DFS system is searching for candidates of input variables all the time from various sources. Each variable, whether sourced from an official or alternative source, goes through the following seven procedures before modeling.

Seasonal adjustment.

Conversion of nominal variables into real[6] ones.

Outlier and missing value detection.

"Ragged edge"[7] [8] solution (Kalman filtering [9]).

Missing early values solution.

Stationarity.[8]

Standardization.[9]

There are great challenges in applying these processes even to highly aggregated time series values from government sources. For some unstructured alternative data, challenges are magnified. For example, transforming satellite images into a normalized format requires paramount data reduction techniques.

Generally, there are five major categories of variables as described below.

Expenditure. This category covers four aspects: Capex, Imports, Housing and Government. Capex includes indices such as heavy trucks and shipment of computers. Imports include indices import aggregates from all major countries to the country in concern, or export aggregates from the country in concern to all major economies. Housing has indices such as new housing starts,[10] new home permits.[11] Government covers indices such as new school building construction and new healthcare building construction.

[6]The real value is the nominal value adjusted for inflation and other related measures, e.g. the nominal value of gross domestic product is usually adjusted by a deflator to derive its real values.

[7]Macroeconomics data is typically not available for today and the immediate past ("ragged edge") and subject to revision.

[8]A stationary time series is one with constant statistical properties such as mean, variance, autocorrelation, etc.

[9]Standardization is the process of scaling variables so that they are comparable.

[10]The number of privately-owned new houses on which construction has been started in a given period.

[11]Permits for new-home construction.

Income. This category covers one aspect: Labor. Labor includes indices such as new job offers, personal income, average hourly earnings, cash earnings scheduled, job offers/seekers ratio, unemployed, job opening vacancies and so forth.

Output. This category covers two aspects: Service and Industry. Service includes indices like real estate, wholesale, finance and insurance and transport. Industry includes indices such as manufacturers new orders, IP[12] motor vehicle and parts, IP headline, computers and capacity utilization rate total industry.[13]

Sentiment. This category covers three aspects: Household, Corporate and Financial. Household has indices such as conference board—present situation and consumer confidence (overall). Corporates have indices such as ISM non-manufacturing employment,[14] Richmond Fed[15] future 6 months, NAHB[16] next 6 months, Small business conditions index, Tankan[17] services for businesses and Economy Watchers Nonmanufacturing current.[18]

Others. This category covers all alternative data. Examples for this category are high-frequency industrial indicator processed from satellite images, hourly web traffic data and online job postings.

2.2 Econometrics: Dynamic Factor Modeling

The premise of a dynamic factor model (DFM) is that a few latent dynamic factors, f_t, drive the comovements of a high-dimensional vector of time series variables, x_t, which is also affected by a vector of mean zero idiosyncratic disturbances, e_t. The latent factors follow a time series process, which is commonly taken to be a vector autoregression (VAR). In equations, the dynamic factor model is,

$$x_t = \lambda(L) f_t + e_t \tag{1}$$

$$f_t = \psi(L) f_{t-1} + \eta t \tag{2}$$

where there are N series, so x_t and e_t are $N \times 1$, there are q dynamic factors, so f_t and η_t are $q \times 1$, L is the lag operator, and the lag polynomial matrices $\lambda(L)$ and

[12]Industrial production.

[13]The percentage of resources used by corporations and factories to produce goods in manufacturing, mining, and electric and gas utilities.

[14]The ISM Non-Manufacturing Index (NMI) is an economic index based on surveys of more than 400 non-manufacturing firms purchasing and supply executives, within 60 sectors across the nation, by the Institute of Supply Management (ISM).

[15]Federal Reserve Bank of Richmond.

[16]National Association of Home Builders.

[17]Short-Term Economic Survey of Enterprises in Japan.

[18]The Economy Watchers Current Index measures the current mood of businesses that directly service consumers, such as barbers, taxi drivers, and waiters.

$\psi(L)$ are $N \times q$ and $q \times q$. The ith lag polynomial $\lambda_i(L)$ is called the dynamic factor loading for the ith series, x_{it}, and $\lambda_i(L) f_t$ is called the common component of the ith series. All the processes in (1) and (2) are assumed to be stationary. The idiosyncratic disturbances are assumed to be uncorrelated with the factor innovations at all leads and lags.

DFM fits very well with macroeconomics, not only because of its ability to deal with sparse high dimensional data, but also its premise of latent factors driving co-movements of high dimensions. As forcefully argued by Lucas [10] and many others, the business cycle is not about any single economic signal but the dynamics and interactions of many signals.

The proposed approach uses DFM to extract a one-dimensional latent factor from all the input economic variables. The resulting latent factor is interpreted as factor of business cycles, or business cycle factor.

2.3 Information Theory: Transfer Entropy

Let $X = (x1, x2, \ldots, xN)$ and $Y = (y1, y2, \ldots, yN)$ denote two time series of simultaneously measured scalar quantities. The state spaces X and Y are reconstructed by the method of time delay embedding in which the state (or embedding) vectors are formed by delayed "past" scalar observations of the time series. The m-dimensional embedding vector is defined as:

$$x_t^{m,\tau} \triangleq \left(x_t, x_{t-\tau}, x_{t-2\tau}, \ldots, x_{t-(m-1)\tau}\right)^T \tag{3}$$

$$y_t^{m,\tau} \triangleq (y_t, y_{t-\tau}, y_{t-2\tau}, \ldots, y_{t-(m-1)\tau})^T \tag{4}$$

m is the embedding dimension, τ is the time lag, and $t = n, \ldots, N$ with embedding window $n = (m-1)\tau$. Note that the state vectors $x_t^{m,\tau}$, $y_t^{m,\tau}$ are points in the m-dimensional spaces X and Y. Essentially, transfer entropy (TE) measures the deviation from the generalized Markov property:

$$p(x_{t+1}|x_t^k) = p(x_{t+1}|x_t^k, y_t^l) \tag{5}$$

p denotes the transition probability. k, l denote time lag, If the deviation from the generalized Markov process is small, then we can assume that the state of space y has little (or no) relevance on the transition probability of the state vectors of space x. However, if the deviation is large, then the assumption of Markov process is not valid. The incorrectness of the assumption can be quantified by the transfer entropy which is formulated as Kullback–Leibler distance [11] between $p(x_{t+1}|x_t^k)$ and $p(x_{t+1}|x_t^k, y_t^l)$:

$$T_{Y \to X} = \sum_{x_{t+1}, x_t^k, y_t^l} p(x_{t+1}, x_t^k, y_t^l) \log \frac{p(x_{t+1} | x_t^k, y_t^l)}{p(x_{t+1} | x_t^k)} \tag{6}$$

TE represents the information about future observations of x_t obtained from the simultaneous observation of past values of both x_t and y_t, after discarding the information about the future of x_t obtained from the past of x_t alone.

The proposed approach applies TE in order to measure the dependency of the US business cycle factor on Japan. As one of the most influential informational methods to detect directed influence between two stochastic time series [12], TE is chosen because of its three attributes or functionalities.

- Nonparametric.
- Asymmetric and directional.
- Able to measure nonlinear coupling effects.

2.4 System Architecture

The proposed approach has the following three phases. Phase One and Two are based on a deployed factor system in DeepMacro (DFS).

Phase One: Vintage Generation. As shown in Fig. 1, Phase One generates vintages of factors on a daily basis. The process starts with integrating input variables from various data sources. Size of data sources could be kilobytes, megabytes or gigabytes. Variables are updated at different frequencies (ranging from streaming to annually) but are normalized on a fixed frequency basis (daily). Normalization has

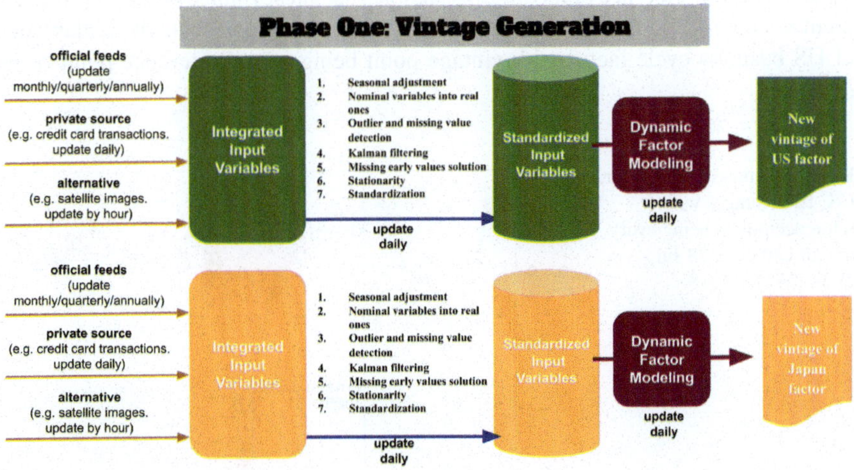

Fig. 1 Phase One, vintage generation

seven steps (as discussed detailly in Sect. 2.1), namely, seasonal adjustment, conversion of nominal variables into real variables, outlier and missing value detection, "ragged edge" solution, missing early values solution, stationarity and lastly standardization. After transformation, variables will be fed into dynamic factor modeling to generate new vintage of business cycle factors.

DFS updates business cycle factors every day, creating a vintage with every passing day as reference point. All vintages are monthly and 10 years' history long.

To give examples of business cycle vintages, Fig. 2 shows the 104 vintages of the US business cycle factor, with reference points being month-end dates between 28 Jul 2009 and 19 Mar 2018. In other words, with every month-end date between 28 Jul 2009 and 19 Mar 2018 being reference point, DFS computed a 10-year-long monthly time series up to the latest month available. As shown in Fig. 2, business cycle factor values on the same vintage points are different among vintages with different reference points. For example, the business cycle factor on Jan 31, 2005 vintage point varies about 1 standard deviation among vintages.

Phase Two: Vintage Storage. As shown in Fig. 3, every day whenever a vintage is created in Phase One, DFS stores it into a real time database where all the vintages are preserved.

Phase Three: Vintage Transformation. Phase Three transforms vintages of US and Japan factor and generates the USDJPY tracking index on a weekly basis (0:00 a.m. NY Eastern Time Every Monday). As shown in Fig. 4, every week, the latest vintages of the US and Japan factors are retrieved respectively from the real time database. Then, transfer entropy is applied on those two vintages with the dependency direction from Japan to the US, creating the USDJPY index.

Equations (6) and (7) illustrate Phase Three in detail. $V_{d_{(t)}}^{US}$ denotes a vintage of the US business cycle factor with reference point being $d_{(t)}$ and $V_{d_{(t)}}^{Japan}$ a vintage of Japan factor for reference point $d_{(t)}$.$d_{(t)m}$ denotes the latest vintage point for reference point d_t and $d_{(t)m-N}$ is the Nth latest vintage point. $F_{d_{(t)m}}^{US}$ represents the scalar value of US business cycle factor with vintage point being m and reference point being $d_{(t)}$.

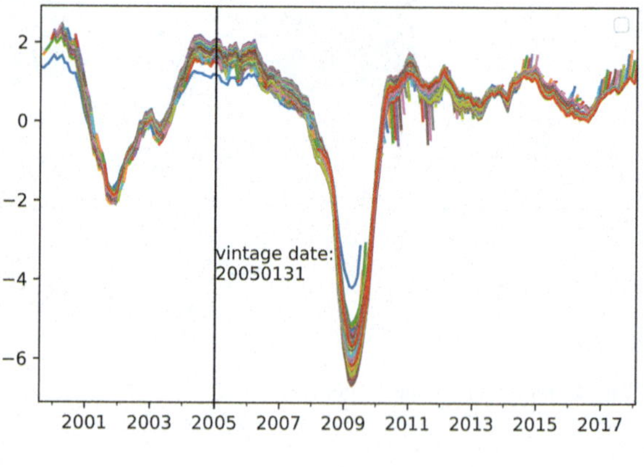

Fig. 2 Business cycle factor, US, 104 vintages with reference points being every month-end date, 28 Jul 2009–19 Mar 2018

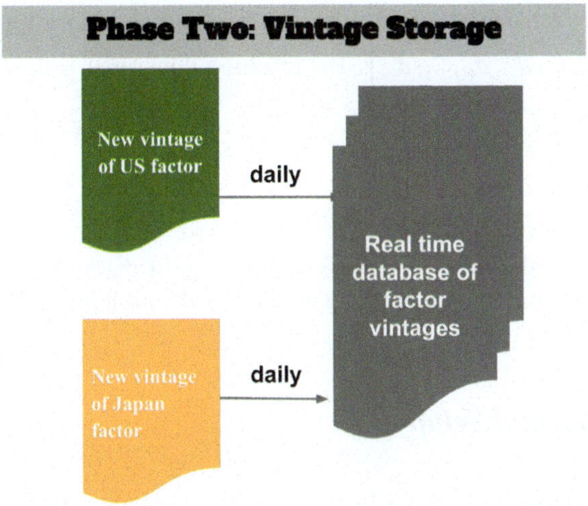

Fig. 3 Phase two, vintage storage

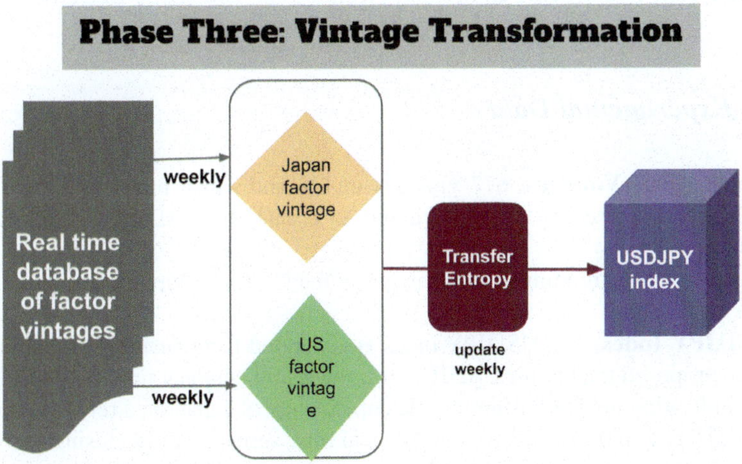

Fig. 4 Phase Three, vintage transformation

$$V_{d_{(t)}}^{US} = \left(F_{d_{(t)m}}^{US}, F_{d_{(t)m-1}}^{US}, F_{d_{(t)m-2}}^{US}, F_{d_{(t)m-3}}^{US} \cdots F_{d_{(t)m-N}}^{US}\right) \tag{7}$$

$$V_{d_{(t)}}^{Japan} = \left(F_{d_{(t)m}}^{Japan}, F_{d_{(t)m-1}}^{Japan}, F_{d_{(t)m-2}}^{Japan}, F_{d_{(t)m-3}}^{Japan} \cdots F_{d_{(t)m-N}}^{Japan}\right) \tag{8}$$

To calculate the USDJPY index for $d_{(t)}$, we apply TE on $V_{d_{(t)}}^{US}$ and $V_{d_{(t)}}^{Japan}$ with the dependency direction from Japan to the US. The resulting scalar, denoted as $T_{d_{(t)}}^{Japan \to US}$, is the US business cycle dependency on Japan on date $d_{(t)}$.

Similarly, we can generate a time series of $T^{Japan \rightarrow US}_{\{d_{t_1}, d_{t_2}, d_{t_3}, \ldots, d_{t_N}\}}$ for a group of specific vintage points $\{d_{t_1}, d_{t_2}, d_{t_3}, \ldots, d_{t_N}\}$. The resulting time series is the USDJPY index.

3 Evaluation

In this section, we describe our experimental setup of the proposed approach. We estimate its performance by evaluating its statistical correlation with the target.

3.1 Experimental Setup

Following the methodology described in Sect. 3, we generate vintages of the US and Japan business cycle factors and transform vintages into the USDJPY index.

For evaluation, we sample the US and Japan vintages of business cycle factors with reference points being all Mondays from 08 Jan 2007 to 30 Dec 2019.

3.2 Experimental Data

Business Cycle Vintages. 677 vintages of US and Japan business cycle factors respectively are used in our experiment to create the USDJPY index tracking USDJPY.

Figure 5 shows trending of the vintages of the US factor (upper) and Japan factor (lower).

USDJPY Index. The USDJPY index is generated from business cycle vintages. It is the proposed tracker for USDJPY. It is a one-dimensional time series with each scalar indicating the US business cycle dependency of Japan on a reference point.

USDJPY. USDJPY is a one-dimensional time series of daily closing (4.59 p.m. NY Eastern Time) spot US Dollar Yen exchange rate on selected reference points (Mondays, from 08 Jan 2007–30 Dec 2019).

3.3 Experimental Results

We examine the results by looking at the correlation between USDJPY and USDJPY index. Figure 6 shows the correlation between the index and the target.

The index inversely correlates the dynamics of USDJPY in a significant linear way most of the time during the examining period. There exist some dates when the relationship deviating from the normal linearity.

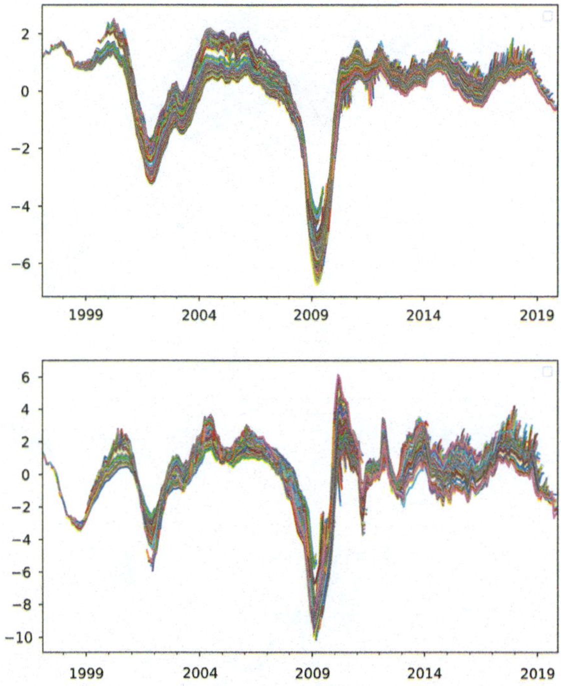

Fig. 5 US (upper) and Japan (lower) Business cycle vintages. Reference points: Mondays, 08 Jan 2007–30 Dec 2019

From the scatter plot in Fig. 6 (upper), we can see that the deviation mainly happens during two periods, early 2011 to late 2012 and late 2014 to early 2015, both coinciding with the Bank of Japan unconventional monetary policies. During 14 Mar 2011 to 20 Dec 2012, the Bank of Japan announced enhancement of monetary easing nine times. On 31 Oct 2014, the Bank of Japan unexpectedly announced expansion of the Quantitative and Qualitative Monetary Easing, sending the Yen to its lowest value against the dollar in almost seven years.

An overall linearity in the examined relationship implies that the index is capable of closely tracking USDJPY in a significantly linear way. Brief deviations during abnormal monetary policy periods make sense considering that the index only factors in the dependency of macroeconomic growth.

To further validate the overall correlation, we exclude those two abnormal periods (14 Mar 2011–20 Dec 2012 and 31 Oct 2014–20 Apr 2015) and calculate the Pearson correlation between the index and USDJPY as well as Pearson correlation between their percent change transformations. Table 2 shows the result.

With the absolute value of coefficient for correlation being over 80% and that for the Pct. Chg. correlation being over 16%, together with significantly small p values, we infer that USDJPY index and USDJPY are highly correlated in significantly linear way.

Furthermore, we run an OLS (ordinary least squares) regression analysis. Table 3 shows the result. We can tell from the skew and kurtosis statistics that residuals are

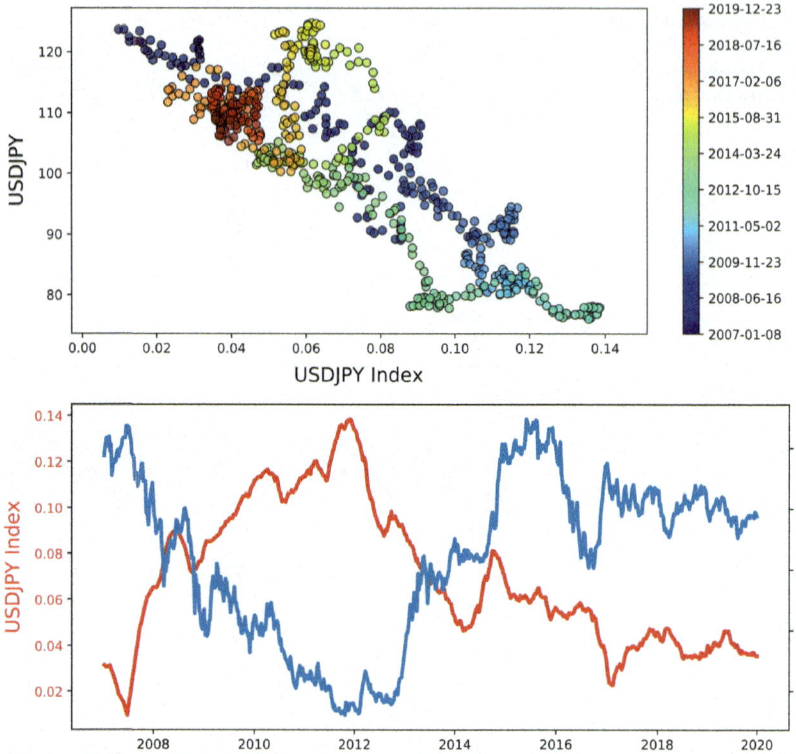

Fig. 6 Scatter Plot (upper) of USDJPY Index (x axis) and USDJPY (y axis). Colorized by dates. Jan 2007–Dec 2019. Trend plot (lower) of USDJPY Index (red) and USDJPY (blue), Jan 2007–Dec 2019

Table 2 Correlation between USDJPY index and USDJPY. Mondays, Jan 2007–Dec 2019. abnormal monetary policy periods excluded

	USDJPY USDJPY index	USDJPY Pct. Chg USDJPY Index Pct. Chg
Sample count	560	559
Correlation coefficient	−0.8081	−0.1693
P Value	<0.001	<0.001

fairly symmetrical, and a Durbin-Watson test statics of 1.9 indicates that the variance of the errors is independent of our independent variable here.

With reasonable amount of caution, we conclude that aside from abnormal monetary policy periods, the USDJPY index derived from mere macroeconomics information demonstrates ability of explaining about 37% variances in the US Dollar Japanese Yen exchange rates over a 13-year period.

Table 3 Regression analysis between USDJPY and USDJPY index

Dependent variable			USDJPY			
Independent variable			USDJPY index			
R squared			0.37			
F statistic			209			
Prob (F-statistic)			<0.001			
	Coef	Std Err	t	P > |t|	[0.025	0.975]
Const	119.124	0.946	125.879	<0.001	117.263	120.985
X1	-228.720	15.805	−14.471	<0.001	−259.803	−197.637
Error analysis						
Skew			0.430			
Kurtosis			2.835			
Durbin-Watson			1.930			

4 Conclusion

We proposed a real-time approach for tracking the US Dollar Japanese Yen exchange rate. The approach is innovative in the sense that it uses a real time database, leverages a wide coverage of macroeconomics information, updates in a timely manner and considers the macro dependency between countries.

The approach has three phases with the first two based on a deployed dynamic factor system dated and the last phase creatively combining econometrics and information theory techniques.

Using macroeconomic information only and with strict forward-looking avoidance, our approach achieved a tracking index able to explain about 37% of the variances of USDJPY dynamics in monetary policy normal days over a 13-year-old period.

We expect the future work to achieve similarly significant results for other major currency pairs, such as EURUSD and GBPUSD, using the same methodology.

References

1. Botman, D., Danninger, S., Schiff, J.: Abenomics: lessons learned from two decades of conventional and unconventional monetary policies. In: Can Abenomics succeed? Overcoming the legacy of Japan's lost decades. IMF (2015)
2. The New York Times. https://archive.nytimes.com/query.nytimes.com/gst/fullpage-9C02E7DA153AF930A15753C1A9679D8B63.html. Last accessed 23 October 2011
3. Haselton, M., Nettle D., Andrews, P.: The evolution of cognitive bias. In: The Handbook of Evolutionary Psychology, pp. 724–746. Wiley, New Jersey (2005)
4. Ito, T., Isard P., Symansky S.: Economic growth and real exchange rate: an overview of the balassa-samuelson hypothesis in Asia. In: Ito, T., Krueger, O (eds.) Changes in Exchange Rates

in Rapidly Developing Countries: Theory, Practice, and Policy Issues, pp. 109–28. University of Chicago Press, Chicago (1999)
5. Sarac, T., Karagoz, K.: Impact of short-term interest rate on exchange rate: the case of Turkey. In: Murat Karagoz, M., Bildirici, M. (eds.) Procedia Economics and Finance 2016, vol. 38, pp. 195–202. https://doi.org/10.1016/S2212-5671(16)30190-3 (2016)
6. Kwofie, C., Ansah, R.: A study of the effect of inflation and exchange rate on stock market returns in Ghana. Int. J. Math. Math. Sci. https://doi.org/10.1155/2018/7016792 (2018)
7. McKenzie, R., Gamba, M.: Data and metadata requirements for building a real-time database to perform revisions analysis. In: OECD/Eurostat Task Force on "Performing Revisions Analysis for Sub-Annual Economic Statistics (2008)
8. Bouwman, K., Jacobs, J.: Forecasting with real-time macroeconomic data: the ragged-edge problem and revisions. J. Macroecon. (2011)
9. Harvey, A.: Forecasting, Structural Time Series Models and the Kalman Filter. Cambridge University Press, Cambridge (1989)
10. Lucas, R.: Understanding business cycles. Carnegie-Rochester Conf. Ser. Publ. Policy **5**, 7–29 (1977)
11. Kullback, S., Leibler, R.: On information and sufficiency. Ann. Math. Stat. **22**, 79–86 (1951)
12. Ito, S.: Backward transfer entropy: Informational measure for detecting hidden Markov models and its interpretations in thermodynamics, gambling and causality. Sci. Rep. **6**, 36831 (2016)

An Economic Model Based on the System of National Accounts and Stock Flow Consistent Theory

Thomas Wang, Tai Young-Taft, and Harold M. Hastings

Abstract We propose a model of a closed economy based upon the System of National Accounts (United Nations [11]) and Stock Flow Consistent theory (Godley and Lavoie [2]) to model observed complex dynamics in real economies. We find that our model displays cyclic behavior while approaching an equilibrium, and simulates the effects of various shocks. We also find that a more egalitarian economy appears to be both yield a higher rate of return and be more stable (Marx [5]-Piketty[9] thesis).

Keywords Stock Flow Consistent · Godley-Lavoie · Marx-Piketty thesis · Minsky instability

1 Introduction

We propose a model of a closed economy with four interacting components based upon the System of National Accounts (United Nations [11]) and Stock Flow Consistent theory (Godley and Lavoie [2]) to model the complex dynamics of the economy. This model is similar to the Leontief model, however it examines the economy from a different perspective. We have added banks as an additional component to the usual three components, namely households, firms, and the government. Their interactions will be modeled with a four-by-four matrix. As Minsky [7] states on page 2, "A concrete real-world capitalist economy is characterized by an interrelated set of balance sheets, income statements and statements about sources and uses of funds." Here we attempt to realize this description of the economy using the above framework, and attempt to apply mathematical stability analysis to explain Minsky's financial instability hypothesis.

After analyzing the reasons for changes in asset prices that lead to business cycles, we also propose an asset pricing model that will capture macroeconomic dynamics in asset markets. We analyze the structure of the economy and its stability by running

T. Wang · T. Young-Taft · H. M. Hastings (✉)
Bard College at Simon's Rock, Great Barrington, USA
e-mail: hhastings@simons-rock.edu

D. Braha et al. (eds.), *Unifying Themes in Complex Systems X*, Springer Proceedings in Complexity, https://doi.org/10.1007/978-3-030-67318-5_12

simulations under various conditions. Furthermore, we discuss potential applications of the model in identifying crises by observing the effects of changes of variables in the model.

We find that our model displays cyclic behavior while approaching an equilibrium, and simulates the effects of various shocks to the economy fairly realistically. In addition, we find that a more egalitarian economy appears to be both yield a higher rate of return and be more stable (Marx [5]-Piketty [9] thesis). The model in its current state is far from complete or realistic, however its dynamics mimics key features of real economies. This might provide some policy insights.

2 Network Model of the Economy

We begin with the United Nations [11] framework for accounting of national economies. To quote the definition of the System of National Accounts [11], "The System of National Accounts (SNA) is the internationally agreed standard set of recommendations on how to compile measures of economic activity in accordance with strict accounting conventions based on economics principles." Based on this statistical framework [11] and Stock Flow Consistency theory [2], we propose a model of the economy with four participants: Households, Firms, Banks and the Government. We will examine their stocks and the flows between the balance sheet of the participants. We define our variables in Table 1. Subscripts h represents households, f firms, c capitalists households and w worker's households. The difference between workers and capitalists is that the former could not possess assets. The subscript o represents an outstanding amount. T represents the maturity on loans. Table 2 shows flows of funds in our model, yielding the 4×4 matrix shown in Eq. 1, below.

As Minsky states [6] "This theory (financial instability hypothesis) leads to the conclusion that economic instability,...results from the normal functioning of such a capitalist economy." As hypothesis indicates, economic crises and business cycles

Table 1 Definition of variables

Item	Variable	Item	Variable	Item	Variable
Consumption	C	Investment	I	Rate of return	r
Profit	P	Wage	W	Interest rate (loans)	i_l
Deposits	D	Loan maturity	N	Taxes	Ta
Transfers	Tr	Quantitative easing	Q	Net worth	N
Interest rate (deposits)	i_d	Government expenditure	G	Asset value	A
Debt	Db	Expectations	ρ	Capital	K

Table 2 System of national accounts for the model of the economy

	Household	Firms	Banks	Government	Sum
Consumption	$-C$	$+C$			0
Investment	$-I$	$+I$			0
Profit	$+P$	$-P$			0
Wage	$+W$	$-W$			0
New loans	$+\Delta L_h$	$+\Delta L_f$	$-\Delta L$		0
Interest payments on loans	$-i_l * L_{ho}$	$-i_l * L_{fo}$	$+i_l * Lo$		0
Deposits households	$-\Delta D_h$	$-\Delta D_f$	+D		0
Interest payments on deposits	$+i_d * D_h$	$+i_d * D_f$	$-i_d * D$		0
Payment on principles of loans	$-L_h/T$	L_f/T	L/T		0
Government expenditure		$+G$		$-G$	0
Taxes	$-Ta_h$	$-Ta_f$		$+Ta$	0
Transfers	$+Tr$			$-Tr$	0
Quantitative easing			$+Q$	$-Q$	0
Net worth	$-N_h$	$-N_f$	$-N_b$	$+N_g$	0
Sum	0	0	0	0	0

must exist endogenously with respect to the economy. Here we attempt to replicate this instability.

$$\begin{array}{c} \\ H \\ F \\ B \\ G \end{array} \begin{array}{c} \begin{array}{cccc} H & F & B & G \end{array} \\ \begin{bmatrix} \Delta A & C+I+\Delta N_h & \Delta D_h+(L_h/N)+i_l * L_{ho} & Ta \\ P+W & r\times F_c+\Delta N_f & \Delta D_f+L_f/T+i_l * L_{fo} & T_f \\ \Delta L_h & \Delta L_f & \Delta N_b & Q \\ Tr & G & Q & \Delta N_g \end{bmatrix} \end{array} \tag{1}$$

3 Asset Pricing Model

We consider constructing an asset pricing model for the model of the economy. After we combine the economic theories of Friedman [1], Keynes [3], Minsky [8], and Marx [5], we propose the following asset pricing model.

$$\Delta A/A_{t-1} = \Delta L/A_{t-1} \times (r+\rho)/i \times K/Db \tag{2}$$

We regard the right hand side of Eq. (2) as the product of three ratios. The first $\Delta L/A_{t-1}$ is the real money balance from a Chicago Monetarist Model, according to Taylor [10] (203). The price of the asset at the previous moment determines how many assets a particular amount of money newly injected into the market could buy.

The second ratio $(r + \rho)/i$ is Minsky's interpretation of the valuation ratio, also according to Taylor. This term is very straightforward; it asks merely whether the firm or asset is producing enough returns to payoff the interest on their debt.

The third ratio is the debt to capital ratio described by Taylor [10] on page 273 as "a natural state variable for dynamic analysis."

We tested this model in the US housing market from 1993 to 2010, a period which saw the bubble of 2007. The important observation here is not the asset itself but price dynamics. We found $r^2 = 0.62$ with $p < 0.0001$.

4 Simulation

4.1 Parameters and Initial Values

Now we put this model into action first under a hypothetical situation. Before that there are certain assumptions of the economy that must be made. The first assumption is that there will be capitalist's households and worker's households. The main difference between the two is that the capitalist households invest in assets whereas the worker households do not. An assumption about the financial sector is that there is only savings bank in the economy, and they will only issue loans according to the exogenous reserve ratio (rr), which means the banks could not issue more than $(1/rr)D$ of loans in total, thus they will stop issuing new loans at that point. We are not considering any derivatives or insurances and interest rate on deposits are zero, because if not, it would considerably complicate the portfolio choice. The government will increase spending during a recession, when growth becomes negative. The exact figures for the first iteration of the model economy are as follows (Table 3) below:

Table 3 Parameters and initial values

Item	Amount	Item	Amount
$r \times K$	75	P	$0.6 \times r \times K$
W	$0.6 \times r \times K$	A	1000
K	1000	ΔA	0.1
Ta	0.02	D_w	1000
D_c	1200	Db_w	150
Db_f	150	Db_c	300
Payment on loan	$-\frac{L_h}{N}$	G	$0.05 \times (300 + Ta)$

There are also some special parameters, such as the interest rate under normal circumstances is 5%, and when loan demand reaches half of loan ceiling, the interest rates rise to 8% and banks would only issue 80% of loan demanded, and when loan demand exceeds the loan ceiling, the bank could not issue any loans and the interest rate rises to 10%. The government would aid the economy when it is in distress or suffering negative growth by increasing their budget to 5% of 2000 plus taxes, and often taxes would be negative. This action could be interpreted as government aid for when people are in distress. Not all funding for corporations could be satisfied, and we are supposing that 80% of their demand is satisfied.

4.2 *Results*

As implied by the rate of change of assets and wages as seen in Fig. 1, all other parts of the economy display cyclic behavior, which periods of high growth are followed by periods of low or negative growth. The explanation of this is that the capitalists have demanded too much loan during the up rise, and could not keep up with the interest payment on loans, thus they are forced to liquidate assets (negative loan demand). However there is another case of negative loan demand, which is that they could finance their allocation of money by themselves. In this case, the savings would increase, in this model the only thing the capitalists can do with extra money is to save.

In this epoch of slow growth, the firms and capitalists are demanding and provided with loans continuously, with slow economic growth, they do not demand much funds, which the bank is capable of providing. However, this tranquil era ends up in a period of negative growth.

In the end, however as the liquidity or bank savings dry up, the bank could not issue more loans, and then the economy enters an equilibrium as we see after period 50. Equilibrium is achieved when no new liquidity is injected into the economy, or that the demand for loans always exceeds the loan ceiling. Also as rate of returns diminishes, the well-being of the economy is increasingly dependent upon external injections of

Fig. 1 Rate of growth and rate of return in simulated economy

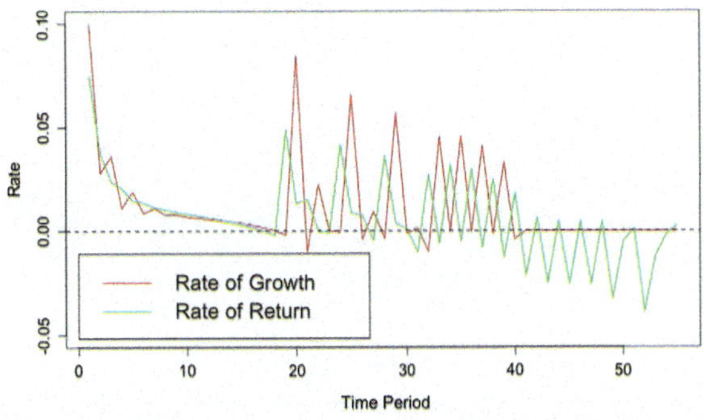

liquidity, either through financial innovations (implying the ability to issue more loans with given amount of money), import of capital or more and persistent government injection. This dependence will weaken the economy, and each crises more severe in scale, eventually ending up somewhere similar to Japan's lost decades [4].

4.3 Effect of Reducing the Time Step

One observes that the model displays approximately cyclic oscillations. One might explore this problem by reducing the time that passes between updates which might smooth the cycle. The other problem we might note from the normal time step version of the model is that there is negative wage implied by negative rate of return for long periods of time, which needs to be addressed somehow.

As shown in Fig. 2, we continue to see rapid cycling; however the problem of persistent negative wage is resolved, and the dynamics is preserved, consistent with the existence of an attractor.

We now explore the effect of some changes to the model.

4.4 Effect of Reduced Government Aid

We had previously set the level of government aid to the economy during recessions to be $0.05 * 2000 + Ta$. We now reduce this to $0.05 * 500 + Ta$. As shown in Fig. 3, the economy descends into a persistent crisis, from which the economy does not recover. This corresponds to Koo's [4] argument that the reason for second dip recessions of America in 1938 and Japan in 1996 is insufficient government aid.

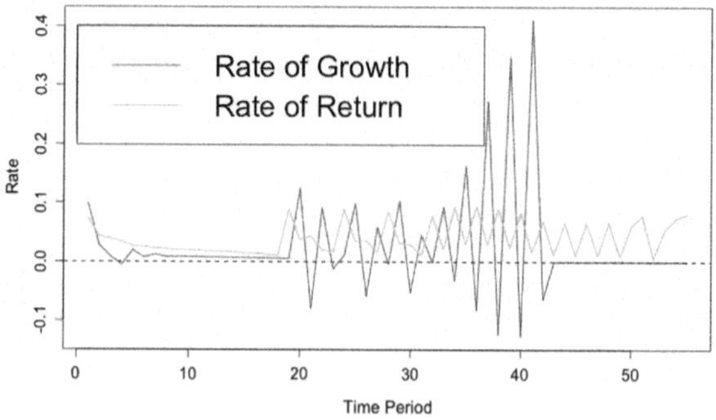

Fig. 2 Rate of growth and rate of return with reduced time step

Fig. 3 Rate of growth and rate of return with reduced government stimulus

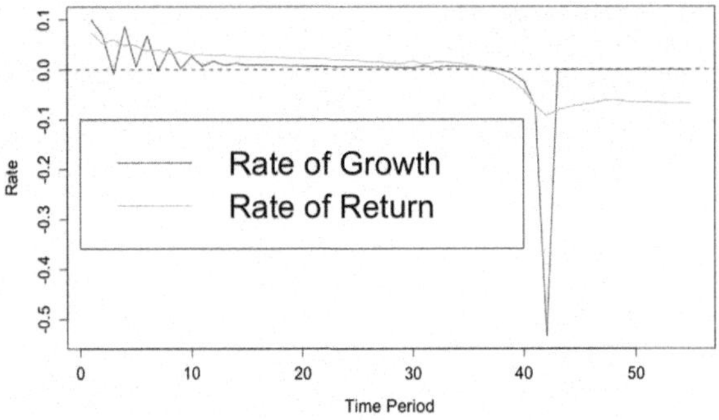

4.5 *Effect of Income Distribution*

Finally we explore the effects of the distribution of between capitalists and workers. As shown in Figs. 4 and 5, allocating 80% of the income to workers yields a higher average rate of return than allocating 80% to capitalists (7.6% vs. 4.8%). This result corresponds to Piketty's observation that the capitalism experienced most growth and stability in the 1950 to the late 1970s period where inequality is the lowest in the century, and growth decreased in the 1980s when inequality increased. This would imply that a more egalitarian economy would be more stable and healthy at equilibrium.

5 Discussion

Limitations. There are other essential aspects of the economy not considered in this model, such as the bond market. Intervening in the bond market is an important way for the government to regulate the economy. One can also simulate the effects of government bankruptcy. Finally, it is necessary to improve the portfolio distribution

Fig. 4 Rate of growth in capitalist-dominated and worker-dominated economies

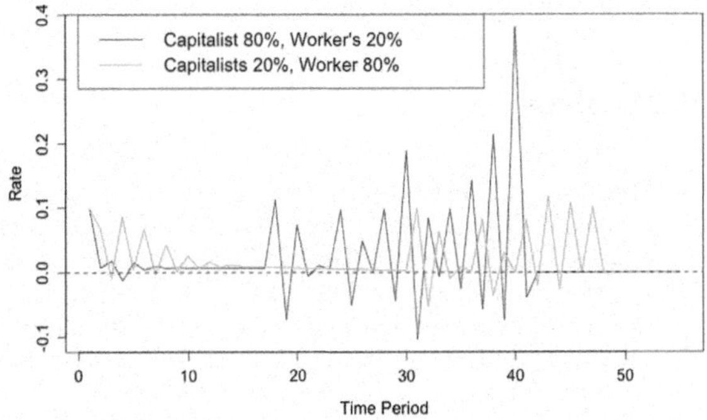

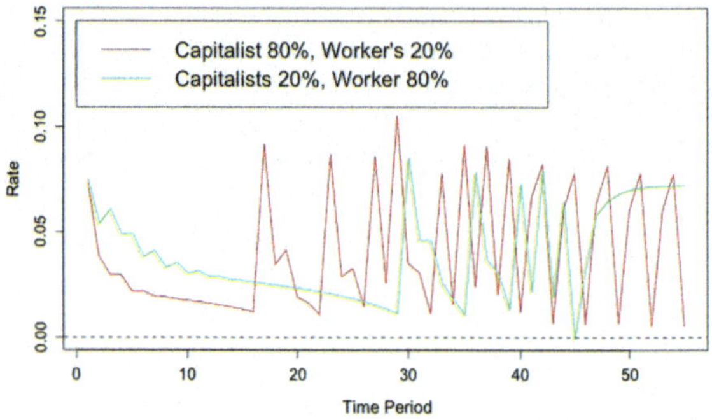

Fig. 5 Rate of return in capitalist-dominated and worker-dominated economies

design for smoother performance of the model, and to possibly incorporate more assets and liquidity preference into the economy.

In this paper, we attempt to explore the dynamics of a Stock Flow Consistent model of a closed economy and briefly explore connections with the Marx-Piketty Thesis and Minsky's financial instability hypothesis. We find that our model displays cyclic behavior while approaching an equilibrium, and simulates the effects of various shocks. We also find that a more egalitarian economy appears to be both yield a higher rate of return and be more stable (the Marx-Pikkety thesis).

As noted above, this framework bears the potential to simulate various scenarios such as transactions within any sector of the economy, trade networks of any scale, and most importantly, networks of economies. By then, one could simulate various events and examine their effect on the economy, and possibly provide insights to controlling the economy in a more direct and effective way. It would also be interesting to further explore reasons for apparently persistent oscillations in the rate of returns and more generally to explain stability of the model.

In the future one could build a network of these (now open) models together to form international asset market and trade dynamics, where the workers and capitalists could purchase products and assets from other economies and vice versa. One might also use this kind of model to analyze the flows within a sector, such as money flows between banks to analyze their stability.

References

1. Friedman, M.: The role of monetary policy. Am. Econ. Rev. **LVIII** (1968)
2. Godley, W., Lavoie, M.: Monetary Economics. Palgrave Macmillan, New York, NY (2007)
3. Keynes, J.M.: The General Theory. The Macmillan Press LTD, London and Basingstoke (1973)
4. Koo, R.: The Escape from Balance Sheet Recession. Wiley, New York, NY (2014)
5. Marx, K.: Grundrisse. Random House, New York, NY (1973)
6. Minsky, H.P.: The financial instability hypothesis: the behavior of a sophisticated capitalist economy. Hyman P. Minsky Arch. **466** (1980)

7. Minsky, H.P.: Marxian economics: a centenary appraisal. Hyman P. Minsky Arch. **170** (1994)
8. Minsky, H.P.: Stabilizing and Unstable Economy. McGraw Hill, New York, NY (2008)
9. Piketty, T.: Capital in the Twenty-First Century. The Belknap Press of Harvard University Press, Cambridge, MA (2017)
10. Taylor, L.: Reconstructing Macroeconomics. Harvard University Press, Cambridge, MA (2004)
11. United Nations: System of National Accounts. United Nations, New York, NY (2008)

Does Exist Gap-Filling Phenomenon in Stock Markets?

Xiaohang Liu, Yan Wang, Ziyan Jiang, Qinghua Chen, and Honggang Li

Abstract To Chartists, there is an axiom about the stock gap which is "gaps always get filled soon" . However, there is a lack of sufficient academic literature to discuss this kind of issue carefully. This paper exhibits some characteristics of the gap by collecting empirical data in Chinese and American stock markets. By applying a detrending and random exchange process on the original data, the results reveal that the real data series has some inherent structure behind the price variation and the overall trend hinders the gaps' generation and slows down the gaps' refilling process to a certain extent. The difference between the original data and the randomly exchanged data after the detrending suggests that the gap-filling phenomenon may exist in the stock markets of China and the United States.

Keywords Price gap · Gap-filling · Stock market · Random shuffling process · Detrending

1 Introduction

Revealing universal law based on analyzing empirical data, as necessary and essential part, has significantly promoted the considerable progress of the study on financial markets. It is well known some stylized facts such as the fat-tailed distribution of

X. Liu · Q. Chen (✉) · H. Li
School of Systems Science, Beijing Normal University, Beijing 100875, China
e-mail: qinghuachen@bnu.edu.cn

H. Li
e-mail: hli@bnu.edu.cn

X. Liu · Z. Jiang
School of Government, Beijing Normal University, Beijing 100875, China

Y. Wang
Department of Mathematics, University of California, Los Angeles 90095, USA

Q. Chen
New England Complex Systems Institute, Cambridge, MA 02139, USA

Department of Chemistry, Brandeis University, Waltham, MA 02453, USA

D. Braha et al. (eds.), *Unifying Themes in Complex Systems X*, Springer Proceedings in Complexity, https://doi.org/10.1007/978-3-030-67318-5_13

returns, volatility clustering of returns, and slow decay of auto correlation in absolute returns etc. [1, 2] have discovered and became useful tools to judge the quality of theoretical models [3]. But, the process of discoveries and tests for new possible facts never stopped. For instance, the power law relation between skewness and kurtosis [4], the universal and stationary price formation mechanism [5], and spurious trend switching [6]. Consistent empirical findings would inspire relevant theoretical work and practical application attempts.

This paper is supposed to discuss some features about gap-filling, the interesting but easily neglected phenomena, empirically. The stock price gaps generally refer to the presence of a "vacuum" on candle chart, which resulted from the difference between the highest and lowest prices. Some scholars have long time noticed that evidences as "the market abhors a vacuum and all gaps will be filled" [7]. It is consistent with the opinion that only when a gap in the stock price be replenished in short time, and then the stock price will continue to move forward [8]. However, until now, there are only a few academic literatures on the gap which mainly focus on the similarity and difference of statistics of the trade volume in transactions period such as before, during and after the gap [9, 10]. In short, it remains as only hearsay because there are insufficient researches in price gaps in stock markets, which deserves to be attached more importance.

For systematically study on gap-filling phenomena, this paper adopts random-shuffle and de-trending technology among different markets in several countries. We attempt to answer whether a market tend to fill price gaps and whether there is a universal law in gap-filling among different markets. What we do is a statistical analysis, which is a very beginning of tremendous work. The sections of the article will be arranged as follows. In Sect. 2, a brief introduction to the gap phenomenon and our data sources will be present. In Sects. 3 and 4, a stochastic exchange process will be presented, and the results will be compared with the empirical ones. In Sect. 5, the statistical results of no trend data will be shown. In the last section, we conclude and discuss.

2 Background and Data Source

2.1 Gap and Gap-Filling Phenomena

The candlestick-graph is a style of financial chart used to describe price movements and is one of the most convenient tools to find the trend and the gap of the stock[11]. Each candlestick represents four important prices for that day: opening price, closing price, maximum price and minimum price. In Fig. 1, it is obvious to find 4 vacancies that appears between some adjacent bars that indicating the gaps.

Price gaps can be divided into up gaps and the down ones by jumping directions. To form an up gap, the lowest price must be higher than the highest price of the previous day, and a down gap is formed when the highest price is lower than the

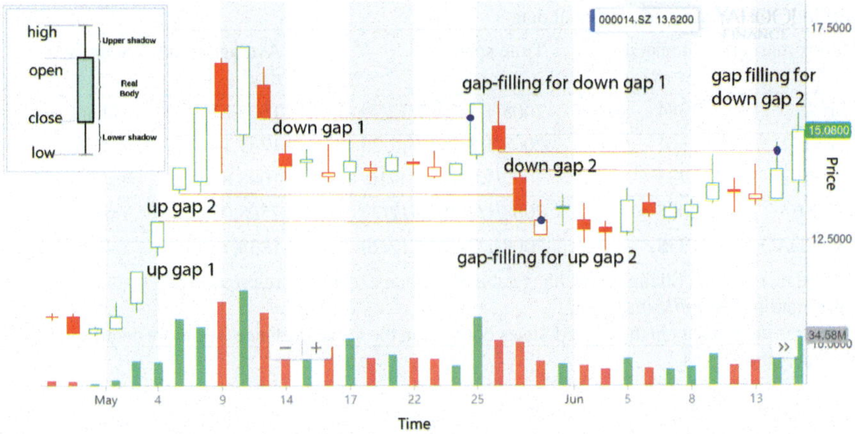

Fig. 1 Phenomena of gap and gap-filling. The data is got from https://sg.finance.yahoo.com

lowest price of the previous day. There are 2 up gaps on 4th May and 7th May and 2 down gaps on 14th May and 29th May in Fig. 1.

A gap is filled or closed means that the price movement cover the space at a later time. For an up gap, the lowest price on the day of filling the gap must lower or equal to the lower boundary of the gap. For a down gap, the highest price on the day of filling the gap must higher or equal to the upper boundary of the gap. In that case, the "up gap 2" was filled on 30th May; the "down gap 1" and "down gap 2" were filled on 25th May and 14th June respectively. Until the end of the period, the "up gap 1" has not closed yet.

Gaps are meaningful for providing clues about price movements. The breakup in price continuity implies that something important has happened to the fundamentals or the psychology of the crowd has triggered for one stock even the whole market. By the inception of technical analysis, these "holes" can be identified as the possibility of its profitable exploitation by traders in the stock markets [10].

2.2 *Data Description*

We collected all the available stock data from Shanghai stock exchange (SHSE), the main board of Shenzhen stock exchange (SZSE-1), the second board of Shenzhen stock exchange (SZSE-2), New York stock exchange (NYSE) and NASDAQ stock exchange (NASDAQ) respectively. We use backward adjusted prices to get rid of the effect of splits and dividends. These stock markets are representative markets in both China and the U.S.

Table 1 Information of empirical data

Stock market	Stocks no.	Time span	Average trade days	Source
SHSE	1384	2008/1/2-2017/12/29	1605.37	tushare
SZSE-1	1361	2008/1/2-2017/12/29	1751.18	tushare
SZSE-2	733	2009/12/30-2018/10/19	1098.85	tushare
NYSE	2005	2008/1/2-2017/12/29	2508.02	Yahoo
NASDAQ	1838	2008/1/2-2017/12/29	1838.76	Yahoo

[a]The data of China Shanghai exchange and Shenzhen exchange are provided by tushare (see more information at http://tushare.org/)
[b]The transaction data in the United States come from the Yahoo Finance (see more details on https://finance.yahoo.com)

As shown in Table 1, a ten-year period is taken into consideration. The data pre-processing includes deleting missing values. Yahoo provides stock data with null values when the stock is suspended. We have deleted stocks with too many missing or abnormal values.

3 Random Shuffling

In this subsection, we will present a randomly shuffling process. As a kind of time series, price features both distribution and time-series characteristics such as volatility aggregation and time-relevant correlation. By replacing original series with a randomly distributed series, Random Shuffle would destroy the time-relevant correlation while reserving the distribution characteristic, which is a regular method broadly applied[12, 13]. Meanwhile, the gap-filling phenomenon reflects short-term correlation behavior of time series, so if there does exist inherent mechanism of gap generation and complement, we would break it by Random Shuffle. In short, by comparing the differences in statistical characteristics of time-series between real data and the randomly generated data, we can reveal that if the real data is dependent in time series and then we can find the features of gap and gap-filling in real data.

As shown in Fig. 2a, the process is as follows

- **Get real data and make statistics**
  Figure 2 is the candlestick plot of original empirical data. There are two gaps, one is an up gap which occurred on day 2, and another one is a down gap on day 4. We define $X_{t,open}$, $X_{t,close}$, $X_{t,high}$ and $X_{t,low}$ to indicate the opening, closing price, highest and lowest price of a stock on the tth trading day respectively. For $t \in [1, 2, 3, \ldots, T]$ and T represents length of the data.
- **Calculate ratios to last close**
  Calculating the ratios of the four prices per day to the previous day's closing price in turn.

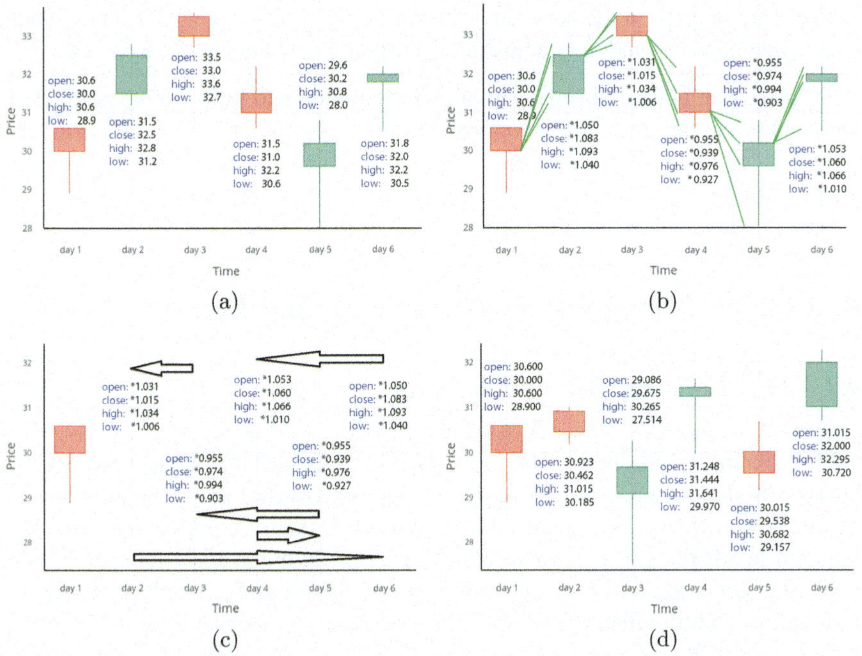

Fig. 2 Randomly switching processing: **a** the original data; **b** calculate the daily ratios to close price on previous trade day; **c** randomly switch daily ratios; and, **d** recalculate the price data from earliest close price with the shuffling daily ratios

$$r_{t+1,open} = \frac{X_{t+1,open}}{X_{t,close}}, \quad r_{t+1,close} = \frac{X_{t+1,close}}{X_{t,close}},$$
$$r_{t+1,high} = \frac{X_{t+1,high}}{X_{t,close}}, \quad r_{t+1,low} = \frac{X_{t+1,low}}{X_{t,close}}. \tag{1}$$

In this way, we can get the relative changes of the price indices of every day. As shown in Fig. 2b, this ratios are given on a daily mark as vector $r_{t+1} = (r_{t+1,open}, r_{t+1,close}, r_{t+1,high}, r_{t+1,low})$.

– **Randomly shuffle ratio vectors**
  Keep the data of the first day, and randomly rearrange the other daily's ratio vectors. In Fig. 2b, new $r_2^*, r_3^*, r_4^*, r_5^*, r_6^*$ are shuffled from old r_3, r_4, r_6, r_4, r_2 respectively. i.e. $r_{2,open}^* = r_{3,open}$, $r_{2,close}^* = r_{3,close}$, $r_{2,high}^* = r_{3,high}$, and $r_{2,low}^* = r_{3,low}$ and so on.
– **Consequently recalculate prices**
  The new daily prices are recalculated from the first day's original data and the rearranged ratios. This process should be carried out in order.

$$X_{t+1,open}^* = r_{t+1,open}^* X_{t,close}^*, \quad X_{t+1,close}^* = r_{t+1,close}^* X_{t,close}^*,$$
$$X_{t+1,high}^* = r_{t+1,high}^* X_{t,close}^*, \quad X_{t+1,low}^* = r_{t+1,low}^* X_{t,close}^* \tag{2}$$

There are no gaps in the new candlestick plot 2d. Notice that it is also possible to generate gaps by randomly adjusting the original data that do not have a gap.

If there is an internal mechanism of gap generation and filling in the time series, the stock prices before and after would have strong correlation. If we switch the price sequences randomly, it will eliminate the internal structure and see different statistics results.

4 Results of Empirical Statistics on Original Data

4.1 The Statistics of Original Data

Table 2 exhibits some meaningful statistical norms, such that (N_{100}) the average number of gaps in the 100 trade date per stock; the average and standard deviation of TD_{fill} and ND_{fill} which presents the number of the trade days and natural days required to fill the gap respectively; and $N_{unfilled}$ is the average number of gaps that were not filled until the end of the period. Besides, S_{gap} defined as the gap size or gap width to distinguish how big one is the gap. For an up gap, the size is $S_{t,gap} = \frac{X_{t,low} - X_{t-1,high}}{X_{t-1,close}}$ and for a down gap $S_{t,gap} = \frac{X_{t-1,low} - X_{t,high}}{X_{t-1,close}}$ respectively. There are some clear evidences that we could find from Table 2.

1. On average, there is a gap of about 3.5–6.8 in 100 trading days. Both kinds of gaps in the US stock exchange markets have a larger amount than Chinese markets do. The less occurrence of the gaps in Chinese markets perhaps due to the "T+1" trading rules and 10% restrictions. Besides, the number of unfilled gaps is also significantly larger in America than that in China. The proportion of down gaps in the US market is almost twice that of up gaps.
2. The average filling time is about 27.6 to 50.5 trading days, and both China and the U.S. have a large variance. In detail, the gap-filling time of down gaps is much longer than that of up gaps in China except SZSE-2, while it is almost the same in the U.S. This may indicate that the development of the Chinese stock market is relatively weak, and it requires more time to recover from the unexpected environment impacts.
3. The average gap size is between 0.9 and 1.6%. No matter in up gaps or down gaps, the order is the same, NASDAQ $\succ$ SZSE-2 $\succ$ SHSE $\succ$ SZSE-1 $\succ$ NYSE.

4.2 Statistical on Shuffling Data

This paper uses randomly shuffling technology to compare how far in difference of the actual data to random series. We believe that the differences can reflect the gap-

Table 2 Gap statistics of actual data from some stock exchanges

Gap type	Index	SHSE	SZSE-1	SZSE-2	NYSE	NASDAQ
Up gap	N_{100}	1.5579	1.6637	1.8118	3.6070	2.4547
	TD_{fill}	30.171 (101.634)	27.626 (89.562)	36.780 (124.639)	33.417 (124.921)	33.452 (119.584)
	ND_{fill}	47.443 (158.724)	44.657 (143.611)	61.423 (210.110)	48.498 (181.419)	48.533 (173.702)
	S_{gap}	0.01464 (0.02501)	0.01446 (0.04997)	0.01577 (0.01889)	0.01021 (0.39952)	0.01638 (0.09832)
	N_{unfilled}	1.3721	1.7355	0.5681	6.6599	4.2136
Down gap	N_{100}	1.9666	2.0534	2.1424	3.2272	2.1723
	TD_{fill}	50.531 (161.448)	47.893 (150.297)	27.767 (90.522)	30.140 (108.777)	33.237 (120.140)
	ND_{fill}	80.175 (254.046)	76.911 (238.986)	44.973 (143.906)	43.789 (158.284)	48.262 (174.473)
	S_{gap}	0.01137 (0.01391)	0.01127 (0.01492)	0.01271 (0.01506)	0.00943 (0.01860)	0.01630 (0.04216)
	N_{unfilled}	1.7211	1.6032	3.4128	1.5172	1.4469
Total	N_{100}	3.5245	3.7171	3.9542	6.8342	4.6270
	TD_{fill}	41.531 (138.606)	38.822 (127.162)	31.896 (107.598)	31.869 (117.585)	33.351 (119.845)
	ND_{fill}	65.707 (217.741)	62.474 (202.580)	52.511 (177.521)	46.274 (170.901)	48.405 (174.064)
	S_{gap}	0.01282 (0.01968)	0.01271 (0.03537)	0.01403 (0.01689)	0.00985 (0.29481)	0.01634 (0.07790)
	N_{unfilled}	3.0932	3.3387	3.9809	8.1771	5.6605

filling phenomenon. The results of statistics on shuffling data are shown in Table 3. Comparing the Table 3 to Table 2, it can be found as follows.

1. Almost in all markets, shuffling data have more gaps than empirical data. Especially, the down gap in each market has considerable increment. The increment degree is larger in the U.S. than that in China. It indicates that the transaction behavior in the actual market would avoid the appearance of gaps, especially in the U.S.
2. The time to fill an up gap change significantly. Almost all the U.S. markets take more time to fill the gaps, which indicates that there is a certain phenomenon of early gap-filling. However, the situation in China is quite complicated. The average gap-filling time of down gaps in SHSE and SZSE-1 has decreased by about 40% and the gap-filling time of up gaps has increased by around 25%. This may indicate that the up gaps tend to be refilled while the refilling of the down gaps is impeded. Overall, the difference on gap-filling time has been narrowed down.

Table 3 Gap statistics on data from randomly shuffling process

Gap type	Index	SHSE	SZSE-1	SZSE-2	NYSE	NASDAQ
up gap	N_{100}	1.6138	1.6660	1.7014	4.4009	3.2939
	TD_{fill}	38.693* (139.629)	35.967* (129.187)	29.788 (96.372)	34.442* (130.213)	34.026* (125.903)
	ND_{fill}	61.118* (221.774)	57.823* (207.958)	48.962 (160.567)	49.998* (189.075)	49.391* (182.776)
	S_{gap}	0.01212* (0.02400)	0.01216* (0.04986)	0.01243* (0.01573)	0.01049* (0.36743)	0.01593* (0.08766)
	$N_{unfilled}$	1.3171	1.6062	1.1744	5.0613	3.6147
down gap	N_{100}	2.3995	2.3731	2.6045	3.7991	2.8963
	TD_{fill}	39.447* (137.351)	39.023* (132.376)	30.937 (98.115)	38.052* (132.823)	35.545* (129.311)
	ND_{fill}	62.368* (217.707)	62.983* (214.584)	50.664 (161.983)	55.295* (193.047)	51.588* (187.646)
	S_{gap}	0.00905* (0.01150)	0.00918* (0.01252)	0.01052* (0.01263)	0.009813* (0.01835)	0.01538 (0.03994)
	$N_{unfilled}$	1.7934	1.6466	1.6485	2.6528	2.2409
total	N_{100}	4.0133	4.0391	4.3060	8.2001	6.1902
	TD_{fill}	39.144* (138.271)	37.763* (131.078)	30.484 (97.431)	36.114* (131.441)	34.737* (127.511)
	ND_{fill}	61.865* (219.351)	60.855* (211.891)	49.992 (161.425)	52.452* (190.943)	50.419* (185.074)
	S_{gap}	0.01028* (0.01770)	0.01042* (0.03537)	0.01128* (0.01397)	0.01018* (0.2708)	0.01567* (0.06971)
	$N_{unfilled}$	3.1105	3.2527	2.8229	7.7142	5.8556

[a] *means we can reject the null hypothesis that the random result is the same with the original one at 5% significance level using Kruskal-Wallis rank sum test
[b] The red letters mean bigger and blue letters mean smaller than the original results

3. Apart from NYSE, the size of the gap has been reduced to varying degrees. The overall size of the gap is slightly reduced.

5 Result of Statistics on No-trend Data

5.1 Detrending Processing

The overall trend of the stock market does have an influence on the time of gap filling. For example, an upward gap is difficult to be filled when prices are at rising trend, so as to a downward gap at down trend. We use the detrending technology to get rid of the overall trend from the individual stock by the following steps.

- Step 1: Take the first trading day of the research period as the base period and calculate the adjustment coefficient.
- Step 2: For the five stock markets, the adjustment coefficients are calculated respectively using its market index.
- Step 3: Divide the daily four prices of each stock by the adjustment coefficient of the corresponding date.

5.2 *Statistical Result on No Trend Data and Shuffling Data*

Table 4 shows the statistic result of no trend data and the statistics on shuffling no-trend data is shown in Table 5.

Comparing Table 4 to Table 2, we can find some changes.

1. The average number of both kinds of gaps increases, while the changes of unfilled gaps are not significant. The increment in the U.S. is somewhat larger than China,

Table 4 Gap statistics of no trend data from some stock exchanges

Gap type	Index	SHSE	SZSE-1	SZSE-2	NYSE	NASDAQ
Up gap	N_{100}	2.1725	3.0157	2.6894	4.2063	3.3948
	TD_{fill}	23.039 (103.802)	17.871 (75.305)	17.976 (75.612)	28.351 (118.555)	25.705 (107.116)
	ND_{fill}	36.464 (164.133)	28.729 (122.013)	29.685 (127.049)	41.165 (172.032)	37.312 (155.528)
	S_{gap}	0.01250 (0.02356)	0.01174 (0.02678)	0.01352 (0.01858)	0.00839 (0.03122)	0.01511 (0.08566)
	N_{unfilled}	1.1185	2.0154	0.8595	2.3461	1.6823
Down gap	N_{100}	2.2705	3.2461	2.8684	4.0237	3.4649
	TD_{fill}	17.818 (73.020)	17.208 (72.198)	14.654 (61.776)	23.315 (88.606)	23.573 (94.618)
	ND_{fill}	28.417 (116.132)	28.034 (116.431)	23.903 (101.689)	33.718 (129.397)	34.215 (137.435)
	S_{gap}	0.00811 (0.01078)	0.00785 (0.01082)	0.00889 (0.01136)	0.00796 (0.01643)	0.01261 (0.03356)
	N_{unfilled}	1.3208	1.4798	1.3656	4.7450	4.8997
Total	N_{100}	4.4430	6.2618	5.5578	8.2300	6.8597
	TD_{fill}	20.371 (89.443)	17.527 (73.711)	16.262 (68.839)	25.889 (105.374)	24.628 (101.002)
	ND_{fill}	32.352 (141.707)	28.369 (119.152)	26.701 (114.698)	37.536 (152.784)	35.748 (146.676)
	S_{gap}	0.01025 (0.01831)	0.00974 (0.02029)	0.01111 (0.01543)	0.00818 (0.02501)	0.01381 (0.06425)
	N_{unfilled}	2.4394	3.4952	2.2251	7.0911	6.5820

Table 5 Gap statistics on data from randomly shuffling process

Gap type	Index	SHSE	SZSE-1	SZSE-2	NYSE	NASDAQ
up gap	N_{100}	2.0299	2.7262	2.5611	4.4780	3.9218
	TD_{fill}	24.220* (98.993)	21.745 (94.686)	16.908 (70.481)	26.276 (110.075)	27.231* (110.232)
	ND_{fill}	38.434* (157.376)	35.261* (155.349)	27.729 (117.035)	38.183 (159.997)	39.522* (159.973)
	S_{gap}	0.01165 (0.02312)	0.01124 (0.02748)	0.01191 (0.01630)	0.00902 (0.03119)	0.01580 (0.08114)
	$N_{unfilled}$	1.4400	1.9662	1.3070	2.6325	1.7134
down gap	N_{100}	2.6938	3.5377	3.3660	4.4492	4.0792
	TD_{fill}	18.310* (90.705)	16.145 (80.146)	12.951 (61.944)	27.634 (116.768)	24.896* (101.430)
	ND_{fill}	29.144* (144.452)	25.987* (129.430)	21.259 (104.044)	40.162 (169.686)	36.149* (147.212)
	S_{gap}	0.007445 (0.009444)	0.007812 (0.01015)	0.008441 (0.009921)	0.008459 (0.01674)	0.01298 (0.03299)
	$N_{unfilled}$	0.7449	0.9104	0.8336	3.4283	4.4272
total	N_{100}	4.7237	6.2639	5.9271	7.1105	5.6352
	TD_{fill}	20.850* (94.401)	18.582* (86.818)	14.661 (65.797)	26.953* (113.461)	26.041* (105.842)
	ND_{fill}	33.136* (150.212)	30.023* (141.370)	26.701* (114.698)	39.169* (164.899)	37.802* (153.608)
	S_{gap}	0.00927 (0.01697)	0.00932 (0.01985)	0.00996 (0.01321)	0.00874 (0.02503)	0.01433 (0.06107)
	$N_{unfilled}$	2.1849	2.8766	2.1426	6.0608	6.1406

[a] *means we can reject the null hypothesis that the random result is the same with the original one at 5% significance level using Kruskal-Wallis rank sum test
[b] The red letters mean bigger and blue letters mean smaller than the original results

whereas the SZSE-1 experience the sharpest increment. These may indicate that the real market will avoid the appearance of gaps.

2. The time to fill a gap is significantly shorter, and the difference between the up gaps and down gaps are narrow. It takes even less time to close a down gap than it does to close an up gap in SHSE and SZSE-1, which means the overall trend in these two markets hinders the refilling process.
3. As to the gap size, the changes in both China and the U.S. are not evident. The changes go in the smaller direction, and the order of the five markets remains the same.

Then we focus on the changes on the no-trend data after the random exchange. Comparing Table 5 to Table 4, it can be found as follows.

1. The changes of almost all statistics after random exchange are reduced. The total number of the gaps did not change too much, even there are less gaps in the U.S.

2. What ever types of up or down, time to fill an gap is changed significantly. They are almost increased after random shuffling except SZSE-2, which implies there is a certain degree of gap filling phenomena in the original detrending data.
3. The changes in the gap sizes are statistically insignificant.

These may indicate that the patterns in the original data series have been broken after the detrending process, so the statistic results are quite the same after the random shuffling process, which means that the gap-filling phenomenon has something to do with the overall trend.

6 Conclusions and Discussions

In this paper, we discussed a gap-filling phenomenon which does not receive much attention from academic world. Specifically, we focused on the statistic characteristics of gap and gap-filling time. Different markets were slightly different in some statistic characteristics, such as the number of gaps, the gap-filling time, the size of the gap and so on, but generally they are similar.

To reveal that the existed gap can be recovered in a specific pattern, we proposed a randomly shuffling process. After the random shuffling process, we found some significant differences in each characteristic, especially for the gap-filling time. We can reject that most of the random results are the same as the original one at a 5% significance level by Kruskal-Wallis rank sum test. In short, we found that the stock market had some internal pattern in terms of gap-filling and may recognize that this phenomena are widespread in these countries.

We use detrending technology to remove the trend of the whole market from individual stocks. We found the changes of statistics after the random exchange are reduced, which means that the gap-filling phenomenon may possibly be related to the overall trend. Moreover, regardless of the trend, the number of gaps in the U.S. market and the number of gaps that have not been filled are larger than those in China, reflecting the high activity of the U.S. stock market, which is related to China's "T+1" trading rules and the daily 10% restrictions. Some differences between real data and no-trend data suggest that the overall trend hinders the gaps' generation and slows down the gaps' refilling process to a certain extent. After the detrending, the difference between the original data and the randomly exchanged data suggests that the gap-filling phenomenon may exist in the stock markets of China and the United States.

Researches about gap-filling phenomenon could have some guiding significance for stock traders. For instance, a trading strategy might be established on the possibility of a stock gap being filled in future time. Therefore, it is a subsequent research direction that to focus on the distribution of gap-filing time. Additionally, the mechanism of how a gap is generated and then filled is also a question worthy of discussion.

Acknowledgements We appreciate comments and helpful suggestions from Prof. Yougui Wang. This work was partly supported by Chinese National Natural Science Foundation (No. 71701018, No. 61673070) and China Scholarship Council.

References

1. Cont, R.: Empirical properties of asset returns: stylized facts and statistical issues. Quant. Financ. **1**(2), 223–236 (2001)
2. Bouchaud, J.P., Mézard, M., Potters, M., et al.: Statistical properties of stock order books: empirical results and models. Quant. Financ. **2**(4), 251–256 (2002)
3. Lehoczky, J., Schervish, M.: Overview and history of statistics for equity markets. Ann. Rev. Stat. Appl. **5**(1), (2018)
4. Cristelli, M.: Universal relation between skewness and kurtosis in complex dynamics. In: Complexity in Financial Markets, pp. 141–150. Springer (2014)
5. Sirignano, J., Cont, R.: Universal features of price formation in financial markets: perspectives from deep learning. Quant. Financ. **19**(9), 1449–1459 (2019)
6. Filimonov, V., Sornette, D.: Spurious trend switching phenomena in financial markets. Eur. Phys. J. B **85**(5), 155 (2012)
7. Peacock, A.T.: The Political Economy of Economic Freedom. Edward Elgar Publishing (1997)
8. Guo, Y.: The guiding significance of the gap theory in portfolio investment technique analysis. Sci/Tech Inform. Dev. Econ. **14**(2), 80–82 (2004)
9. Dahlquist, J.R., Bauer, R.J.: Technical Analysis of Gaps: Identifying Profitable Gaps for Trading. FT Press (2012)
10. Caporale, G.M., Plastun, A.: Price gaps: another market anomaly? Investment Anal. J. **46**(4), 279–293 (2017)
11. Morris, G.L.: Candlestick Charting Explained: Timeless Techniques for Trading Stocks and Futures: Timeless Techniques for Trading Stocks and Sutures. McGraw Hill Professional (2006)
12. Kim, H., Oh, G., Kim, S.: Multifractal analysis of the Korean agricultural market. Phys. A Stat. Mech. Appl. **390**(23–24), 4286–4292 (2011)
13. Bolgorian, M., Gharli, Z.: A multifractal detrended fluctuation analysis of gold price fluctuations. Acta Phys. Polon. B **42**(1), 159–169 (2011)

Empirical Scaling and Dynamical Regimes for GDP: Challenges and Opportunities

Harold M. Hastings and Tai Young-Taft

Abstract We explore the distribution of gross domestic product (GDP) and per capita GDP among different countries in order to elucidate differences in the dynamics of their economies. An initial analysis of GDP data and per capita GDP data from 1980 and 2016 (and many years in between) typically finds three scaling regions—signatures of likely differences in dynamics. The GDP of the largest ~25 economies (nations, EU) follows a power law GDP ~1/rank (c.f. Garlaschelli et al. in Eur Phys J B 57:159–164 [1]); followed by a second scaling region in which GDP falls off exponentially with rank and finally a third scaling region in which the GDP falls off exponentially with the square of rank. This broad pattern holds despite significant changes in technology (enormous growth in computing power, "intelligent" automation, the Internet), the size of the world economy, emergence of new economic powers such as China, and world trade (almost free communication, containerized shipping yielding sharp declines in shipping costs, trade partnerships, growth of the EU, multinationals displacing the traditional economic role of nation-states). Thus, empirically, these patterns may be universal in which case one approach to growth of less developed economies (in the second and third scaling regions of per capita GDP) may be to identify and target causative differences between these economies and those in the first (power law) scaling region.

Keywords Power law · Lognormal · Scaling regions · Economic growth

1 Introduction

Many scaling laws have been observed in economic systems [2], and references therein. Although specific metabolism of biological organisms scales as $(\text{mass})^{-0.25}$, per capita productivity of cities grows with population as $(\text{population})^{0.15}$. Thus cities,

H. M. Hastings (✉) · T. Young-Taft
Bard College at Simon's Rock, Great Barrington, MA 01230, USA
e-mail: hhastings@simons-rock.edu

T. Young-Taft
Levy Economics Institute of Bard College, Annandale-on-Hudson, NY 12504, USA

D. Braha et al. (eds.), *Unifying Themes in Complex Systems X*, Springer Proceedings in Complexity, https://doi.org/10.1007/978-3-030-67318-5_14

organizing centers for of economic activity serve as engines of economic growth. as do many similar activities. See also [2–7].

Here we consider scaling of GDP relative to rank, which implies a scaling rule for trade links given by the gravity law [8, 9], and also scaling of per capita GDP. We find that GDP data 1980 and 2016 (and many years in between, not discussed further here) displays three scaling regions. The GDP of the largest ~25 economies (nations, EU) follows a power law GDP ~1/rank (c.f. [1]); this is followed by a second scaling region in which GDP falls off exponentially with rank and finally a third scaling region in which the GDP falls off exponentially with the square of rank. The distribution of per capita GDP also displays these three scaling regions.

This broad pattern holds despite significant changes in technology (enormous growth in computing power, "intelligent" automation, the Internet), the size of the world economy, emergence of new economic powers such as China, and world trade (almost free communication, containerized shipping yielding sharp declines in shipping costs, trade partnerships, growth of the EU, the effect of multinationals displacing the traditional economic role of the nations-state [10]).

Thus, empirically, these patterns may be universal [11–17], in which case one of the targets for growth of potentially less developed economies (those in the second and third scaling regions) may be to identify and target causative differences between these economies and those in the first (power law) scaling region. For example, Montroll and Shlesinger [11] suggest a basic lognormal distribution which may be a consequence of multiplying many independent random variables, with a power law high-end tail because "the very wealthy generally achieve their superwealth through amplification processes that are not available to most." On the other hand, Reed and Hughes [14] show how power law behavior can arise if "stochastic processes with exponential growth in expectation are killed (or observed) randomly."

The rest of this paper is organized as follows. The observed scaling behavior is discussed in Sect. 2, and its persistence through time in Sect. 3. A few potential causes are briefly discussed in Sect. 4, followed by a concluding Sect. 5.

2 The Data

Here we consider GDP (US$, purchasing power parity (PPP)) and Per Capita GDP (US$, PPP) as functions of rank of country (highest to lowest) in 1980 and 2016. We observed the existence of three apparent scaling regions: a power law for the highest ranking countries, followed by a region of exponential fall-off, and finally a region of more rapid fall off ($\sim \exp(-const \times rank^2)$). We located their approximate boundaries empirically by maximizing R-squared.

Although a lognormal distribution appears similar, the Kolmogorov-Smirnoff (KS) test shows these distributions of GDP are unlikely to be lognormal (Table 1). Thus, differences in scaling likely represent differences in dynamics.

Table 1 Kolmogorov-Smirnoff statistics for distributions of GDP and per capita GDP

Year	Overall GDP	Per capita GDP
1980	$p = 0.0448$	$p = 0.0555$
2016	$p = 0.1505$	$p = 0.0022$

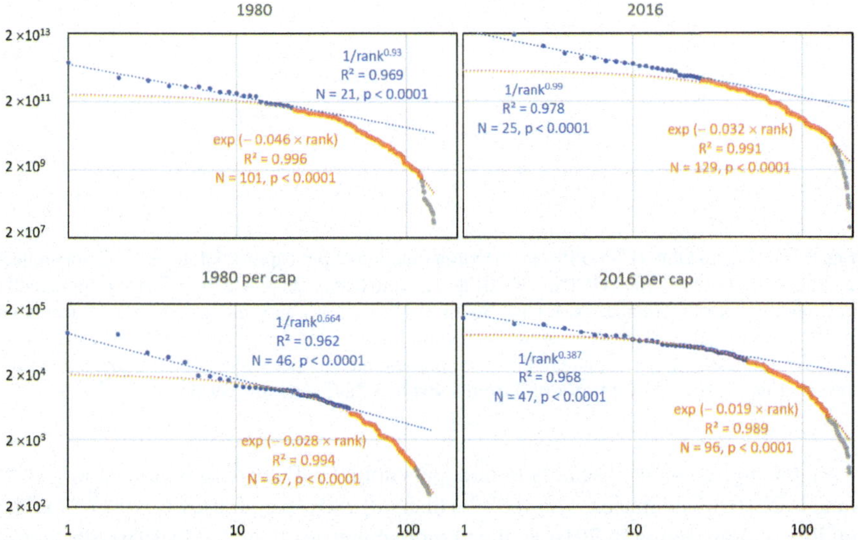

Fig. 1 GDP (US$, PPP) and Per Capita GDP (US$, PPP) as functions of rank of country (highest to lowest). Note the existence of three empirical scaling regions: a power law for the highest ranking countries (blue dots), followed by a region of exponential fall-off (orange dots), and the finally a region of more rapid fall-off ($\sim \exp(-const_1 \times (rank - const_2)^2)$, gray dots). In particular, the GDP of high-ranking economies scales as $\sim 1/rank$, a Zipf-Pareto distribution. The per capita GDP of high-ranking economies has a slower dependence on rank, $\sim 1/rank^{\beta}$, $\beta < 1$. Data from IMF, as reported in easy-to-use tabular form in [18, 19].

Although correlations between 1980 and 2016 data for the KS test are difficult to determine, the 2016 data meets the $p = 0.01$ cutoff, and the 1980 data is close to the $p = 0.05$ cutoff (Fig. 1).

3 Rank Tends to Persist

As shown in Fig. 2, the distribution of per capita GDP across countries is remarkably persistent over the 36-year interval from 1980 to 2016. This that the economic dynamics of most countries are surprisingly stable, with two notable general trends and a few exceptions. This has important consequences for reducing poverty in countries in the lowest scaling region.

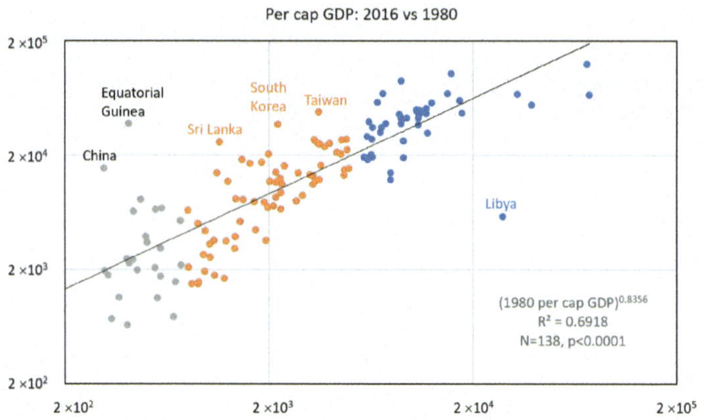

Fig. 2 Per Capita GDP (US$, PPP) in 2016 as a function of per capita GDP in 1980. Colors reflect scaling regions of per capita GDP in 1980 (Fig. 1). Note that per capita GDP increased significantly in China, Equatorial Guinca, Sri Lanka, South Korca and Taiwan (among) othcrs, and dccrcascd significantly in Libya. The slope of the regression line in a log–log plot (0.8356) also suggests a slight compression in per capita GDP's. Note that the median per capita GDP of countries in the lowest group increased by a factor of 5 from $ 464 to $ 2228 over this period

In the highest-ranked scaling region, per capita GDP decreased more slowly with rank in 2016 (47 countries, $1/rank^{0.387}$) than in 1980 (46 countries $1/rank^{0.664}$). The middle scaling region in 2016 included more countries than that in 1980 (96 vs. 67) and also a slower rate of decrease as function of rank ($\exp(-0.019 \times rank)$ versus $\exp(-0.028 \times rank)$). Economic development which moves countries from the lowest-ranked scaling region to the middle scaling region or from the middle scaling region to the highest scaling region is thus associated with significant increases in per capita GDP. As seen in the case of China, South Korea and Taiwan in Fig. 2, altering a country's dynamics appropriate industrialization provides one way to significantly increase per capita GDP, thus a useful change in economic "metabolic function." is a relevant question to pose with respect to the questions of growth and development. As suggested above, the stability of the dynamic regimes overtime suggests an aggregate metabolic function.

While the data appears remarkably consistent ($p < 0.0001$), there are some notable exceptions. China, South Korea and Taiwan rapidly industrialized (c.f. [20, 21]). Similarly, Sri Lanka progressed from a rural to an industrialized economy [22]. Equatorial Guinea's growth has been largely fueled by the discovery of oil, and thus has seen economic deterioration with the recent oil price drops [23]. [https://www.worldbank.org/en/country/equatorialguinea/overview]. At the same time, we have seen economic collapses in Libya and South Africa over this period.

4 Causes and Questions

We explore data which may shed light on some of the factors behind apparently persistent scaling laws: location, level of corruption, connectedness (as a surrogate for effective size), human development, and GINI index. The following example suggests that population itself does not drive per capita productivity, as West [2] shows for cities: Denmark, a member of the EU had a 2016 per capita GDP of $ 44,339 and 2019 population of 5.8 million; Pakistan, had a $ 2016 per capita GDP of 5095 and a 2018 population of 212 million.

As shown in Fig. 3, per capita GDP depends on location: higher levels are seen in North America and Europe; much lower levels in Africa, and the per capita GDP of a country is similar to that of its neighbors. Per capita GDP appears inversely related to level of corruption (the corruption perception index [24, 25] is inversely related to corruption, thus per capita GDP is directly related to this index), and directly related to the level of logistics.

There is a complex relationship between inequality as measured by the GINI index and economic growth [28]. At low initial incomes (less than $ 3000), increasing inequality is associated with greater economic growth; at high initial incomes, increasing inequality is associated with slower growth. A change of 0.01 in the GINI index implies a ~1% decrease in growth rate and an ultimate ~5% decrease in per capita GDP.

According to data in the UN human development index [29], education level appears closely related with per capita income, both before and after log-transforming income (Eq. 1). However, the question of cause and effect is complex: more education likely yields a more productive economy, and a more productive economy can afford more education.

$$
\begin{aligned}
&income = \$4382 \times (years\ of\ education) - \$29,725,\ R^2 = 0.411,\ N = 189,\ p < 0.0001,\\
&income = \$217 \times exp(0.350 \times years\ of\ education),\ R^2 = 0.700,\ N = 189,\ p < 0.0001.
\end{aligned} \tag{1}
$$

There are many other potential correlates of economic development, such as internet access, type of economy, type of government, etc. We plan to analyze factors associated with economic development and their historical evolution more extensively in future work.

5 Discussion

We have seen that both overall GDP data and per capita GDP data tends to display three distinct scaling regions. The large GDP tail appears to follow a power law; the middle exponential decay $GDP = exp(-const \times rank)$, and the low GDP tail decay as $GDP = exp\left(-const \times rank^2\right)$.

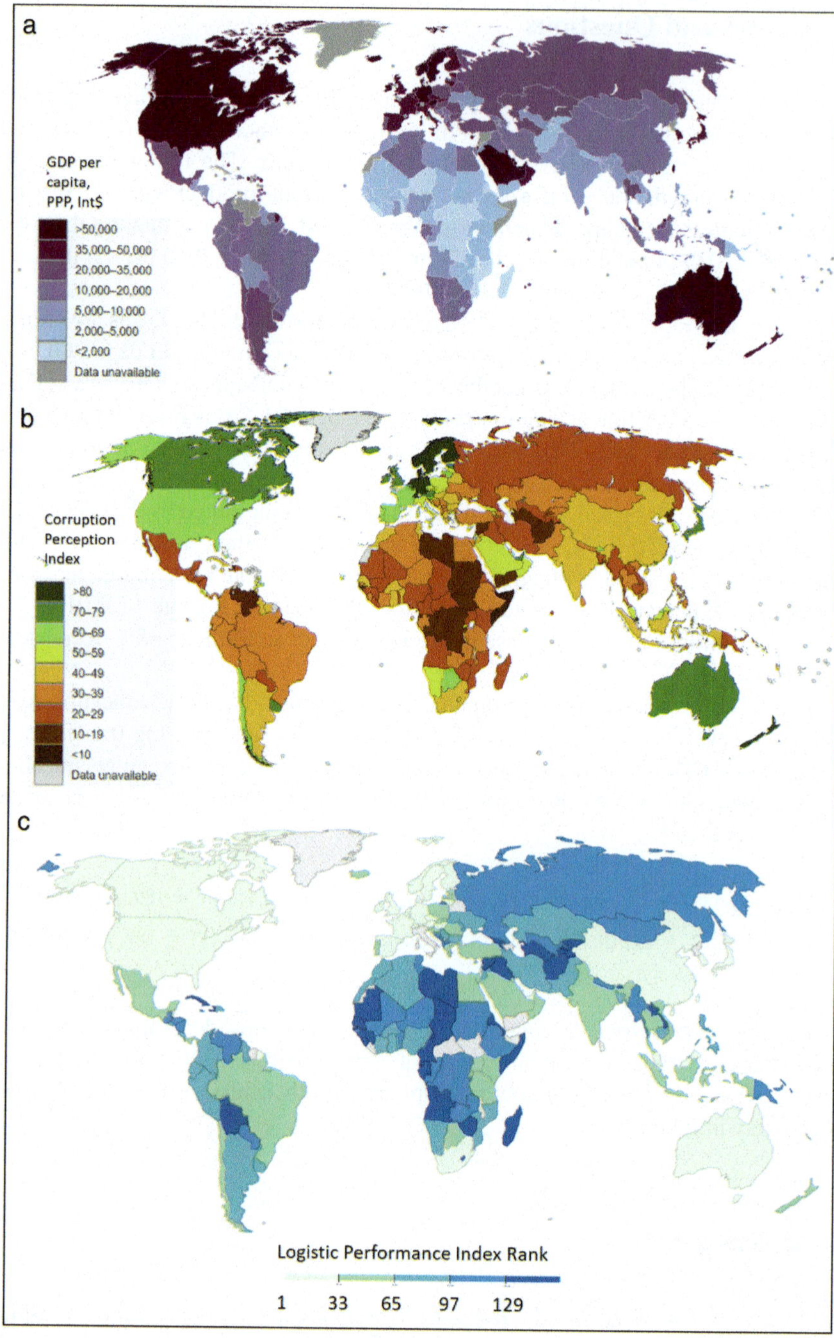

Fig. 3 **a** Per Capita GDP (Int$, PPP) according to IMF, reprinted from [26], CC license. **b** Corruptions perception index, as defined by [24], *source* Reprinted from [25], CC license. **c** World bank logistic performance index, reprinted from [27], CC license

In addition, there appears to be distinct dynamics in distinct scaling regions. For example, the high GDP tail includes both large, industrialized economies and trading economies such as Qatar and Luxembourg. Four countries which moved from the low of middle GDP scaling regions to the high per capita GDP scaling region, China, South Korea, Sri Lanka and Taiwan developed highly industrial economies. Equatorial Guinea apparently moved from the low per capita GDP scale to the middle scaling region from oil exports. Notable collapses included Gabon, Venezuela and Libya [30].

Although we have not identified an economic explanation, we note that addressing causes for break points between scaling regions may lead to identification of opportunities for economic growth.

In future work, we plan to consider evolution of the distribution of per capita GDP over time and identify potential changes due for example to industrial revolutions, wars, the structure of innovations, and pandemics.

Acknowledgements We thank our students Cole Peterson, Thomas Wang and Peter Xu for help in examining the data.

References

1. Garlaschelli, D., DiMatteo, T., Aste, T., Caldarelli, G., Loffredo, M.L.: Interplay between topology and dynamics in the world trade web. Eur. Phys. J. B **57**, 159–164 (2007)
2. West, G.: Scale: The Universal Laws of Growth, Innovation, Sustainability, and the Pace of Life in Organisms, Cities, Economies, and Companies. Penguin Press, New York (2017)
3. Duboff, R.: Accumulation and Power: Economic History of the United States. Routledge, New York (1989)
4. Dos Santos, P.: The Principal of Social Scaling. Complexity **2017**, 8358909 (2017)
5. Smith, A.: An Inquiry into the Nature and Causes of the Wealth of Nations. University of Chicago Press, Chicago (1977)
6. Braudel, F.: Capitalism and Civilization. Harper & Row, New York (1984)
7. Arrighi, G.: The Long Twentieth Century: Money, Power, and the Origins of our Times. Verso, Brooklyn, New York (2010)
8. Tinbergen, J.: Shaping the World Economy: Suggestions for an International Economic Policy. Twentieth Century Fund, New York (1962)
9. Poyhonen, P.: A tentative model for the volume of trade between countries. Weltwirtschafriches Archiv. **90**, 93–99 (1963)
10. Weber, M.: General Economic History. Dover, New York (2003)
11. Montroll, E.W., Shlesinger, M.F.: On 1/f noise and other distributions with long tails. Proc. Natl. Acad. Sci. **79**(10), 3380–3383 (1982)
12. Solomon, S., Richmond, P.: Power laws of wealth, market order volumes and market returns. Phys. A **299**, 188–197 (2001)
13. Solomon, S., Richmond, P.: Stable power laws in variable economies; Lotka-Volterra implies Pareto-Zipf. Eur. Phys. J. B-Condens. Matter Complex Syst. **27**, 257–261 (2002)
14. Reed, W.J., Hughes, B.D.: From gene families and genera to incomes and internet file sizes: why power laws are so common in nature. Phys. Rev. E **66**(6), 067103 (2002)
15. Mitzenmacher, M.: A brief history of generative models for power law and lognormal distributions. Internet Math. **1**, 226–251 (2004)

16. Yakovenko, V.M.: Econophysics, statistical mechanics approach to. In: Encyclopedia of Complexity and Systems Science, pp. 2800–2826. Springer, New York (2009)
17. Yakovenko, V.M., Rosser, J.B., Jr.: Colloquium: Statistical mechanics of money, wealth, and income. Rev. Mod. Phys. **81**, 1703 (2009)
18. List of countries by past and projected GDP (PPP) home page. https://en.wikipedia.org/wiki/List_of_countries_by_past_and_projected_GDP_(PPP). Accessed 31 May 2020
19. List of countries by past and projected GDP (PPP) home page. https://en.wikipedia.org/wiki/List_of_countries_by_past_and_projected_GDP_(PPP)_per_capita. Accessed 31 May 2020
20. Zhang, J. What's the secret of China's growth? https://www.weforum.org/agenda/2014/12/whats-the-secret-of-chinas-growth. Accessed 31 May 2020 (2014)
21. The World Bank in China. https://www.worldbank.org/en/country/china/overview. Accessed 31 May 2020
22. The World Bank in Sri Lanka. https://www.worldbank.org/en/country/srilanka/overview. Accessed 31 May 2020
23. The World Bank In Equatorial Guinea. https://www.worldbank.org/en/country/equatorialguinea/overview. Accessed 31 May 2020
24. Transparency International home page. https://www.transparency.org/en/cpi
25. Corruption Perceptions Index map. https://en.wikipedia.org/wiki/Corruption_Perceptions_Index#/media/File:Countries_by_Corruption_Perceptions_Index_score_(2019).png, Accessed 31 May 2020
26. Countries by GDP map. https://upload.wikimedia.org/wikipedia/commons/f/fb/Countries_by_GDP_%28PPP%29_per_capita_in_2019.png, Accessed 31 May 2020
27. Logistics Performance Index The World Bank. https://datacatalog.worldbank.org/dataset/logistics-performance-index. Accessed 31 May 2020
28. Inequality and GDP per capita, Pubdocs.worldbank.org. https://pubdocs.worldbank.org/en/755201504498011731/inequality-and-growth-3-september-2017.pdf. Accessed 31 May 2020
29. Human Development Index and its components. https://hdr.undp.org/en/content/table-1-human-development-index-and-its-components-1. Accessed 29 September 2018 (newer similar data 2020/5/31)
30. Kuperman, A.J.: Obama's Libya debacle: how a well-meaning intervention ended in failure. Foreign Aff **94**(2), 66–77 (2015)

Process Mining Organization Email Data and National Security Implications

John Bicknell and Werner Krebs

Abstract Many think of processes as sequential, deliberate activities which sustain businesses and government agencies; employees integrate themselves into defined organizational processes. From an ecosystem vantage, however, emergent processes exist and are discoverable. Emergent ecosystems form without human intention and may be especially influenceable. If emergent organizational processes–especially critical infrastructure processes–were explicit, they may be exploited. Tremendous intelligence is contained within semi-structured and unstructured organizational data sources. Properly analyzed, these data provide government and private organizations with actionable management and risk mitigation insights. Using explainable process technologies combined with natural language processing, a private critical infrastructure participant's organizational process model is discovered from semi-structured email data. Data derived from the process model are presented which elucidate internal operations and contribute to automated situational awareness of dynamically evolving events. National security implications and future research needs are described.

Keywords Process mining · Information warfare · Critical infrastructure · Organization modelling

1 Introduction

Many people think of processes as intentional groups of activities which serve profit maximizing businesses or mission driven government agencies. While this is true, information ecosystems and complex processes are also emergent and self-organizing with no human intention [1, 2]. Processes underlie all complex naturally occurring phenomena; indeed, the Earth is a process [3]. Process technologies

J. Bicknell (✉)
More Cowbell Unlimited, Inc., Lake Oswego, OR 97034, USA
e-mail: john@morecowbellunlimited.com

W. Krebs
Acculation, Inc., Los Angeles, CA 90036, USA
e-mail: wkrebs@acculation.com

D. Braha et al. (eds.), *Unifying Themes in Complex Systems X*, Springer Proceedings in Complexity, https://doi.org/10.1007/978-3-030-67318-5_15

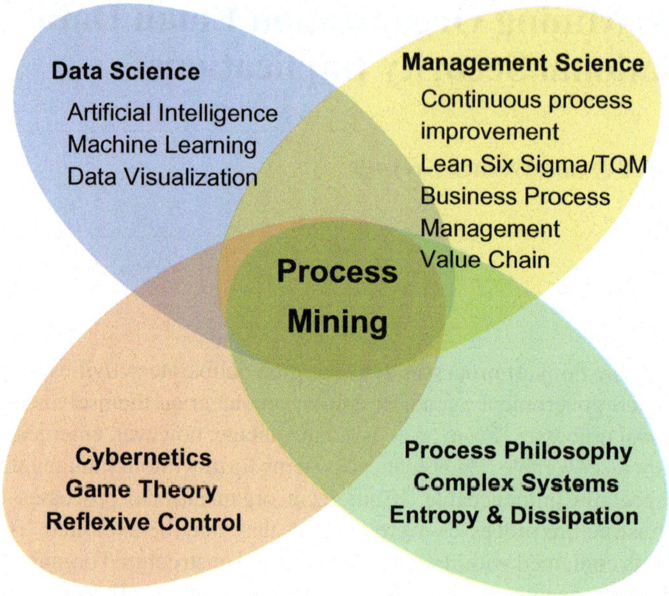

Fig. 1 Process mining is highly interdisciplinary. *Source* Authors

exist today which are an effective means to analyze time-domain relationships. As Fig. 1 suggests, process mining is inherently interdisciplinary–providing additive capability to data scientists, process improvement experts, operations researchers, organizational design practitioners, and national security analysts.

Process mining has its modern roots in Western Europe and was conceived originally in a business process improvement context. Data science and management science practitioners, Fig. 1, use process mining to speed the discovery of business process models while improving accuracy and preparing organizations for digital transformation [4]. Process mining is a highly versatile technique with vast utility beyond corporate process improvement. Poorly understood or opaque processes are a national security intelligence gap. If corporate, critical infrastructure, government, and societal processes were visible and explicit, opportunities for exploitation become available. Combined with other analytic disciplines such as game theory and complex systems theory (again, Fig. 1), process mining may inform critical infrastructure defense, information warfare (IW) vulnerability detection, network fragility analysis, geosurveillance, and many other analyses of national security importance.

Organizational process mining projects typically use structured enterprise resource planning system logs or other structured system data which record explicit activities, by specific employees, and at specific times. For example, a talent hiring process might include the following activities: Create Job Requisition; Post Job Announcement; Conduct Phone Screening; Conduct Phone Interview; Conduct On-site Interview; Extend Offer; Accept Offer. These process steps are defined frequently in structured system tables and are suitable for process mining with relatively little pre-processing.

In this study, we use a simple natural language processing (NLP) AI classifier to convert non-event email data into an IEEE standards compliant event log XES format [5], which facilitates process mining. Unstructured or semi-structured data must be pre-processed using an AI adapter. Previous email process mining efforts sought to understand business processes with a desire to improve business operations [6, 7] or to automate workflow discovery [8]; our research focuses on using business email data in national security contexts such as IW and critical infrastructure vulnerability detection.

Many defined business processes, as described earlier, produce Markov models which are absorbing. Our analysis is considerably different, although it uses the same IEEE data standard. We analyze an informal, emergent email communication process which is non-absorbing and ergodic; in other words, the process has no definitive beginning nor ending, and it is characterized by cybernetic looping behavior. Unlike previous email process mining efforts which required a tag in the subject line to identify a process activity [6] or which examined the entire email text for keywords [7], our intent is to show that an adversary may discern process information merely from untagged subject lines. Email subject lines often are considered less-sensitive than the full email and appear in SMTP log files.

Unstructured corporate data–including emails–contain significant business intelligence. It is therefore desirable to analyze email data in order to understand how a business or government agency might improve operations, identify vulnerabilities, and automate situational awareness of dynamically evolving events. In this paper, we present a process mining project which modelled an emergent corporate process using email data and process simulation technologies [9, 10]. We then discuss implications and suggest follow-on efforts.

1.1 Methodology

This study uses process mining, a nascent and powerful technique, to analyze the publicly available Enron email dataset [11].

The goal of process mining is to turn event data into insights and actions [12]. Process technologies elucidate ecosystem information flows, decision-making probabilities and temporal measures. Machine-readable process outputs and models of complex natural phenomena enable numerous applications. As the name implies, process mining AIs "mine" data and surface (e.g. Markov or Bayesian) models of decision-making processes from various input formats with no a priori knowledge. Process mining is also a human-understandable, human-verifiable, and human-explainable AI.

Structured or unstructured data which chronicle events are usable–cyber security logs, telemetry, information systems, multispectral imaging, social media data [13], etc. Three pieces of information are needed to discover processes; additional features enrichen the analysis:

- Case ID: An identifier that represents a specific execution of a process.
- Activity: One of several steps performed within a process. For example, cyber exploit attempts or tweet hashtags.
- Time Stamp: This orders the activities within each case and enables sophisticated modeling.

Figure 2 contains a trivial example where an event log is converted into a temporal process model. Process cases 1 through 3 all contain the same activities, labeled "A" through "E." In Case 1, the activities happen in natural order. In Case 2, Activity "C" precedes "B." Finally, in Case 3, Activity "D" is repeated before concluding with Activity "E." The discovered process accounts for these process variations. Real world processes are significantly more complicated.

Process models are temporal representations of events organized by process case. The process mining software used in this analysis creates a machine-readable business process model notation (BPMN) diagram, calculates various Markov transition probability matrices, descriptive statistics associated with the process, capacity estimates for process activities, and distribution probabilities associated with activity state arrival data.

CASE	ACTIVITY
1	A
2	A
1	B
1	C
3	A
2	C
3	B
2	B
1	D
2	D
2	E
3	C
3	D
1	E
3	D
3	E

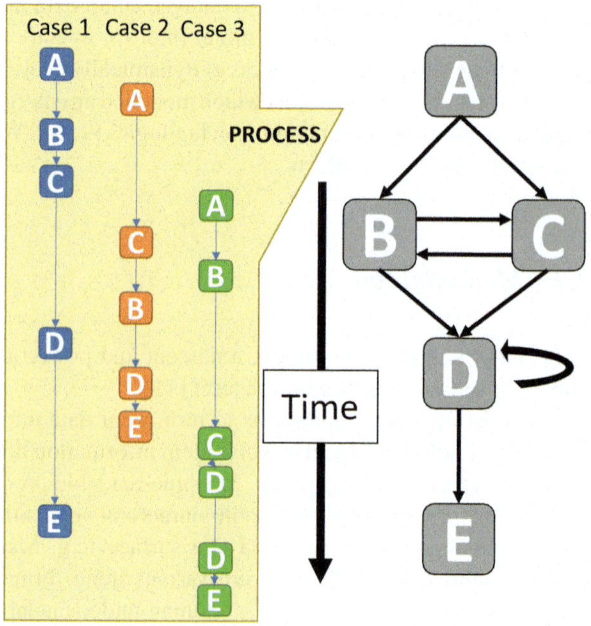

Fig. 2 Trivial process mining example

Fig. 3 Transition probability matrix state 1 → 2. *Source* Authors

State 1 \ State 2	A	B	C	D	E
A		0.67	0.33		
B			0.67	0.33	
C		0.33		0.67	
D				0.25	0.75
E					

The model presented in Fig. 2 may also be represented algebraically as a Markov transition probability matrix, Fig. 3. Information flows from beginning process activity states are expressed as probabilities. The beginning state is labelled "State 1" and is contained in the matrix rows; the ending state is labeled "State 2" represented by the matrix columns. Referencing the same simple event log contained in Fig. 2, information flows from Activity "A" to Activity "B" two-thirds of the time, and from Activity "A" to Activity "C" one-third of the time. When starting at Activity "D," there is a 25% probability that Activity "D" repeats and a 75% likelihood that Activity "E" follows Activity "D." The software used in this analysis uses a heatmap color coding scheme to visualize relative values and to suggest noteworthy findings.

Event log quality and richness varies greatly. The concept of a process sojourn is important. A sojourn comprises the path of a case between activities. Sojourn time may be measured from an event log and represents the entire activity wait and activity service times combined, Fig. 4. A sojourn for a case may proceed from Activity "A" to Activity "B", and so the sojourn time may be calculated as the Activity "B" End Timestamp minus the Activity "A" End Timestamp. If service start times are available, then explicit wait and services time calculations are possible. In a business context, sojourn measures describe waiting queues and service times for defined operational processes such as: parts manufacturing, loan processing, ecommerce webform user experience, or employee onboarding. Since this analysis uses email data, the only available sojourn measure is each email's send timestamp; therefore, it is impossible to calculate directly the sojourn wait and service times. So, we use the time between emails as a rough proxy for how much time is being spent on a

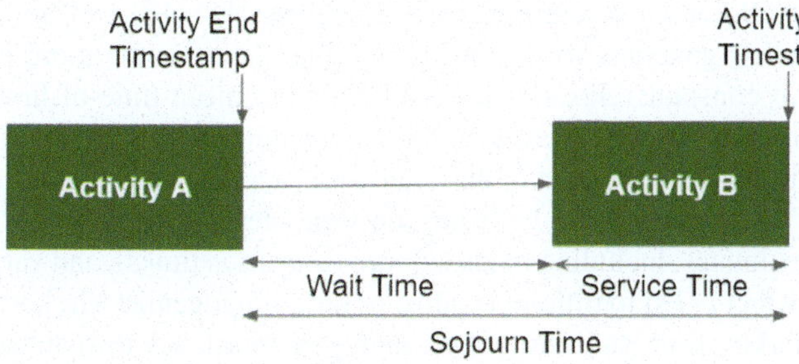

Fig. 4 Deconstructed process activity

task related to that email. This is an important consideration for estimating capacity of process activities, explained next.

Capacities for process activities may be estimated one of two ways. First, activity capacity may be estimated using simple queuing theory (SQT) comprising arrival rate and service rate estimates for activities derived from the event log; an alternative method for estimating capacity of an activity from the event log involves computing fractional spectral downshift in a frequency-domain representation of the event log. In this paper, we use the first method for estimating process activity capacities. Capacity is estimated using heuristics derived from SQT. Making a number of hand-waving assumptions, SQT-based heuristics look at the average number of tasks running in parallel together with sojourn times to estimate the number of process activities likely in a wait state at any given time; from this queue lengths, wait states, service times, and capacity may be estimated. We hypothesized that our very simple NLP email subject line event analysis, although noisy, would nevertheless serve as a rough proxy for the tasks employees were engaged in at any time. The resulting timestamps, in turn, when analyzed with SQT, in turn, provide a useful rough ranking of where employee bandwidth might be at a premium.

Descriptive statistics such as average, median, mode, skew, and kurtosis describe process sojourn times associated with state transitions. Probability distribution estimates with labels such as "Gaussian distribution," "extreme value distribution," "uniform distribution," "Poisson distribution," "platykurtic normal distribution," and so on are derived by applying heuristics to the statistical characteristics.

These process measures may be used as data-driven input parameters to simulate the organization. The analytic expression for the capacity parameter, defined above, is accurate under the assumption that there is only a single worker or single automated process processing events in the workflow. The expression, however, provides at least an estimate of capacity or utilization that is often directionally accurate as a simulation input parameter. Similarly, probability distribution descriptions are assumed to be approximately accurate.

Process mining and social network analysis (SNA) are conflated easily. While both techniques have a strong network component, process mining is inherently temporal, as explained earlier. Process models, therefore, may be analyzed using many of the same centrality, distance, and entropy measures associated with SNA. We describe future research possibilities which should explore these possibilities. A number of well-known visualizations and analyses [14–16] can be applied to the Markov models generated from process mining. Some of the more interesting outputs (Kemeny constant, relaxation time, mixing time, mean-time-of-first-passage, entropy, and eigenvalues, when present) have time units in terms of Markov steps. The process mining software used in this analysis generates "time-inhomogeneous" or "multiscale" Markov models with uneven time steps along with various time measures and estimates, including mean sojourn (transition times), and various techniques are known [17–19] to time-normalize or time-homogenize Markov matrices, after which Markov time step units are roughly proportional to real world time. Any method that generates XES process log data, including data from previous process mining email efforts [6–8] can be subjected to our time-homogeneous, time-of-first-passage, capacity estimates, and resulting simulations.

Enron Corporation was an American energy, commodities, and services company based in Houston, Texas. It was founded in 1985. During the late 1990s, Enron was heralded as an American success story. Famously, Enron went bankrupt in late 2001 after accounting fraud and corporate corruption became public knowledge [20]. The Enron email dataset contains data from about 150 users, mostly senior management. The data contains a total of about 0.5 M messages and primarily spanned late 1998 through mid-2002 as Enron's collapse unfolded (although some spurious emails as well as incoming emails have timestamps outside this range). These data were originally made public, and posted to the web, by the Federal Energy Regulatory Commission during its investigation [11].

There are numerous ways to organize unstructured or semi-structured data for process mining. Different process event log configurations offer various temporal perspectives of information flow through an organization. For example, external email subject lines may be used as the process case in order to understand how external information flowing into an organization is handled. If department, team, or supervisory work group information is available, then these data provide aggregate views which offer comparisons between organizational sub-groups. For this analysis, the sender's and recipient's email addresses were used as CaseIDs (so that each retained email appeared as at least two events; bulk emails would appear as multiple events). A portion of the event log is contained in Fig. 5.

Process activities were derived using a simple, case-insensitive bag-of-words model [21]. We constructed a simple NLP classifier for the emails in Enron's dataset by looking at approximately the top-10, top-20, and top-40 non-stop-word email subject keywords. Emails were classified into unique activity categories (assumed to be approximately correlated with employee tasks) based on the first such top-10, top-20, or top-40 keywords that appear in the subject line. Emails with no non-stop-word subject lines, or emails which had subject lines not containing a top keyword, were dropped. An alternative, which was also attempted, is to classify such emails

	A	B	C
1	**Case ID**	**Start Time**	**Activity**
2	a..[redacted]@enron.com	2001-04-19T15:32:27-07:00	meeting
3	a..[redacted]@enron.com	2001-05-04T10:15:30-07:00	customer
4	a..[redacted]@enron.com	2001-05-21T10:23:25-07:00	meeting
5	a..[redacted]@enron.com	2001-05-29T15:30:24-07:00	SHOUT
6	a..[redacted]@enron.com	2001-05-31T05:39:42-07:00	SHOUT
7	a..[redacted]@enron.com	2001-05-31T09:03:24-07:00	SHOUT
8	a..[redacted]@enron.com	2001-06-01T14:29:55-07:00	address
9	a..[redacted]@enron.com	2001-06-05T13:15:46-07:00	meeting
10	a..[redacted]@enron.com	2001-09-28T14:11:27-07:00	tickets

Fig. 5 Enron email event log excerpt

as ‘other’; this resulted in data which was deemed too noisy. In some of our data, we made one exception to this rule: noting that Enron frequently sent out emails with ALL-CAPS subject lines as a way of indicating an urgent email, if an email subject line was in ALL-CAPS (except for common control stop words like ‘Re:’), this was assumed to be a proxy for a crisis or urgent matter, and designated as belonging to the “SHOUT” process activity.

Process mining techniques include a number of filtering strategies for dealing with noisy datasets. Not all “SHOUT” Enron emails were crisis related; some were merely social events. Filtering process logs to retain only a top set of the most common activities is a heuristic referred to as an “activity filter” in the process mining literature. The subject line keyword in the last email sent or received by an employee with the top-keyword-set was assumed to be a proxy for that employee’s current activity in our very simple NLP model. The idea was to demonstrate that even very simple, easily automatable NLP analyses of unstructured corporate data could yield interesting process mining results. If an employee sent or received multiple emails containing that keyword in a row (even if this activity was interspersed with emails not containing subject lines of interest which were removed, as described previously) the employee was assumed to still be engaged with the subject keyword line activity, and the duplicate records were deleted.

The Enron dataset includes emails from outside the company, a small but significant percentage of corrupted emails, and emails from different email software systems (Enron transitioned off of Lotus Notes during this period). To ensure our dataset only looked at emails originating from within Enron and avoided bulk mail duplicates, we limited our data to emails found in ‘sent-mail’ folders, which may have been specific to one email system. By limiting to emails in employee ‘sent-mail’ folders, we initially processed only a subset of the 0.5 M Enron emails, looking at 37,921 emails, of which 37,830 were deemed non-corrupt, resulting (after ‘activity filtering’ on top keywords), in 4,994 events (which represented emails spanning multiple employee/CaseIDs as described above). Since ‘sent-mail’ may have been an artifact of one email system used by Enron, for comparison purposes we also did a second analysis of the entire Enron email archive (including emails sent from outside Enron). This comparison run looked at 517,401 emails, of which 495,466 were deemed non-corrupt. After “activity filtering” to limit events to approximately the top-10 subject line keywords as described above, this dataset had 63,306 events. Despite including emails from outside Enron and a different date range, this larger but presumably noisier dataset resulted in reasonably similar process mining statistics as the smaller ‘sent-mail’-only dataset.

2 Results

Derived process models combined with post hoc manual analysis revealed insights about Enron corporate culture and how management dealt in email with recurring corporate issues.

The Markov model contained in Fig. 6 represents Enron corporate decision-making or information flow through the lens of general email exchanges. Decision patterns emerge among Enron employees which are unintended, yet observable. The matrix includes a "re-work" perspective which is observable as the diagonal values. Re-work in this context means that an employee sent consecutive emails with a similar contextual subject, which is interpreted as engagement.

Emails with all capitalizations ("SHOUT") in the subject line take up a significant portion of employee email dialogue, as evidenced by the high transition probabilities going into the "SHOUT" state from other states. Moreover, there is a 55% probability that Enron employees remain in a "SHOUT" email re-work state. Responding to "meeting" emails is another large resource expenditure as evidenced with relatively high inflow probabilities and 47% probability of sending consecutive meeting-related emails (re-work). "Urgent," "Vacation," and "Please" are all activities with relatively little re-work.

The software used in this analysis auto-generates a narrative summary of the process:

> This process model was created by analyzing 172 cases. The average case time is 116.48 days, and the median case time is 93.07 days. The min case time is 0.00 s. The max case time is 431.99 days. It appears that the case durations are normal skewed to the right. There were 4994 total events processed. The activities which were estimated to take the longest amount of time are "address" at 7.45 days and "hi" at 7.26 days. There was an average of 29.03 events per case, and a median of 15.0 events per case. There were a max of 292 events per case and a min of 1. The most frequent activities are "SHOUT" at 1780 sojourns and "meeting" at 1265 sojourns. The process activities with the highest capacity estimates are "SHOUT" and "meeting."

	Activity Counts	hello	hi	lunch	SHOUT	address	conference	customer	game	holiday	imbalance	meeting	office	party	please	settlement	tickets	training	urgent	vacation
hello	81	.27	.04		.36	.01	.03		.01		.01	.11	.04	.04	.03		.01	.01	.01	.01
hi	58	.06	.31	.02	.31	.07	.02					.09		.02	.04	.02		.02		.02
lunch	69		.05	.32	.21	.02	.09	.03	.02	.02	.05	.08	.02	.03	.02		.02	.03		.03
SHOUT	1780	.02	.01	.01	.55	.02	.04	.02	.01	.01	.01	.16	.02	.02	.03	.01	.01	.01	.01	.02
address	115		.01	.03	.32	.20	.05	.05	.02			.18	.02	.05	.04		.02		.01	.01
conference	329		.01	.01	.24	.03	.28	.04	.01	.01	.01	.23	.02	.03	.03	.01	.02	.01	.01	.02
customer	134	.02		.01	.33	.05	.05	.18		.01	.01	.22	.01	.02	.03	.02	.01		.02	.02
game	70		.03		.22	.06	.06		.28	.01		.18	.03	.04	.04			.01		.03
holiday	54	.06			.20		.04			.26	.02	.20	.06	.04	.04		.02		.02	.04
imbalance	107				.22	.02	.11		.01		.42	.14	.01	.02	.02		.01			.02
meeting	1265	.01			.21	.01	.07	.02	.01	.01	.02	.47	.02	.01	.03	.02	.02	.02	.01	.02
office	114	.04		.01	.24	.01	.04	.02	.01	.03	.03	.23	.17	.01	.07	.04	.03	.01	.01	.02
party	168	.01	.01	.02	.29	.01	.02	.03	.01	.01		.10		.42	.04	.01	.02	.01		.01
please	195	.02	.03	.01	.31	.02	.05	.02	.01	.01	.03	.22	.01	.03	.17	.03	.01		.04	.02
settlement	94	.01	.01	.01	.23		.07	.02	.02		.01	.26	.01		.04	.26	.01		.01	.01
tickets	107	.02			.20	.01	.06	.01	.02	.01	.02	.26	.03	.01	.04	.01	.25	.01	.01	.05
training	78		.01	.03	.29			.03	.03		.03	.25			.03	.01	.05	.22	.03	
urgent	64		.02	.02	.32	.02	.02	.05				.24	.02	.02	.05	.02	.03	.02	.16	.03
vacation	111	.01		.03	.22	.03	.05	.06	.01	.01	.03	.23	.01	.05	.04	.03	.03		.01	.18

Transition probabilities less than .01 are not displayed.

Fig. 6 Enron email transition probability matrix with re-work

By removing re-work process events, transition probabilities for the "SHOUT" and "Meeting" activity states are more pronounced, Fig. 7. In fact, all of the matrix transition probabilities are slightly higher than the results contained in Fig. 6. This makes sense, as the process case activity event counts are reduced by the amount of re-work and the matrix row divisors are adjusted accordingly. Comparing the results from the model with re-work removed with the previous matrix (Fig. 7), we observe significant percent changes in information flows into "SHOUT" from the "Meeting" activity state, and vice versa. There is also a 22% change in probability that Enron employees send a "SHOUT" email following a "Party" email.

> This process model was created by analyzing 172 cases. The average case time is 116.48 days, and the median case time is 93.07 days. The min case time is 0.00 s. The max case time is 431.99 days. It appears that the case durations are normal skewed to the right. There were 2985 total events processed. The activities which were estimated to take the longest amount of time are "settlement" at 15.10 days and "game" at 12.93 days. There was an average of 29.03 events per case, and a median of 15.0 events per case. There were a max of 292 events per case and a min of 1. The most frequent events are "SHOUT" at 843 observations and "meeting" at 677 observations. The process activities with the highest capacity estimates are "meeting" and "SHOUT."

We hypothesized that our very simple NLP email subject line event analysis, although noisy, would nevertheless serve as a rough proxy for the tasks employees were engaged in at any time and provide a useful ranking of where employee bandwidth might be at a premium. Figure 8 compares the capacity estimates for Enron's email corporate model with activity event counts. "SHOUT" and "Meeting" have the largest capacity estimates at 0.92 and 0.85, respectively. The lowest capacity estimates include: "Urgent" at 0.23, "Training" at 0.24, "Settlement" at 0.26, and

	Activity Counts	hello	hi	lunch	SHOUT	address	conference	customer	game	holiday	imbalance	meeting	office	party	please	settlement	tickets	training	urgent	vacation
hello	61		.05		.49	.02	.04		.02		.02	.15	.05	.05	.04		.02	.02	.02	.02
hi	41	.08		.03	.46	.11	.03					.14		.03	.05	.03		.03		.03
lunch	48		.07		.31	.02	.13	.04	.02	.02	.07	.11	.02	.04	.02		.02	.04		.04
SHOUT	843	.04	.02	.03		.05	.10	.04	.02	.03	.02	.35	.04	.05	.07	.03	.03	.02	.02	.04
address	93		.01	.03	.40		.07	.06	.02			.23	.02	.07	.05		.02		.01	.01
conference	240		.01	.01	.32	.04		.06	.01	.01	.01	.31	.03	.04	.04	.01	.03	.02	.01	.02
customer	111	.03		.01	.41	.06	.06			.01	.01	.26	.01	.03	.04	.03	.01		.03	.02
game	51		.04		.31	.08	.08			.02		.24	.04	.06	.06			.02		.04
holiday	41	.08			.27		.05				.03	.27	.08	.05	.05		.03		.03	.05
imbalance	63				.38	.03	.20		.02			.25	.02	.03	.03		.02			.03
meeting	677	.01	.01	.01	.41	.02	.14	.04	.02	.01	.03		.05	.02	.06	.04	.04	.04	.02	.04
office	96	.04		.01	.29	.01	.04	.02	.01	.03	.03	.28		.01	.08	.04	.03	.01	.01	.02
party	100	.01	.01	.03	.51	.02	.03	.05	.01	.01		.17			.06	.01	.04	.02		.01
please	163	.02	.03	.01	.37	.02	.07	.02	.01	.01	.03	.26	.01	.04		.03	.01		.05	.02
settlement	70	.01	.01	.01	.31		.09	.03	.03		.01	.36	.01		.06		.01		.01	.01
tickets	81	.03			.27	.01	.08	.01	.03	.01	.03	.34	.04	.01	.05	.01		.01	.01	.06
training	61		.02	.03	.37			.03	.03		.03	.32			.03	.02	.07		.03	
urgent	54		.02	.02	.38	.02	.02	.06				.28	.02	.02	.06	.02	.04	.02		.04
vacation	91	.01		.03	.26	.03	.05	.08	.01	.01	.03	.29	.01	.05	.04	.03	.03		.01	

Transition probabilities less than .01 are not displayed.

Fig. 7 Enron s email transition probability matrix re-work removed

Fig. 8 Enron email process activity capacity estimates compared to event counts

	Activity Counts	Capacity Estimate
SHOUT	1780	.92
meeting	1265	.85
please	195	.61
address	115	.57
conference	329	.56
party	168	.56
office	114	.51
tickets	107	.49
vacation	111	.49
customer	134	.41
hi	58	.40
game	70	.39
imbalance	107	.37
hello	81	.36
holiday	54	.31
lunch	69	.27
settlement	94	.26
training	78	.24
urgent	64	.23

"Lunch" at 0.27. Although there is a significant positive correlation between event counts and capacity estimates, $r = 0.83$, $n = 19$, $p < 0.001$, capacity estimates also looks at event timings and number of parallel tasks, so that "training" appears at significantly lower capacity than various social activities ("hi", "holiday", "lunch", "game") despite the former having higher or equal event counts (Enron was not noted for its compliance culture, so we might expect social activities to run at a higher capacity than "training" and "settlement", although it is even more surprising so many social-related keywords made it into the top keywords list. Given that "party" is above the median for both activity counts and capacity, it is perhaps not surprising that Fig. 7 shows a "SHOUT" response frequently following a "party" email, but not the reverse).

Skew, kurtosis, mean, variance and similar distribution statistics were applied together with simple heuristics to classify activity time interval probability distributions in the event logs, Fig. 9. For example, distributions with an absolute value or skew or excess kurtosis greater than 2 that are not classified as log normal are classified above as 'Extreme Value.' Similar heuristics classify distributions as likely 'Normal or Poisson', "Log normal', 'Leptokurtic normal', 'Platykurtic normal', 'Discrete', or 'Uniform.' These classifications can provide insights into the nature of processes and serve as data-driven simulation parameters. For example, processes with normally distributed sojourn times likely involve manual labor,

	hello	hi	lunch	SHOUT	address	conference	customer	game	holiday	imbalance	meeting	office	party	please	settlement	tickets	training	urgent	vacation
hello	4	4		4	7	1		7		7	4	6	6	1		7	7	7	7
hi	4	4	7	4	6	7					6		7	1	7		7		7
lunch		6	4	4	7	6	4	7	7	4	6	7	1	7		7	1		4
SHOUT	4	4	6	2	4	4	4	4	4	4	4	4	4	4	4	4	4	2	4
address		7	4	4	4	3	5	1			4	1	5	6		1		7	7
conference		1	4	4	4	4	5	4	6	4	4	6	5	4	6	4	6	6	5
customer	6		7	4	5	6	4		7	7	4	7	6	6	4	7		6	1
game		1		4	6	4		4	7		3	1	4	6			7		1
holiday	6			4		4			4	7	4	6	1	1		7		7	4
imbalance				4	1	4		7		4	4	7	1	1		7			1
meeting	4	6	4	4	4	4	4	4	4	4	2	4	6	4	4	4	4	6	4
office	6		7	4	7	4	1	7	6	6	4	4	7	4	4	4	7	7	1
party	7	7	6	4	1	6	5	7	7		4		4	3	7	4	1		7
please	6	4	7	4	6	4	4	7	7	4	4	1	4	4	5	7		4	4
settlement	7	7	7	4		4	1	1		7	4	7		4	4	7		7	7
tickets	1			5	7	6	7	1	7	1	4	4	7	6	7	4	7	7	6
training		7	1	4			1	4		1	4			1	7	6	4	1	
urgent		7	7	4	7	7	6				4	7	7	4	7	1	7	3	1
vacation	7		4	4	6	5	3	7	7	4	4	7	6	6	4	4		7	4

Legend:
1 = Discrete, 2 = Extreme value, 3 = Leptokurtic normal, 4 = Log normal, 5 = Normal, 6 = Platykurtic normal, 7 = Uniform

Fig. 9 Enron email probability distribution matrix

whereas processes with uniform or discrete time distributions may involve automated processes or indicate insufficient sample sizes. Processes near capacity often show log normal, leptokurtic or extreme value time distributions, likely due to employees utilizing triage to prioritize work queues. The Enron email process activities with the highest capacity estimates, "SHOUT" and "Meeting," are both estimated to have "Extreme Value" distributions.

"Normal modes" are a useful classifying characteristic in a number of dynamical systems [22, 23] and are associated with Markov processes when eigenvalues are present [23]. It is therefore reasonable to expect corporate processes with interesting eigenvectors to have normal modes, which might correspond to a steady-state-response to crisis (e.g. "SHOUT") and more relaxed ("game") corporate modes of operation. Eigenvalues have units of frequency, or inverse Markov step time, so a time-homogeneous Markov matrix may be more relevant to eigenvalue analyses. Not all corporate event traces will have interesting normal modes; typical corporate order flows have "absorbing" Markov Matrices without interesting loops and thus without interesting eigenvalues. However, our Enron-derived Markov chains have many looping processes, consequently they are ergodic and appear to have multiple interesting eigenvalues, which can be illustrated on a (time-inhomogeneous) eigen plot, Fig. 10.

Fig. 10 Enron email process activity eigen plot

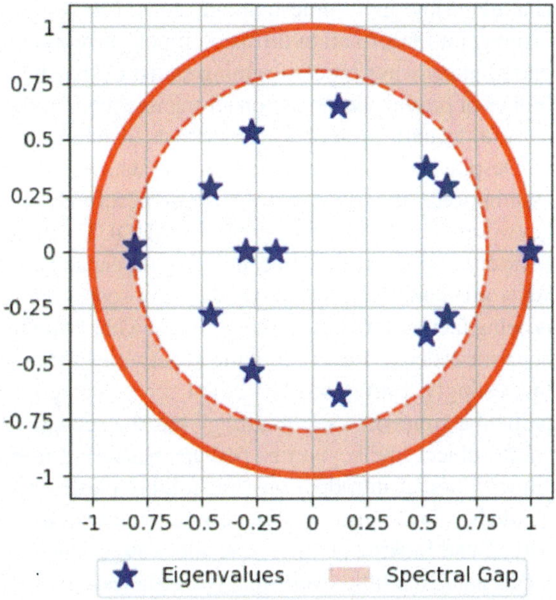

3 Discussion

Email process mining results using simple NLP AI model the flow of information and suggest Enron has a "meetings" culture and presumably has limited bandwidth for its total meetings and events. Based on our email subject line analysis, Enron meetings fall into different categories: trainings, client calls, conferences, networking or social event planning, and what appear to be crisis management and internal escalation meetings, all of which spawn each other at different rates. A crisis management meeting makes a training or conference call email, as well as future crisis management or internal escalation meetings more likely. Similarly, friendly meetings around networking or social events seem to make other social event-related meetings more likely. Thus, observation of the different categories of meetings impacts the probabilities of future instances of other meeting categories. A crisis-related meeting makes crisis-response and training meetings more likely as the company continues to respond to an ongoing crisis, whereas a social event meeting makes follow up social event meetings more likely as the company presumably relaxes and shifts "meeting bandwidth" towards these friendlier events.

Multiple process loops are present, resulting in a Markov matrix that is non-absorbing and ergodic with chain and topological entropies, relaxation time, and eigenvalues present. Since the dataset often shows switching between urgent ALL-CAPs ("SHOUT") subject lines and more relaxed social tasks, the Markov matrix presumably encodes data about Enron's response to various corporate crises and subsequent relaxation from those crises. Thus, eigenvectors, relaxation times, and chain-entropy in a time-homogeneous Markov Matrix can potentially provide insights into information diffusion times and corporate response to external stimuli, one of many ways process-mining statistics may be used to infer corporate vulnerabilities or provide business consulting insights. Markov chain walk plots also provide a useful visualization of corporate task switching within our model.

Despite the noise in the underlying dataset and violations of single server SQT assumptions, the capacity algorithm nevertheless tends to identify work-related keywords such as "meeting", urgent "SHOUT"-related tasks, and other work-related keywords as among the top candidates for capacity drains, whereas the social-related keywords tend to receive lower capacity scores. Note that, because task parallelism is being factored in, this ranking is somewhat different than one achieved via simpler heuristics such as merely looking at activities with top counts or long-running tasks. Interestingly, Enron was not noted for its compliance culture, and at least in this particular time-slice set "training" has the second lowest capacity score of the top email subject line keywords, perhaps because it is lower priority than "settlement", "lunch," social introductions, or holiday parties. (In fairness to Enron, "training" would be a lower priority at many companies compared to "customer" if not "[social] party", and looking at other time slices suggest Enron's priorities in this regard may have changed a little over time. Nevertheless, given Enron's history, it is interesting that "training" scored quite low here despite considerable activity counts and sojourn times spent on the activity). Despite the low capacity score for "training", it is worth noting that many other keyword-activities in this same set would be lower ranked when merely looking at simpler criteria, such as raw activity counts or sojourn times. These results suggest that SQT analysis may add value as a useful ranking heuristic even on noisy data where SQT assumptions are violated. The software used in this analysis is also able to combine the Markov transition results with its activity timing estimates to automatically generate more accurate agent-based simulations free from the restrictions of SQT assumptions.

Conceptually, Enron employees are likely doing many tasks in parallel, but due to lack of data granularity and because this project was a first effort, we modeled Enron employees as working on one email task (process activity) at a time, even if there were minor distractions in between. Despite this simplification, our consistent approach appears to present a correct approximate picture of Enron corporate operations. Using a simple NLP classifier and simplifying assumptions about employee processing of work, this project constructed a process mining event log from semi-structured corporate email data. Post hoc analysis suggests promise for future research.

4 Implications and Future Work

In this section, we describe national security implications and future research considerations.

Government agencies and American corporations–especially critical infrastructure participants–must assume intelligent, persistent, and resourced adversaries are gathering intelligence in order to launch insidious and potentially devastating IW attacks on the homeland [24–27]. Since the 1960s, Russia has enhanced IW with systematic psychological understandings of adversary reflexive processes and continues honing the technique. Known as reflexive control (RC), this highly analytical method has roots in cybernetics and game theory [24, 28, 29] and is a means of conveying to a partner or an opponent specially prepared information to incline him to voluntarily (or reflexively) make a predetermined decision desired by the initiator of the action [30].

IW enhanced with RC extends beyond human-to-human interactions; in fact, human-to-human IW RC is a minority use case in the Information Age. Today, entire societies and all public/private critical infrastructure participants must thwart human–machine and machine-machine IW attacks [25]. IW RC may be deployed with devastating effect against machines, information systems, and physical infrastructures [30]. Mapping of decision-making patterns, also known as "information cartography" [31], is a challenging but still achievable task. It is the knowledge of patterns within the decision-making process that allows an adversary to insert information into the process that would ultimately allow manipulation of the decision [32]. As such, all aspects of critical infrastructure ecosystems should be analyzed using a process-centric lens.

The informal, emergent corporate process contained in this study, for example, suggests ways information may be injected into the Enron ecosystem in order to disrupt operations, distract executives, or drain resources with frivolous activity. The processes discovered in this study represent probabilistic flows of information in the Enron corporate ecosystem. Future work should build upon this study and develop creative measures and simulations for elucidating process ecosystems further. Understanding ecosystem dynamics empirically answers operational questions of interest such as: how long will it take for specific information to reach a specific person, what is the organizational rhythm or tempo, how will the ecosystem react when information is injected at various entry points [33], how do adjacent ecosystems interact, and how long does it take for an ecosystem to recover? Measures such as the Kemeny constant, relaxation time, mixing time, mean-time-of-first-passage, entropy, and eigenvalues should all be considered. Time-normalized or time-homogeneous Markov matrices will likely provide additive ecosystem intelligence, as well.

Process models representing contextual temporal decision flows through organizations contain substantial intelligence from which organizational vulnerabilities may be inferred and mitigation strategies developed. Derived vulnerabilities may include reflexive reactions to deliberate or incidental stimuli (red team cyber-attack provocations, major financial upheavals, or weather disasters, for example), decision network fragilities/network points of failure, and diffusion of information both contextually and temporally. Future research may reveal important trends with profound national security implications. If enough organizations are studied and processes catalogued, generalized reflexive response trends to certain stimuli may emerge and be exploited in various ways such as using information as a maneuver element in a larger physical or cyber kill chain. New social engineering techniques may be developed as well which are, again, generalizable. Future research should include manipulating

Markov models in various ways to understand ecosystem perturbations empirically, simulating ecosystems to understand and predict information flows, investigating concepts like entropy warfare, entropy transfer, or directed entropy [34], developing cyber-attack/defense software which accepts machine-readable process model outputs [24, 25] to understand vulnerabilities through a new lens, and developing mitigation strategies which do not affect organizational operations.

As discussed, pre-processing tools are required for creating event logs; this is especially true for unstructured or semi-structured data. The simple bag-of-words NLP classification combined with human analyst adjustments methodology used in this project extracted keywords from email subject lines. New pre-processing AI adapters are necessary, however, in order to maximize future research. We hypothesize that useful process mining insights could be derived from the email-derived event logs using fully-automated contextual NLP AI adapters. If simple NLP techniques yield interesting results, imagine what can be done with more sophisticated natural language understanding (NLU) technologies, such as unsupervised document clustering, TensorFlow-based open source document NLU classification systems like Rasa, or proprietary systems like IBM Watson. For example, email bodies, email attachments, and entire corporate data systems are a much richer and untapped data source for process technologies. Process mining which combines customized contextual NLP from various perspectives (human resources, operations, finance, logistics, marketing, legal, etc.) enables comprehensive multi-dimensional organizational elucidation with greatly enhanced possibilities for whole-business optimization simulations and critical infrastructure vulnerability analyses.

5 Conclusion

This paper reported the findings of a novel application of organizational process mining. Enron corporate emails were analyzed after applying a simple NLP methodology to extract keywords from the email subject lines. The technique yielded interesting results and tells a logical post hoc story of internal operations which warrant further research to understand national security and organizational management implications.

References

1. Prigogine, I., Nicolis, G., Babloyantz, A.: Thermodynamics of evolution. Phys. Today **25**(11), 23–28 (1972). https://doi.org/10.1063/1.3071090
2. Hidalgo, C.: Why Information Grows: The Evolution of Order, from Atoms to Economies. Basic Books (2015)
3. Whitehead, A.N.: Process and Reality, 2nd edn. Free Press, New York (1979)
4. Bicknell, J.W.: Process mining technologies. ORMS Today **46**(5) (2019). https://doi.org/10.1287/orms.2019.05.01

5. IEEE: IEEE Standard for eXtensible Event Stream (XES) for Achieving Interoperability in Event Logs and Event Streams. IEEE Std 1849-2016, pp. 1–50, Nov 2016. https://doi.org/10.1109/IEEESTD.2016.7740858
6. van der Aalst, W.M.P., Nikolov, A.: EMailAnalyzer : an e-mail mining plug-in for the ProM framework (2007)
7. Jlailaty, D., Grigori, D., Belhajjame, K.: A framework for mining process models from emails logs. ArXiv. abs/1609.06127 (2016)
8. Allard, T., Alvino, P., Shing, L., Wollaber, A., Yuen, J.: A dataset to facilitate automated workflow analysis. PLoS ONE **14**(2), e0211486 (2019). https://doi.org/10.1371/journal.pone.0211486
9. Bicknell, J.W., Krebs, W.G.: Methods and systems for estimating process capacity. United States 10,846,194, issued November 24, (2020)
10. Bicknell, J.W., Krebs, W.G.: Methods and systems for inferring behavior and vulnerabilities from process models. U.S. Patent Application No. 16/440,639. Washington, DC: U.S. Patent and Trademark Office
11. Enron Email Dataset, 18 May 2015. https://www.cs.cmu.edu/~./enron/. Accessed 12 Mar 2019
12. van der Aalst, W.M.P.: Process Mining: Data Science in Action, 2nd edn, 2016 edn. Springer, New York, NY (2016)
13. Bicknell, J.W., Krebs, W.G.: Detecting botnet signals using process mining (2019)
14. Khinchin, A.Y.: Mathematical Foundations of Information Theory, 1st Dover edn. Dover Publications, Mineola, NY (1957)
15. Kemeny, J.G., Snell, J.L.: Finite Markov Chains: With a New Appendix "Generalization of a Fundamental Matrix." Springer, New York (1976)
16. Belluzzo, T.: A framework for discrete-time Markov chains analysis (2019)
17. Feldman, J.F., Roberge, F.A.: The normalized transition matrix. A method for the measure of dependence between inter-spike intervals. Electroencephalogr. Clin. Neurophysiol. **30**(1), 87–90 (1971). https://doi.org/10.1016/0013-4694(71)90209-4
18. Hanks, E.M., Hooten, M.B., Alldredge, M.W.: Continuous-time discrete-space models for animal movement. Ann. Appl. Stat. **9**(1), 145–165 (2015). https://doi.org/10.1214/14-AOAS803
19. Bielecki, T.R., Cialenco, I., Gong, R., Huang, Y.: Wiener-Hopf factorization for time-inhomogeneous Markov chains and its application (2018). https://arxiv.org/abs/1801.05553. Accessed 02 Oct 2019 [Online]
20. McLean, B., Elkind, P.: The Smartest Guys in the Room: The Amazing Rise and Scandalous Fall of Enron, Reprint Portfolio Trade, New York (2004)
21. Bag-of-words model. Wikipedia, 07 Sep 2019. https://en.wikipedia.org/w/index.php?title=Bag-of-words_model&oldid=914476224. Accessed 18 Sep 2019 [Online]
22. Krebs, W.G., Alexandrov, V., Wilson, C.A., Echols, N., Yu, H., Gerstein, M.: Normal mode analysis of macromolecular motions in a database framework: Developing mode concentration as a useful classifying statistic. Proteins Struct. Funct. Bioinforma. **48**(4), 682–695 (2002). https://doi.org/10.1002/prot.10168
23. Normal mode. Wikipedia, 08 Sep 2019. https://en.wikipedia.org/w/index.php?title=Normal_mode&oldid=914576448. Accessed 18 Sep 2019 [Online]
24. Bicknell, J.W., Krebs, W.G.: Process mining: the missing piece in information warfare. ResearchGate (2019). https://doi.org/10.13140/RG.2.2.23584.94722/1
25. Bicknell, J.W., Krebs, W.G.: FOCAL information warfare defense standard. ResearchGate, June 2019. https://doi.org/10.13140/RG.2.2.12672.07687.
26. Sviridova, A.: Vectors of the development of military strategy, 04 Mar 2019
27. Mueller, R.S.: Report on the investigation into Russian interference in the 2016 presidential election, Mar 2019. https://www.hsdl.org/?abstract&did=824221. Accessed 01 Oct 2019 [Online]
28. Chotikul, D.: The Soviet Theory of Reflexive Control in Historical and Psychocultural Perspective: Preliminary Study. Naval Postgraduate School, Monterey, California (1986)

29. Novikov, D.A., Chkhartishvili, A.G.: Reflexion and Control: Mathematical Models. CRC Press (2014)
30. Thomas, T.: Russia's reflexive control theory and the military. J. Slav. Mil. Stud. **17**(2), 237–256 (2004). https://doi.org/10.1080/13518040490450529
31. Waltzman, R.: SASC Testimony: The Weaponization of Information (2017)
32. Jaitner, M., Kantola, H.: Applying principles of reflexive control in information and cyber operations. J. Inf. Warf. **15**(4), 27–38 Fall (2016)
33. Ruocco, A., Buchheit, N., Ragsdale, D.: A combined offensive/defensive network model. In: 1st Annual IEEE Systems, Man, and Cybernetics Information Assurance Workshop, West Point, NY, June 2000, pp. 14–18. Accessed 22 Aug 2019 [Online]
34. Dobson, T.K.: Entropy and self-organization—an open system approach to the origins of homeland security threats. Thesis, Naval Postgraduate School, Monterey, California (2015)

Efficient Self-organized Feature Maps Through Developmental Learning

Adit Chawdhary and Ali A. Minai

Abstract Learning and adaptation occur at several temporal and spatial scales in animals with central nervous systems. In the broadest sense, these include evolution, development, synaptic learning, and real-time modulation. Of these, artificial neural networks have focused mainly on learning at the synaptic level with fixed network sizes and architectures, thus excluding some very important dimensions of adaptation available to biological systems. With the increasing scale of problems to which neural networks are being applied, and the quest to achieve more robust and flexible general intelligence in artificial systems, it is important to explore learning at other scales as well, of which developmental learning is the least explored. In this paper, we study a simple model of developmental learning in self-organized feature maps with the narrow goal of demonstrating that adding a developmental aspect to synaptic learning can enhance the performance of neural networks while reducing computational load. Though simplistic, the approach described represents a general paradigm for neural learning that mimics the gradual complexification that biological neural networks undergo in animals. We also argue that this is a crucial step for neural learning to move beyond narrow applications towards a more general intelligence framework with continuous learning.

Keywords Neural networks · Unsupervised learning · Self-organized feature maps · Developmental learning · Brain-inspired computation

A. Chawdhary · A. A. Minai (✉)
Department of Electrical Engineering and Computer Science, University of Cincinnati, Cincinnati, OH 45221, USA
e-mail: Ali.Minai@uc.edu

A. Chawdhary
e-mail: chawdhat@mail.uc.edu

D. Braha et al. (eds.), *Unifying Themes in Complex Systems X*, Springer Proceedings in Complexity, https://doi.org/10.1007/978-3-030-67318-5_16

1 Introduction

Ever since the beginning of the AI project in the 1950s, the goal has been to achieve intelligence that operates at the same level as humans, or at least at the level of intelligent mammals. However, in spite of repeated predictions about imminent success [24], the goal has proved to be elusive [27, 30]. The recent explosive growth in AI systems, led mainly by deep neural networks, has been focused on narrow tasks and pre-defined goals—often with impressive but brittle performance. It is also based for the most part on supervised (or at least gradient-based) learning, uses very large amounts of data, and requires exorbitant computational effort for training. All of these features make even the most successful current AI very different from natural intelligence, which is versatile and robust, learns mainly through self-organization or reinforcement, needs relatively little data, and often learns in very few trials. The question must therefore be asked: What is missing? The answer is likely to be very complicated, but one crucial feature that is missing from most neural network learning today is the exclusion of evolution and development from the learning process.

Two important reasons why animal nervous systems can learn very complex tasks with relative ease are:

1. *Evolution* pre-configures useful processing architectures, feature detectors, and constraints that act as powerful priors for subsequent learning.
2. *Development* learns complex tasks requiring many degrees of freedom gradually, beginning with simple learning in a simple (infant) body, and systematically adding more complexity of task and body in tandem while using the previous simpler learning as a biasing prior.

In contrast, most current AI systems—both neural networks and robots—usually begin with a fully complex architecture and use learning to acquire complex capabilities, necessitating computationally intensive, time-consuming learning. This paper focuses on the second item listed above, making the case that adding a developmental component to neural learning can make it more efficient and scalable.

2 Background

While the vast majority of work in neural networks has focused on adaptation of fixed architectures, there has also been significant research in on applying evolutionary algorithms [38]—primarily in the context of evolutionary robotics [31]. The results show that this approach can often be very promising, but the ability to truly capture the power of evolution in actual implementation or even simulation is limited due to the fact that evolution is fundamentally a slow process that leverages extremely large populations. These conditions are impossible to replicate artificially, and ultimately, the evolutionary aspects of biological learning have to be incorporated mostly into the design process. However, the biological processes of development are much more amenable to simulation and implementation because they occur at the level of the individual animal. But surprisingly, work on incorporating developmental processes into neural learning or robotics has been more limited (see [3] for a notable exception in the area of robotics).

In neural networks, development usually involves combining learning with growth—mainly to support incremental learning [33, 35, 39]. Algorithms proposed for growth or pruning in neural networks [5, 17, 19, 25] are not especially developmental in their approach. Most importantly, they are grounded in the supervised learning paradigm whereas biological development occurs in unsupervised or reinforcement learning situations. The theme of development in reinforcement learning, however, has been investigated in particular by Asada and collaborators [2] in the context of robotics.

Interest in enabling neural networks to show life-long learning has resulted in several incremental learning strategies. Some of these rely on growing networks in a manner similar to development. Miller and colleagues [28, 29] have proposed a model where neurons are endowed with the capacity to grow new connections and neurons in ways that mimic development in the nervous system. However, the "developmental" process is actually simulated within an evolutionary algorithm. Perhaps the most systematic growth-based approaches to incremental learning are the growing neural gas (GNG) model of Fritzke [7–9] and the subsequently developed self-organized incremental neural network (SOINN) model of Hasegawa and colleagues [10–12, 20, 36]. In both these models, a SOFM network is grown (and occasionally pruned) adaptively based on the distribution of neuron activity across the training data. The goal is to balance the representational load across neurons, to identify the number of clusters in the data adaptively, or to handle non-stationarity in the data.

The motivation and approach for the study we present are very different. First, the focus is on scalability and efficiency of learning, so the dataset is neither growing nor changing. The growth in the network is not driven by any calculated metric (as in GNG, SOINN and other systems), but rather represents the expanding degrees of freedom in a growing organism (though we do not claim biological plausibility for the mechanism used.) The "incremental" aspect in our model is thus hierarchical: What is increasing is the system's capacity to perceive and learn the *same* data with increasing resolution and in more complex ways, not the ability to keep learning new data. This is discussed in more detail in the next section.

3 Motivation and General Framework

The supervised learning paradigm will always need large amounts of labeled data and training iterations. Moving towards more natural general intelligence requires focusing on unsupervised and reinforcement learning, which are the basis of learning in animals. The long-term goal of our research is to investigate how developmental methods can help address complex learning tasks in animals, robots, and AI systems. The *Developmental Self-Organized Map* (DevSOM) model studied in this paper represents the simplest step in that investigation: Showing that a simple, developmentally-inspired growth process can significantly improve the learning of feature maps.

The self-organized feature map (SOFM) model first proposed by Kohonen [21–23] has become one of the most influential and profoundly useful paradigms in neural networks. In a very simple framework, it captures the cognitively critical function of learning informative spatial maps from data in complex, high-dimensional feature spaces [32]. Such maps are seen explicitly in many parts of the mammalian brain [15]—notably the visual [13, 18], somatosensory [6, 34, 37] and motor cortices [1, 14, 16, 34]—and very likely occur in more abstract form throughout the nervous system. It is one of the brain's primary ways of organizing data and learning meaningful representations based on feature similarity. Thus, we choose the SOFM paradigm to demonstrate the utility of a developmental approach. A key aspect of building feature map representations is to cluster data based on feature similarity, so that distinguishable categories of data are represented in distinct regions of the map. We use this requirement as a way to quantify the performance of learning in this paper.

A complex domain with a high-dimensional feature space and a large amount of data requires a network large enough to represent all the categories, including small, nuanced differences between samples. But when a large network is trained on a large amount of high-dimensional data, the process is difficult. It often requires very slow training with careful tuning of parameters, and even then, it is often difficult to achieve a sufficient performance level. This is an instance of the standard situation encountered in most AI paradigms today, where complicated systems are forced to learn complicated information directly, thus setting up an inherently very complex learning task. We propose that this can be alleviated by solving the problem developmentally in a hierarchical fashion as follows:

1. Start with a small network that quickly learns to make gross distinctions in feature space, thus laying down the foundations of the map at an approximate level.
2. Then grow the network in stages where the neurons in each stage inherit the imperfect but useful learning of the neurons from the prior stage, but at the same time, add some variation so that finer features of the data can be learned.

The results in this paper show that significantly better results can be achieved with this approach with a much lower computational cost.

This profound aspect of developmental learning was explored in an early work by Elman [4] in the area of learning a language model, though that model still used supervised learning. Elman showed that a system that starts with initially limited but gradually increasing information processing capacity can learn complex functions better than one that is forced to learn the same data at full capacity from the outset. As he states: "Limited capacity acts like a protective veil, shielding the infant from stimuli which may either be irrelevant or require prior learning to be interpreted. Limited capacity reduces the search space, so that the young learner may be able to entertain a small number of hypotheses about the world." The DevSOM model applies the same principle in a more biologically plausible—and potentially more broadly applicable—unsupervised setting. The work presented in this paper is intended as a "proof of concept" for this more general approach.

4 The DevSOM Model

As in all SOFM models, the goal in the DevSOM model is to organize data in a high-dimensional feature space onto a low-dimensional neural surface (or hypersurface) such that data points with similar features are mapped close together. If the data has clusters or categories defined by feature similarity, an SOFM would map data points of the same category to a compact region on the map, and data points of distinct categories to non-overlapping regions. Thus, an SOFM also functions as a clustering network if the underlying data has clustered structure. Furthermore, if there are variations within a category or data points that bridge categories, the structure of the map would also reflect that. This case provides a convenient way of studying the capabilities of SOFMs, and is adopted in this paper. In particular, we use data from the well-known MNIST digit dataset [26] to pose the learning problems on which the DevSOM model is evaluated.

DevSOM is a hierarchical growth model, where successively larger standard SOFMs are trained in combination with a simple growth process. At each stage, the larger network inherits weights from its predecessor as described below. All networks are trained using the standard SOFM training algorithm with a shrinking Gaussian neighborhood function. Each neuron, i, in the network has a topographical location r_i on the map, and receives input from all m feature inputs through the weight vector $w_i = [w_{ij}]$. When a data point $x^q = [x_1^q, \ldots x_m^q]$ is presented to the network, the winner neuron is determined by $i^* = argmin_i ||x^q - w_i||$, and the weights of the neurons are updated using:

$$\Delta w_{ij}(t) = \eta(t)\Lambda(i, i^*, \sigma)(x_j^q - w_{ij}(t-1)) \quad (1)$$

with

$$\Lambda(i, i^*, t) = \exp\left[-(r_i - r_{i^*})^2\right]/2\sigma^2(t) \quad (2)$$

$$\sigma(t) = \bar{\sigma}/(1 + (2t/T)) \quad (3)$$

$$\eta(t) = \bar{\eta}/(1 + (2t/T)) \quad (4)$$

where $\bar{\sigma}$ is the initial width of the neighborhood function (in grid cells), $\bar{\eta}$ is the initial learning rate, and T is the total number of iterations, which equals number of training epochs times the number of datapoints in training dataset. Both the hyperparameters undergo hyperbolic decay.

The process begins with a small *infant network*, N_0, which is a standard SOFM with randomly initialized weights. In the simulations for this paper, we assume that the network is a two dimensional square $n_0 \times n_0$ lattice. The network is trained only for a very small number of iterations through the data, which may represent only a small subset of all available data. This network is not large enough to capture the details of the complex data, and training results in a relatively low-quality gross map where categories are only vaguely discriminated and nuances within categories are not captured at all.

Once the infant network has been trained, it is expanded into the Level-1 network, N_1, with $n_1 \times n_1$ neurons. For our simulations, we use $n_k = 2n_{k-1}$, so $n1 = 2n_0$. The neurons of N_1 are obtained by replicating each neuron of N_0 into a 2×2 sub-lattice of four neurons that inherit their weights from their parent neuron in N_0, but with $\mathcal{N}(0, 0.2)$ independent Gaussian noise added to each weight. Thus, this network inherits the information learned by N_0, but with twice the potential resolution in both dimensions of the map. This network is again trained for a small number of epochs, after which the same process is repeated to generate the N_2 network with $n_2 \times n_2$ neurons, and so on. After K-steps of developmental growth, the feature map is of size $2^K n_0 \times 2^K n_0$. Each stage is trained only for the same small number of epochs (5 in the simulations).

5 Simulations and Results

5.1 Experimental Protocols

To investigate the potential benefits of the DevSOM model, we perform experiments under four protocols as described below.

The MNIST dataset has 70,000 samples of handwritten digits from 0 through 9, with 7,000 samples of each digit. Of these, 1,000 samples of each digit are used for testing and the remaining 6,000 for training in various configurations. The test set for all protocols is the same. The two main hyper-parameters in SOFM are the $\bar{\sigma}$ and $\bar{\eta}$, namely the initial width parameter of the neighborhood function and the initial learning rate, respectively. All weights are initialized randomly.

The protocols are:

- **Direct Training (D)**: In this case, a 24×24 network is trained on the full 60,000 sample training set for 15 epochs, i.e., a total of 900,000 steps. To find good hyper-parameters, all combinations of $\bar{\sigma} = \{1.5, 3.0, 6.0\}$ and $\bar{\eta} = \{0.025, 0.05, 0.1\}$ are evaluated in a grid search to find the best combination, which is reported below.
- **DevSOM with Full Data (DevFull)**: This is the baseline developmental case consisting of three stages: i) A 6×6 network is trained with the full training set for 5 epochs; ii) The network is developed to a 12×12 network and trained on the full training set for 5 epochs; and iii) The network is again developed to 24×24 and trained on the full training set for 5 epochs. Thus, the training in this whole protocol is 15 epochs, as is the case in Direct Training, but 10 of those epochs are for smaller networks, so the overall computation time is much lower. Since

Table 1 Results for 24 × 24 networks for all protocols on MNIST dataset

Size of network	σ	η	Normalized metric		Classification metric	
			Train split	Test split	Train split	Test split
D	3.0	0.025	0.7453	0.7693	0.8206	0.8232
DevFull	1.5	0.025	0.8135	0.8355	0.8759	0.8872
DevGrow	1.5	0.025	**0.8173**	**0.8371**	**0.8799**	**0.8906**
DevShrink	1.5	0.025	0.7279	0.7475	0.8082	0.8210

each stage requires its own hyperparameters, doing a grid search is impractical. However, good values were found by trial and error (see Table 1).

- **DevSOM with Growing Data (DevGrow)**: This protocol has the same three stages as DevFull, but the first stage (6 × 6) uses only 600 datapoints for training (60 from each digit), the middle (12 × 12) stage uses 6,000 datapoints including the 600 from stage (i), and the final (24 × 24) stage uses the full training set. This reduces the computational effort dramatically over the D and DevFull protocols. This protocol corresponds best to natural developmental learning, where learning in early stages is done on more limited data than learning in later stages.
- **DevSOM with Shrinking Data (DevShrink)**: Like DevGrow, this protocol uses different amounts of training data in each stage, but in the reverse order: The 6 × 6 network is trained on the full training set, the 12 × 12 on 6,000 points, and the 24 × 24 network on only 600 points. The computational effort here is much lower than DevGrow, because the smallest network is using the most data. However, this is an extreme protocol that is not as consistent with development as DevGrow, since intelligent agents typically learn from more data as they become more complex.

The purpose of DevFull is to show that development can match or outperform Direct learning with less computation. DevGrow shows that a small amount of early learning can facilitate more complicated learning later. Finally, DevShrink demonstrates that a large amount of early learning with a small system can be extended developmentally to a larger system with much less further training.

5.2 Performance Metrics

While the performance of self-organized maps can be evaluated in many ways, we measure it in terms of category representation, i.e., to what extent the ten classes are mapped distinctly in the final network. This is done using a visualization and two numerical metrics, calculated separately on the training and testing sets:

1. **Visualization of Neuron Tuning**: This metric quantifies the degree to which each neuron becomes tuned to data from a specific class. Let M be the number of classes C_i, $i = 1, \ldots, M$ ($M = 10$ is these experiments). Consider the set of

all neurons that win for any sample in the dataset (i.e., neurons that do not win at all are excluded). Given a neuron with coordinates (x, y), let $m(x, y)$ be the total number of datapoints for which the neuron is the winner, and $n_i(x, y)$ the number of these that come from class C_i. Then $p_i(x, y) = n_i(x, y)/m(x, y)$ is the fraction of the datapoints for which (x, y) wins that come from class C_i. The vector $p(x, y) = [p_1(x, y) \ \dots \ p_M(x, y)]$ is then the probability distribution of class representation by (x, y). If only one $p_i(x, y) = 1$ and the rest are all 0, the neuron is responding purely to one class. At the other extreme, if $p_i(x, y) = 1/M$ for all i, then the neuron responds equally to all classes, and thus does not represent any class. This is quantified using the entropy of $p(x, y)$:

$$E(x, y) = -\sum_{i=1}^{M} p_i(x, y) \log(p_i(x, y)) \tag{5}$$

A neuron that represents only one class has entropy 0, and a neuron that responds equally to all classes has entropy $log M > 0$. Thus, looking at the entropy of all neurons in the network through a heatmap gives a visual impression of how well-tuned the map has become to the classes in the data. A well-tuned map will have most neurons with entropy near 0.

2. **Normalized Tuning Metric**: The entropy metric above captures the specificity of tuning in individual neurons, but not the localization of each category to specific neurons. To get a measure of the latter for the whole network, a normalized global performance metric is calculated which would be 1 if each class is represented by neurons that respond only to that class. Let N be the size of the dataset and N_i be the number of samples for class C_i. Define $q_i(x, y) = \frac{n_i(x,y)}{N_i}$ as the fraction of datapoints in C_i for which neuron (x, y) is the winner. Then the class-wise metric Q_i is defined as:

$$Q_i = \sum_x \sum_y q_i(x, y) p_i(x, y) = \frac{1}{N_i} \sum_x \sum_y \frac{n_i^2(x, y)}{m(x, y)} \tag{6}$$

Thus, if all the neurons that win for any datapoint in class C_i win only for that class, $Q_i = 1$, indicating that class C_i is very well-discriminated.
Finally, the normalized global performance metric Q is defined as the mean of the tuning quality over all classes:

$$Q = \frac{1}{M} \sum_{i=1}^{M} Q_i \tag{7}$$

3. **Categorization**: This metric quantifies how well the trained map can function as a simple classifier. To do this, every winner neuron (x, y) is assigned class label i if $n_i(x, y) > n_j(x, y)\ \forall j \neq i$, i.e., the class for which it wins the most *in the training set*. Then, each sample *in the test set* is applied to the network, and assigned the class label of the winning neuron. If the winning neuron had not won for any training sample (and thus has no class label), the datapoint is assigned a random class label. After all test samples have been labeled, the fraction correct (hit-rate) is calculated as the categorization metric. While the information captured by this metric is closely related to that of the previous one, it is a more direct test of categorization capability and—very importantly—checks for generalization from the training set to the test set explicitly.

5.3 Results

The performance results for all four protocols are summarized in Tables 1 through 5. Table 1 provides a comparison of both metrics on the 24 × 24 networks of each protocol. It also lists the hyperparameters used in each case. It should be noted that, for the D protocol, the network was run with these hyperparameters—obtained via a grid search—for 15 epochs with all training data. For the three developmental protocols, the hyperparameters for the smaller networks were different, as listed in Tables 2, 3, 4 and 5. The comparisons in Table 1 show the following:

Table 2 Results for the D protocol

Size of network	σ	η	Normalized metric		Classification metric	
			Train split	Test split	Train split	Test split
6 × 6	3.0	0.1	0.4536	0.4577	0.5163	0.5152
12 × 12	2.25	0.05	0.6582	0.6701	0.7349	0.737
24 × 24	3.0	0.025	0.7453	0.7693	0.8206	0.8232

Table 3 Results for the DevFull protocol

Size of network	σ	η	Normalized metric		Classification metric	
			Train split	Test split	Train split	Test split
6 × 6	3.0	0.1	0.4551	0.4657	0.5137	0.5267
12 × 12	2.25	0.05	0.6526	0.6674	0.7469	0.7666
24 × 24	1.5	0.025	0.8135	0.8355	0.8759	0.8872

Table 4 Results for the DevGrow protocol

Size of network	σ	η	Normalized metric		Classification metric	
			Train split	Test split	Train split	Test split
6 × 6	3.0	0.1	0.4752	0.4718	0.5368	0.5247
12 × 12	2.25	0.05	0.6551	0.6480	0.7434	0.7482
24 × 24	1.5	0.025	0.8173	0.8371	0.8799	0.8906

Table 5 Results for DevShrink protocol

Size of network	σ	η	Normalized metric		Classification metric	
			Train split	Test split	Train split	Test split
6 × 6	3.0	0.1	0.4757	0.4852	0.5340	0.5436
12 × 12	2.25	0.05	0.6553	0.6415	0.7378	0.7405
24 × 24	1.5	0.025	0.7279	0.7475	0.8082	0.821

Table 6 Time taken in training the networks on MNIST dataset

Protocol	Size of network	Time taken (sec)	Total time taken (sec)
D	6 × 6	179.21	4926.36
	12 × 12	941.58	
	24 × 24	4926.36	
DevFull	6 × 6	170.28	2266.43
	12 × 12	450.77	
	24 × 24	1645.38	
DevGrow	6 × 6	**1.18**	1414.55
	12 × 12	**33.86**	
	24 × 24	1379.51	
DevShrink	6 × 6	181.17	**227.26**
	12 × 12	33.94	
	24 × 24	**12.15**	

1. The DevFull protocol does significantly better than the base D protocol on both metrics and both datasets (training and test), while using far less computation time (see Table 6).
2. The DevGrow protocol shows a slight improvement even over DevFull, even though it uses even less computation.
3. The DevShrink protocol has performance only slightly worse than D but significantly worse than DevFull and DevGrow, but uses far less computation time.

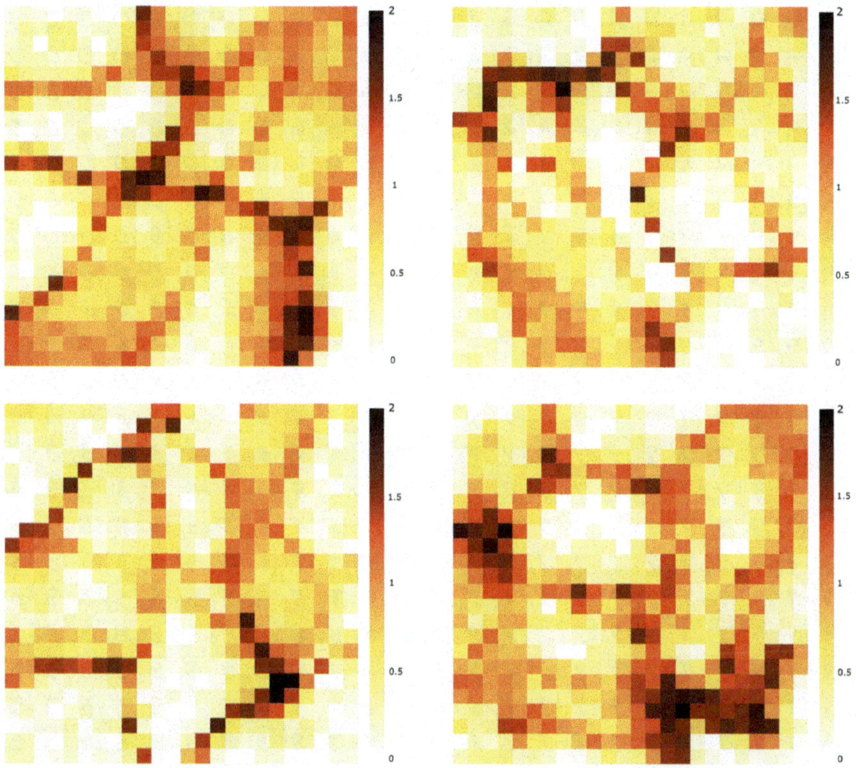

Fig. 1 Map of entropy values for the final 24 × 24 network obtained via the four protocols: Direct training (upper left); DevFull (upper right); DevGrow (lower left); and DevShrink (lower right)

Since the three developmental protocols required training of 6 × 6 and 12 × 12 networks, these were also simulated in the D case. Tables 2, 3, 4 and 5 show the results and hyperparameters for networks of all three sizes for each of the protocols. This gives a good idea of the progression of developmental learning.

Figure 1 shows the neuron entropy distributions over the final 24 × 24 networks obtained in each of the protocols. Recalling that well-tuned neurons have entropy near 0 (lighter colors), it can be seen visually that the DevFull and DevGrow maps are better tuned than the D and DevShrink networks—a fact reflected in their performance metrics as well. It is also interesting to observe the formation of boundaries between regions tuned to distinct classes in each case. These boundaries—which show up as darker neurons—represent regions of overlap between classes, and are narrower in the DevFull and DevGrow maps.

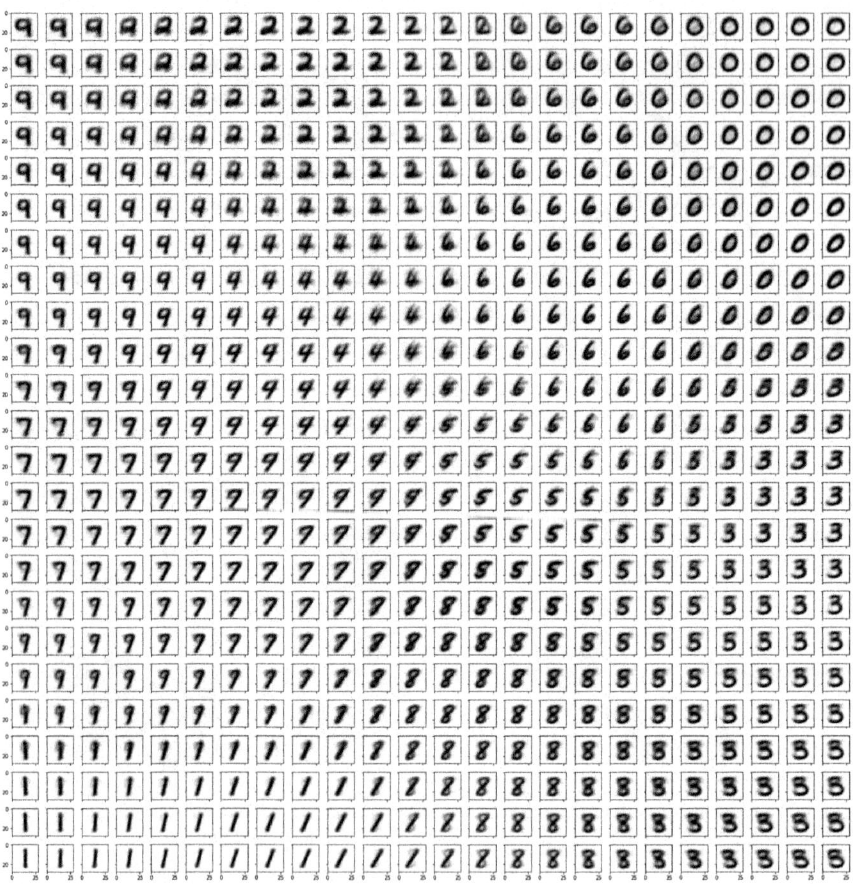

Fig. 2 Feature map for the final 24 × 24 network obtained via the D protocol

Another informative way to assess what the networks have learned is to look at the features to which each neuron has become maximally tuned. These can be obtained simply by looking at the 784 weights of each neuron as a 28 × 28 image. Figures 2, 3, 4 and 5 show these feature maps for the final 24 × 24 networks for each protocol, which demonstrate which region became tuned to which digit. Issues of interest here are: (a) Were all the digits well-represented? (b) What are the sizes of the regions representing each digit? and (c) Which digits are represented close to or far from which ones? As can be seen from the maps, most digits get good representations, except the digit 4 which is very narrowly represented in all four cases.

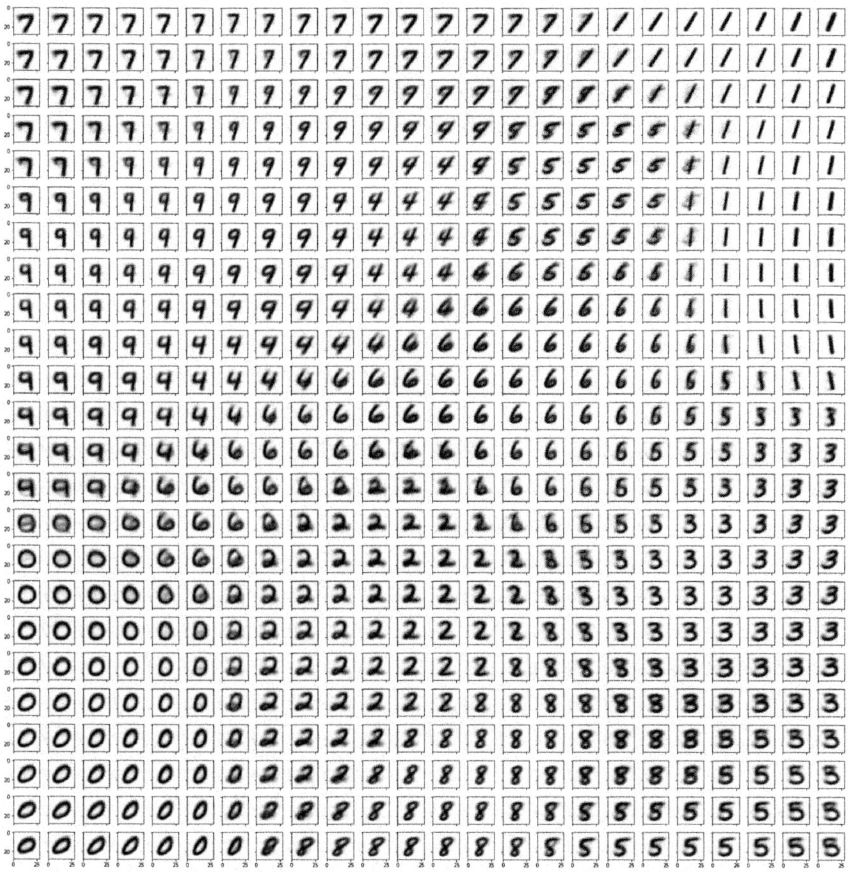

Fig. 3 Feature map for the final 24 × 24 network obtained via the DevFull protocol

As the number of epochs over the training dataset remained constant in all protocols, there was a significant reduction in training time of the network in the developmental cases. Table 6 shows that DevFull starting with a 6 × 6 network and with 2-step developmental learning achieves an approximately 50% reduction in training time for a 24 × 24 network over the D Protocol, and DevGrow 70% reduction—both with a significant performance improvement over D. While DevShrink uses only about 5% of the training time needed for D, it comes at the cost of a very small loss in performance (see Table 1).

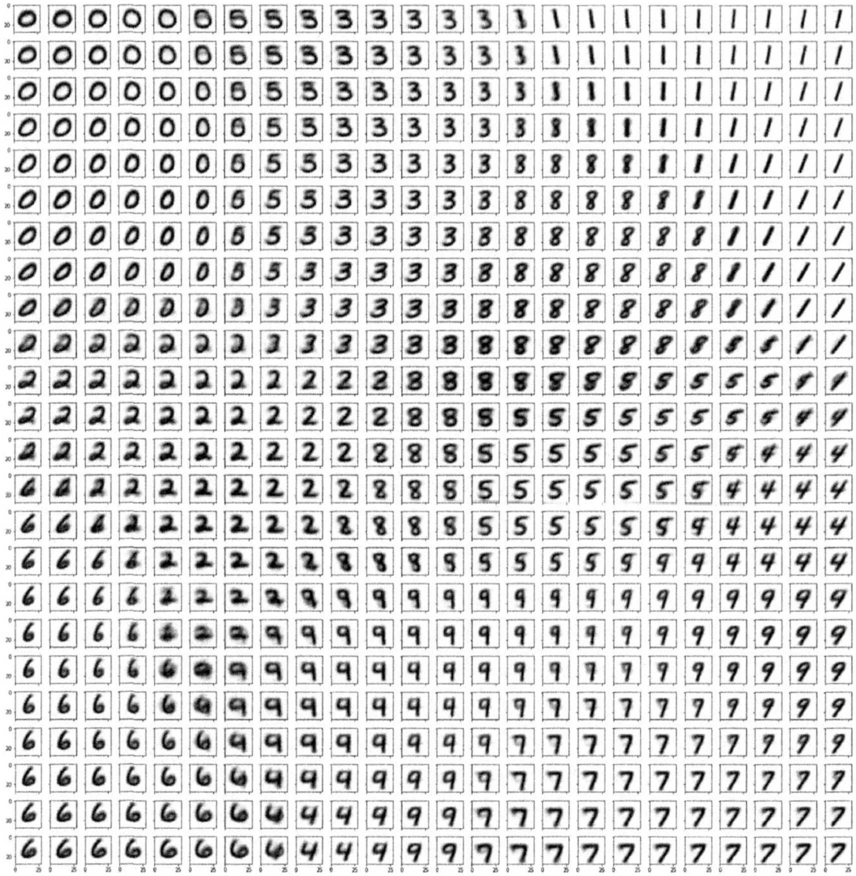

Fig. 4 Feature map for the final 24 × 24 network obtained via the DevGrow protocol

6 Conclusions

This paper has reported a pilot study of using a developmental approach for unsupervised learning in self-organized feature maps. The results show that, with this developmental approach, superior results can be obtained with significantly lower computational load. This is achieved by letting limited learning in small networks to create the gross organization on which larger networks in subsequent stages can build more effectively. In contrast, training a naive large network directly requires more training and may still fail to match the results of the developmental approach.

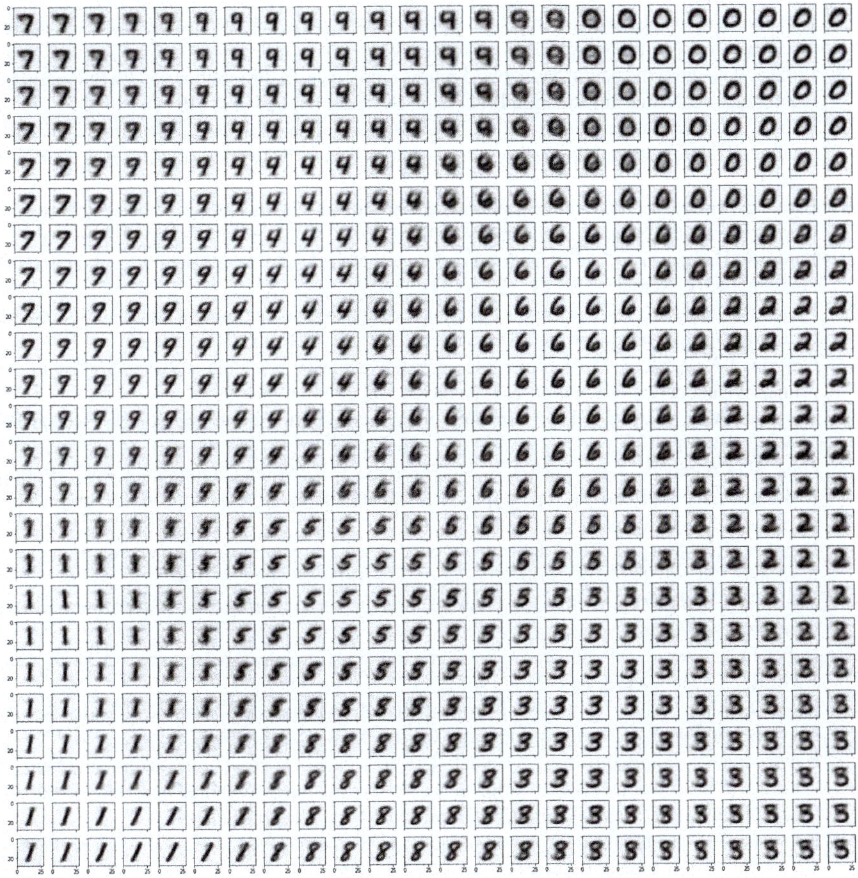

Fig. 5 Feature map for the final 24 × 24 network obtained via the DevShrink protocol

References

1. Aflalo, T.A., Graziano, M.S.A.: Possible origins of the complex topographic organization of motor cortex: reduction of a multidimensional space onto a two-dimensional array. J. Neurosci. **26**, 6288–6297 (2006)
2. Asada, M., Noda, S., Tawaratsumida, S., Hosoda, K.: Purposive behavior acquisition for a real robot by vision-based reinforcement learning. Mach. Learn. **23**, 279–303 (1996)
3. Cangelosi, A., Schlesinger, M., Smith, L.B.: Developmental Robotics: From Babies to Robots. MIT Press (2000)
4. Elman, J.L.: Learning and development in neural networks: the importance of starting small. Cognition **48**, 71–99 (1993)
5. Fahlman, S.E., Lebiere, C.: The cascade-correlation learning architecture. In: Touretzky, D.S. (ed.) Advances in Neural Information Processing Systems 2. Morgan-Kaufmann (1990)
6. Feldman, D.E., Brecht, M.: Map plasticity in somatosensory cortex. Science **310**, 810–815 (2005)

7. Fritzke, B.: Growing cell structures—a self-organizing network for unsupervised and supervised learning. Neural Netw. **7**, 1441–1460 (1994)
8. Fritzke, B.: A growing neural gas network learns topologies. In: Proceedings of 7th International Conference on Neural Information Processing Systems, pp. 625–632 (1994)
9. Fritzke, B.: A self-organizing network that can follow non-stationary distributions. In: Proceedings of International Conference on Artificial Neural Networks, pp. 613–618. Springer (1997)
10. Furao, S., Hasegawa, O.: An incremental network for on-line unsupervised classification and topology learning. Neural Netw. **19**, 90–106 (2006)
11. Furao, S., Hasegawa, O.: A fast nearest neighbor classifier based on self-organizing incremental neural network. Neural Netw. **21**, 1537–1547 (2008)
12. Furao, S., Ogura, T., Hasegawa, O.: An enhanced self-organizing incremental neural network for online unsupervised learning. Neural Netw. **20**, 893–903 (2007)
13. Gordon, J.A., Stryker, M.P.: Experience-dependent plasticity of binocular responses in the primary visual cortex of the mouse. J. Neurosci. **16**, 3274–3286 (1996)
14. Graziano, M.S.A., Aflalo, T.A.: Mapping behavioral repertoire onto the cortex. Neuron **56**, 239–251 (2007)
15. Guo, W., Chambers, A.R., Darrow, K.N., Hancock, K.E., Shinn-Cunningham, B.G., Polley, D.B.: Robustness of cortical topography across fields, laminae, anesthetic states, and neurophysiological signal types. J. Neurosci. **32**, 9159–9172 (2012)
16. Harrison, T.C., Murphy, T.H.: Motor maps and the cortical control of movement. Current Opin. Neurobiol. **24**, 88–94 (2014). https://doi.org/10.1016/j.conb.2013.08.018
17. Hassibi, B., Stork, D.G., Wolff, G.J.: Optimal brain surgeon. In: Proceedings of the International Conference on Neural Networks (1993)
18. Hubel, D.H., Wiesel, T.N.: Receptive fields and functional architecture of monkey striate cortex. J. Physiol. (London) **195**, 215–243 (1968)
19. Irsoy, O., Alpaydin, E.: Continuously constructive deep neural networks. IEEE Trans. Neural Netw. Learn. Syst. (2019). https://doi.org/10.1109/TNNLS.2019.2918225
20. Kawewong, A., Tangruamsub, S., Kankuekul, P., Hasegawa, O.: Fast online incremental transfer learning for unseen object classification using self-organizing incremental neural networks. In: Proceedings of International Joint Conference on Neural Networks (2011)
21. Kohonen, T.: Self-organized formation of topologically correct feature maps. Biolog. Cybernet. **43**, 59–69 (1982). https://doi.org/10.1007/BF00337288
22. Kohonen, T.: Self-Organization and Associative Memory, 3rd edn. Springer (1989). https://doi.org/10.1007/978-3-642-88163-3
23. Kohonen, T.: Self-Organizing Maps. Springer (1995). https://doi.org/10.1007/978-3-642-97610-0
24. Kurzweil, R.: The Singularity Is Near: When Humans Transcend Biology. Penguin Books (2006)
25. Le Cun, Y., Denker, J.S., Solla, S.A.: Optimal brain damage. In: Touretzky, D.S. (ed.) Advances in Neural Information Processing Systems 2. Morgan-Kaufmann (1990)
26. LeCun, Y., Bottou, L., Bengio, Y., Haffner, P.: Gradient-based learning applied to document recognition. Proc. IEEE **86**, 2278–2324 (1998)
27. Marcus, G., Davis, E.: Rebooting AI: Building Artificial Intelligence We Can Trust. Pantheon Books (2019)
28. Miller, J.F.: Evolving an artificial brain to solve multiple problems. In: Proceedings of ALife'19 (2019)
29. Miller, J.F., Wilson, D.G.: A developmental artificial neural network model for solving multiple problems. In: Proceedings of GECCO'17 Companion (2017). https://doi.org/10.1145/3067695.3075976
30. Mitchell, M.: Artificial Intelligence: A Guide for Thinking Humans. Farrar, Straus and Giroux (2019)
31. Nolfi, S., Floreano, D.: Evolutionary Robotics: The Biology, Intelligence, and Technology of Self-Organizing Machines. MIT Press (2015)

32. Obermayer, M.K.: Self-organizing maps. In: Squire, L.R. (ed.) Encyclopedia of Neuroscience. Elsevier (2009). https://doi.org/10.1016/B978-008045046-9.01431-5
33. Part, J.L., Lemon, O.: Incremental online learning of objects for robots operating in real environments. In: Proceedings of 2017 Joint IEEE International Conference on Development and Learning and Epigenetic Robotics (ICDL-EpiRob) (2017)
34. Penfield, W., Boldrey, E.: Somatic motor and sensory representation in the cerebral cortex of man as studied by electrical stimulation. Brain **60**, 389–443 (1937)
35. Rebuffi, S.A., Kolesnikov, A., Sperl, G., Lampert, C.H.: icarl: incremental classifier and representation learning. In: Proceedings of 2017 IEEE Conference on Computer Vision and Pattern Recognition (CVPR'2017) (2017). https://doi.org/10.1109/cvpr.2017.587
36. Shen, F., Yu, H., Kasai, W., Hasegawa, O.: An associative memory system for incremental learning and temporal sequence. In: Proceedings of International Joint Conference on Neural Networks (2010)
37. Sherrington, C.S.: Observations on the scratch-reflex in the spinal dog. J. Physiol. (London) **34**, 1–50 (1906)
38. Stanley, K.O., Miikkulainen, R.: Evolving neural networks through augmenting topologies. Evolution. Comput. **10**, 99–127 (2002)
39. Xiao, T., Zhang, J., Yang, K., Peng, Y., Zhang, Z.: Error-driven incremental learning in deep convolutional neural network for large-scale image classification. In: Proceedings of 22nd ACM International Conference on Multimedia (2014)

Modeling and Simulation Analysis of a Complex Adaptive Systems of Systems Approach to Naval Tactical Warfare

Bonnie Johnson

Abstract A complex adaptive systems of systems approach is an engineered solution to a highly complex problem domain. Highly complex problems are comprised of many diverse objects and events that are interrelated, rapidly changing, and give rise to unpredictable, severe, and often destructive consequences. A recent study produced a grounded theory that defines complex adaptive systems of systems as a new class of engineered systems that can address highly complex problem domains. Naval tactical warfare was identified as a highly complex problem domain and used as a use case to study the complex adaptive systems of systems theory. A modeling and simulation analysis of a complex naval tactical scenario produced a comparative analysis of the existing traditional approach to naval warfare with the new theoretical complex adaptive systems of systems approach. This paper describes the naval tactical modeling and simulation approach and presents the results of the comparative analysis—providing insight into engineered complex adaptive systems of systems solutions and demonstrating theory validity.

Keywords Complex adaptive systems of systems · Naval tactical warfare · Modeling and simulation

1 Introduction

A recent study of highly complex problem domains produced a theory for engineering complex adaptive systems of systems (CASoS) as solutions to address these domains [1]. This paper presents the results of a case study modeling and simulation (M&S) analysis that applied the CASoS approach to naval tactical warfare as an example of a highly complex problem domain. The M&S analysis provided insight into the behavioral characteristics of a CASoS approach and provided a venue for comparing the CASoS approach with the existing approach to managing naval warfare assets. The quantitative results of the M&S analysis support validation of the CASoS theory.

B. Johnson (✉)
Naval Postgraduate School, Monterey, CA 94044, USA
e-mail: bwjohnson@nps.edu

D. Braha et al. (eds.), *Unifying Themes in Complex Systems X*, Springer Proceedings in Complexity, https://doi.org/10.1007/978-3-030-67318-5_17

Highly complex problem domains present a large number of diverse, distributed, interrelated, and unexpected events that are changing states rapidly, are difficult to understand or gain awareness of, and result in potentially severe consequences [1]. According to Calvano and John [2] they arise from two phenomena: an unprecedented level of integration and processes and the rapid pace of technological change. Alberts [3] asserts that humans have entered the Age of Interactions in which decisions and events are inextricably linked with many other decisions and events, and these complex interactions produce many emergent outcomes that are often unpredictable. Bar-Yam [4] describes many examples of highly complex problem spaces including military conflict, health care, education, international development, large scale natural disasters, ethnic violence, and terrorism.

Legacy systems that are traditionally engineered and often monolithic, are ill-equipped to deal with the complexity of these problem spaces. Complex problems require engineered solutions that adapt and even evolve over time as the environment will be unpredictably changing with a volatile tempo at times. System solutions must anticipate change and be able to quickly ascertain when adaptation is needed. Solutions must have situational awareness and self-awareness and must be able to self-organize to produce desired emergent behavior to address the changing problem space.

This paper describes how naval tactical combat presents an example of a highly complex problem domain and describes how naval warfare resources could be engineered to behave as a CASoS solution capability. The paper explains how a modeling and simulation analysis was conducted to study how a CASoS solution approach could address a complex naval threat scenario. The paper presents the results of the analysis showing how warfare assets can self-organize and adapt collaboratively to produce desired emergent defense actions.

2 The Highly Complex Tactical Environment

The naval tactical domain encompasses aspects of the maritime environment that pose a combative threat. A tactical situation can very quickly and unexpectedly escalate into a highly complex environment, as in the case of an unconventional or surprise attack, or more slowly as tension builds with a known adversary. The offensive use of naval warfighting assets also initiates a tactical situation. The naval tactical domain can occur in deep water or in a littoral region and can involve threats and assets in the sea, air, space, cyberspace, underwater, and on land. The domain can include affected civilians and participating coalition partners. Actions and events in the tactical domain include threats, countermeasures, evasion, retaliation, defensive and offensive measures, stealth, sensing and tracking, jamming, blinding, cyber-attacks and combat readiness. The tactical domain includes military assets that are in a state of combat readiness—either in a defensive or offensive posture—preparing for a warfighting situation. In this tactical domain, the actual warfighting is a fraction in duration of strategic campaigns and often spans a smaller area that is part of a larger theater of war.

Naval tactical warfare is highly complex [4]. The domain is comprised of potentially large numbers of diverse, distributed, and often interrelated threat objects and events. Threats include ships, aircraft, missiles, countermeasures, submarines, decoys, surveillance systems, and cyber means. Threat events include weapon deployment, asset placement, sensing, jamming, and hacking. The enemy is generally attempting to increase its tactical advantage by avoiding detection, presenting a false depiction of its location and capabilities, and attempting to outmaneuver and overwhelm our military forces [5].

The tactical domain is constantly changing and presenting a continuum of unique operational environments. Many tactical objects are in motion, have dynamic interactions, and are surrounded by both fixed and changing environmental features. The climate, weather, atmosphere, humidity, sea states, hydrography, and topography can affect sensor and weapon performance and affect the ability to gain situational awareness. Nearby urban areas and other civilians in the area can confuse combat identification and change the dynamics of combat. These features contribute to creating a dynamic and complex environment.

The tactical domain is unpredictable and often challenging in terms of gaining complete and accurate situational knowledge. Adversaries purposely attempt to create a fog of war to confuse naval forces [6]. Methods include unexpected attacks, stealth, denied access, distributed assets, decoys countermeasures, and cyber actions. Gaining and maintaining shared battlespace awareness, or "collective consciousness among the elements of the warfighting ecosystem" [7] is a critical enabler of tactical effectiveness. Human warfighters must manage uncertainty in the data collected by sensors and processed. Often the uncertainty, coupled with the amount and diversity of information and the shortened decision timescales, overwhelms human cognitive abilities [8].

Complexity in the tactical environment often results in severe and dire consequences such as warfighter casualties, civilian casualties, and destruction of military and civilian assets. The severity of consequences places a criticality on making effective tactical decisions.

The Navy has the mission of preparing its forces to achieve and maintain tactical superiority. In practice, this involves understanding and anticipating the tactical environment and adversarial threat. The Navy has recognized the tactical environment's growing complexity and actively seeks engineered solutions to address this challenge. Figure 1 highlights some of the limitations of current naval tactical systems when faced with highly complex tactical problems. Examples of these limitations include when naval forces cannot defend against threats that are unexpected, too fast moving, or too-numerous; or friendly fire incidents when civilians or blue force partners are mistaken for threats. These situations arise when the reaction time is too short, or the decision space is too complex for an effective decision.

Two types of naval capabilities limit overall tactical ability: the performance of individual resources (sensor detection range, sensor resolution, sensor multi-function ability, weapon range, weapon accuracy, weapon destructive ability, etc.)

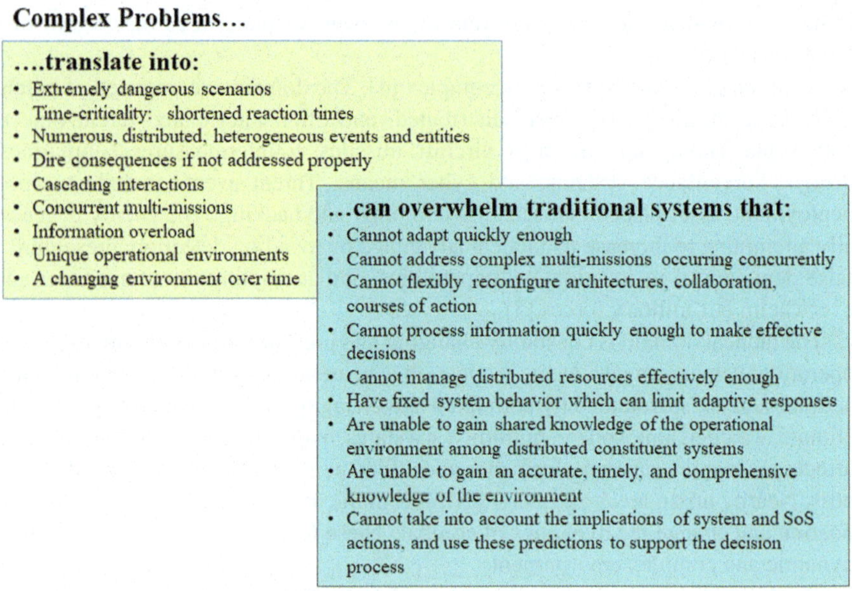

Fig. 1 Limitations of existing traditional naval approaches to address the complex tactical domain [1]

and the decision making process, which relies on the naval architecture, to use these individual assets most effectively. Improving these capabilities is a critical part of achieving tactical success.

Current naval approaches rely on a combination of human decision-making and automated processes to make tactical decisions. The process of threat detection, involving collecting and processing sensor data, is largely automated. However, humans play a role in threat identification. Fully automated engagement decisions are possible for air and missile self-defense on some naval ships—this capability (using the Aegis weapon system) is platform-centric, relying on only resident (or organic) sensors and weapons [9]. However, in general, engagement decisions rely on manual decision-making, groups of decision-makers, and significant negotiation between humans on the multiple platforms [10]. The Navy is working on future capabilities to enable collaborative engagements involving distributed platforms, such as the ability for a weapon to be fired based on remote data (the Navy Integrated Fire Control—Counter Air (NIFC-CA) program plans to use Cooperative Engagement Capability (CEC) data from a remote aircraft sensor to provide tracking data to support a weapon system on a ship) [11]. However, these capabilities are currently limited to threat cues [12]. These cues provide an alert, identifying the existence of an incoming threat. The weapon platform must still use its resident (organic) sensor to detect and track the threat to provide this data to the weapon system [12]. Thus, the current use of naval weapon systems for air and missile defense is a largely platform-centric capability. As tactical warfare missions become more complex, human decision-makers become overwhelmed by information uncertainty, diversity, and overload and by the challenge of identifying effective decision options involving distributed warfare assets [13]. The time-critical nature of many tactical warfare missions often provides only minutes to make these complex decisions [10].

3 Complex Adaptive Systems of Systems Approach to Naval Tactical Warfare

This section describes how a CASoS approach would be implemented in the naval tactical domain. The CASoS approach would identify the systems in the tactical realm that constitute constituent systems. For this domain, the constituent systems would be the distributed warfare assets: platforms (ships, aircraft, submarines, helicopters, etc.), weapons, sensors, jammers, decoys, countermeasures, and other resources that contribute to tactical operations. The CASoS approach would reengineer these existing assets by implementing an adaptive architecture to support enhanced interaction and collaboration and embedding each platform with an intelligent agent (software and computing environment) for enhanced decision-making. Figure 2 illustrates a future CASoS approach with an adaptive architecture connecting distributed naval assets that are embedded with intelligent agents. The CASoS approach would maximize the use of the distributed warfare assets by requiring the means for purposeful, adaptive, collaborative, and emergent behavior, thus improving overall tactical operations.

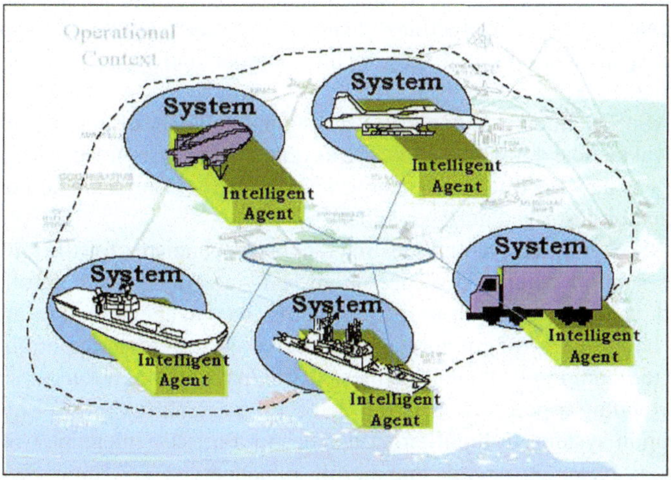

Fig. 2 Illustration of the CASoS architecture and intelligent agents embedded in naval tactical systems [1]

The mission-oriented SoS approach designs constituent systems to meet the required capabilities of the SoS, as a whole, to address different missions [14, 15]. The CASoS approach enables a mission-oriented approach during the design phase as well as during operations. The naval CASoS approach, as illustrated in Fig. 2, provides the capabilities desired in mission-oriented SoS—to identify what desired behaviors are needed from the constituent systems according to overall mission needs. The CASoS intelligent agents and adaptive architecture enable the constituent systems to individually and collectively self-organize to address naval missions adaptively and in near-real-time during operations.

3.1 Naval Tactical CASoS Adaptive Architecture

The CASoS adaptive architecture connects the distributed warfare platforms and assets through their embedded intelligent agents. The architecture shares data, information, and knowledge so each intelligent agent can develop and maintain both situational awareness and self-awareness. The architecture shares data between them to support synchronization of these internal models and synchronization of the COA decision options. The architecture enables the distributed warfare assets to collaborate to an automated degree that is not currently possible. As a CASoS, the distributed warfare assets are effectively managed as if they were all collocated on one ship or aircraft. The CASoS architecture, as illustrated in Fig. 3, enables the distributed ships and aircraft to truly become a fully integrated system of systems.

The CASoS architecture enables the naval battle group to adapt as a whole to the dynamic threat environment through the interaction of the distributed ships and aircraft. Enabling these interactions creates many more permutations of possible force-level behaviors—limited only by the number and types of warfare assets participating.

The CASoS architecture supports the changing boundary of this solution system. For the naval tactical domain, warfare assets (constituent systems) can join (or leave) the CASoS as new aircraft or ships move near (or away from) a CASoS or as assets are damaged or destroyed during combat.

The CASoS architecture provides interaction mechanisms for the naval tactical domain to support collaborative operations. For integrated fire control, in which distributed weapons and sensor participate in an engagement, the architecture provides "handshakes" of commitment between these assets to support the collaboration for the duration of the engagement. Participation from a remote sensor may include providing a precision cue that detects the threat, fire control quality data to the weapon system, in-flight-target-updates to a missile interceptor in flight, or illumination of a target for endgame missile guidance.

The CASoS architecture must support the following types of adaptation:

- Adaptation in the relationships of the warfare resources, resulting in multi-level and multi-minded responsiveness (at system level and force level).

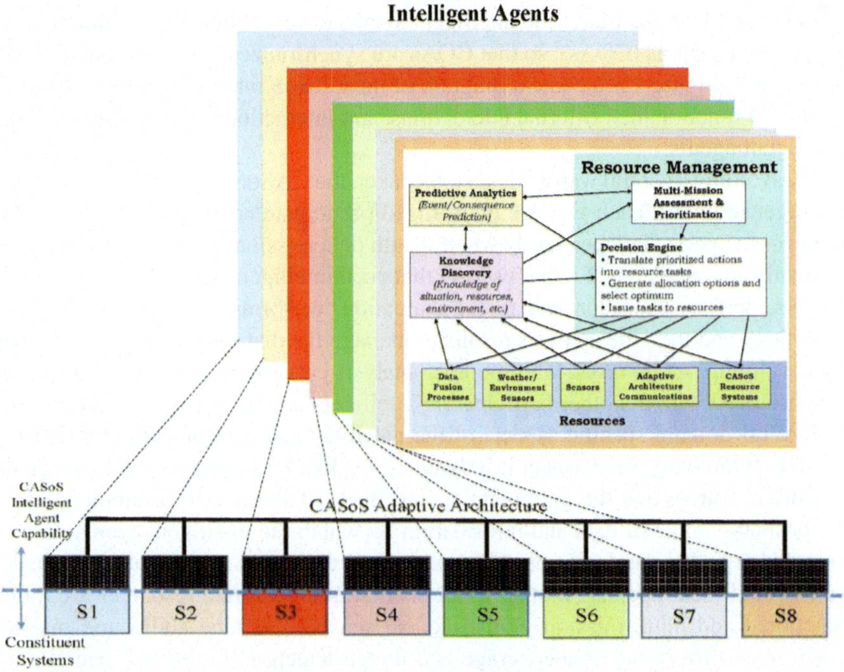

Fig. 3 Adaptive architecture with embedded intelligent agents [1]

- Adaptation as changes occur for the rules, doctrine, plans, and polices governing tactical missions and engagements.
- Adaptation in the level of collaboration: formation of new SoSs, addition or deletion of systems from a collaborative SoS, different levels of collaboration within a SoS.

3.2 *Naval Tactical CASoS System of Intelligent Constituent Systems*

In the naval tactical CASoS solution, an identical intelligent agent is embedded in each distributed warfare platform (ship, aircraft, submarine, etc.). The intelligent agents work together to develop shared knowledge and collaborative decisions regarding tactical courses of action (COA). Through this system of intelligent constituent systems approach, the CASoS empowers each platform—giving each a "god's eye" view of the operational environment (shared situational awareness) and a "god's eye" view of the distributed warfare resources (shared self-awareness). With this knowledge, each platform develops force-level COAs—meaning they can collectively use their distributed warfare resources from a holistic perspective. The adaptive architecture allows the intelligent agents to share data, information, knowledge, and decisions (COAs), so that COAs are synchronized and coordinated across the CASoS. Figure 4 is a context diagram of the CASoS intelligent agent—showing high level functionality, external interactions, and interactions with the other CASoS intelligent agents.

By avoiding a central warfare decision-maker, the CASoS approach of distributing intelligence allows each warfare platform to participate collaboratively as a strike group or to operate independently when stealth or "emissions control" operations are tactically advantageous. Figure 4 shows that each intelligent agent develops internal models, performs mission and threat evaluation, performs wargaming (predictive analytics), and uses the analysis results to manage the distributed warfare resources by developing COA tasks. The models, analysis, wargaming, and COA tasks are synchronized with the other participating intelligent agents within the CASoS.

Several domain specific aspects of the naval tactical environment contribute to CASoS complexity. One aspect is a dynamic feedback loop that exists between the warfare resources and the intelligent agents. Tactical sensors, for example, provide the primary source of data and information by which the intelligent agents develop knowledge and make decisions. These sensors are also resources that are managed by the intelligent agents. Another aspect is that many warfare resources are multi-functional and multi-mission. Some tactical sensors have the ability to sense the environment for broad area coverage and threat detection, focus their energy on a

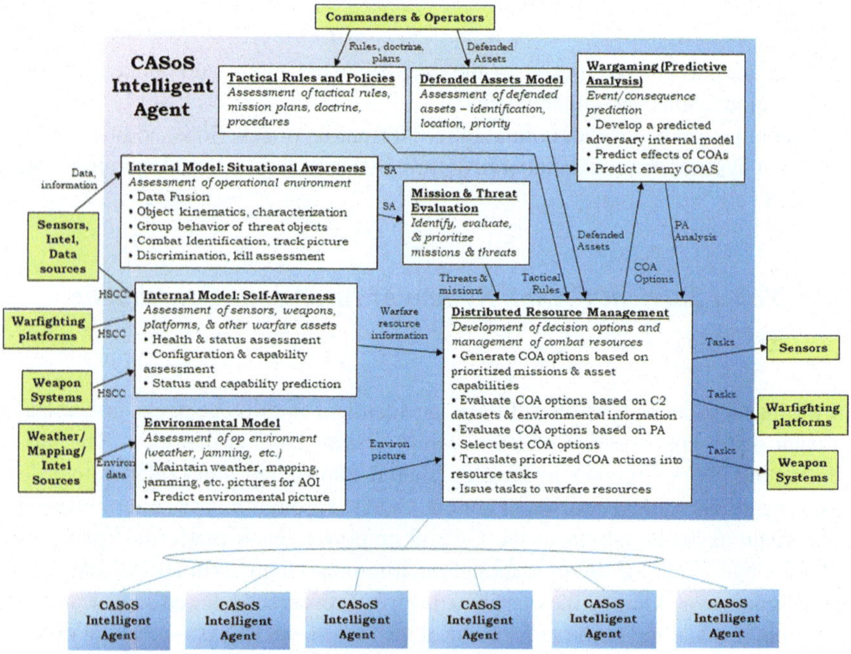

Fig. 4 CASoS intelligent agent context diagram [1]

specific target for higher resolution and higher update rate tracking, and illuminate threat targets to support endgame weapons guidance. In a highly complex operational environment, warfare resources may be in high demand and choices will have to be made about how best to use them.

3.3 Naval Tactical CASoS Knowledge Discovery and Predictive Analytics

Discovering knowledge (or gaining battlespace awareness) and predicting outcomes of actions are both key to a CASoS solution to the naval tactical domain. CASoS knowledge discovery is the attainment and management of knowledge of the entire domain (both problem space and solution space) within the naval tactical decision space. This includes situational awareness (knowledge of the operational environment or problem space) and self-awareness (knowledge of all the assets and resources comprising the solution space). The CASoS knowledge is shared and synchronized among constituent systems to ensure that each constituent has the same knowledge. Figure 5 illustrates examples of shared and self-awareness information in the naval tactical domain.

The four blocks on the left side of Fig. 5 describe types of situational awareness information concerning the problem domain. For the naval tactical domain, this includes: a "track picture" or representation of objects (ships, aircraft, drones, submarines, missiles, etc.) in the environment. CASoS intelligent agents would process, fuse, and analyze different types of data to determine the location, kinematics, and identification of the objects. The intelligent agents would perform object identification (i.e., friendly, neutral, or adversarial), object intent (i.e., neutral or hostile), and object attribution (identifying which country or organization it belongs to). The intelligent agents would develop an internal model of the environment for the region of interest to be considered for calculating environmental effects on resource

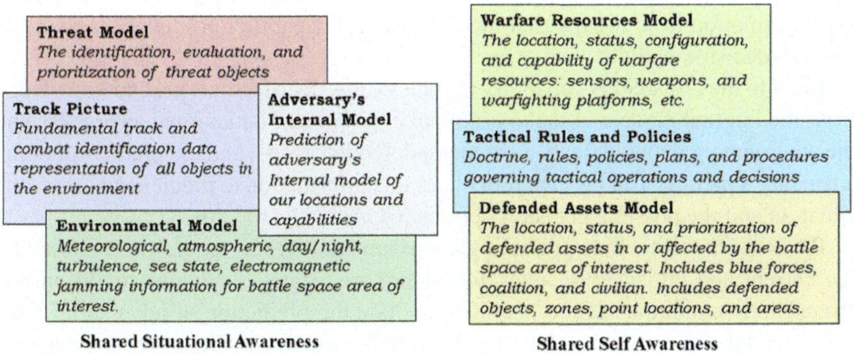

Fig. 5 CASoS internal models for knowledge discovery [1]

capability predictions. The intelligent agents would develop a predictive model of the adversary's situational awareness—the adversary's internal model—to support wargaming analysis.

The right side of Fig. 5 shows three types of information that contribute to shared self-awareness: knowledge of the distributed warfare assets within the CASoS, knowledge of rules and policies affecting the CASoS, and knowledge of defended assets within the area of interest. By developing an internal model of the participating distributed warfare resources, each intelligent agent is able to manage CASoS resources with a "god's eye" perspective. This opens up a much larger selection of behavior based on the many combinations of interactions between distributed resources that can produce desired emergent behavior. Managing knowledge pertaining to rules and policies that guide allowable courses of action ensures that the tactical operations are compliant. Keeping track of defended assets within the area of interest supports courses of action that maximize objectives. In a highly contested area, if resources are limited, the CASoS can prioritize engagements to defend the highest value targets.

Conceptually, the CASoS acquires data from CASoS sensor assets as well as from data and information sources external to the CASoS. This data is shared among the constituent system intelligent agents. The intelligent agents perform data fusion and processing and the adaptive architecture supports synchronization among the agents to develop the shared internal models. The intelligent agents analyze these models for uncertainty and incompleteness. The results of this analysis are used to update the tasking of sensor assets—to collect more data for specific objects or regions to improve the uncertainty or incompleteness in the situational awareness. Thus, the CASoS process of developing situational awareness is an adaptive process of managing knowledge uncertainty through continuous analysis and sensor feedback tasking.

CASoS predictive analytics (PA) provides a real-time wargaming capability to assess decision options during all phases of tactical operations: for force readiness, offensive operations, and even during defensive and combat operations when the decision reaction time is very short. The CASoS PA concept is illustrated in Fig. 6. The PA is a set of data analytics that assess decision options based on knowledge and information from the internal models. The PA capability produces assessments of COA decision options.

Input to the CASoS PA capability includes COA decision options and knowledge from the internal models of the operational environment (situational awareness and the environmental model), self-awareness, defended assets, and the prediction of the adversary's picture. The PA capability uses this information to predict and assess the situation and the possible decision options (COAs).

PA assessments include predicting environmental effects on the COAs based on the current and projected weather, turbulence, sea states, day/night condition and atmospheric conditions. They may also include the prediction of adversarial electronic warfare, jamming, and clutter conditions based on possible adversary capabilities. Environmental effects can greatly influence sensor performance and can therefore be used to assess the sensor tasking options. Environmental effects can also affect

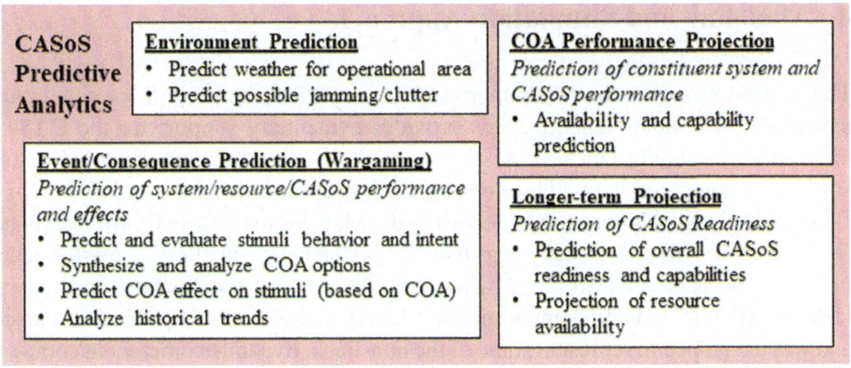

Fig. 6 Predictive analytics for a naval tactical CASoS [1]

weapon selection—weapons that depend on sensors may diminished performance in certain conditions. Directed energy weapons performance is largely affected by environmental effects. Weapons may be affected by possible adversarial jamming or countermeasures.

PA assessment includes the performance projection of warfare assets. The PA capability would assess COA options based on the warfare resource internal model of self-awareness. For example, from this knowledge, the PA could calculate the probability of detection or probability of kill based on sensor or weapon status, location, and expected capability performance.

The PA capability could perform longer-term assessment of CASoS force readiness. This prediction capability would coordinate tactical readiness with planning and strategic goals. The PA could predict the adversary's projected capabilities and intent and then develop and assess longer-term tactical plans and strategies to prepare warfare resources (sensor coverage, weapons load-out), platform locations, stealth and emission control operations, and overall force readiness.

The PA capability could also provide a real-time tactical wargaming capability that adds strategic thinking to time critical tactical operations. PA wargaming would predict the effects of COA options—predicting adversarial responses based on the estimated adversarial internal model (situational awareness of the blue forces), strategies, and capabilities. Predicted adversarial responses could be taken into account in the selection of COA alternatives. PA wargaming would also assess COA alternatives to determine which has the highest probability of success and the best chance of protecting defended assets. Consequence prediction and assessment would provide additional insight into the selection of COAs for tactical operations.

4 Modeling and Simulation Approach

This section describes the M&S approach taken to study the CASoS solution to the naval tactical domain. This analysis provides evidentiary support for the CASoS theory by presenting data results showing tactical improvements using a CASoS solution as compared with a baseline non-CASoS approach.

The M&S analysis was conducted using the Map Aware Non-Uniform Automata (MANA) tool. MANA is an agent-based, time-stepped, stochastic mission-level modeling environment developed by the New Zealand Defense Technology Agency [16]. The MANA modeling environment is based on the ideas of complexity science and was developed to represent some of the non-linear dynamics inherent in complex combat environments. It accomplishes this by treating many aspects of combat behavior as simple rules subject to stochastic random probabilistic processes.

The model represented abstractions of a complex tactical problem domain and naval tactical solution to highlight the differences between a baseline (non-collaborative) approach with a CASoS approach. Many aspects of a real-world tactical scenario were simplified in the model; however, the model provides further understanding into how the collaborative and adaptive behavior of distributed constituent systems produce desired emergence. The model did not provide an exact and exhaustive solution to the tactical problem domain; but instead, used this complex domain as a use case to explore the CASoS approach.

In order to validate the CASoS theory, the M&S scenario had to contain the characteristics of a highly complex operational environment. Table 1 lists the characteristics of highly complex operational environments and describes how these characteristics are represented in the model's scenario.

The resulting scenario in MANA is a littoral A2/AD missile threat environment able to demonstrate the characteristics of a highly complex operational environment. Incorporating sea-based and shore-based components enabled the scenario to easily represent geographically-distributed red force ships and land-based launchers. This allowed the simulation to randomly generate two different types of missile threats being fired from distributed locations in relation to the blue force strike group. As listed in Table 1, this created a model scenario with objects, features, and events that were distributed, randomly-generated, unpredictable, kinematically diverse, rapidly occurring, and lethal. The incoming missiles were given an initial set of parameters: number of missiles (32 anti-ship and 12 anti-aircraft), lethality (0.8), and speed (1111 km/hr anti-ship and 5186 km/hr anti-aircraft); which could be increased to represent higher levels of complexity. The initial parameters were set so they would present a complex situation without overwhelming and quickly annihilating the blue force. The intent was to study how the blue force's systems behavior could address the threat environment as a comparison of a baseline non-CASoS set of blue force actions with a CASoS approach of blue force collaborative interactions. The randomness of red force missile launches presented unique and unexpected threat scenarios for each simulation run. This created an environment that was challenging in terms of the blue force's ability to gain adequate situational awareness and defend with engagement

Table 1 Characteristics of a highly complex problem domain represented in the model [1]

Characteristics of a highly complex environment	How the characteristics are represented in the model
Large numbers of objects/events/features	A relatively large number of red force missile threats
Heterogeneity and/or diversity of environment objects/events/features	Multiple types of threats: anti-ship missiles and anti-aircraft missiles
Geographically distributed	The red force launch sites are distributed in the scenario
Diverse kinematics among objects in the environment	Threat missiles are launched from different locations and at different (random) times—results in diverse threat kinematics
Environment's objects/events/features are highly interrelated and/or highly related to the solution	The threats are aimed at the blue forces, thus creating a highly interrelated situation between the blue force and its environment
Highly dynamic/rapid tempo of change	The kinematics and speed of the threats creates a rapidly dynamic tempo of events
Uniqueness of situations or states	The randomly generated threats create a unique set of states in each run of the model
Severe consequences of environment behaviors and events	The threats can kill blue forces if not successfully engaged
Unexpected and rapid shifts in states; behaviorally unpredictable	The threats are unexpected and can only be known by the blue forces once they enter the sensor detection range
Unknowable—difficult to gain accurate situational awareness	The threats are unknown until they enter the detection range of the blue force sensors
Accompanied by constraints, rules, and parameters	The threats cannot be engaged until they are within the allowable weapons engagement range

weapons. The blue forces were equipped with sensors with set detection ranges and weapons with set parameters for engagements. The sensor and weapon constraints represented real world limits on the abilities to detect and engage missile threats.

The M&S scenario, illustrated in Fig. 7, contains a blue force high value unit (HVU) ship escorted by a strike group consisting of six destroyers (DDG), and two airborne early warning (AEW) aircraft. Two blue force fighter aircraft happen to be nearby. The blue force strike group must safely escort the HVU through highly contested waters. The A2/AD littoral region contains a red force with threats consisting of 32 anti-ship missiles launched from ships and eight anti-aircraft missiles launched from land.

The blue force strike group must traverse the operational area, reach land, and then traverse back out of the operational area. The blue force strike group consists of a HVU ship, accompanied by six DDGs and two surveillance aircraft (AEW). The blue force is also aided by two fighter aircraft that are not initially part of the force

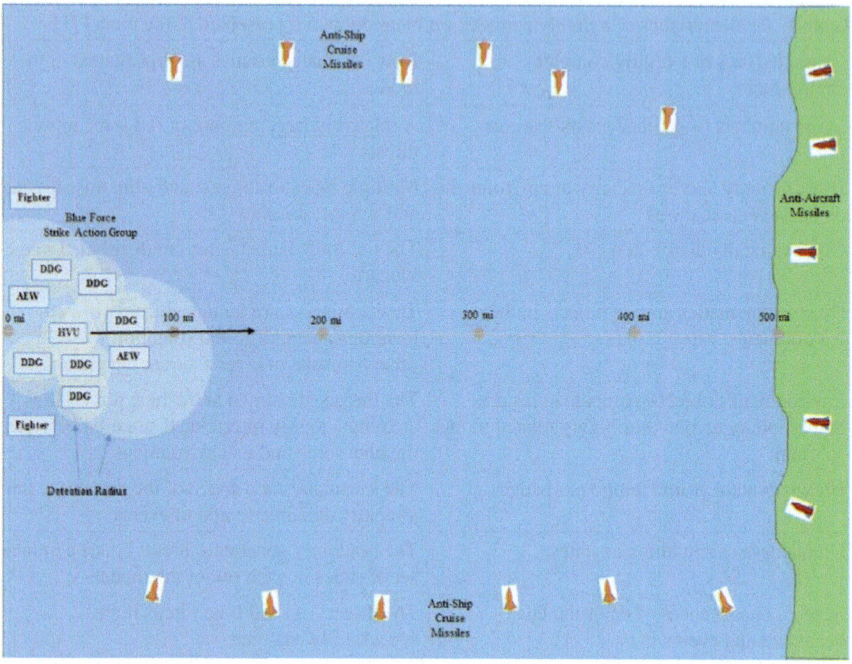

Fig. 7 Naval tactical modeling and simulation scenario [1]

group but join in the battle. The overall mission objective is for the HVU to safely reach the land and then safely traverse back out of the A2/AD region.

The threat consists of anti-ship missiles which only attack the HVU and DDGs, and anti-aircraft missiles which are land-based and only attack the AEWs and fighter aircraft. The HVU and destroyers can engage both anti-ship and anti-aircraft red force missiles. The fighter aircraft can only engage anti-aircraft red force missiles.

The blue force strike group moves together at a constant speed and in a constant, fixed formation (distances between the ship and aircraft platforms are set and with an approximate separation of 20 nautical miles). They are traveling at a higher-than-normal cruising speed of 30 nm/hour as they are in an A2/AD environment. They approach the land along a perpendicular trajectory, and then leave the land along the same trajectory. The two fighter aircraft move at a higher speed and each enters the scenario at a random time after the start.

Tables 2 and 3 contain the initial model parameters for the blue and red force assets. These parameters include the sensor detection ranges, the weapon ranges and lethality (probability of kill), and the speeds of the blue force assets. The threat parameters include the numbers of missiles, the missile lethality (probability of kill), and the missile speeds.

Two approaches were modeled: (1) a baseline (or non-CASoS approach) representing the current approach to naval tactical operations, and (2) a CASoS alternative representing a CASoS approach to naval tactical operations.

Table 2 Initial model parameters for blue force assets [1]

Warfare asset	#	Detection capability	Weapon	Speed (nmi/hour)
Surveillance aircraft—airborne early warning (AEW)	2	100 mi range (all directions)	None	300
High value unit (HVU) ship	1	30 mi range	Short range weapon: 20 mi range $P_{kill} = 0.9$ 1 launch per second	30
Destroyer ship (DDG)	6	30 mi range	Long range weapon: 100 mi range $P_{kill} = 0.9$ 1 launch per 10 s	30
			Short range weapon: 30 mi range $P_{kill} = 0.9$ 1 launch per second	
Fighter aircraft	2	30 mi range	AMRAAM: 30 nmi range $P_{kill} = 0.9$ 1 launch per 2 s	400
			JDAM: 15 nmi range $P_{kill} = 0.7$ 1 launch per 2 s	

Table 3 Initial model parameters for red force assets [1]

Threats	Lethality	Number	Speed (km/hr)
Anti-ship missiles	$P_{kill} = 0.8$	32 (8 per each of 4 red ships)	1111
Anti-aircraft missiles	$P_{kill} = 0.8$	12 (4 per each of 3 launchers)	5186

4.1 Baseline (Non-CASoS Approach)

Currently, naval ship and aircraft assets are connected through Link-16, over-the-horizon message types, chat, message traffic and voice communications [10]. Miller [13] explains that a variety of naval decision-makers, including the Joint Interface Control Officer (JICO), the Command, Control, Communications, and Computer (C4) Officer and the N2 (Naval Intelligence) officer, are equally responsible for ensuring the assets work together. The current approach is highly manual and slow in response and decision time; and, miscommunication and misunderstanding are rampant [13].

The model represented the naval baseline approach as a set of distributed blue force assets without shared situational awareness for this real-time scenario. Each asset used only its own organic (or resident) sensor to gain detection of the threat. Further, the distributed blue force assets did not coordinate their defensive engagements. Each individual ship and fighter aircraft was configured to fire at an enemy threat missile after detection and when the threat was within range of its defensive weapon. In the baseline model, the two fighter aircraft were configured with random movement, independent situational awareness and engagements with red force anti-aircraft missiles.

4.2 CASoS Alternative

In the CASoS alternative, the model was configured to represent a CASoS approach to naval warfare. In this model, the distributed blue force assets functioned collaboratively as a CASoS. The blue force ships and aircraft have shared situational awareness, meaning that as soon as one of their sensors detects a threat, all of the blue force assets gain this knowledge. The blue force also shared self-awareness, or knowledge of each other's weapon capabilities. This is represented in the model as coordinated engagements against the threats. This was modeled by selecting the blue force asset closest to the threat to fire an engagement shot first and then allowing another asset to fire if the first shot failed.

In the CASoS alternative model, the two fighter aircraft are initially not part of the CASoS. As soon as each fighter aircraft gets within 30 miles of one of the blue force strike group's assets, it becomes part of the CASoS force structure network and gains shared situational awareness (both as a contributor of detected threats and as a recipient of the blue force's detected threats). The aircraft joins the formation and changes its speed to cover the strike group. Its movement is no longer random, but instead matches the strike group movement. The fighter aircraft are equipped with Joint Direct Attack Munitions (JDAM) for bombing red force missile launchers and Advanced Medium-Range Air-to-Air Missiles (AMRAAM) for engaging anti-aircraft missiles.

5 Modeling and Simulation Analysis Results

Two M&S analyses were conducted. The first M&S analysis compared the behavior of the baseline approach with the behavior of the CASoS approach using the initial threat scenario parameters. The second M&S analysis studied the effects of increased complexity in the environment by comparing its effect on the baseline and CASoS alternatives. In the second M&S analysis, an increase in the operational environment's complexity was modeled by increasing the number, lethality, and speed of the red force threats. The same set of evaluation metrics were used in both analyses. Each

M&S experiment involved 200 simulation runs (100 for the baseline and 100 for the CASoS). The metrics for the analyses are shown in Table 4.

The evaluation metrics provide insight into the interactions of the blue force solution approaches with their complex operational environment. Collecting data showing the number of blue force casualties, the time required to defend against (or kill) red forces, and the expenditure of weapons resources, provided evidence of the differences between a CASoS and non-CASoS approach to implementing a set of distributed constituent systems, or in this case, blue force assets.

The number of blue force casualties is an overarching indicator of the performance of the layered defense—of how well and how quickly the distributed and heterogeneous weapons resources are used. This metric indicated the contributions of an adaptive and collaborative architecture and purposeful emergent behavior, which the CASoS approach provided. It also showed implementation of important CASoS principles such as the information principle, the contextual principle, the goal principle, and the principle of holism.

The time required to kill 50% of the red forces provided a means to evaluate whether the CASoS approach would be able to decrease the decision reaction time, thereby defending assets more quickly. This evaluation measure demonstrated the contribution of shared and improved situational awareness, which is enabled by the CASoS collaborative architecture, information principle, and contextual principle.

The amount of weapons resources expended is indicated by the evaluation metrics for numbers of DDG long and short range weapons fired and fighter aircraft AMRAAMs and JDAMs launched. These metrics provided insight into the ability of the two approaches to provide a layered defense. They measured how well the two approaches were able to make use of the different types of distributed weapons resources. The results indicated the contributions of CASoS characteristics (adaptiveness, collaboration, purposefulness, self-organization, emergence, constituent variety, changing boundary) and principles (holism, high flux, information, contextual, requisite variety).

Table 4 Evaluation metrics for M&S analysis [1]

Metric	Metric description
Blue force casualties	Total blue assets killed
Time required to kill 50% of the red threat	Time, in time steps, that it took to kill 50% of the red force
Number of DDG long range weapons fired	Total DDG long-range weapons used (sum over all 6 DDGs)
Number of DDG short range weapons fired	Total DDG short-range weapons used (sum over all 6 DDGs)
Number of fighter AMRAAMs fired	Total Fighter AMRAAMs used (sum over both fighters) (AMRAAM = Advanced medium range air-to-air missiles)
Number of Fighter JDAMs Fired	Total Fighter JDAMs used (sum over both fighters) (JDAM = Joint direct attack munition)

The following two subsections contain the results of the two M&S analyses.

5.1 M&S Analysis #1—Comparison of the Current (Baseline) Approach with a CASoS Approach

The first M&S analysis compared the baseline and CASoS alternatives to study the similarities and differences of naval tactical behavior in the two approaches. Each model variant was run 100 times and data was collected according to the evaluation metrics shown previously in Table 4.

For the case of the CASoS alternative, there were significantly fewer blue force casualties and the time required to kill 50% of the red force was significantly lower. The ability for the distributed blue force assets to function collaboratively and exhibit desired emergent behavior in terms of shared and increased situational awareness and cooperative engagements, resulted in these tactical improvements.

In the CASoS alternative, the DDGs used many fewer short range weapons and many more longer range weapons because they had greater situational awareness and could detect threats in time to support the longer range weapons engagement requirements. The CASoS approach was, in effect, able to employ a layered defense tactic; whereas the baseline approach was only able to employ a short-range close-in defense tactic.

In the CASoS alternative, the fighter aircraft used more of the AMRAAM missiles because they were called into the fight much sooner and could respond adaptively to the threat. This indicated that the fighter aircraft took many more shots (or exhibited more purposeful and participatory behavior) when they were able to collaborate as part of the CASoS.

Table 5 contains a summary of the M&S #1 analysis results. The analysis demonstrated that the CASoS improved layered defense options, reduced casualties, and improved the engagement decision reaction time.

The improvements in blue force defense and decision reaction time demonstrate the benefits of a CASoS approach. The adaptive architecture connecting the distributed blue force assets enable the strike group to function as a CASoS. This enables solution behavior that is multi-level, emergent, adaptive, purposeful, and collaborative. The model demonstrates that enabling a changing boundary and maximizing the use of constituent variety (a diverse set of warfare resources) provided a layered defense approach that resulted in tactical improvements.

Table 5 M&S #1 analysis results [1]

Evaluation metrics	Baseline mean	CASoS mean	Analysis results
Blue force casualties	2.01	0.78	Alternative hypothesis accepted: Mean number of blue casualties was lower in the CASoS alternative
Implications for the CASoS theory validation:	Indicates that the distributed, collaborative, and adaptive architecture, which enables shared situational awareness and behavior that is collaborative, adaptive, and purposeful led to fewer blue force casualties		
Time required to kill 50% of the red threat	40.9 min	25.53 min	Alternative hypothesis accepted: Mean time required to kill 50% of the red forces was lower in the CASoS alternative
Implications for the CASoS theory validation:	Indicates that the CASoS architecture and behavior leads to increased and shared situational awareness which decreases the time required to make engagement decisions and increases the effective engagement range of the blue force weapons. Earlier and more effective engagement shots lessens the amount of time required to defend against the red force threats		
Number of DDG long range weapons fired	26.43	174.7	Alternative hypothesis accepted: Mean number of DDG long range weapons was higher in the CASoS alternative
Implications for the CASoS theory validation:	Indicates that the CASoS architecture and behavior enables increased and shared situational awareness. This earlier and shared detection of red force threats increases the number of long range weapons that can be fired—giving the blue force a significantly improved layered defense		
Number of DDG short range weapons fired	60.02	48.83	Alternative hypothesis accepted: Mean number of DDG short range weapons fired was lower in the CASoS alternative
Implications for the CASoS theory validation:	Indicates that the CASoS's increased use of long range weapons decreases the number of short range weapons that need to be used. This is due to the fact that the red force threats are engaged by the long range weapons at larger ranges from the blue force assets. Fewer red force threats get close enough to the blue force assets to require the use of the short range weapons		

(continued)

Table 5 (continued)

Evaluation metrics	Baseline mean	CASoS mean	Analysis results
Number of fighter AMRAAMs fired	3.96	21.8	Alternative hypothesis accepted: Mean number of fighter aircraft AMRAAMs fired was higher in the CASoS alternative
Implications for the CASoS theory validation:	This indicates that the ability of the fighter aircraft to join the Blue Force CASoS enables them to participate in the CASoS mission of defending the blue force HVU. Upon joining the CASoS, the fighter aircraft have increased and shared situational awareness with the rest of the blue force assets. They have significantly improved awareness of red force threat location and can purposefully coordinate their use of AMRAAM weapons as part of the blue force layered defense		
Number of fighter JDAMs fired	7.64	7.38	Null hypothesis accepted: The mean numbers of fighter aircraft JDAMs fired was equivalent in the baseline and CASoS alternatives
Implications for the CASoS theory validation:	There was not a significant difference in the number of JDAMs fired in the baseline and CASoS alternatives. In the baseline alternative, the fighter aircraft motion was random. In the CASoS alternative, the fighter aircraft moved in conjunction with the blue force strike group. The blue force movement and random movement created approximately the same number of opportunities for the fighter aircraft to launch JDAMs		

5.2 *M&S Analysis #2—Effects of Increasing the Complexity of the Scenario Environment*

The second M&S analysis increased the complexity of the threat scenario by increasing the number, speed, and lethality of the red force threat to analyze these effects on the baseline and CASoS alternatives. The purpose of this analysis was to study the behavior of the two approaches (baseline and CASoS) as they encountered a heightened level of complexity in the operational environment with a greater number of events (more missiles), a faster tempo of events (with greater missile speed), and more deadly consequences (greater missile lethality). The same evaluation metrics (listed in Table 4) were used in this second M&S analysis.

A design of experiments (DOE) was conducted to vary the number, speed, and lethality of the red force threats to identify values for these threat parameters that would provide a more stressing environment without completely overwhelming the blue force strike group. The DOE crossed a model alternative factor (2 levels, CASoS/baseline) with a 33 Design Point (DP) Nearly Orthogonal Latin Hypercube (NOLH) for the three threat factors, resulting in $2 \times 33 = 66$ DPs, each run with 100 stochastic replications. The analysis was based upon 6600 runs. The evaluation produced the following parameters to represent the more stressing environment: a threat number of 12 missiles per launcher, a threat speed of 6000 km/hr, and a threat lethality (probability of kill) of 0.9. Therefore, there were 48 anti-ship missiles and 36 anti-aircraft missiles in this second M&S scenario, as compared with 32 anti-ship missiles and eight anti-aircraft missiles in the first M&S scenario. The threat speeds in the first M&S scenario were 1111 km/hr for the anti-aircraft missiles and 5186 km/hr for the anti-aircraft missiles. These speeds were increased to 6000 km/hr for both the anti-aircraft and anti-ship missiles in the second scenario. Finally, the probability of kill was raised from 0.8 in the first M&S scenario to 0.9 in the second M&S scenario. These values increased the complexity of the threat environment without completely destroying the blue force assets as represented in both the baseline and CASoS approaches.

For the more stressing threat scenario of M&S #2, the CASoS solution alternative continued to have fewer blue force casualties, and a shorter time required to kill 50% of the red forces than the baseline approach. This demonstrates the tactical benefits of implementing the distributed warfare assets as a CASoS—a collaborative and adaptive system of distributed constituent systems with desired (purposeful) emergent behavior.

In M&S #2, the CASoS used a significantly greater number of long-range DDG weapons and fighter aircraft AMRAAMS than the baseline approach. However, the CASoS used approximately the same number of short-range DDG weapons and fighter aircraft JDAMs, on average, as the baseline approach. The use of the fighter aircraft weapons resources followed a similar behavior as in M&S #1. Upon joining the CASoS, in the CASoS alternative, the fighter aircraft were able to fire many more AMRAAMs, benefitting from the collaborative, adaptive architecture with improved and shared situational awareness (earlier threat detection), and changing boundary to allow the fighters to join the CASoS. This clearly demonstrates the utility of these CASoS characteristics and principles.

The use of fighter aircraft JDAMs followed a similar behavior pattern as in M&S #1. In both the first and second M&S analyses, the fighter aircraft fired a little over seven JDAMs on average in both the baseline and CASoS solution approaches. This coincided with the number of red force launch sites, which in both M&S #1 and M&S #2, was seven. On average, the fighter aircraft destroyed the seven red force launch sites in the simulation runs and then ran out of targets. In some cases, the JDAMs missed their target and fired again—resulting in more than seven total JDAMs fired. In some cases, fewer than seven JDAMs were fired because the fighter aircraft were not close enough proximity to the launch sites. The use of JDAMs in the modeling analyses did not end up being a distinguishing factor between the baseline and CASoS alternatives; however, it provided more insight into the behavior and utility of having constituent variety in the available resources. A closer examination of individual simulation runs, showed that in both the baseline and CASoS variants, there were instances in which the JDAMs destroyed red launch sites early in the run, resulting in fewer blue force casualties and decreasing the time required to kill 50% of the red forces. The CASoS approach could evolve to improve the use of the JDAMs with more sophisticated predictive analytics and intelligent agents.

Table 6 contains a summary of the M&S #2 analysis results. The analysis demonstrated that the CASoS continued to decrease blue force casualties, improve decision reaction time, and maximize the use of distributed heterogeneous assets as a layered defense; even as the threat situation grew in complexity.

The improvements in blue force defense and decision reaction time demonstrate the benefits of a CASoS approach. The adaptive architecture connecting the distributed blue force assets enable the strike group to function as a CASoS. This enables solution behavior that is multi-level, emergent, adaptive, purposeful and collaborative. Analyzing how the CASoS solution approach behaved in a more complex environment, provided more insight into the value of the CASoS characteristics and principles. The model demonstrated that the CASoS solution approach was able to adapt to the changing threat environment by doubling its use of short range weapons to continue to provide tactical advantages over the baseline approach. The CASoS adapted its behavior and maximized the use of its constituent variety (diverse warfare resources) to provide an improved layered defense strategy.

6 Conclusion

In summary, the CASoS solution alternative demonstrated tactical improvements over the baseline alternative in an initially complex threat environment; and continued to demonstrate improvements in a more stressing threat environment. The results showed significant improvements in (1) blue force defense (lowering the number of casualties), (2) decision reaction time (destroying red forces more quickly), and (3) layered defense (maximizing the use of different warfare weapon assets).

As the threat environment grew in complexity (from M&S #1 to M&S #2), the mean number of blue force casualties increased for both the baseline and CASoS approaches. However, there were fewer blue force casualties using the CASoS approach; and therefore, the CASoS was better able to address the increased threat than the baseline approach.

The overall time required to kill 50% of red forces, for both the baseline and CASoS alternatives, decreased as the threat environment became more complex. The scenario had a much faster tempo as the speeds of the incoming red threat missiles were much faster in the second scenario. The CASoS alternative had a significantly higher speed of defense—killing 50% of red forces much faster than the baseline, even as the threat environment was more stressful.

The CASoS alternative employed a layered defense—making better use of the variety of weapons than the baseline alternative. The baseline approach used a close-in defense, depending primarily on the DDG short range weapons to defend the blue force ships. The baseline used only roughly 15% of the amount of long range DDG weapons and 18% of the AMRAAMs used by the CASoS. These percentages stayed the same as the threat environment became more complex. The baseline also took advantage of the random path of the nearby fighter aircraft to launch JDAMs at the red force launch sites if they happened to be within proximity. The CASoS

Table 6 M&S #2 analysis results [1]

Evaluation metrics	Baseline mean	CASoS mean	Analysis results
Blue force casualties	3.56	2.74	Alternative hypothesis accepted: Mean number of blue casualties was lower in the CASoS alternative
Implications for the CASoS theory validation:	Demonstrates that the distributed, collaborative, and adaptive architecture, which enables shared situational awareness and behavior that is collaborative, adaptive, and purposeful led to fewer blue force casualties		
Time required to kill 50% of the red threat	28.3 min	20.84 min	Alternative hypothesis accepted: Mean time required to kill 50% of the red forces was lower in the CASoS alternative
Implications for the CASoS theory validation:	Demonstrates that the CASoS architecture and behavior leads to increased and shared situational awareness which decreases the time required to make engagement decisions and increases the effective engagement range of the blue force weapons. Earlier and more effective engagement shots lessens the amount of time required to defend against the red force threats		
Number of DDG long range weapons fired	22.14	167.42	Alternative hypothesis accepted: Mean number of DDG long range weapons was higher in the CASoS alternative
Implications for the CASoS theory validation:	Demonstrates that the CASoS architecture and behavior enables increased and shared situational awareness. This earlier and shared detection of red force threats increases the number of long range weapons that can be fired—giving the blue force a significantly improved layered defense		
Number of DDG short range weapons fired	77.38	82.16	Null hypothesis accepted: The mean number of DDG short range weapons fired were not dissimilar enough between the two alternatives to reject the null hypothesis
Implications for the CASoS theory validation:	Demonstrates that given a more stressing threat environment, the CASoS relied as heavily on short range DDG weapons as the baseline approach. This indicates that the CASoS adapts its behavior to use both long range and short range weapons in a collaborative way as the threat grows in complexity		

(continued)

Table 6 (continued)

Evaluation metrics	Baseline mean	CASoS mean	Analysis results
Number of fighter AMRAAMs fired	4.31	23.19	Alternative hypothesis accepted: Mean number of fighter aircraft AMRAAMs fired was higher in the CASoS alternative
Implications for the CASoS theory validation:	Demonstrates that the ability of the fighter aircraft to join the Blue Force CASoS enables them to participate in the CASoS mission of defending the blue force HVU. Upon joining the CASoS, the fighter aircraft have increased and shared situational awareness with the rest of the blue force assets. They have significantly improved awareness of red force threat location and can purposefully coordinate their use of AMRAAM weapons as part of the blue force layered defense		
Number of fighter JDAMs fired	7.54	7.86	Null hypothesis accepted: The mean numbers of fighter aircraft JDAMs fired was equivalent in the baseline and CASoS alternatives
Implications for the CASoS theory validation:	There was not a significant difference in the number of JDAMs fired in the baseline and CASoS alternatives. In the baseline alternative, the fighter aircraft motion was random. In the CASoS alternative, the fighter aircraft moved in conjunction with the blue force strike group. The blue force movement and random movement created approximately the same number of opportunities for the fighter aircraft to launch JDAMs		

approach used a layered defense strategy, making the most of the different types of weapons available. In the initial scenario, the CASoS primarily depended on long-range DDG weapons as well as fighter aircraft AMRAAMs and JDAMs. As the scenario became more complex, the CASoS approach doubled the number of short-range DDG weapons fired while still using about the same numbers of long-range weapons, AMRAAMs, and JDAMs. This demonstrated the ability of the CASoS to adapt to the changing circumstances and to take advantage of the constituent system variety.

Table 7 compiles the results of M&S #1 and #2 to summarize the mean values computed for each evaluation metric. The table compares the results gathered in the initial complex scenario of M&S #1 with the more stressing scenario of M&S #2. The last column in the table contains descriptions of how the results demonstrate the CASoS theory.

Table 7 Summary of M&S analysis results [1]

Evaluation metrics	Initial complex scenario: baseline mean	More stressing scenario: baseline mean	Initial complex scenario: CASoS mean	More stressing scenario: CASoS mean
Blue force casualties	2.01	3.56	0.78	2.74
Implication for the CASoS approach:	Demonstrates that the CASoS distributed, collaborative, and adaptive architecture, which enables shared situational awareness and behavior that is collaborative, adaptive, and purposeful, led to fewer blue force casualties; even with an increase in threat complexity			
Time required to kill 50% of the red threat	40.9 min	28.3 min	25.53 min	20.84 min
Implication for the CASoS approach:	Demonstrates that the CASoS architecture and behavior leads to increased and shared situational awareness which decreases the time required to make engagement decisions and increases the effective engagement range of the blue force weapons. The CASoS's earlier and more effective engagement shots decrease the amount of time required to defend against the red force threats			
Number of DDG long range weapons fired	26.43	22.14	174.7	167.42
Implication for the CASoS approach:	Demonstrates that the CASoS architecture and behavior enables increased and shared situational awareness. This earlier and shared detection of red force threats increases the number of long range weapons that can be fired—giving the blue force a significantly improved layered defense			
Number of DDG short range weapons fired	60.02	77.38	48.83	82.16
Implication for the CASoS approach:	Demonstrates that the CASoS approach enabled adaptive behavior by doubling the number of short-range weapons fired as the scenario became more complex. The CASoS used its adaptive behavior and constituent variety to employ a more layered defense to address the increased threat			
Number of fighter AMRAAMs fired	3.96	4.31	21.8	23.19
Implication for the CASoS approach:	Demonstrates the CASoS changing boundary, adaptive architecture, purposeful collaborative behavior, and constituent variety. The CASoS continued to take advantage of the fighter aircrafts' contributions even as the threat increased			
Number of fighter JDAMs fired	7.64	7.54	7.38	7.86
Implication for the CASoS approach:	Demonstrates that the baseline and CASoS alternatives took advantage of the fighter aircraft JDAMs. In the baseline, the fighter aircraft's random movement placed them in proximity of red force launch sites. In the CASoS, the fighter aircraft joined the CASoS and then fired JDAMs as the strike group as a whole approached red force launch sites			

References

1. Johnson, B.: A Framework for Engineered Complex Adaptive Systems of Systems. Naval Postgraduate School (2019)
2. Calvano, C., John, P.: Systems engineering in an age of complexity. IEEE Eng. Manage. Rev. **32**(4), 29–38 (2004)
3. Alberts, D.: Agility Advantage. Command Control Press, Washington, DC (2011)
4. Bar-Yam, Y.: Making Things Work: Solving Complex Problems in a Complex World. New England Complex Systems Institute Knowledge Press, St. Louis, MO (2004)
5. Hughes, W., Girrier, R.: Fleet Tactics and Naval Operations, 3rd edn. Naval Institute Press, Annapolis, MD (2018)
6. Hughes, W.: Fleet Tactics and Coastal Combat, 2nd edn. Naval Institute Press, Annapolis, MD (2000)
7. Alberts, D., Garstka, J., Stein, F.: Network Centric Warfare: Developing and Leveraging Information Superiority. Command and Control Research Press, Washington, DC (2000)
8. Talbot, P., Ellis, D.: Applications of Artificial Intelligence for Decision-Making. Amazon Digital Services, LLC (2015)
9. Young, B.: Integrated Fire Control for Future Aerospace Warfare. Presented at the National Fire Control Symposium, Kauai, HI (2004)
10. Treadway, W.: OPNAV N2/N6 discussion. Washington, DC (2019)
11. Eckstein, M.: NIFC-CA Advances Could Allow the Navy to use Cheaper 'Dumb' Weapons. USNI News, 22 February (2017)
12. Young, B.: Future integrated fire control. In: Proceedings of the International Command and Control Research and Technology Symposium. Alexandria, VA (2005)
13. Miller, S.: NPS discussion concerning current naval tactical practices. Monterey, CA (2019)
14. Silva, E., Batista, T., Oquendo, F.: A mission-oriented approach for designing system-of-systems. In: Proceeding of the 10th System of Systems Engineering Conference, pp. 346–351 (2015)
15. Giachetti, R.: System of systems capability needs analysis via a stochastic network model. Naval Eng. J. **127**(4), 67–79 (2015)
16. Lauren, M., Stephen, R.: Map-aware non-uniform automata (MANA)—a New Zealand approach to scenario modeling. J Battlefield Technol **5**(1), 27–31 (2002)

Deterministic Chaos Constraints for Control of Massive Swarms

Josef Schaff

Abstract Problems such as the control of massive swarms or MANET radios in a Battlespace on the order of tens of thousands of entities is challenging with conventional approaches. Entities that enter or leave this space need to have either some means of coordination or known constraints. Moreover, each entity should have some limited awareness of the relative positioning of all the others, without the massive computational expense incurred by a combinatorial explosion. This may rule out the quantum state space type of approach where each entity is a computationally-unique unit vector. To control such behaviors, and allow self-organizing MANET relay nodes as well as collective swarm control, the author uses a topological structure based upon deterministic chaos—the fractal. Constraining the entities to "believe" that they exist only within the boundaries of an adaptive fractal forces their topological layout to map to topological clusters. The invariant features of a fractal topology allow each node to compute the IFS (Iterative Function System) to effectively "know" the relative positions of all the other entities, and form adaptive, self-healing MANET or swarm clusters. One of many possible equations for the adaptive fractal is given, which computationally is of O(n). Since it uses one floating point number, the computations for each entity/node to determine the relative positions of all Battlespace entities is trivial, therefore each entity can achieve situational awareness in near-realtime. A simulation of the dynamics for different adaptive fractal topologies written in Mathematica, can be demonstrated during the paper presentation.

Keywords Large-scale swarm dynamics · Adaptive fractal · Large-scale Manet

---

J. Schaff (✉)
Naval Air Systems Command, Patuxent River, USA
e-mail: josef.schaff@navy.mil

D. Braha et al. (eds.), *Unifying Themes in Complex Systems X*, Springer Proceedings in Complexity, https://doi.org/10.1007/978-3-030-67318-5_18

1 Problem: Scaling in Spatially Distributed Large Dynamic Networks

The Congressional Research Report [1] showed scaling limitations impacting the large numbers of networked nodes needed for large-scale dynamic networks, such as a Battlespace, with the loss of edge node communications. Current methods to determine dynamically changing network connectivity result in a combinatorial explosion from the massive numbers of route permutations. To increase availability and resiliency in network-centric clouds and intelligent swarms, ad-hoc nodes must rapidly self-organize by using shared topology data. Additionally, topology can affect the network failures or successes for cyber offense and defense. The effective sharing of topological information between dynamic nodes on massive networks allows rapid self-organization to take place, thereby providing network resiliency. An emergent macroscopic (i.e., multi-agent world) perspective could address computationally intractable solutions—e.g., the large scale (10,000 + nodes) that impact the self-organization of ad-hoc networks and adaptive swarms. A method needs to be defined to accomplish such a solution.

1.1 Large Ad-Hoc Network Routing Issues and a Possible Path to a Solution

Some limitations for large-scale system of systems (SoS) may be caused by emergent phenomena that are common in complex systems. Bar-Yam [2] discussed the potential impact of complexity science upon large network-centric topological environments. While these emergent behaviors of complex systems may cause issues if not constrained, they may also present a solution to emergent network topologies. Since the topologically distributed nodes need to share both local and distant information, as described by Bar-Yam [2], it would be prudent to observe how nature creates self-organization. One method used in nature is called stigmergy [3]. This is the sharing of sparse amounts of dense information [4], where behaviors are impacted by different topologies in this instance specifically for distributed network nodes [5].

The internet and smaller network topologies have been well studied to understand their positive attributes and potential weaknesses, as described by Kleinberg et al. as resilience to random failures while being vulnerable to attacks targeted at a few critical nodes [6]. Additionally, this provides a rationale for developing a resilient topology, unlike the internet [7], while mitigating unwanted emergent behaviors due to the magnitude of the number of nodes [2]. The mathematical descriptions of complex networks [8], their construction [2], and the emergent order [9] have helped to direct us to look for a mathematical topology for resilient ad-hoc networks.

1.2 From Chaos to Iterative Order: Applying Fractals and the Iterated Function System (IFS)

Many real-world phenomena are self-similar (i.e., a smaller section of the whole can be scaled iteratively to represent the whole). Examples include trees, broccoli, cauliflower, and their respective branches; planets with multiple moons that can scale to a solar system with multiple planets; coastlines that can scale both upward and downward; and many other phenomena. Many phenomena can be represented by recursive generating functions, or Iterated Function Systems (IFS), which typically generate fractals [10–12]. The IFS generates a fractal shape by constructing many copies of itself and overlaying these to produce the fractal pattern. The copies are affine transformations of the original shape, and are typically contractive-affine, or a reduced size copy of the original. This IFS is the result of a set of contractive affine elements, by use of the Hutchinson operator, which converges to a unique attractor [13]. The overlays of these affine transformations are mathematically the union of these functionally transformed shapes, which create a fractal topology.

These generating functions can be iterated to the desired degree of behavioral fidelity with respect to real-world observations. This means that although the true fractal may be infinite, we can take the IFS computation to almost any arbitrary level and still see some of the resulting emergent pattern (similar to a hologram, where each small element of the hologram still contains enough critical attributes to reconstruct the complete picture).

Static or Dynamic?

The above approach will give us a static picture, but what about dynamical systems in general? That is indeed the nature of our world—dynamic. What if we consider another approach, i.e., to use an adaptive, stand-alone IFS to generate unique self-organizing sets of patterns, and let these evolve into a representation of an ordered structure? This dynamic IFS would adapt, morph, and distort the fractal to include new points as they are acquired (e.g., entities could join or leave swarms or ad-hoc networks). Because fractals are self-similar, their elementary parts can each represent an aspect of the real world. Such a "rubber" fractal can stretch and adapt, and by its nature coalesce or contract to a stable set of patterns. The vertices on this fractal would be related by analogy to real-world parameters. This produces a paradigm that allows a machine to behave more like the dynamics of natural phenomena.

The Dynamic IFS—What Is It?

The dynamic IFS described below is just one of possibly infinitely many new or unknown mathematical functions that, instead of being derived from formal proofs and existing mathematical functions, is a product of research using experimental mathematics. In this case, it was discovered by the author after deriving it from one aspect of Barnsley's "chaos game" [11, 14], and is a simple linear algorithm that takes minimal computational power to generate a repeatable and predictable complex system. This specific algorithm, named by the author Non-predetermined

Parametric Random (NPPR) IFS [15], creates an adaptive fractal whose shape is formed from many points in self-similar or repeating patterns, analogous to the way that tree branches are similar to a twig at the smallest scale, and trunks at the largest end.

This dynamic IFS is created from points that topologically represent the positions of nodes or simple agents. The positions of these nodes are mapped to the IFS equation, which is dependent upon several parameters, specifically a small subset of initial points or vertices that become attractors (e.g., targets or waypoints for mobile entities), and a contractive scaling factor. When the scaling factor approaches either $+1$ or -1, the system displays what appears to be a random scattering of node points. When the scaling factor approaches 0, the points coalesce in clusters around an arbitrary attractor, covering all the attractor vertices. These emergent clusters provide the topology needed for point-to-point node communication links, as well as a choice of inter-cluster routing topologies.

This solves several problems, including those associated with a topologically distributed network or swarm. It also solves some of the currently intractable distributed information problems by creating emergent solutions through sharing of information. A stigmergic solution results when the small amount of data that is information-dense is shared among all nodes—in this case, the vertex positions and the scaling factor value. By each node knowing this sparse amount of data, they can each reconstruct the entire topology of the SoS.

This has narrowed down our approach to using a dynamic IFS, which produces an emergent solution that depends upon a relatively small change in one or more variables. The solution would also need to represent the random changes in a large ad-hoc network. Finally, the results would need to be repeatable each time that the dependent parameters are set to the same values. By adding networked sensor data into the system, the NPPR IFS could modify (adapt) itself to a dynamically changing environment or changing threat behavior. Networked inter-agent communication would allow shared knowledge to enhance and update each agent's knowledge map [13], promoting a consistent shared knowledge of the networked environment.

NPPR Adaptive Fractal—How It Works, and What Is Displayed by the Software Tool

Similar to the appearance of the simple point-slope equation for line, the deterministic chaos equation is given as:

$$X(n) = M * X(n - 1) + Z \tag{1}$$

$X(n - 1)$ is the current point, $X(n)$ is the next point, M is the scale parameter, which controls the distance that the next point is generated from the current point, and Z corresponds to one of the randomly selected "vertices." These vertices are the set of initial points that constrain all network node points, and for our example can represent network hubs. The number of initial points can range from three [16] to over 100 if needed, although experimental results show that most unique emergent behaviors are observed with 12 or less initial points. The specific Z is randomly selected out of this set of initial points. The value of M is $0 < |M| < 1$, and it takes experimental discovery to determine what values of M give the desired emergent behaviors. Once these values are determined, any random run of this equation with the selected M and the same number of initial points in similar topologies will produce the same topological structure overall. This is particularly interesting, since changing the random number seed produces almost no visible change to the topologies generated, despite the fact that all the points are newly generated, and justifies this mathematical function belonging to the area of deterministic chaos. Both variables M and Z share interdependencies that affect the overall network topologies, including controlling the degree of clustering, and implicit mappings between specific cluster subsets at different scales.

The algorithm has "tunable" values that control the repeatable transition from order to chaos. This is controlled by the variable M coupled with the initial set of vertices. The resulting data points, in addition to producing a topological map for network routing, may also be used to create several categories of information. The information may result from sensor data that consists of several features to be associated with feature categories and can be represented by topological relationships. If several sensors at different locations detect something from a common location, but categorize it differently, the sensors' data maps into multiple interrelated categories. The "topological map" would contain the sensor positions and data and correlate the relationships between them while preserving the differences. This allows shared information to define distinct yet common features. A small amount of data transferred between sensor elements could be used to share the global picture for the sensor nodes, i.e., a stigmergic relationship. Nature uses a similar stigmergic approach when ants and other social insects mark a location that is of interest. As other ants pass the marked location, they too mark it, and a sufficient number of marks cause all ants to follow that trail of landmarks. This small chunk of information can become extremely powerful and is analogous to the sparse data elements used for defining behaviors in the NPPR algorithm. For the investigation of the NPPR algorithm, a tool was built in Mathematica® to display parametric changes as they are made dynamically by the user.

Comparing the two pictures in Fig. 1, one can see the relationship between the initial points, which can also be considered as topological representation for five categories, with each attractor/vertex representing a category. As described previously, the top left (smaller) rectangle in each picture contains five points that represent the initial categories (in this data analytics example) that are spatially separated. The set of clustered points in the large rectangle in each picture represents the data. The left picture shows symmetric clusters resulting from a symmetric layout of categories. When the lower left category vertex is moved so that it is closer to some of the other categories, the clustering data points change to match the relative category positions.

To see the impact of parameter M on the topology, in Fig. 2, M is chosen as a value greater than 0.5, and the resulting pattern for five initial points is random scattering, which would be the default pattern guess anyone would make for a simple function with the random aspect of NPPR. Now M is changed to around 0.33 in Fig. 3, and the resulting pattern resembles the classic video game of Space Invaders. Other example patterns are included at the end of this chapter.

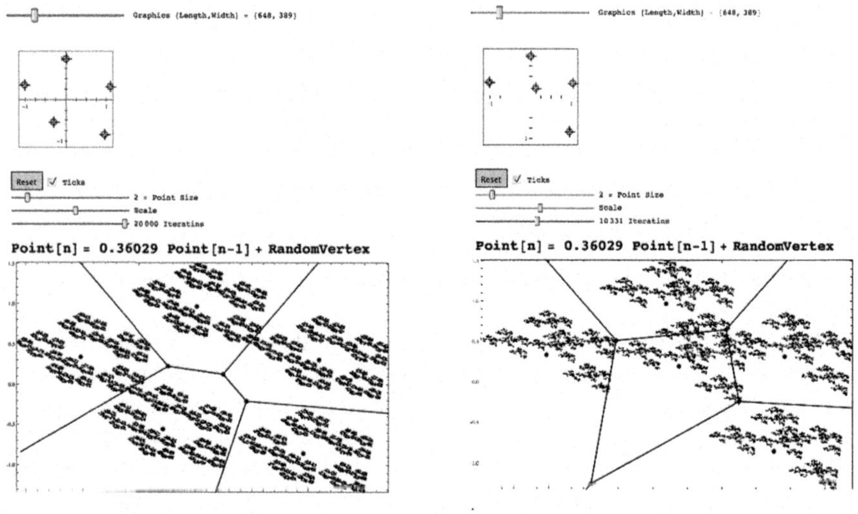

Fig. 1 A vertex is shifted

From above, we can see that the emergent pattern of clusters in Fig. 3 resulted from the seemingly random cloud of points in Fig. 2. Therefore, in this case for 5 vertices, changing the scaling parameter M from about 0.5 or more to about 0.33 causes the order to emerge. The lines that divide the pictures into tile-like sections are the result of applying a Voronoi function mapping to the points on the screen. The Voronoi tessellations provide an unbiased method of determining which points map to a particular cluster. This avoids the human errors or biases that would result from estimating just where a cluster begins and ends, or the illusion of a cluster being at a location, i.e., pareidolia. The Voronoi is not infallible, as occasionally some parts of a cluster will appear on the line between two tessellations. It provides a general metric for when clusters are readily separable.

1.3 A Proposed Solution: Apply NPPR to Solve the Ad-Hoc Routing Problem

We now have enough information to provide a solution to our large ad-hoc network routing problem. Using the previous highly structured topology shown in Fig. 3, we now add routing, and define a communication distance. This result is shown in the marked-up version of Fig. 3, labeled as Fig. 4.

The five major clusters in the picture, i.e., one major cluster per Voronoi tile, is divided into five sub-clusters, one of which, as an example, is circled in purple. These are again divided into sub-sub-clusters, and onwards due to the self-similar nature of the IFS. Let's assume that the window with the clusters is a two-dimensional map

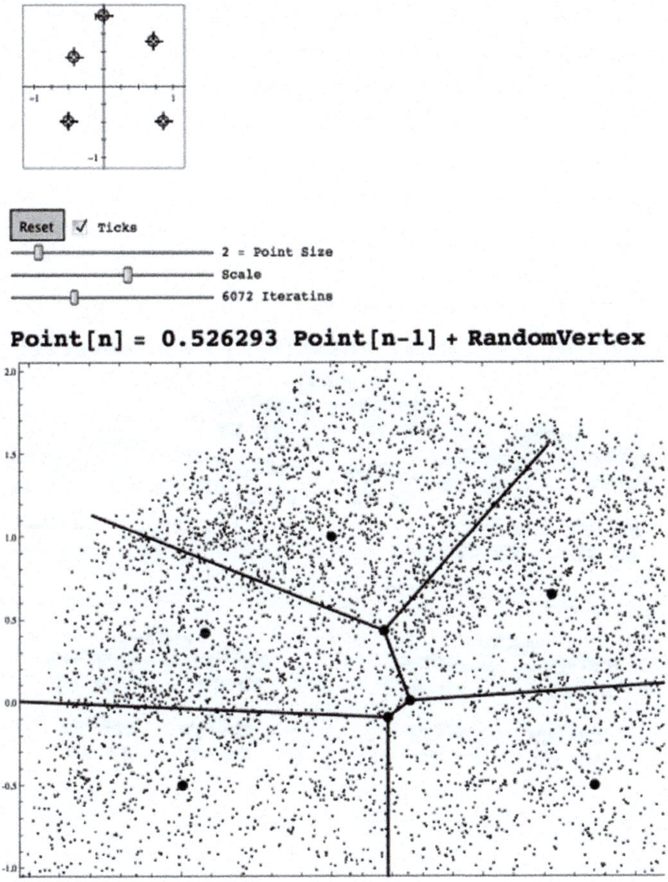

Fig. 2 M > 0.5

of a region within which these points are located. Let's assume that each one of the smallest points on the screen corresponds to an ad-hoc radio (or IoT device) possibly within an agent or autonomous vehicle. Note that each individual point in this picture is not explicitly visible since this run of the tool plotted 20,000 unique points, many of which are tightly clustered. For the sub-cluster circled in purple, let's assume that the circle's diameter is the typical range of a point-to-point communication between two nodes. Finally, two nodes have been selected (actually for ease of viewing, two sub-sub-clusters were selected) and are circled in green. For the purpose of showing the routing, the green circled nodes were selected at diagonally opposite ends of the window.

Since each node would be capable of calculating the general layout of all the other nodes, it would be able to determine which direction to forward its message for the next leg or "hop" of its route, shown as red arcs in Fig. 4. The emergent self-similarity makes it easier since despite the nodes being randomly iterated, the overall topology

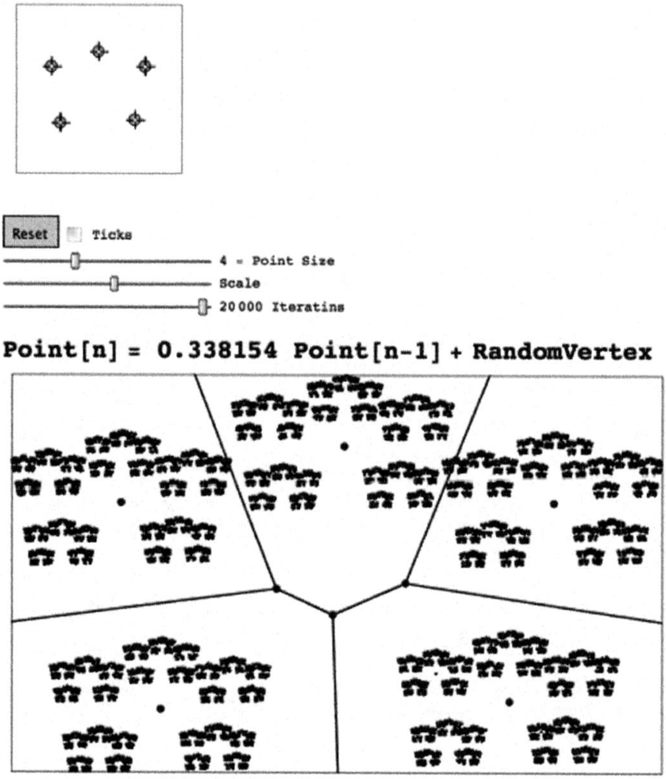

Fig. 3 M ~ 0.33

will remain self-similar, at least for certain values of the scaling parameter M. If each node knows its range limitations, i.e., the diameter of the purple circle, then it would broadcast to the sub-sub-clusters at the edge of its range and in the appropriate direction in order to reach the destination node.

Each node has the information needed to calculate the positions of all other nodes, which in this example consists of the positions of the five vertices and the value of the scaling factor M. As positions of vertices are changed, or M changes, these values can be broadcasted to all nodes in a short transmission. The recalculation would be trivial for any modern processing node, since it consists of a linear addition with a single floating-point number as multiplier. These internal calculations of node topologies provide all nodes with the ability to create their internal world model that is consistent with the overall topology, hence the ability to derive a shared world model. This imparts a shared understanding for communication between the nodes, even when parameters dynamically change.

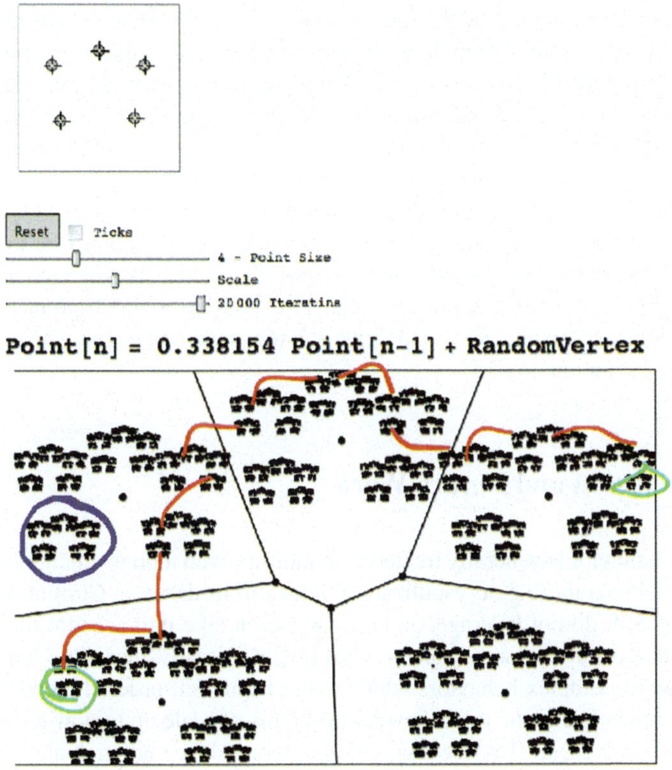

Fig. 4 Routing mark-up of Fig. 3

Walking Through the Point-to-Point Network Connection

As we walk through the route from the source node to the destination, it would be easiest to number the clusters and sub-clusters in some order. Arbitrarily, let's go clockwise within each cluster and sub-cluster, starting with the lower left, and ending with the lower right. So, for the green circled nodes in the lower left, we can refer to these cluster elements by three numbers (cluster, sub, and sub-sub) as 1, 2, 1. The red arcs represent the paths for each hop that is within the range limit of each node. The approximate sequence of hops from the green circled nodes in 1, 2, 1 to the destination green circled nodes in 4, 4, 5 is as follows:

1, 2, 1 to 1, 3, 3 to 2, 5, 1 to 2, 4, {3, 4, or 5} to 3, 2, 1 to 3, 3, 2 to 3, 4, {1 or 2} to 4, 2, 2 to 4, 3, 2 to 4, 4, {3, 4, 5}.

The braces represent the approximate number of hops, since all the nodes within braces are in range, so if some are unavailable, others nodes can take their place.

While the exact routes chosen may not be optimal every time, this type of routing topology provides a measure of resilience for both edge-nodes and more centrally located ones, as the communication routes can work even if some sub-clusters and sub-sub-clusters lose up to half of their nodes. This has a clear advantage over current ad-hoc routing protocols, e.g., AODV (Ad-hoc On-Demand Distance Vector Routing), which suffers from large latency time in route discovery. Instead, this approach using NPPR consists of calculating just a few route permutations to find an efficient, albeit not optimal, route.

This adaptive fractal topology can also apply to the emergent organization of massive swarms, since a change in one parameter can cause the entire swarm of autonomous vehicles to either produce emergent flocking behaviors [17], or scatter into a random pattern. A shared stigmergic world model can still be known, and additional messages can also be routed despite the randomly scattered appearance of the nodes. The swarm application can also provide a single-point control for many vehicles whose objective is to arrive at a few destinations (i.e., the vertices can represent destinations).

2 Conclusion and Future Work

Emergent behavior is generally treated as a liability, even though emergence is effectively used in nature (e.g., crystallization, flocks of birds, etc.). Current methods of addressing SoS do not leverage, and may not even take into account the emergent behavior that occurs in large-scale systems [18]. Generally, these SoS will exhibit a wide range of complex behaviors which range from well-ordered predicted behaviors to completely chaotic ones. Several problems of scale up to now have not been successfully solved, such as large swarm behaviors and the routing dynamics of large-scale ad-hoc communication networks. Using the appropriately selected constraints, random IFS such as NPPR can converge to ordered patterns [19] and dynamically adapt to changes in the environment. Such emergent topologies provide insight into solutions for massively distributed agents awareness [20], and apply to cognitive sensors, communication nodes, and intelligent swarms.

References

1. Library of Congress, Congressional Research Service. Network centric operations: background and oversight issues for Congress (Report Order Code RL32411) (2007). https://www.fas.org/sgp/crs/natsec/RL32411.pdf. Accessed 02 Jan 2015
2. Bar-Yam, Y.: About engineering complex systems: multiscale analysis and evolutionary engineering (2005). https://necsi.edu/research/multiscale/ESOA04.pdf. Accessed 01 Apr 2017
3. Stygmergy. https://en.wikipedia.org/wiki/Stygmergy. Accessed 01 May 2016
4. Masoumi, B., Meybodi, M.R.: Speeding up learning automata based multi agent systems using the concepts of stigmergy and entropy. Expert Syst. Appl. **38**(7), 8105–8118 (2011)
5. Dagon, D., Guofei, G., Lee, C.P., Wenke, L.: A taxonomy of botnet structures. In: Computer Security Applications Conference, 2007, ACSAC 2007, pp. 325–339. ACSAC (2007)
6. Kleinberg, J., Sandler, M., Slivkins, A.: Network failure detection and graph connectivity. In: SODA '04 Proceedings of the Fifteenth Annual ACM-SIAM Symposium on Discrete Algorithms, pp. 76–85. Society for Industrial and Applied Mathematics, Philadelphia (2004)

7. Dorogovtsev, S., Mendes, J.: The shortest path to complex networks. Am. J. Med. Genet. **71**, 47–53 (2004)
8. Shargel, B., Sayama, H., Epstein, I., Bar-Yam, Y.: Optimization of robustness and connectivity in complex networks. Phys. Rev. Lett. **90**(6) (2003)
9. Bak, P., Tang, C., Wiesenfeld, K.: Self-organized criticality: an explanation of 1/f noise. Phys. Rev. Lett. **59**, 381–384 (1987)
10. Park, K., Willinger, W.: The internet as a large-scale complex system. In: Santa Fe Institute Studies on the Sciences of Complexity, Oxford University Press, New York (2005)
11. Barnsley, M.F.: Fractals Everywhere. Academic Press, San Diego, CA (1988)
12. Demko, S., Hodges, S., Naylor, B.: Construction of fractal objects with IFS. Comput. Graph. **19**(3), 271–278 (1985)
13. Peitgen, H., Jurgens, H., Saupe, D.: Chaos and Fractals: New Frontiers of Science. Springer (1992)
14. Barnsley, M.F., Vince, A.: The chaos game on a general iterated function system. Ergodic Theory Dyn. Syst. **31**(4), 1073–1079 (2010)
15. Schaff, J.: Creating human-like behavior in a UAV: a counter-terrorism model. In: Proceedings of the Fourteenth Annual Software Technology Conference (2002)
16. Diacu, F.: The solution of the n-body problem. Math. Intell. **18**(3), 66–70 (1996)
17. Reynolds, C.: Flocks, herds and schools: a distributed behavioral model. In: SIGGRAPH '87: Proceedings of the 14th Annual Conference on Computer Graphics and Interactive Techniques, pp. 25–34. Association for Computing Machinery (1987)
18. Office of the Deputy Under Secretary of Defense for Acquisition and Technology: Systems and software engineering. Systems engineering guide for systems of systems, version 1.0. ODUSD(A&T) SSE, Washington, DC (2008)
19. McKelvey, B.: Complexity science as order-creation science: new theory, new method. ECO **6**(4), 2–27 (2004)
20. Brooks, R.A.: Intelligence without reason. In: IJCAI'91 Proceedings of the 12th International Joint Conference on Artificial Intelligence, vol. 1, pp. 569–595. Sydney (1991)

Complex Systems and Classical Military Theory

David J. Lyle

Abstract War has always been an emergent phenomenon, comprised of countless, constantly interacting physical and cognitive elements of individuals, groups, societies, and nation states competing violently for power, influence, and access to resources. If war is the sum total of these multiple levels of competition, can war be any less complex than any of the phenomena that play a part in it? Schools of military theorists have largely followed the same tensions that schools of science have followed between "positivist" schools of thought born during the Enlightenment who believed that the world is inherently ordered and controllable via by formal scientific description and method, and those "romantics" who suspected that randomness played a much greater role in the universe than our scientific tools and reason could handle on their own. Various military writers throughout the centuries have sought to offer prescriptive principles that can be used to make warfare more predictable and manageable. But modern understandings of complex systems can help us better understand the true degrees of efficacy we can hope to achieve through our calculations of war, and help us to avoid coming to false conclusions about what we can hope to achieve through the force of arms alone.

Keywords Complex systems · Military · Theory

1 First Section

War has always been an emergent phenomenon, comprised of countless, constantly interacting physical and cognitive elements of individuals, groups, societies, and nation states competing violently for power, influence, and access to resources. If war is the sum total of these multiple levels of competition, can war be any less complex than any of the phenomena that play a part in it?

D. J. Lyle (✉)
Princeton University, Princeton, NJ 08544, USA
e-mail: lncs@springer.com

Springer Heidelberg, Tiergartenstr. 17, 69121 Heidelberg, Germany

D. Braha et al. (eds.), *Unifying Themes in Complex Systems X*, Springer Proceedings in Complexity, https://doi.org/10.1007/978-3-030-67318-5_19

Any new theory should not only replicate the success of the old ones, but should also explain things that the previous theories could not account for, and also do it in a more elegant manner, a concept described as "Occam's Razor" [1]. Claims that any new theory will "change the nature of war" should rightfully be looked upon with suspicion. But complex systems theory makes no such promises, complex systems theory gives us a better way to look at the nature of war that has always existed, with the nature defined as the basic interaction rules of human systems that don't change, and character defining the interactions themselves which can change. Complex systems theory explains many of the key insights of the classical military theories, and also offers new ways to both expand on their original insights and correct some of their deficiencies. Using complex systems concepts as the foundation of military theory is nothing new, but our dawning recognition of this currently constitutes the leading edge in military thought.

1.1 The Ancient Military Classics

Thucydides' History of the Peloponnesian War and Sun Tzu's The Art of War, both written approximately 2400 years ago, are ancient classics of military strategy that have maintained their relevance in modern times. Among Thucydides' most famous contributions is his proposition that the actions of individuals, groups, societies, and states can be understood in the context of an attempt to simultaneously balance the prioritized social forces of fear, honor, and interest [2]. In complex systems terms, these are the powerful psychological attractors that guide all basic individual decisions, with fear being the primary motivator geared toward physical survival, honor being focused on the survival of one's sense of identity within the contexts of groups, and interest guiding decisions when the previous two imperatives are satisfied. While an understanding of these attractors cannot guarantee accurate prediction of individual decisions in specific instances, they can often be used to describe and predict what psychological behaviors will drive human behavior from the bottom up in aggregate, as well as in the long term. This serves as the foundation for prediction that all sound strategy relies on.

The story of Athens and Sparta itself is a classic case of two complex adaptive systems competing against each other within the larger complex adaptive system of the Ancient Mediterranean and Persian worlds. For decades, the sea power Athens and land power Sparta two sides clashed with futility, neither side being able to overcome the strengths of each other, in what students of military strategy describe as "elephant vs. whale" stalemate. Then, both sides began the process of adaptation, each adopting the means of the other in order to break through the defenses of the other, and ending ironically when the land power, Sparta, defeated the sea power, Athens, in a decisive naval engagement, after they allied with their traditional enemies, the Persians. But in winning, Sparta illustrated another key takeaway from complex systems theory: in complex adaptive systems, you can "never do merely one thing." [3]. In making an alliance with Persia to defeat Athens, Sparta weakened the entire alliance of Greek city states that had traditionally banded together to defend the region from foreign invasion. Only decades later, after even more Greek internecine warfare, all of the weakened Greek city states eventually fell separately to Phillip of Macedon and his son Alexander the Great. Thus, the real admonition of Thucydides in writing is book may be his warning echoed by the most important principle of complex systems. If your own definition of the system is too narrow, as was both the Athenian and Spartan definitions of victory, your solution set will be to narrow, and if you fail to understand how competition and cooperation work at various levels of scale, you cannot hope to adapt successfully within the context of the larger system.

Sun Tzu (or the group of scholars represented by him) is very possibly the earliest known advocate for systems thinking in warfare, as a holistic understanding of the system and the environment in which a military commander competes is crucial to success according to his theory. If the acme of skill is to win without fighting—by first defeating the enemy's strategy, as Sun Tzu suggests—then one must not only be able to adapt oneself to the situation and the enemy, but also be able to predict how the enemy is adapting, and factor that into your own model for adaptation [4]. Sun Tzu seeks to use knowledge of the potentialities inherent within the total system in his favor, using deception to maneuver the enemy into positions of geographic and material disadvantage that can then be exploited with minimal effort and risk, a concept described by Francois Julien as seeking "efficacy." [5]. Successful generals recognize the fundamental nature of the system and the environment, and pattern their actions (or inaction) to take advantage of three different potentials: moral potential, topographic potential, and potential of adaptation. It is the third of these aspects of potential that the maneuver of military forces can influence. The potential of the situation cannot be anticipated, as it "proceeds from continuous adaptation…" [6].

1.2 Theories of War During the Enlightenment

Schools of military theorists have largely followed the same tensions that schools of science have followed between "positivist" schools of thought born during the Enlightenment who believed that the world is inherently ordered and controllable via by formal scientific description and method, and those "romantics" who suspected that randomness played a much greater role in the universe than our scientific tools and reason could handle on their own. Various military writers throughout the centuries have sought to offer prescriptive principles that can be used to make warfare more predictable and manageable. In the emerging "scientific" theories of war, some even believed that "Bellona, the furious goddess of war, was to be rendered tractable, tamed like a kitten on the hearth."

Theorist after theorist designed systems of war built upon principles that if followed, would practically ensure victory for those who had the diligence to study and master them. This school of thought was a natural follow-on from the era of "cabinet wars" in which fortification had dominated the character of war, and was represented by a geometric style of calculation best represented by the theoretical works of Sébastien Le Prestre de Vauban, Maurice De Saxe, Louis Pierre de Chastenet, comte de Puységur, Count Turpin de Crisse, Paul Gideon Joly de Maizeroy, and Dietrich Heinrich Von Bulow. This positivist school described by Gerhard Ritter in The Sword and the Scepter, "Many people hoped that war would ultimately eliminate itself, having become a pure, universal science by means of mathematical equations that ruled out chance and the fortunes of war." And perhaps the most famous of the positivists, Baron Antoine Jomini advocated prescriptive theories based on principles of war that would help you recognize "decisive points" that have a greater impact on the system than others, and should therefore be the focus of one's efforts in warfare [7].

But others were not so sure. Carl Von Clausewitz—would write in his magnum opus On War that "…all military action is intertwined with psychological forces and effects", and elsewhere that "One might say that the physical seem little more than the wooden hilt, while the moral factors are the precious metal, the real weapon, the finely honed blade." Those unquantifiable moral factors "will not yield to academic wisdom. They cannot be classified or counted. They have to be seen or felt." [8]. While Clausewitz did acknowledge the usefulness of principles of war and the study of history to search a for a useful general theory, he was firmly against the idea that war could ever be subjugated or controlled mathematically, as Clausewitz described war's main driving forces as psychological passion that drove the combatants to conflict, irreducible chance that creates both challenge and opportunity for military leaders, and the attempt to subordinate war to reason and rationality by those directing the war [9].

1.3 A Modern Diagnosis of the Old Debates

The deeper insight Clausewitz and the others above intuited on—the core issue of misapplied scientific philosophy driven by positivist thinking—can today be described as technical rationalism, a variant of the positivism that fueled the French Enlightenment, the sense that the world was inherently ordered and would eventually be brought under control by a combination of good science and deliberate effort. As described by MIT's Donald Schön, Technical Rationality is the "dominant epistemology of practice," an "instrumental problem solving made rigorous by the application of scientific theory and technique", a scientific philosophy that is "implicit in the institutionalized relations of research and practice, and in the normative curricula of professional education." Under this positivist paradigm, "real knowledge lies in the theories and techniques of basic and applied science." [10]. But what if those scientific theories and techniques imply a degree of order or stability of outcomes that don't exist in real world?

Technical rationalism falls far short when it comes to evaluating competing moral factors, or evaluating the tradeoffs between competing value-laden options or balancing irreducible ethical dilemmas. These are the intangible and often capricious social dynamics that describe the difference between knowledge and wisdom, a distinction that cannot be described in formal or universal logical functions minus specific contexts, but must be fit to the unique contexts of specific situations. Technical rationalism offers a seductive but false sense of certainty when applied to problems of strategy, and despite several near misses and certain disasters in the recent past caused by its application in war, technical rationalism continues to ensnare senior leaders who fail to understand its obvious inadequacies above the level of tactics, as it did with Robert McNamara's quantitatively focused "Whiz Kids" in Vietnam, who later confessed that "I had always been confident that every problem could be solved, but now I found myself confronting one – involving national pride and human life – that could not." [11].

Technical rationalism continues to seduce those not steeped in the realities of complex systems, as was demonstrated in the 1990s with the overreaches of Network Centric Warfare and Effects Based Operations which sought to make war tamable by digitized, algorithmic logic run by increasingly powerful computer and communications systems [12]. More recently, the "Third Offset", led by former Deputy Secretary of Defense Robert Work sought to describe war as a series of grids, hearkening back to the technical rationalists of the French Enlightenment [13]. Taken metaphorically, these descriptions are valuable, but too many continue to literally believe that war can be reduced to formal logic, and "gonculated" via algorithm, neglecting the true difficulties of formally describing the connected, contingent, and ever changing human value functions that accompany real social intercourse—the same point Clausewitz made in the early 1800s. This "chessmaster" tactical focus has been commented on repeatedly, and can perhaps be summed up by the following observation on the US military by the late éminence grise of strategic studies, Dr. Colin Gray: "The problem, to repeat, is that the United States has a severe strategy deficit. It is, and has long been, guilty of what is known as the "tacticization" of strategy. US military power does tactics well and tends to expect success at that level to translate automatically into strategic victory." [14].

The good news is that complex systems theory does indeed help to bolster the military theories that have resonated with generations of military thinkers, giving us new insights into why these ideas indeed seem to be timeless. As more and more military thinkers become familiar with complex systems concepts, and relate the new ideas to the old ones, there is a much better chance of larger organizational acceptance of intellectual models that better approximate war as it really is, reducing surprise even when it cannot eliminate uncertainty. But it's extremely critical that we adopt the conceptual frameworks of complex systems to complement those classical scientific rationalist approaches that work we with well-structured problems—if we fail to discern the difference between complicated and complex tasks, we'll apply the wrong types of tools to the wrong types of problems, and risk creating "Weapons of Math Destruction" that will force us to learn the lessons of complex systems far too late to achieve our intended goals.

References

1. Mitchell, M.: Complexity: A Guided Tour, pp. 99–100. Oxford University Press, New York (2009)
2. Thucydides: The Landmark Thucydides: A Comprehensive Guide to the Peloponnesian War. In: Crawley, R. (trans.), p. 43. Simon and Schuster, New York (1996)
3. Hardin, G.: Filters Against Folly: How to Survive Despite Economists, Ecologists and the Merely Eloquent. Penguin Books, SA (1986)
4. Tzu, S.: The Illustrated Art of War. In: Griffith, S.B. (trans.), p. 115. Oxford University Press, New York (2005)
5. Jullien, F.: A Treatise on Efficacy: Between Western and Chinese Thinking. In: Lloyd, J. (trans.), p. 8. University of Hawaii Press, Honolulu (2004)
6. Jullien, F.: A Treatise on Efficacy: Between Western and Chinese Thinking. In: Lloyd, J. (trans.), p. 23. University of Hawaii Press, Honolulu (2004)
7. De Jomini, B.: The Art of War. In: Mendell, C.G.H., Craigshill, L.W.P. (trans.), p. 63. Wilder Publications, LLC, Radford (2008)
8. De Jomini, B.: The Art of War. In: Mendell, C.G.H., Craigshill, L.W.P. (trans.), p. 184. Wilder Publications, LLC, Radford (2008)
9. This is the "wondrous trinity" that Clausewitz presents as arguably the conceptual centerpieces of his theory of war. Von Clausewitz, C.: On War. Howard, Paret (trans.). Princeton University Press, Princeton (1989). Contained in Book One, Chapter One, section 28, "The Consequences for Theory" (page 89 of the Howard and Paret 1976 translation). See also "Tiptoe through the Trinity" by Christopher Bassford. https://www.clausewitz.com/mobile/trinity8.htm
10. Schön, D.A.: The Reflective Practitioner: How Professionals Think in Action, p. 27. Basic Books, New York (1984)
11. McNamara, R.: In Retrospect: The Tragedy and Lessons of Vietnam, p. 207. Vintage Books, New York (1996)
12. Wilson, C.: Congressional Research Service Report to Congress: Network Centric Operations: Background and Oversight Issues for Congress. Congressional Research Service, Washington DC. Updated March 15, 2007
13. Work: U.S., NATO Must Use 21st-Century Approaches for Deterrence, Dominance. US Department of Defense News Service, 29 Apr 2016. https://dod.defense.gov/News/Article/Article/746336/work-us-nato-must-use-21st-century-approaches-for-deterrence-dominance/
14. Gray, C.: The Airpower Advantage in Future Warfare: The Need for Strategy, p. 4. Air University Press (2007). https://apps.dtic.mil/dtic/tr/fulltext/u2/a477043.pdf

On the Middleware Design, Cyber-Security, Self-monitoring and Self-healing for the Next-Generation IoT

Predrag T. Tošić

Abstract Internet-of-Things (IoT) is one of the most important technological advances in the realm of Internet-powered cyber-physical and cyber-secure systems of the last decade and likely for many years to come. As such, IoT poses a number of research challenges when it comes to system scalability, availability, inter-operability, security and reliability. We study the programming abstractions, "smart middleware", multi-agent systems and cyber-security aspects of what it will take to enable a reliable, secure, inter-operable and human-friendly next-generation IoT. In particular, three important aspects of the NG IoT are discussed in this paper. One, we address software design challenges and appropriate programming abstractions for the software agents that would provide IoT's inter-operability, enabling different users, devices and platforms to effectively communicate and coordinate with each other. We revisit agent-based computing, and especially the Actor model as a potentially perfectly suited programming paradigm for an inherently open, distributed and heterogeneous infrastructure such as NG IoT. Second, we discuss some critical capabilities of the next-generation IoT middleware. Last but not least, to address cyber-security aspects of IoT holistically, we outline several key elements of computational intelligence that could enable the self-healing and self-recovery capabilities of the NG IoT. With such self-defense/self-healing/self-recovery capabilities, future cyber-attacks would be less likely to cause a major disruption to the IoT infrastructure and users.

Keywords Internet-of-Things · Smart middleware · Distributed intelligence · Multi-agent systems · Cyber-security · Self-healing systems

P. T. Tošić (✉)
Independent AI and Computer Science Researcher, Spokane, WA, USA
e-mail: predrag.tosic@ieee.org

Department of Mathematics & Statistics (Adjunct Faculty), Washington State University, Pullman, WA, USA

D. Braha et al. (eds.), *Unifying Themes in Complex Systems X*, Springer Proceedings in Complexity, https://doi.org/10.1007/978-3-030-67318-5_20

1 Introduction: Future of Internet-of-Things

Internet-of-Things (IoT) is arguably one of the most significant and impactful technological advances in the realm of the "consumer", Internet-powered cyber-physical systems of the last decade. Moreover, IoT technologies are expected to become even more pervasive and of practical importance to how we work, live, play and interact with technology in general in the years to come. This is why IoT and its many R&D challenges have been a focus of both industry leaders in the Internet technologies, smart devices, platforms, and applications, as well as increasingly among the academic researchers, as well.

Due to inherently open and heterogeneous nature of IoT, as well as the needs for preserving user's data privacy, confidentiality and integrity, IoT architectures and deployments need to address security and privacy from the get-go, as first-order design and implementation concerns [1]. On the other hand, a large (and ever growing) number of heterogeneous devices necessitates the design that will ensure interoperability, integration and scalability across a variety of users, devices and compute platforms [2]. From a system architecture standpoint, these challenges of IoT design and deployment imply that the IoT middleware needs to be more flexible, layered/modular and complex than the familiar middleware platforms deployed with "traditional" distributed and cyber-physical applications.

The list of technological and research challenges behind enabling the next-generation IoT is quite long and versatile. We are particularly interested in the programming abstractions, the middleware architecture aspects, as well as the applied AI (especially distributed intelligence and multi-agent systems) aspects, of what it will take to enable the reliable, secure, inter-operable and human-friendly next-generation IoT.

The promise of IoT lies in enabling different users, devices, platforms and technologies to effectively interact with each other, including communication, coordination, and various forms of cooperation. Since different agents in this "soup" of users and devices called IoT in general represent different human users or organizations, these agents will in general have different goals from each other [3]. Sometimes, these goals might be mutually exclusive or conflicting. Further, just like in human societies, certain groups of agents will have better aligned interests and goals with each other than with agents outside of their "circle". Similarly, again just like in human societies, certain agents will be more reputable or trustworthy than others when it comes to delivering on commitments made as a part of a multi-agent cooperation, or with respect to the quality of expertise they provide about a particular domain [4].

Some research prototypes of IoT architectures have attempted to incorporate cognitive-like capabilities into the IoT middleware design; one such prominent example is the iCORE project, based on the Virtual Objects (VO) paradigm [5]. These VOs are semantically rich virtual representations of the capabilities and resources provided by real-world objects and devices. As such, the VO paradigm and prototype design based on it build on the past research on distributed resource and task allocation in heterogeneous multi-agent systems, a classic example of how computational intelligence and Distributed AI contribute to advancing the IoT state-of-the-arts.

These observations suggest some of the key elements, perhaps a partial roadmap, for the next-generation AI/MAS research and design of effective autonomous software agents for the up-and-coming *Internet of Agents* (IoA). For instance, the focus in [3] is on the desiderata pertaining to intelligent agents' abilities to coordinate and cooperate with other agents and in particular create coalitions, while taking into the account those other agents' capabilities, "expertise" and reputations.

The rest of the paper organizes our main ideas into three main parts. First, we discuss the key middleware desiderata and, related to that, appropriate programming abstractions for the NG IoT; in that context, we revisit Agent-Oriented Programming, and argue that an extended Actor model is a suitable programming paradigm for the next-generation IoT. We then discuss some key general desiderata for the NG IoT middleware design. Third, we briefly address the self-monitoring/healing/recovery aspects of IoT, and argue that reinforcement learning techniques combined with appropriate computational "capture" of the self-defending and self-recovery capabilities of biological organisms, hold a promise of providing the right high-level design guidelines for making an intrinsically open, heterogeneous and very unpredictable system such as NG IoT capable of adapting to and surviving various types of future cyber-attacks.

2 Agent-Oriented Programming for IoT and the ACTOR Model

Agent-oriented programming (AOP) [6] emerged in the 1990s as a novel programming paradigm spurred by the growth in intelligent agent software technologies and applications at that time. Tracing history of programming languages and how the nature of dominant applications has been driving the most popular programming paradigms during different eras of computing, AOP can be contextualized (in hind-sight) as a very natural successor to the object-oriented programming (OOP) paradigm; see, e.g., [7]. In particular, the transition from object-oriented towards agent-oriented programming was motivated by the emergence of open distributed computing platforms in the late 1980s and throughout the 1990s, which naturally required that concurrency and resource sharing be explicitly controllable and therefore easy to capture and exploit in the underlying programming model.

As computing started becoming increasingly distributed both physically and logically, and these distributed systems getting increasingly heterogeneous, complex and open, the individual components of complex software systems started being transitioning from non-autonomous components of typically either monolithic or else hierarchical architectures towards each such component becoming increasingly sophisticated, autonomous and complex (sub)system on its own, with the system architecture entailing only a loose coupling of many such components into an overarching larger system. Further, such systems were growing rapidly in scale, becoming highly decentralized, and would usually be based on a peer-to-peer (P2P) paradigm, as opposed to being either centralized or hierarchical. This is how the AOP paradigm rose to prominence, striving to meet the increasing demands in terms of autonomy and flexibility of the individual components in such open distributed systems [6].

We emphasize some key characteristics of (what were back then, emerging) agent-based applications of the 1990s that required new programming paradigms with respect to the "old school" procedural imperative programming (and, to some extent, also with respect to the traditional object-oriented programming the way it was done in the 1980s and early 1990s, with the rise of first Object-C and then C++):

- larger scales and higher degree of system complexity;
- physical and logical decentralization of compute tasks, resources and data;
- increasing heterogeneity of computing resources, platforms, tasks and users;
- inherent openness, where new "actors" come and go, and it cannot be anticipated at the design time, how many or how diverse tasks, resources and/or users may "join" the infrastructure at various points of time in the future;
- great need for fault-tolerance, reliability, and availability in the likely presence of individual "compute node" and/or communication link failures;
- concerns about data privacy, confidentiality and integrity, as well as access control to resources;
- preparedness for and resilience to cyber-attacks of known and unknown varieties, from familiar and unfamiliar "bad guys" trying to disrupt the infrastructure, compromise or steal data of individual users, etc.

Reflecting upon the above list, it is immediate that today's and tomorrow's IoT has virtually all of the above characteristics—if anything, the "features" listed above are even more pronounced in the IoT context than what was the case with agent-oriented and service-oriented software architectures of the 1990s and early 2000s. This is why, in our view, revisiting and especially identifying how to appropriately adapt those methodologies and programming abstractions based on the *Service-Oriented Architecture* (SOA) and *Agent-Oriented Programming* (AOP) paradigms will be of utmost importance for NG IoT, specifically for the IoT software architecture and middleware design, in the years to come. Due to space constraints, we discuss here only one such programming abstraction—the Actor model of distributed computing.

The *Actor model* was introduced with a particular goal of specifying distributed computation in large-scale, open distributed systems [8]. An actor encapsulates its internal state, as well as its behavior, which includes both data on which to compute, and procedures or methods to be applied to that data. Each actor has a unique name and a "mail box" to which other actors can send messages; actors communicate with each other via asynchronous message passing. In OOP, an object has a state and a set of procedures or methods that can change that state. An actor, additionally, encapsulates a thread of control thereby enabling concurrency (based on asynchronous exchange of messages among actors). A new actor can be created as desired, and various actors might be entering or leaving a given actor system, that is, this "soup" of actors, at various times. All these characteristics of the actor model make it very versatile and suitable for open distributed systems.

We note the Actor model was originally introduced in the 1980s, with the main goal of extending the OOP to capture multi-threading and concurrency as the first-order entities. The Actor model was later expanded to provide a simple yet versatile communication and concurrency programming model for Distributed AI and designing multi-agent software systems (see [9, 10] and references therein). Many years later, we find the general actor-based computing paradigm quite appealing for the middleware as well as "application software" (running on top of that shared middleware) for the next-generation IoT, precisely because the model was designed with large-scale, highly heterogeneous and intrinsically open distributed systems in mind—i.e., systems such as IoT.

In summary, the Actor model has many highly desirable properties from the standpoint of middleware and software design for the next-generation Internet-of-Things. What the original actors do not have, however, are mechanisms explicitly addressing the (actor data and/or state) integrity, privacy, confidentiality and related aspects; likewise, the actor model does not provide for self-healing defensive mechanisms in the presence of cyber-attacks. An open distributed software system based on actors will be fault tolerant with respect to individual actors joining or leaving the system, becoming too slow or unresponsive, or dying.

However, the original Actor model can do little about more Byzantine types of "bad behavior"—for example, dealing with malicious actors in the system. Therefore, should the future software infrastructures for IoT be based on some variation or extension of the Actor model, the cyber-security, self-monitoring and self-healing capabilities for such software infrastructure would need to be provided as "extras". One possible direction for future R&D on the programming abstractions and models for the NG IoT, therefore, is how to extend the original Actor model so that certain "core" cyber-security aspects too are captures in the extended model as "first order entities". An alternative approach would be to design an actor-based architecture which would include "non-user", special-purpose actors ensuring data and system integrity, user privacy and confidentiality, providing monitoring and defensive mechanisms in case of cyber-attacks, and so on. Some ideas along this latter direction will be discussed in the subsequent sections of this paper. Of course, we are well-aware that how to actually design a scalable cyber-secure system deploying such "protection actors" will pose significant research, technology development and real-world deployment challenges.

3 Smart Middleware for the NG IoT

The rise of middleware technologies and underpinning R&D was prompted by the evolution in computing and a paradigm shift from tightly coupled multi-processor systems towards distributed, loosely coupled systems in which each node is usually a standalone computer, and all these "nodes" are interconnected in a local or wide-area or other network, communicating and cooperating with each other on various data and/or compute/CPU intensive tasks [10]. Once deployments of "clustered" systems involving "nodes" that run different software including different underlying Operating Systems (OS) became commonplace (in the 1990s), an additional layer of abstraction between the applications and the underlying hardware became necessary—a layer enabling different compute nodes in the system, especially when running on different operating systems, being able to communicate, coordinate, share tasks and resources with each other, etc.

However, the traditional clustered distributed systems were not characterized by the levels of openness and unpredictability found in IoT: addition of new nodes (computers or other interconnected devices) to the system, for example, was in general infrequent and predictable. In stark contrast, in IoT context, new devices or even platforms (with potentially many "attached" devices) can join or leave the existing "soup" of devices, users and software agents at any time, with arbitrary frequency, and in unpredictable time patterns. Hence, designing a highly flexible, portable, adaptable and capable of self-reconfiguration middleware layer for the IoT kind of distributed systems is much more challenging, than for the more predictable and structured (and less open) traditional distributed clustered systems.

3.1 Some Examples of IoT Middleware Prototypes

Recent years have seen an upsurge in research and prototype design on middleware for IoT. Some prominent examples of R&D efforts include [1, 2, 5, 11–14]. One frequently encountered theme in this pursuit of the "right" kind of middleware for IoT, is the SOA design as the guiding principle: abstracting users and devices into "smart objects" and web services, and then providing a semantics-rich layer on top of that abstraction, that in turn enables those smart objects and services to exchange data, knowledge, tasks and resources with each other. Given the heterogeneity of devices found in IoT, the need for this kind of abstraction providing a uniform interfacing framework of interaction appears quite natural. What is tricky here, of course, is finding the right kind(s) of abstraction and interfacing, meeting the needs not only of the existing devices and applications but also predicting new types of devices and especially applications that will likely "want to join" IoT in the upcoming years.

The EU-sponsored *ebbits project* [12] has focused on allowing businesses to semantically integrate the IoT into the mainstream enterprise IT systems, thus moving a step closer to interoperable, real-world (as opposed to research prototype), end-to-end interaction among different business applications. The ebbits architecture is an attempt to bridge different backend enterprise applications with each other, as well as with people, various services and the physical world. The semantically rich information is generated from the tags, smart sensors, and other devices and data streams involved in IoT. In particular, "the ebbits platform will feature a Service oriented Architecture (SOA) based on open protocols and middleware, effectively transforming every [organization's] subsystem or [individual] device into a Web service with semantic resolution. The ebbits platform does enables the convergence of the Internet of People (IoP), the Internet of Things (IoT) and the Internet of Services (IoS) into the [integrated] 'Internet of People, Things and Services' (IoPTS) for business purposes" [12]. We note that any platform striving to enable interoperability of different "backends" needs to be aware, that the organizations to whom those backends belong need not trust each other, and that any such shared platform needs to provide security, reliability, privacy and confidentiality guarantees, as well as agreed-upon policies, to all stakeholders in order to be adopted and actually for important business interactions among different organizations.

Some researchers have realized early on the utmost importance of dealing with security, privacy, confidentiality and trust issues in the IoT context. In [13], the key idea is to integrate the (human) user directly into the "trust chain", ensuring visibility and transparency in the IoT security and reliability properties. The prototype described in [13] also provides manufacturers of devices and platforms with a unified interface of conveying the security, reliability and trust considerations to the users in simple, non-technical language—so that the users can reason about the tradeoffs or potential issues, and make decisions accordingly, without being cyber-security experts. The downside of this approach is fairly obvious: while human users are indeed given more direct, explicit control, the cognitive burden on the human is increased; and most people want to go about their daily lives and have their various devices and technology doing the work for them, without having to worry much about how or why something may malfunction, and how to fix it.

The Butler IoT middleware prototype [14] allows its users to manage their individual profiles by allowing data replication and identity control (incl. access control) across different applications "belonging" to the same user. The objectives behind this are two-fold: to give the users sense of control, and to provide a middleware framework capable of integrating various user data (e.g., location, interests, current activity) in the overarching privacy and security protocols. From the standpoint of those designing such privacy and security protocols, this is good news: highly individualized, contextual information about users and their behaviors may provide information to those protocols helping them detect anomalous or suspicious behavior, trigger cyber-protection mechanisms as appropriate, and similar. However, again, the proposed approach does place an additional burden on the users, majority of whom can be safely assumed not to be particularly "cyber-savvy".

In a nutshell, based on the common themes among these research prototypes, our conclusion is that finding the right tradeoff between (near-)full automation and flexible computational intelligence that humans can trust on one hand, and still leaving humans in control but without cognitive or physical "overload", will likely remain a major challenge in this area for many years to come.

4 Towards the Design of Self-*Next-Generation IoT

An intrinsically open, heterogeneous and complex decentralized such as IoT can be expected to be inherently difficult to control, formally reason about, or be able to predict the behavior of. Additionally, such complex open distributed cyber-physical systems are hard to protect from various types of attacks from the outside, and especially from malicious behaviors by "intruders" or "bad apples" from the inside. Well-known *Distributed Denial-of-Service* (DDOS) attacks on the IoT infrastructure in the eastern parts of the United States, that took place in the fall of 2016, exposed some of the vulnerabilities of the present-day IoT, as well as demonstrated how disruptive and harmful a cyber-attack of a fairly modest sophistication can be [15].

While educating the users is very important, relying IoT's security on human users of smart phones, smart ambient sensors and other devices to always do the "right thing" won't suffice: IoT technologies are meant to enter every household, and their user base will keep growing, including children, the elderly and adults with little or no IT or "cyber-anything" training or understanding. Technologies supposed to be an invisible "helping hand" with daily lives, yet that put too big a cognitive or other burden on users, likely either won't be used correctly or may get jettisoned altogether. Hence, IoT needs to be made cyber-secure without excessive reliance on the users going an extra mile in changing all the passwords on multiple devices, monitoring and validating appropriate access or other settings, etc. Instead, cyber-security of IoT devices, services and platforms should be seamlessly integrated, and made user-friendly, low-maintenance and as little cumbersome to end-users as possible.

Further, while the traditional cyber-security measures applied to wireless communications, software infrastructures and individual devices will continue being crucial components of the overall IoT's cyber-security, in our view these measures and methodologies won't suffice by themselves. Given the intrinsically open and unpredictable nature of IoT, it will be practically impossible to predict all types of future users, devices and applications—and consequently, all possible kinds of malicious or other undesirable behaviors—at the design time (or even, for that matter, at the time of initial deployment).

Hence, the IoT systems need to be enabled with adaptation, self-monitoring, self-recovery and self-healing. For those objectives to be achieved, taking advantage of advanced AI, machine learning and computational intelligence will be of essence. Some particular aspects in which machine learning, data mining and other AI techniques are needed to enable the self-*capabilities of IoT, include the following:

- Collecting, storing and analyzing behavioral patterns from a variety of devices, services, applications and users (including which devices "belong" to each user);
- Identifying anomalous behavior patterns, and as appropriate either recommending to the human user(s), or in some cases automatically and autonomously performing corrective actions;

- Pattern mining including inferring and predicting new types of malicious or abnormal behavior; for example, abnormal patterns of some of the agents' interactions that have not seen before;
- Being able (re)-act similarly to the immune system of a biological organism with respect to identified intruders or other causes of trouble, without necessarily having an a priori knowledge in what forms intruders may appear;
- Ability to reason and meta-reason about devices, applications, users, and behaviors;
- Reinforcement Learning and Meta-Learning over the good policies, behaviors, and corrective actions to "fix problems" in the past, in order to improve over time in how are various cyber-attacks to be dealt with;
- Generating appropriate triggers and alerts, and notifying as appropriate either other agents within the IoT/IoA, or the human end-user; and more.

Securing wireless communications, protecting the integrity, privacy and confidentiality of users' data, applying a variety of existing digital forensics, information assurance and related techniques all already have important roles in cyber-security of IoT, and those roles—and importance of quality solutions in those areas—will only grow in importance in the years to come.

Our main message, however, is that the "traditional" cyber-security methodologies and solutions almost certainly won't suffice for an intrinsically open and highly complex system such as the NG IoT. Further, we argue that the role of AI, computational intelligence and data analytics (each of those, broadly defined) will be absolutely crucial to make this NG IoT (more) secure, and to keep it as safe, reliable, available and secure as possible in the years and decades to come.

4.1 Self-monitoring, Self-healing and Smart Threat Detection

Our discussion so far of some of the anticipated key cyber-security challenges, as well as the desired properties of the adaptable, "smart" middleware for the next-generation IoT, brings us to the final sub-topic, which is, how to use some AI and Machine Learning techniques, to enable such middleware to learn over time, so it can more effectively predict and prevent future cyber-security threats and attacks.

In [16], we argued how certain biological paradigms, more specifically, the workings of the human immune system, can serve as a foundation of the self-healing NG IoT. The immune system does not have a "crystal ball" (or, in the computing context, a knowledge base) with where will attacks on the organism occur, or by what kind of attacking agents. It adapts and reacts to all kinds of attacks, by known and unknown agents; it learns about new types of attacks, and "stores" effective strategies in combatting them. (This, in essence, is why an individual can get sick from certain types of viral diseases only once in his or her lifetime, becoming immune to future attacks by the same agents.) Biological organisms and the external environments in which they operate are intrinsically open systems, with those environments (to use the lingo of AI) being only partially observable and to a considerable extent unknown. Yet, biological organisms succeed in surviving for a while, and, in that process, they successfully combat all kinds of "attackers", due to resilience, adaptability, and capability of "striking back" of their immune systems. These observations were what motivated us to seek the core design principles for IoT cyber-security—specifically, how to enable the NG IoT platforms (in particular, IoT middleware) to effectively self-monitor, self-recover and self-heal in the presence of various malicious agents and behaviors—in the immune systems of biological organisms.

However, how does one design immune system-like resilient IoT middleware, capable of persistently monitoring system's "health", identifying various threats, responding to them appropriately, and providing recovery and self-healing capabilities to the IoT infrastructure? Clearly, "hard-coding" all the desired capabilities, especially with respect to being able to identify and effectively respond to previously unknown types of malicious agents and attacks (the so-*called zero-day attacks*), won't work. While a knowledge base of malicious behaviors should be maintained and expanded as new attacks are discovered and learned about, a resilient NG IoT system needs to be able to pro-actively learn, adapt and react to new types of attacks and other malicious behaviors that have not been previously observed and "recorded" in the said knowledge base.

Should, then, those active, real-time learning capabilities also be based on biological paradigms? This is a harder proposition to support, even at the conceptual level, for biological evolution can certainly be argued to be very slow and inefficient, insofar as how many generations it takes, to create organisms resilient to certain types of viruses or other diseases. With the pace of technology development in general, and how fast different devices, communication protocols, IT platforms and use scenarios keep changing and entering the aforementioned "soup" that is Internet of Things, in particular, clearly a cyber-secure IoT middleware needs to be able to learn and adapt many orders of magnitude faster than what we observe in the biological world.

Is there, then, a better metaphor or learning paradigm for the IoT's learning and adaptation to new threats and attacks? In particular, since not all possible threats and attacks can be "hard-wired", how do we enable a smart, adaptable self-*NG IoT middleware to learn about new threats and how to respond to them quickly? And further, going from the very abstract towards the more concrete system design principles, how would this quick learning, adaptation, self-defense and self-healing of IoT middleware be designed using an appropriately extended Actor-based model for open large-scale distributed infrastructures? A possible answer may be found in the kind of reinforcement learning that has recently been found to be highly effective in mastering complex games of strategy, in particular, learning rapidly how to best deal with previously unseen scenarios (concretely, game positions).

4.2 Reinforcement Learning for Adaptable NG IoT Middleware

In many complex cyber-physical, cyber-secure and multi-agent systems, adapting to various changes in the system's environment and, in particular, learning over time is necessary to ensure the underlying system's effectiveness or even mere survival. This general observation applies to a broad range of socio-economical, socio-technical, biological and physical complex systems. It has been argued in various contexts, that this learning and adaptation oftentimes needs to take place at different "granularity levels" and time scales. By different granularity levels, we mean learning at the level of individual "agents" (whatever may correspond to an individual agent in a concrete decentralized complex systems), small groups of agents (for example, in the form of so-called co-learning, a well-known paradigm from the biological world that has been gaining popularity in Machine Learning and AI worlds, as well), sizable sub-systems made of many agents, or even at the level of the entire large-scale distributed system; see, for example, [17] for a discussion of multi-tiered learning in the context of micro-UAVs (small autonomous unmanned aerial vehicles). By different time-scales, we cf. mean online, real-time (or near-real-time) learning and adaptation, as differentiated from offline learning, usually based on complex symbolic learning and reasoning using a domain-specific knowledge base (KB), and updating the knowledge in that KB as appropriate (see, e.g., [17, 18]).

Biologically-inspired learning and adaptation for the purposes of protecting self-*NG IoT from threats old and new, discussed briefly in the previous subsection and elaborated upon in much more detail in [16], remains a promising paradigm for the offline, long-term learning of agent-based IoT middleware on how to effectively detect and fight against various threats and attacks, as well as enable IoT's recovery from cyber-attacks that were not prevented sufficiently fast. However, biological evolution is a rather slow process. While "simulated evolution" is indeed many orders of magnitude faster than the biological evolution (as is well-known from many experiments using, for example, genetic algorithms and other "generational", biologically-inspired computational techniques), its ability to learn about and adequately respond to "zero-day" attacks likely still won't be "fast enough" for the purposes of protecting NG IoT from previously unseen types of cyber-attacks.

What would, then, be some provably faster ways of predicting, learning about and responding to novel cyber-attacks? One promising such paradigm is reinforcement learning based on what has been dubbed "self-play", stemming from how advanced AI-based programs were trained to very quickly master highly complex games of strategy, such as Chess or Go. In that context, we find the AI approach adopted in the design of *AlphaGo* and its successors particularly promising. In 2016, *AlphaGo* succeeded in convincingly defeating one of the best Go players, Korean Lee Sedol. Since then, *AlphaGo* and its successors have made further major breakthroughs both for the game of Go (as likely the most complex 2-player game of strategy that is being played by millions of people across the globe; in particular, combinatorial complexity of Go on its standard 19×19 board by far exceeds that of chess) and for

Artificial Intelligence. For instance, in addition to defeating one of the best human players in the world, *AlphaGo* achieved an astonishing win rate of 99.8% over other top Go programs [19].

What we find even more fascinating, however, is that this program's successors, *AlphaGoZero* and *AlphaZero,* have rapidly achieved high-level mastery of Go (way above the best human players, or the best other Go programs, so far), using *reinforcement learning* (RL) based on "self-play": generating millions of new games and positions, playing against themselves, evaluating bad versus good versus best strategies based on the outcome of those self-played games, and rapidly iterating that process [20]. Namely, once Google acquired the UK company DeepMinds (the original creators of *AlphaGo*), Google's researchers decided to test the AI's ability to master the game of Go without "prior knowledge", that is, without an appropriate knowledge base (KB) containing the existing "wisdom" on how to play Go well. (Note, such KBs have been at the foundation of the first successful checkers and later on chess programs back in the 1990s.) In particular, *AlphaGoZero* did not have any prior knowledge about how to play Go well, and was trained solely on self-play reinforcement, starting from essentially random play, and then rapidly learning from there [20]. It turned out, that not having "hard-coded" prior knowledge of Go did not slow *AlphaGoZero* much at all. Advanced artificial neural network and Monte Carlo search tree techniques were used to facilitate effective training and reinforcement learning, which took place at a fascinating speed.

Another key aspect of DeepMind's stellar success with *AlphaGoZero* was the ability to generate millions and millions of self-played games, including many positions and particular games (sequences of legal Go moves by the two players) never previously played by the humans (or for that matter, other computers). Self-play and learning from these previously unseen positions provided AlphaGoZero and its variants a major advantage over even the best human Go players. While many millions of self-played games were generated as a part of *AlphaGoZero*'s training process, what is quite remarkable is that the original training only took about 72 h [20]. In a sense, an AI playing against a previously unseen opponent's strategy is rather akin to a "smart" adaptable middleware being tasked to figure out and prevent (or else, quickly heal and recover from) previously unknown, "zero-day" cyber-attacks.

Since 2017, Google/DeepMind have released, in addition to superb Go-playing AI *AlphaGoZero,* also its de facto generalization called *AlphaZero,* that in addition to Go can play other strategy games such as Chess and Shogi [21]. *AlphaZero* is based on the same foundational principles as its predecessor, rapidly learning from self-play and reinforcement how to play these various games of strategy basically from the scratch [21, 22]. *AlphaZero* has proven itself quite successful against the other best AI programs for all games of strategy it can play, showing versatility and universality of the approach adopted in its design—approach based on seemingly very simple idea of neural network implemented reinforcement learning from rapidly and massively iterated self-play [21].

Turning back to self-*IoT, we envision an adaptable, distributed IoT middleware, made of many software agents or actors, that would be able to adapt and learn dynamically, in near-real-time, based on the paradigms behind *AlphaGoZero* and

AlphaGo. Namely, the software foundation of IoT would be designed as a large-scale multi-agent system made of many (millions or more) of autonomous software agents (akin to actors outlined in Sect. 2 and discussed in this IoT context in much more detail in Ref. [16]). Since not all possible future threats and attacks can be "hard-coded" or anticipated and hence captured in the IoT middleware's "knowledge base", the idea of reinforcement learning from self-play would be applied instead.

How could this be made to work, in the context of designing adaptable, scalable, distributed, and above all, resilient agent-based middleware for the next-generation IoT? In particular, how can the ideas of iterated self-play and reinforcement learning be applied to the design of adaptable and capable of rapidly learning IoT software infrastructure? Let us outline briefly a general scenario: while some of the software agents (that constitute the IoT's software platform and provide various "middleware services") are providing resources to various IoT devices and doing other "productive work" on behalf of either the IoT infrastructure or directly of some of its many end-users, a designated subset of agents would "play games" in order to learn about possible future threats and attacks. In particular, some of those agents would play a role of "the bad guys", thus enabling "the good guys" involved in these simulated scenarios (or, in the parlance of *AlphaGo*, engaging in "self-play"), and ultimately the entire IoT infrastructure, to learn about new behaviors (in particular, new *malicious* behaviors) from such "self-play"—rather akin to how *AlphaZero* (or its variants) rather quickly and effectively masters the highly complex strategy games solely from rapidly iterated reinforcement learning applied to "self-play", more specifically, from generating great many possible games and positions according to the applicable rules of Go or Chess or another strategy game (including the many new ones, not previously seen or recorded in the existing "knowledge bases" for those respective games), and then iteratively learning how to most effectively "deal with" all those possible game scenarios.

In summary, reinforcement learning based on simulated self-play appears ideally suited for an intrinsically open, very large-scale, heterogeneous and decentralized system such as IoT. In particular, it is the case with IoT today and will remain so in the future, that it's virtually impossible to predict ahead all possible behaviors of the individual agents or of various other parts or sub-units of the large-scale, highly complex system, and in particular to anticipate and prepare for all possible future threats and attacks. Therefore, being able to quickly and effectively "learn from the scratch" will be the key to designing resilient, secure, self-healing and self-recovering NT IoT.

5 Summary

We have overviewed some of the key challenges in a prominent next-generation cyber-physical/cyber-secure system, the (NG) Internet-of-Things. IoT offers a promise of making many aspects of our daily lives easier, more comfortable and enjoyable, and less encumbered with menial tasks and worries—but it also poses a number of technological, R&D and even policy-making challenges.

In this paper, we discuss some of the key R&D challenges pertaining to the software design and computational intelligence to enable the NG IoT, with an emphasis on its openness, decentralization, cyber-security, self-monitoring and self-healing needs. In particular, among a rich set of those research challenges, we have focused on three: what are the suitable programming abstractions and models for the software infrastructure for IoT, what are some key capabilities and also challenges when it comes to the design of flexible and adaptable middleware for IoT, and last but not least, how can AI, Machine Learning and computational intelligence improve IoT security—in particular, how to enable the NG IoT to become a self-healing, self-correcting autonomous system capable of protecting itself from cyber-threats of both known and unknown varieties.

Our goal in this paper is to highlight the above-mentioned research and system design challenges, focusing on the software design (especially when it comes to the NG IoT's middleware). We have also offered a partial roadmap of what will be required, in our opinion, to adequately address those challenges in order to fulfill the promise of the NG IoT. And, that promise is indeed grandiose: to become a reliable and secure ambient intelligence system that helps us improve how we work, live, play and interact with each other, thus improving quality of people's lives at a scale not seen since the world-wide adoption of the Internet.

References

1. Sicari, S., et al.: A secure and quality-aware prototypical architecture for the IoT. Inf. Syst. **58**, 43–55 (2016). Elsevier
2. Sicari, S., et al.: Security, privacy and trust in Internet of Things: the road ahead. Comput. Netw. **76**, 146–164 (2015)
3. Tosic, P.: Reputation-based distributed coordination for heterogeneous autonomous agents: towards effective coordination, cooperation & coalition formation for autonomous software & device agents belonging to different end-users. In: Proceedings of Internet of Agents (IoA-2016), Web Intelligence Workshops (WIW-16), Omaha, Nebraska, USA (2016)
4. Tosic, P., Wu, Y.: Towards networks of search engines and other digital experts: a distributed intelligence approach. In: Proceedings of 8th International Conference u- & e-Service, Science & Technology (UNESST-15), pp. 35–38, Seoul, S. Korea, 2015
5. iCORE Project. https://www.iot-icore.eu
6. Shoham, Y.: Agent-oriented programming. Artif. Intell. **60** (1993)
7. Jennings, N.R.: An agent-based approach for building complex software systems. Comm. ACM **44**(4) (2001)
8. Agha, G.: Concurrent object-oriented programming. Comm. ACM **33**(9) (1990)
9. Agha, G., Jamali, N.: Concurrent programming for DAI. In: Weiss, G. (ed.) Multiagent Systems: A Modern Approach to Distributed Artificial Intelligence. The MIT Press, Cambridge, MA, USA (1999)
10. Tosic, P. Understanding autonomous agents: a cybernetics & systems science perspective. In: Proceedings of SAI Future Technologies Conference (FTC-2016), San Francisco, CA, USA (2016)
11. IoT-EST project. https://ict-iotest.eu/

12. Ebbits project. https://www.ebbits-project.eu/
13. uTRUST: Usable Trust in the Internet of Things project. https://utrust.pjwstk.edu.pl/
14. BUTLER Project. www.iot-butler.eu
15. Cobb, S.: Ten things to know about the October 21 IoT DDoS attacks. https://www.welivesecurity.com/2016/10/24/10-things-know-october-21-iot-ddos-attacks/ (2016)
16. Tosic, P., Sheldon, F.: On programming models, smart middleware, cyber-security and self-healing for the next-generation internet-of-things. In: Proceedings of IEEE/SAI Computing Conference, vol. 1, London, UK (2018)
17. Tosic, P.T., Vilalta, R.: Learning and meta-learning for coordination of autonomous unmanned vehicles: a preliminary analysis. In: Proceedings of European Conferences on Artificial Intelligence (ECAI-2010), pp. 163–168, Lisbon, Portugal (2010)
18. Tosic, P.T., Vilalta, R.: A unified framework for reinforcement learning, co-learning and meta-learning how to coordinate in collaborative multi-agent systems. In: International Conferenc. on Computational Science (ICCS-2010), Track on Cognitive Agents: Theory and Practice, Amsterdam, The Netherlands, May–June 2010; in Procedia CS, vol. 1, no. 1, pp. 2217–2226 (2010)
19. Silver, D., et al.: Mastering the game of go with deep neural networks and tree search. Nature **529**(7587), 1–37 (2016)
20. Silver, D., et al.: Mastering the game of go without human knowledge. Nature **550**(7676), 1–42 (2017)
21. For more on DeepMind's *AlphaGo,* visit https://deepmind.com/research/casestudies/alphago-the-story-so-far
22. Alphago's next move (visit https://deepmind.com/blog/article/alphagos-next-move)

A Complex Systems Approach to Develop a Multilayer Network Model for High Consequence Facility Security

Adam D. Williams, Gabriel C. Birch, Susan Caskey, Thushara Gunda, Jamie Wingo, and Thomas Adams

Abstract Protecting high consequence facilities (HCF) from malicious attacks is challenged by today's increasingly complex, multi-faceted, and interdependent operational environments and threat domains. Building on current approaches, insights from complex systems and network science can better incorporate multidomain interactions observed in HCF security operations. These observations and qualitative HCF security expert data support invoking a multilayer modeling approach for HCF security to shift from a "reactive" to a "proactive" paradigm that better explores HCF security dynamics and resilience not captured in traditional approaches. After exploring these multi-domain interactions, this paper introduces how systems theory and network science insights can be leveraged to describe HCF security as complex, interdependent multilayer directed networks. A hypothetical example then demonstrates the utility of such an approach, followed by a discussion on key insights and implications of incorporating multilayer network analytical performance measures into HCF security.

Keywords High consequence facility security · Complex systems · Multilayer network

1 Introduction

According to the Cybersecurity and Infrastructure Security Agency (CISA) of the U.S. Department of Homeland Security, critical infrastructure are those assets, systems, or networks whose incapacitation or destruction would have a debilitating effect on national security, economic prosperity, or public health/safety [1]. Similarly, CISA identifies several sectors of critical infrastructure that consists of "high consequence facilities (HCF)" whose benefits are tied to materials, processes, or systems that pose unique risks, including the chemical, energy, and the nuclear reactors,

A. D. Williams (✉) · G. C. Birch · S. Caskey · T. Gunda · J. Wingo · T. Adams
Sandia National Laboratories, Albuquerque, NM 87123, USA
e-mail: adwilli@sandia.gov

D. Braha et al. (eds.), *Unifying Themes in Complex Systems X*, Springer Proceedings in Complexity, https://doi.org/10.1007/978-3-030-67318-5_21

materials, and waste sectors. Protecting HCFs from malicious attacks is challenged by increasing complexity observed in today's operational environments and threat domains.

This complexity is driven by a multi-faceted–and interdependent–set of trends. The first is the analytical challenge of modeling the intentionality of possible adversary actions–which can range from unwitting or accidental (almost random) acts to actions tailored to optimize chances of successfully completing a malicious act. Second, potential adversaries are demonstrating both evolutionary and revolutionary-type improvements in their resources, capabilities, and tactics. Consider the use of advanced unmanned aerial systems by Yemeni rebels to attack Saudi Arabian oil facilities in September 2019 [2], the cyber-attack on the Kudankulam nuclear power plant in India in October 2019 [3], and a 2011 DHS memo stating that "violent extremists have, in fact, obtained insider positions" for damaging physical and cyber-attacks [4] as representative examples. Third, the operating environment in which HCFs must exist is changing, including increased digitization [5], interdependency [6], and operations in non-traditional (e.g., remote) locations [7]. Whether considered individually or taken together, these changes reduce levels of control over the operational environments–which has been a long-standing pillar of security for HCFs. Thus, there is a need to better incorporate elements of operating environments into HCF security [8]. Lastly, as these challenges engender more stringent and comprehensive security responses, levels of organizational (and individual) inertia grow as such changes are seen as opposing normal workplace activities. Thus, in addition to a more complete understanding of threats facing HCFs, there is an additional need to better understand how organizational and human influences impact HCF security performance [9].

Current approaches to HCF security are primarily based on a solution developed by Sandia National Laboratories (Sandia) in the late 1970s and early 1980s. This approach, called the Design Evaluation Process Outline (DEPO), was a response to a 1973 Congressional mandate to improve the security of nuclear materials [10]. Borrowing from the success of probabilistic risk assessment in nuclear safety, DEPO calculates the ability of a collection of security components to achieve a defined probability of defeating a specific adversary along a specific attack path [11, 12]. This methodology defines system effectiveness as the product of the conditional probability that detection and delay system components assess an adversary in time for response forces to arrive onsite to engage (e.g., the probability of interruption) and the conditional probability that, upon arriving, the response force can kill, capture or cause the adversary to flee (e.g., the probability of neutralization). DEPO uses this detect, delay, respond paradigm to fully describe the necessary functions of HCF security.

Yet, DEPO and its derivatives struggle to account for the previously discussed set of challenges to HCF security. At the same time, DEPO provides a strong legacy on which to explore next generation HCF security approaches. Previous arguments [13] have made the case for exploring the multidomain interactions often observed in HCF security operations in terms of complex system behaviors and models. Such interactions include, but are not limited to, the relationships between adversary mitigation mechanisms (e.g., physical security systems and cybersecurity architectures); facility personnel and security procedures; security mechanisms and facility infrastructure; or enhanced adversary capabilities and HCF security upgrades.

2 Multidomain Interactions in HCF Security

To further understand these challenges to HCF security, interviews and focus groups with subject matter experts (SMEs) were conducted to capture the dynamics and perceptions underlying these interactions in HCF security. Using a semi-structured approach, pre-determined, open-ended questions were used to elicit more coherence, depth, and density with which to understand and unpack interviewee responses [14]. This qualitative data collection probed perceptions of the current state of security assessments; strengths and weaknesses of current HCF-related security approaches; and, depictions of ideal future states for HCF security. These interviews and focus groups (summarized in Table 1)–with participants from a diverse set of HCF security-related backgrounds–provided data for describing and better understanding the interactions and dynamics that need to be captured in next generation HCF security approaches.

Analysis of this qualitative data relied "less on counting and correlating and more on interpretation, summary and integration" [14]. Evaluation focused primarily on finding themes in the data and identifying insightful outlier comments germane to better understanding how multidomain interactions impact HCF security. For example, though most of our interviewees supported our research goal, one outlier comment came from Interviewee Q who strongly expressed that it was time to move past current HCF analytical paradigms that are "static and dated."

One theme that emerged was the current siloed nature of HCF security-related activities. The data identified limited (if existing) interactions between security and facility personnel, security designers and implementers, and different domain security personnel (e.g.., physical versus cyber), crystallized by Interviewee P who said "stovepipes kill us." The data also suggests that bridging these siloes extends beyond technical alignment and includes overcoming challenges related to different cultures among different perspectives of HCF. In response was the consensus that a systems approach could help bridge some of these siloes by clarifying specific role(s) and objective(s) for each perspective (e.g., security operations personnel focus on ensuring power supply versus cyber security personnel focus on preventing hackers) and illustrating how these role(s) work together to achieve HCF security.

A second theme from the data related to the inadequate consideration of the role of human actors in HCF security. Whether a need to better address complacency in security personnel, overcoming the "prevalence effect" [15], better understanding impacts from nuisance/false alarm rates, or mitigating insider threats, our data suggests the role of human actors needs to be better understood and more explicitly accounted for in HCF security approaches. For example, Interviewee Q discussed an anecdote relating to HCF security that concluded with the sentiment that having the

Table 1 Summary description of interviews and focus groups with HCF security professionals

Int.	HCF security-related role	Training in current HCF approaches	Years involved in HCF security-related role(s)*	Formal analytic background
A	1	Formal	>10	No
B	1	Formal	>10	No
C	2	Informal	>2	Yes
D	1	Informal	>10	No
E	3	–	>6	Yes
F	4	Formal	>2	Yes
G	5	Informal	>5	No
H	5	Formal	>10	No
I	3, 4	–	>5	Yes
J	4	–	>10	Yes
K	1	Formal	>20	No
L	5	–	>10	No
M	4, 6	Formal	>30	Yes
N	7	–	>10	Yes
O	4, 6	Informal	>30	No
P	3, 4	Formal	>15	Yes
Q	1, 4	Informal	>5	Yes
Focus group 1	6	Informal	2 to 30+	Some
Focus group 2	1, 5	Informal	0.5 to 7	No

1. HCF Security Engineering; 2. Cyber Security Analysis; 3. HCF Resilience Analysis; 4. HCF Security System Analysis; 5. HCF Security Technology Development; 6. HCF Security Operations; 7. Human Cognition in HCF Security

*This refers to **cumulative** years in HCF security-related roles, not just the current role

best security system is still not beneficial if the human elements within the system do not use or access the information provided by technological elements.

A third theme from our data is the impact of the incomplete characterization of HCF-related threats. For example, Interviewee D described how many in the HCF security community still have to "see it to believe it," while Interviewee P depicted a prevailing attitude of "what's happening overseas could not possibly happen in the U.S." Rather than relying on purely quantitative descriptions of threats (like the popular design basis threat [12]), our interview data identified benefits in a threat-agnostic paradigm that would improve HCF security capabilities for responding to various emerging–including un-anticipatable–future threats. Such a paradigmatic shift moves the HCF security design objective, according to one interviewee, from identifying "what is the probability of adversary success" to "influenc[ing] the attacker's deterrence" (Interviewee P).

Evaluation of our interview and focus group data highlighted the need for HCF security solutions to not only meet desired security objectives, but also to consider the impact on overall HCF safety and operations. Across our data, there was a positive response to the goals of this research to more formally identify and incorporate multidomain interactions into HCF security analysis. Moreover, our interviewees further supported taking a systems approach to HCF security and recasting traditional thinking on security performance in broader, more interdependent terms to describe HCF security system stability and dynamics in emerging environments and against adapting adversaries.

3 Multilayer Network Model for HCF Security

Leveraging insights from the complex systems and network theory domains provides potential pathways for closing the gaps identified in our data and current best practices for HCF security. From a complex systems theory perspective, linear concepts of cause and effect are replaced with a non-linear understanding of cause and effect as parallel processes because "meaning is achieved through connections" [16]. Yet, these many, simple interacting components can produce performance different from what would be expected from individual components. This suggests an opportunity to identify more dynamic performance measures better able to describe the efficacy and efficiency of HCF security amidst previously described challenges.

Building on the argument that relationships (or, interactions) can characterize relative priority between nodes (or, system components) to describe emergent behaviors [16], network theory provides additional insights useful to overcome current challenges to HCF security [17]. There has long been a link between network theory and complex systems [18], but more recent efforts (e.g., [19, 20]) are investigating the impacts of multiple, interacting layers within a single network. For example, applying traditional network theoretic approaches to multiple layers provides performance measures such as multilink community detection, versatility, and multilayer communicability [21] that may be able to capture complex characteristics underlying HCF security system design.

This breadth of performance measures effectively expands the definition of HCF security and moves closer to meeting the calls for "threat-agnostic" approaches highlighted in the interview and focus group data. From an analytical perspective, shifting toward threat agnosticism provides greater flexibility in responding to revolutionary changes in security threats than traditional approaches that are tied to smaller, evolutionary threat changes captured in more quantitative models like the design basis threat. For example, multilayer communicability–defined as "a centrality measure which quantifies the number of paths taking both intralinks and interlinks that join a given node of a given layer to the other nodes of the multilayer structure" in [21]–helps describe the potential for digital manipulations to cascade through other domains of HCF security performance. As implied by this example, a multilayer network-based approach also helps take advantage of a larger portion of generated data and better positions HCF security analysis to incorporate dynamic measures of performance.

Coordinating these rich insights lays the foundation for describing the multi-domain interdependencies necessary to align next-generation HCF security capabilities with current dynamic trends. From experience, observation, and confirmed by our data, applying such a "layer-based" approach is a promising solution for capturing the interdependencies, dynamism, and non-linearity that challenges current approaches. In earlier work [13], these layers were described in terms of traditional HCF terms, namely: physical protection systems, cybersecurity architectures, human/organizational actors, and facility infrastructure. (It is of note, however, that these latter two layers–let alone their interactions with the former two–receive minimal consideration in current approaches.) In this work, we propose a function-based layer framework that enables capturing various capabilities within the HCF and framing them in a manner that enables calculating and evaluating related interactions using network analysis techniques. A representative set of notional layers to describe HCF security include:

- **Data and Communications Layer**: Consisting primarily of the components that support data flows for HCF security, these components are modeled as data generators (which collect or manipulate data, like exterior microwave sensors) and data receivers (which typically process, store, and display information to human operators, like command control display system (CCDE) monitors). Many of the traditional detection-related aspects of HCF security are captured in the relationships between these nodes.
- **Supporting Infrastructure Layer**: Though often overlooked in traditional approaches, strong HCF security is dependent on physical infrastructure components that act as an underlying skeleton to provide the necessary operating conditions (e.g., adequate power, temperature control, or fluid flow) for other nodes (and layers).
- **Human Actor(s) Layer**: Perhaps most challenging, this layer is intended to more explicitly describe the role(s) of human in meeting HCF security operations. While the "nodes" in this layer consist of human actors, the structure of this layer helps describe their relative impact in terms of connectedness between nodes. Many of the traditional delay and response-related aspects of HCF security are captured in the relationships between these nodes.

4 Demonstration Case: HCF Security Scenarios

A hypothetical high consequence facility (H2CF) security system was generated to explore the effectiveness and efficacy of a multilayer network representation of the components and relationships described in the previous section. A modified version of the hypothetical facility used in [13], this H2CF security system is composed of ten distinct perimeter intrusion detection zones (i.e., sectors). Figure 1 shows

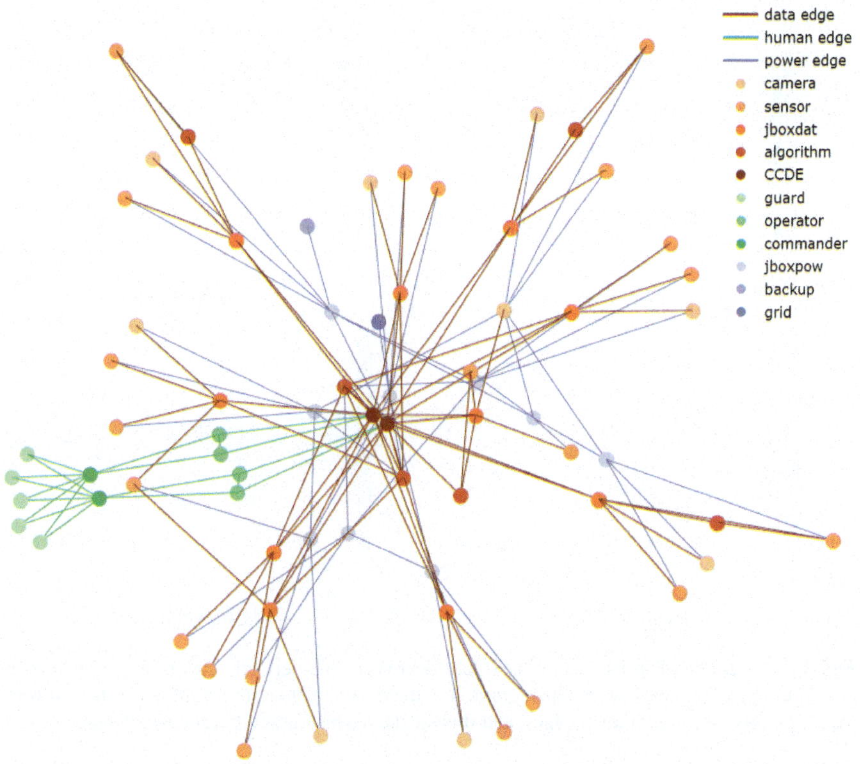

Fig. 1 Example multilayer representation of a ten-sector hypothetical high consequence facility. Red nodes correspond to elements within the data and communications layer, green nodes correspond to elements within the human actors layer, and blue nodes correspond to elements within the supporting infrastructure layer

the multilayer representation of this ten-sector perimeter intrusion detection system. Following HCF security best practices (e.g., [11]), each sector contains sensor devices and cameras that capture data from their areas of responsibility. In addition, these sensors and cameras are connected to junction boxes that exist within the data and power layer, and serve as aggregating points within each sector for communication and power. All related sensor and video data is directed towards the command control display systems (CCDE), which acts as the primary human interface to determine the state of the H2CF security system.

For the multilayer model of the H2CF in Fig. 1, nodes in the data layer consist of forty data generators (e.g., twenty sensors, ten cameras, ten algorithmic processors), ten data transfer elements (e.g., network switches), and two data aggregators (e.g., the CCDE and video management system (VMS)). This representative H2CF security system embeds redundancy in both the data communication and power layers. For example, network components within junction boxes are connected in a ring configuration. Such a design ensures that both communication and power continually

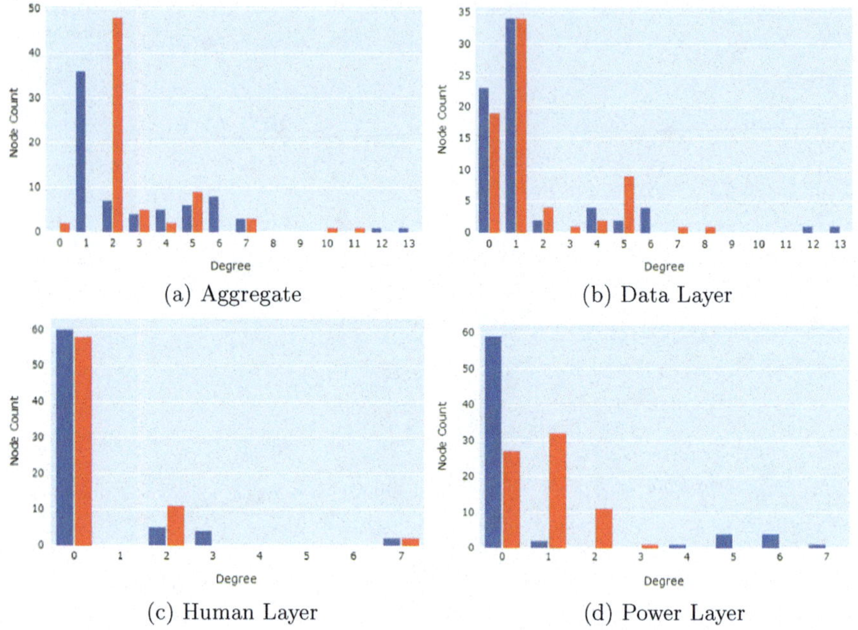

(a) Aggregate (b) Data Layer (c) Human Layer (d) Power Layer

Fig. 2 Histograms of in (red) and out (blue) degree for the network shown in Fig. 1. **a** is the aggregate (i.e., flattened) degree histogram, **b** is the degree histogram for all data edges, **c** is the degree histogram for all human edges, and **d** is the degree histogram for all power edges

operate even when one link is broken. Similarly, the supporting infrastructure layer focuses on power generation and delivery. A backup power supply is connected to the network, adding additional redundant power if a connection to the larger power grid is severed. These elements are represented within the power layer. Lastly, the human layer illustrates organizational relationships between human actors in the H2CF–including the CCDE operator, HCF security supervisors, and HCF security officers/HCF response forces.

The in and out degree distribution was calculated for the aggregate (i.e., flattened) network, as well as for different edge types across the multiple H2CF security system layers (e.g., power, data, human). The resulting degree distribution histograms are shown in Fig. 2 and illustrate several interesting characteristics. First, the in and out degree is generally unequal across layers, indicating an increased importance of the directionality in relationships between elements. For example, more uniformity between in and out degree would suggest a higher measure of redundancy or decentralization in the network. Yet, the large number of in degree and small number of out degrees in the power layer points towards highly centralized and directional infrastructure relationships with respect to power within the H2CF–indicating higher levels of system dependency. Second, the large number of small degree nodes and small number of high degree hub nodes indicates the potential emergence of power-law relationships among elements within the H2CF. For example, several nodes in

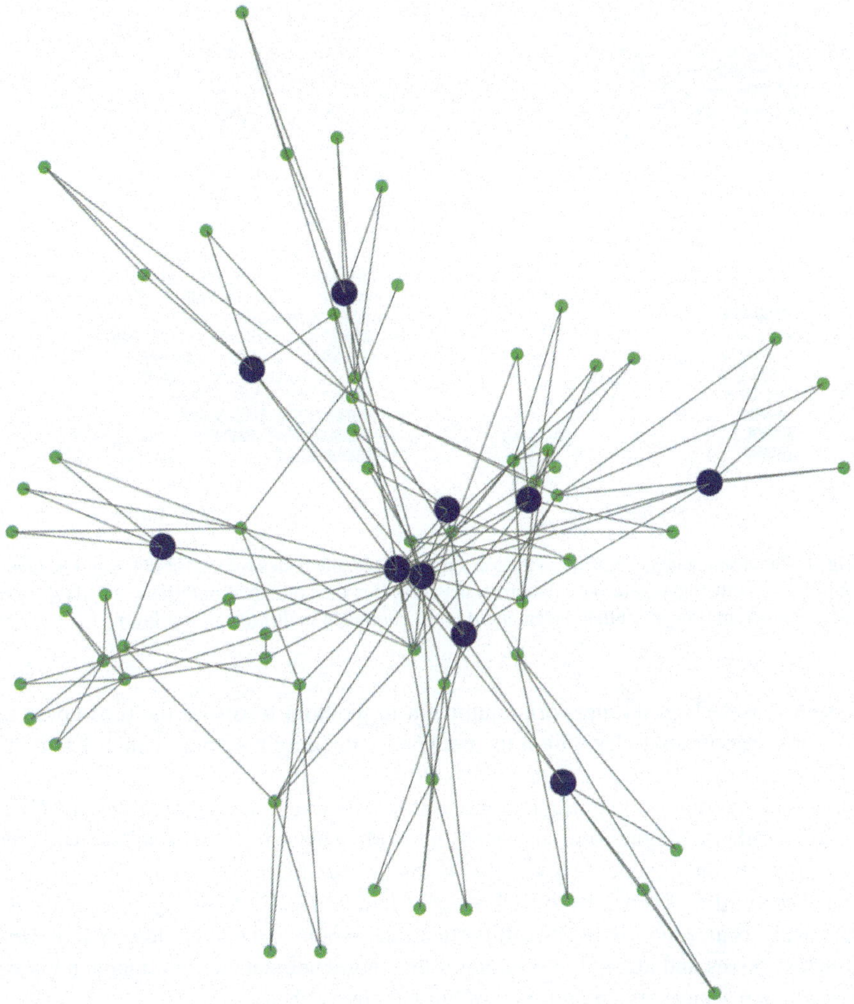

Fig. 3 PageRank values displayed as node size for the H2CF shown in Fig. 1. Blue circles represent the highest page rank values of the flattened network, with the top ten valued nodes being CCDE1, CCDE2, jboxdata10, jboxdata7, jboxdata5, jboxdata4, jboxdata2, jboxdata1, algorithm6, and jboxdata8 (which also matches Fig. 4a)

the data layer–like the CCDE systems–act as network hubs. Identifying hub nodes can aid in making economical design decisions to optimize system resiliency, data processing, and system efficiency.

The PageRank algorithm was performed on this multilayer network, and considers both the combined multilayer PageRank [22], as well as the traditional PageRank algorithm applied to the flattened multilayer network. In the ten sector H2CF, PageRank applied to the flattened network identified the primary and secondary CCDE

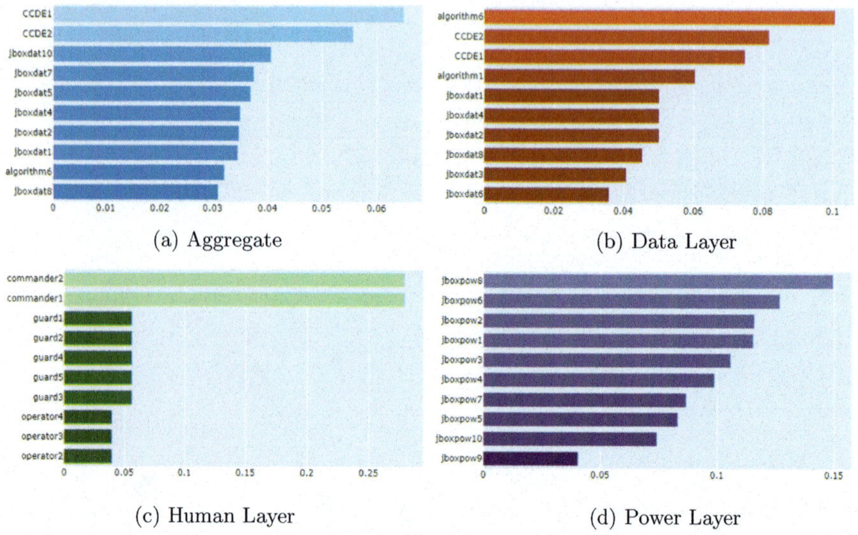

(a) Aggregate (b) Data Layer

(c) Human Layer (d) Power Layer

Fig. 4 PageRank analysis results as the top ten most valuable nodes for the for the network shown in Fig. 1. **a** is the aggregate (i.e., flattened) network, **b** is the combined multilayer data layer, **c** is the combined multilayer human layer, and **d** is the combined multilayer power layer

systems, as well as the network equipment in junction boxes on the data layer, as the most important nodes within the network–indicating they play vital roles in the H2CF security system. The flattened PageRank network is shown in Fig. 3. Figure 4a shows the ten highest PageRank nodes within the system across the different H2CF layers for the aggregate (i.e., flattened) network. This PageRank analysis indicates the primary and secondary CCDE systems are the highest ranking nodes–which matches intuition from current HCF security approaches. Yet, this aggregate PageRank analysis also highlighted the important role of other data layer elements–like the junction boxes and algorithmic switches that aggregate data across data generators within each sector–not captured in traditional approaches.

Figure 4b–d also shows the ten highest PageRank nodes for the combined multilayer PageRank values with respect to the data, human, and power layers. The combined multilayer PageRank is a combination of the additive and multiplicative multilayer page ranks. It measures the influence of a node's centrality in one layer on the centrality of that node in other layers. Additional, more nuanced insights are gained from the application of this multilayer algorithm to the H2CF. For example, the combined multilayer PageRank analysis for the data layer identifies the highly connected algorithms, the primary and secondary CCDE systems, networking equipment among its highest value nodes. This again matches intuition from traditional HCF approaches, though the high rank of certain algorithmic elements may indicate opportunities to increase system resiliency through alternative data layer connections to these elements. The human layer combined multilayer PageRank identified the operational commanders as overwhelmingly valuable nodes as compared to the response force members and security operators that round out the top ten. This matches intuition, considering the highly connected nature of operational commanders within the organizational structure and concept of operations (CONOPS) for HCF security. Finally, the power layer combined multilayer PageRank analysis clearly indicates the importance of oft-overlooked junction boxes to ensuring adequate HCF security system performance.

5 Insights and Implications

Constructing a multilayer model-based approach enabled investigation of traditionally qualitative insights and design best practices for HCF security in a more quantitative manner. The PageRank algorithm applied to the aggregate multilayer network indicated that the central data aggregating points were of high importance. This maps closely to qualitative insights held by traditional HCF security approaches, such as the previously described DEPO method. However, this analysis quantitatively identified the critical nature of the junction box elements within the H2CF, as well as the potentially critical nature of highly interconnected algorithmic elements within an H2CF not explicitly captured in current approaches. These insights could point towards new strategies to decentralize, protect, or optimize sub-components within the HCF for a more balanced distribution of importance within the network.

The themes and insights from the data set supported this approach to security and helped establish the conceptual translation between network-based performance measures and HCF security behaviors. For example, the ability to calculate quantitative network-metric comparisons–based on PageRank in this preliminary analysis–of components in an HCF enable analyses and evaluations that move away from a prescribed threat-based paradigm towards a threat-agnostic, holistic system architecture perspective. Modeling HCF as multilayer systems with many interdependent components enables capturing relational impacts between traditionally segregated domains of security design (e.g., cyber, physical) and results in deeper insights into HCF security behaviors.

These insights imply the potential to expand the multilayer network representation from a static system and move towards dynamic representation. This would enable the calculation of network metrics and analyses that capture the interactivity observed within HCF security operations. Evolving this modeling capability to look at HCF security would shift underlying design, implementation, evaluation, and inspection perspectives from a "reactive" to a "proactive" ethos. Further study and development is needed to adequately incorporate such dynamism and resilience. Yet, this multilayer network model provides a strong start to better address the role(s) of human actors, multidomain interactions, and non-linear operational environments that impact HCF security performance against real-world complexities, innovative adversaries, and disruptive technologies.

Disclaimer: This paper describes objective technical results and analysis. Any subjective views or opinions that might be expressed in the paper do not necessarily represent the views of Sandia National Laboratories, the National Nuclear Security Administration, the U.S. Department of Energy or the United States Government.

Acknowledgements Supported by the Laboratory Directed Research and Development program at Sandia National Laboratories, a multimission laboratory managed and operated by National Technology and Engineering Solutions of Sandia LLC, a wholly owned subsidiary of Honeywell International Inc. for the U.S. Department of Energy's National Nuclear Security Administration under contract DE-NA0003525. SAND2020-5335 C.

References

1. Cyber Infrastructure Security Agency, U.S. Department of Homeland Security: Critical Infrastructure Sectors. https://www.cisa.gov/critical-infrastructure-sectors. Last accessed 4 Mar 2020
2. Hubbard, B., Karasz, P., Reed, S.: Two major saudi oil installations hit by Drone Strike, and US Blames Iran. The New York Times (14 Sept 2019). https://www.nytimes.com/2019/09/14/world/middleeast/saudi-arabia-refineries-drone-attack.html. Last accessed 4 Mar 2020
3. Campbell, A., Singh, V.: Lessons from the cyberattack on India's largest nuclear power plant. Bull. Atomic Sci. (2019) https://thebulletin.org/2019/11/lessons-from-the-cyberattack-on-indias-largest-nuclear-power-plant/. Last accessed 4 Mar 2020
4. Homeland Security News Wire: DHS warns utilities at risk from insider threats. http://www.homelandsecuritynewswire.com/dhs-warns-utilities-risk-insider-threats. Last accessed 4 Mar 2020
5. Gheorghe, A.V., Schlapfer, M.: Ubiquity of digitalization and risks of interdependent critical infrastructures. In: 2006 IEEE International Conference on Systems, Man and Cybernetics, vol. 1, pp. 580–584. IEEE (2006)
6. Rinaldi, S.M., Peerenboom, J.P., Kelly, T.K.: Identifying, understanding, and analyzing critical infrastructure interdependencies. IEEE Control Syst. Mag. **21**(6), 11–25 (2001)
7. U.S. Department of Homeland Security: The Power of Testing Critical Infrastructure in Operational Settings. https://www.dhs.gov/science-and-technology/blog/2018/11/19/power-testing-critical-infrastructure-operational-settings. Last accessed 4 Mar 2020
8. Williams, A.D.: The importance of context in advanced systems engineering. Syst. Eng. Fourth Ind. Revolut. 45–75 (2019)
9. Williams, A.D., et al.: Beyond gates, guards and guns: the systems-theoretic framework for security at nuclear facilities. Massachusetts Institute of Technology, Dissertation (2018)
10. Desmond, W.J., Zack, N.R., Tape, J.W.: The first fifty years: a review of the department of energy domestic safeguards and security program. J. Nucl. Mater. Manag. **26** (1998)
11. Biringer, B., Danneels, J.J.: Risk assessment methodology for protecting our critical physical infrastructures. In: Risk-Based Decision Making in Water Resources IX, pp. 33–43 (2001)
12. Garcia, M.L.: Design and evaluation of physical protection systems. Elsevier (2007)
13. Williams, A.D., Birch, G.C.: A multiplex complex systems model for engineering security systems. IEEE Syst. Secur. Symp. (2019). *In-press*
14. Weiss, R.S.: Learning from strangers: The art and method of qualitative interview studies. Simon and Schuster (1995)
15. Wolfe, J.M., Brunelli, D.N., Rubinstein, J., Horowitz, T.S.: Prevalence effects in newly trained airport checkpoint screeners: trained observers miss rare targets, too. J. Vis. **13**(3) (2013)
16. Tranquilo, J.: An Introduction to Complex Systems: Making Sense of a Changing World Springer International Publishing (2019)

17. Williams, A.D., Jones, K.A.: Invoking network & system theory to improve security risk management in international transport of spent nuclear fuel. In: Proceedings of the 58th INMM Annual Meeting (2017)
18. Liu, Y.Y., Slotine, J.J., Barabási, A.L.: Control centrality and hierarchical structure in complex networks. Plos one **7**(9), e44459 (2012)
19. Gomez, S., Diaz-Guilera, A., Gomez-Gardenes, J., Perez-Vicente, C.J., Moreno, Y., Arenas, A.: Diffusion dynamics on multiplex networks. Phys. Rev. Lett. **110**(2), 028701 (2013)
20. Kivelä, M., Arenas, A., Barthelemy, M., Gleeson, J.P., Moreno, Y., Porter, M.A.: Multilayer networks. J. Complex Netw. **2**(3), 203–271 (2014)
21. Bianconi, G.: Multilayer networks: structure and function. Oxford University Press (2018)
22. Halu, A., Mondragón, R.J., Panzarasa, P., Bianconi, G.: Multiplex pagerank. PloS one **8**(10) (2013)

The Darkweb: A Social Network Anomaly

Kevin P. O'Keeffe, Virgil Griffith, Yang Xu, Paolo Santi, and Carlo Ratti

Abstract We analyse the darkweb and find its structure is unusual. For example, ∼87% of darkweb sites *never* link to another site. To call the darkweb a "web" is thus a misnomer—it's better described as a set of largely isolated dark silos. As we show through a detailed comparison to the World Wide Web (www), this siloed structure is highly dissimilar to other social networks and indicates the social behavior of darkweb users is much different to that of www users. We show a generalized preferential attachment model can partially explain the strange topology of the darkweb, but an understanding of the anomalous behavior of its users remains out of reach. Our results are relevant to network scientists, social scientists, and other researchers interested in the social interactions of large numbers of agents.

Keywords Social networks · Graph theory · Computational network science

1 Introduction

Studies of the World Wide Web (www) have had much impact. An understanding of its topology have let us better understand how its information is stored and can be retrieved [1, 2]. Insight into its paradoxical resilience [3] have allowed us to design more fault tolerant networks [4], and the universal dynamics of it growth patterns [5] have shed light on the behaviors of many others classes of networks [6]. And perhaps

---

K. P. O'Keeffe (✉) · Y. Xu · P. Santi · C. Ratti
SENSEable City Laboratory Massachusetts Institute of Technology,
Cambridge, MA 02139, USA
e-mail: kevin.p.okeeffe@gmail.com

V. Griffith · Y. Xu
Ethereum Foundation, Zug, Switzerland

K. P. O'Keeffe · V. Griffith · Y. Xu · P. Santi · C. Ratti
Hong Kong Polytechnic University, Hong Kong, China

D. Braha et al. (eds.), *Unifying Themes in Complex Systems X*, Springer Proceedings in Complexity, https://doi.org/10.1007/978-3-030-67318-5_22

most importantly, being one of the earliest studied networks[1] [1, 2], the www helped give rise to the flame of attention that network science enjoys today.

This paper is about the www's shady twin: the darkweb. Though the darkweb is similar to the www, both being navigated through a web browser, a key difference the two is that the identities of darkweb users are hidden—that's what makes it 'dark'. This gives the darkweb an infamous air of mystery, which, along with the sparse academic attention it has received [8, 9], makes it ripe for analysis. And beyond satisfying curiosity, its reasonable to think studies of the darkweb could have as much applied impact as the studies of the www.

There are many natural questions to ask about the darkweb. Does it have the same topology as the www, and therefore hold and distribute information in the same way? Is it resilient to attack? Social questions can also be posed. Results from psychology and game theory show people behave more socially when they are watched [10, 11] or have reputations to uphold [12]. Do darkweb users, with whose identities masked, therefore behave more selfishly than www users, whose faces are bare? Finally, beyond the academic impact that answering these questions might generate, insight about the structure or dynamics of the darkweb could have societal benefit, potentially allowing policy-makers to better police its more sinister aspects.

Here we address some of these questions by performing a graph theoretical analysis of the darkweb (we define exactly what we mean by the darkweb below) and comparing it to the www. We find the topologies of the darkweb and the www are starkly different—in fact, the darkweb is much different to many social networks—and conjecture this is due to their users' differing social behaviour. We hope our work stimulates further research on the darkweb's structure as well as the social dynamics of its users.

2 Data Collection

There is no single definition of the darkweb. It is sometimes loosely defined as "anything seedy on the Internet", but in this work we define the darkweb as all domains underneath the ".onion" psuedo-top-level-domain [13] (which is sometimes called the onionweb). Specifically, we mean the subset of the web where websites are identified not by a human-readable hostname (e.g., www.yahoo.com) or by a IP number (e.g., 206.190.36.45), but by a randomly generated 16-character address (specifically, a hash fingerprint). Each website can be accessed via its hash, but it is very difficult to learn the IP number of a website from its hash—this is what makes the web 'dark'; without an IP number, it is exceedingly difficult to trace the geographical origin of a communication.

Crawling the darkweb is not much harder than crawling the regular web. In our case, we crawled the darkweb through the popular tor2web proxy onion.link. Onion.link does all of the interfacing with Tor, and one can crawl all darkweb pages

[1] Alongside smallworld networks [7].

without a login simply by setting a standard crawler to specifically crawl the domain `*.onion.link`. Darkweb pages are written in the same HTML language as the regular web which means we could crawl onion.link using the commercial service scrapinghub.com. Starting from two popular lists of darkweb sites,[2] we accessed each page and crawled all linked pages using breadth-first search.

Most analyses of the www are at the page-level, where each node is an individual URL. One could adopt this model for the darkweb; it would be a natural choice for engineering motivated research question, such as studying crawlability. But the page-level model is not natural for socially motivated research question, which we are interested in in this study, because the page-level graph is influenced more by the various choices of content management system, rather than by social dynamics. So we instead follow the modeling choice in [14] and *aggregate by second-level domain* (for the onionweb the second-level domain is equivalent to [14]'s "pay-level domain"). This means that links within a second-level domain are ignored as socially irrelevant self-connections. In this formulation, each vertex in the darkweb is a *domain* and every directed edge from $u \rightarrow v$ means there exists a page within domain u linking to a page within domain v. The weight of the edge from $u \rightarrow v$ is the number of pages on domain u linking to pages on domain v.

The Tor Project Inc. (the creators and custodians of the darkweb) state there are ~60,000 distinct, active .onion addresses [15]. In our analysis, however, we found merely 7,178 active .onion domains. We attribute this high-discrepancy to various messaging services—particularly TorChat [16], Tor Messenger [17], and Ricochet [18]. In all of these services, each user is identified by a unique .onion domain.

The darkweb has a notoriously high attrition rate; its sites regularly appear and disappear. This complicates our analysis because it creates dead links, links to pages which have been removed. We do not want the dead links in our datasets so we collected responsive sites only; if we discover a link to a page on domain v, but domain v could not be reached after >10 attempts across November 2016–February 2017, we delete node v and all edges to node v. Before pruning nonresponding domains, our constructed graph had 13,117 nodes and 39,283 edges. After pruning, it has 7,178 nodes and 25,104 edges (55% and 64% respectively). The pruned graph is the one used in the rest of this paper, which from now on we call "the darkweb graph".

We note that the darkweb as defined above is different from the network described in [9]. There, the network represents volunteer-operated nodes that could be connected in the sense that they could appear as consecutive nodes in a path through the Tor network. This is completely different from our network.

[2]http://directoryvi6plzm.onion and https://ahmia.fi/onions/.

Table 1 Summarized network level properties between the www and the darkweb. Asterisk for the entries requiring conversion to an undirected graph

Measure	www [14]	darkweb
Num nodes	43M	7,178
Num edges	623M	25,104
Prop. connected pairs	~42%	8.11%
Average SPL	4.27	4.35
Edges per node	14.5	3.50
Network diameter*	48	5
Harmonic diameter	~9.55	232.49

3 Graph-Theoretic Results

Table 1 reports summary statistics of the darkweb and www. In what follows, we discuss these and other statistics.

3.1 Degree Distribution

We begin with statistics on degree distributions, reported in Fig. 1. Panels (Fig. 1a, b) show the in and out degree distributions of the darkweb resemble power laws,[3] just like the www, but with one crucial difference: the location of the y-intercept. In Fig. 1a we see ~30% of domains have *exactly one* incoming link ($k_{in} = 1$), with 62% come from one of the five largest out-degree hubs. In Fig. 1b, we see a whopping 87% of sites *do not link to any other site* ($k_{out} = 0$)!—these are the dark silos that we mentioned in the abstract.

These findings tell us the darkweb is a sparse hub-and-spoke place. The bulk of its sites live in seclusion, not wanting to connect with the outside world. Panel (Fig. 1c) confirms this picture by showing the vast majority of pages have low pagerank (and so are isolated). Panel (Fig. 1d) shows that when a site does link to another, 32% of the time it's only a single page linking out.

As we show in the next section, this siloed structure is not shared by many social networks.

[3]Unfortunately however, we were able to confirm the power laws quantitatively. Following [19], which describes how to fit to power laws rigorously, we tried to fit the degree distribution data using the python package `plfit`, but the fitting failed on account of there not being enough data (the range of the data didn't cover enough orders of magnitude.).

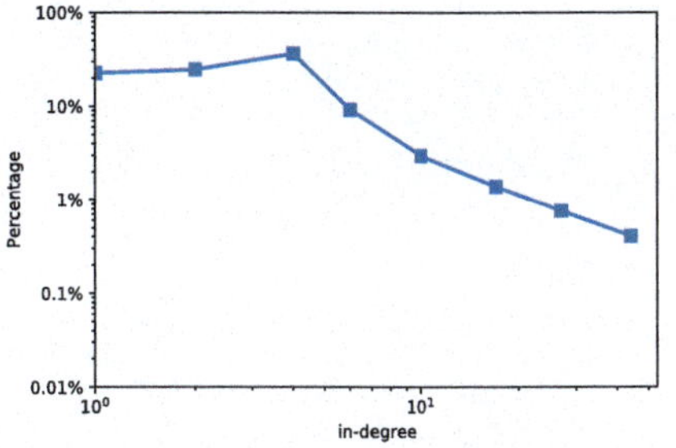

(a) darkweb in-degree

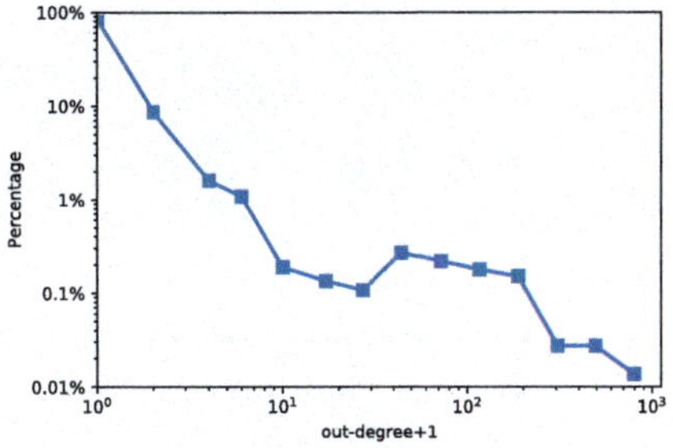

(b) darkweb out-degree

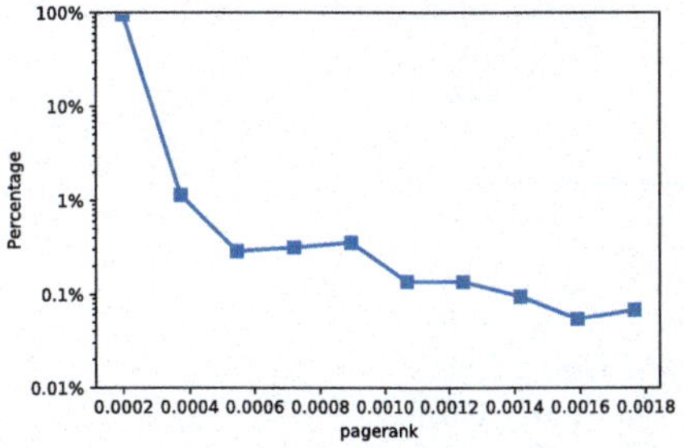

(c) darkweb pagerank

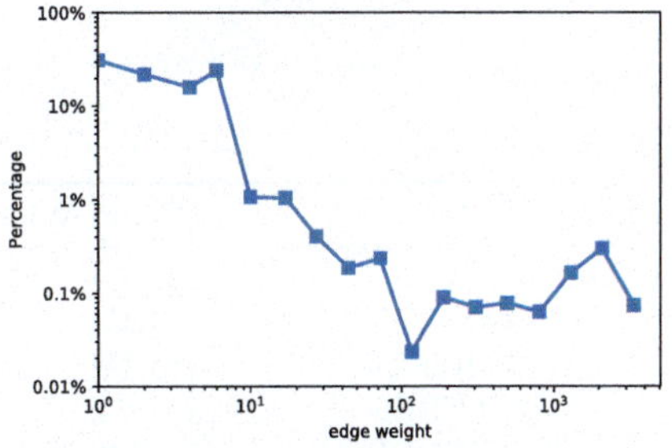

(d) darkweb edgeweights

Fig. 1 The distribution of the in-degree, out-degree, pagerank, and edgeweights. In **c** we exclude the three domains with the highest pagerank because they are such extreme outliers. For all plots with a log-scale axis, we follow following [14, 20] to use the Fibonacci binning from [21]

3.2 Bow-Tie Decomposition

A useful way to describe directed graphs is via *bow-tie decomposition*, where the nodes are divided into six disjoint parts [22]:

1. CORE—Also called the "Largest Strongly Connected Component". It is defined as the largest subset of nodes such that there exists a directed path (in both directions, from $u \rightarrow \cdots \rightarrow v$ as well as $v \rightarrow \cdots \rightarrow u$) between every pair of nodes in the set.
2. IN—The set of nodes, excluding those in the CORE, that are ancestors of a CORE node.
3. OUT—The set of nodes, excluding those in the CORE, that are descendants of a CORE node.
4. TUBES—The set of nodes, excluding those in the CORE, IN, and OUT, who have an ancestor in IN as well as a descendant in OUT.
5. TENDRILS—Nodes that have an ancestor in IN but do not have a descendant in OUT. Also, nodes that have a descendant in OUT but do not have an ancestor in IN.
6. DISCONNECTED—Everything else.

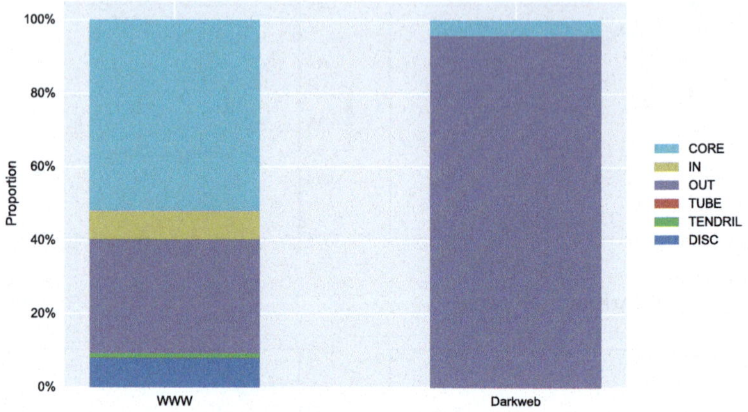

	World-Wide-Web		Darkweb	
CORE	22.3M	51.94%	297	4.14%
IN	3.3M	7.65%	0	0.0%
OUT	13.3M	30.98%	6881	95.86%
TUBES	17k	.04%	0	0.0%
TENDRILS	514k	1.2%	0	0.0%
DISCONNECTED	3.5M	8.2%	0	0.0%

Fig. 2 Bow-tie decompositions of the www and darkweb. The figures for the www were taken from [14]

Figure 2 compares the bowtie decompositions of the darkweb and www. The www numbers are taken from [14, 20, 23]. We chose these works because of the size of their crawls and the rigor of their analyses. Notice the www has each of one the 6 components of the bow-tie decomposition, whereas the darkweb only has a CORE and an OUT component. Moreover, the OUT component of the darkweb contains $\sim$96% (these are the dark silos), leaving the CORE with just $\sim$4% of the mass. This is unusual for a social network; most have large COREs. The www's CORE has $>$50% of the mass, while the cores of Orkut, YouTube, Flickr [24] and Twitter [25] are even larger. In this sense, the darkweb is a social network anomaly.

3.3 *Diameter Analysis*

Next we examine the darkweb's internal connectivity by computing the shortest-path-length (SPL) between nodes. Figure 3a, b shows the SPL's for the www and darkweb. For all pairs of nodes $\{u, v\}$ in the darkweb, only 8.11% are connected by a directed path from $u \to \cdots \to v$ or $v \to \cdots \to u$. This is drastically lower than the $\sim$43.42% found in the www [14]. We again attribute this to the low out-degree

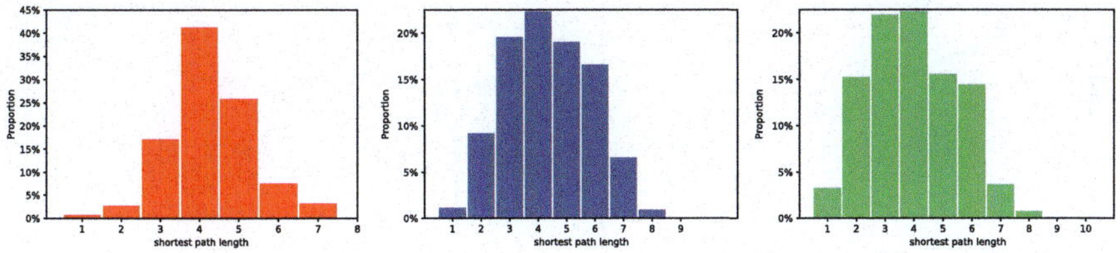

(a) wordwideweb μ=4.27

(b) darkweb μ=4.35

(c) darkweb CORE μ=3.97

Fig. 3 Comparing shortest path lengths between the world-wide-web and the darkweb considering directed edges. Whereas in the www 56.65% of node pairs have have ∞ path-length (no path connecting them), in the darkweb 91.89% of node-pairs have no path connecting them. Moreover, even within that 8.11% of pairs with a directed path between them, the darkweb's average SPL ($\mu = 4.35$) is *higher* than that of the www ($\mu = 4.27$)

per Fig. 1. Of the connected pairs, the darkweb's average shortest path length is 4.35 compared to the 4.27 in the world-wide-web [14]. It's surprising to see a graph as small as the darkweb have a higher mean SPL than the entire world-wide-web, and is a testament to how sparse the darkweb graph really is. Figure 3c plots the distribution of SPLs for the 297 nodes of the CORE. To our surprise, the mean SPL within the CORE is 3.97, only 9% less than the entire darkweb. From this we conclude the CORE is not densely interconnected.

3.4 Robustness and Fragility

Does the peculiar structure of the darkweb make it resilient to attack? Fig. 4 shows it does not; as seen, the entire network (WCC) as well as the CORE quickly disintegrates under node removal. In Fig. 4a, b we see the familiar resistance to random failure yoked with fragility to targeted attacks, in keeping with the findings of [26]. Figures 4b shows that, unlike the www [14], the WCC is *more susceptible to high in-degree deletions than the* CORE. This elaborates the view from Fig. 4c that the CORE is—in addition to not being strongly interconnected—is also not any kind of high in-degree nexus.

Figures 4c, d show the breakdown when removing central nodes. In Fig. 4c, the CORE is largely unaffected by low centrality deletions. In Fig. 4c we see that although the CORE isn't disproportionately held together by high in-degree nodes, it is dominated by very central nodes.

Comparing Fig. 4b, f we see the CORE relative to the entire network consists of more high-pagerank nodes than high in-degree nodes. This implies CORE nodes are not created by their high-indegree (Fig. 4b), but by their high centrality, amply corroborated by Fig. 4c, d. Likewise, Fig. 4e recapitulates Fig. 4a, that nodes with especially low in-degree or centrality are, unsurprisingly, not in the CORE.

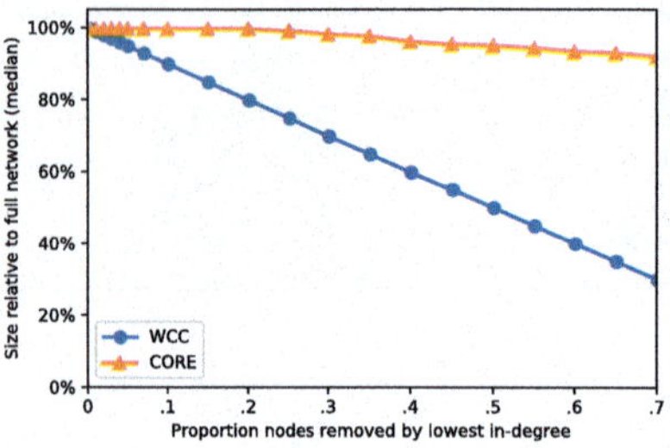

(a) removing nodes by lowest in-degree

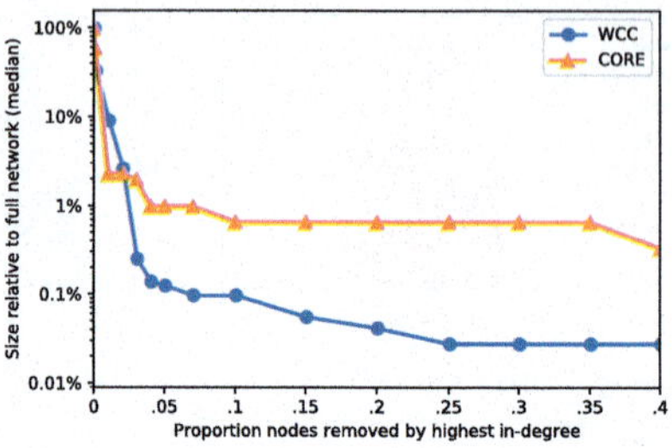

(b) removing nodes by highest in-degree

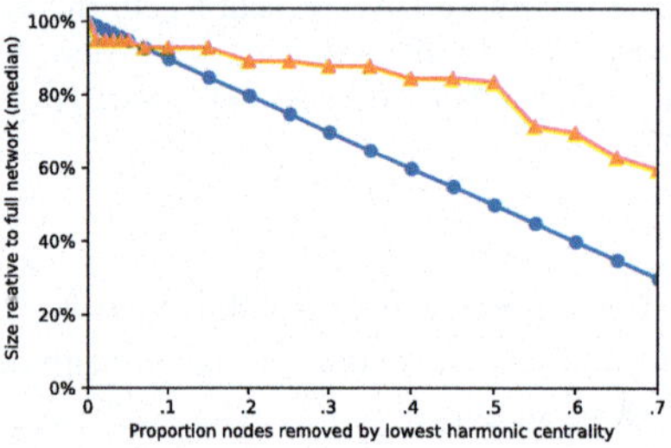

(c) removing nodes by lowest harmonic centrality

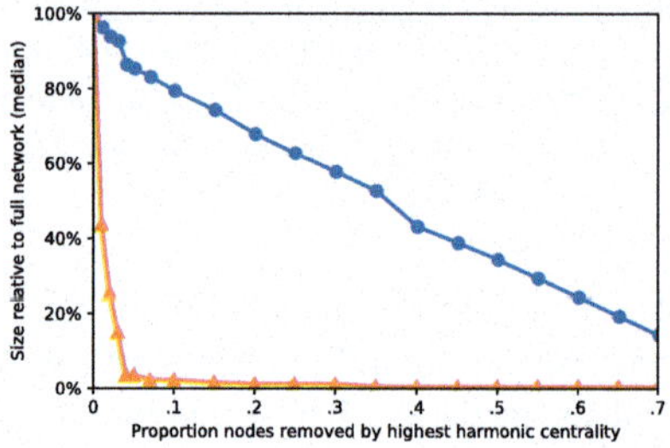

(d) removing nodes by highest harmonic centrality

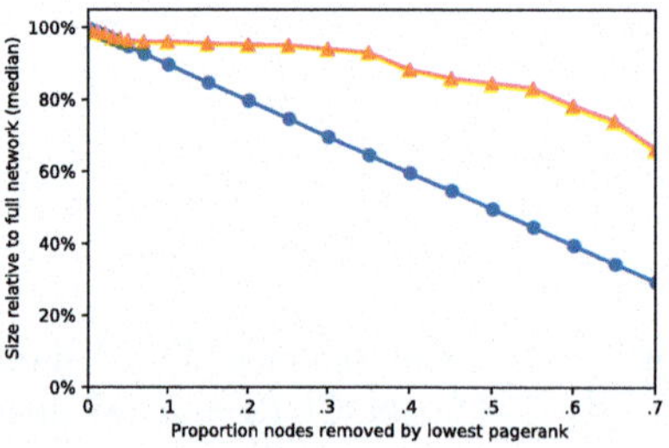

(e) removing nodes by lowest pagerank

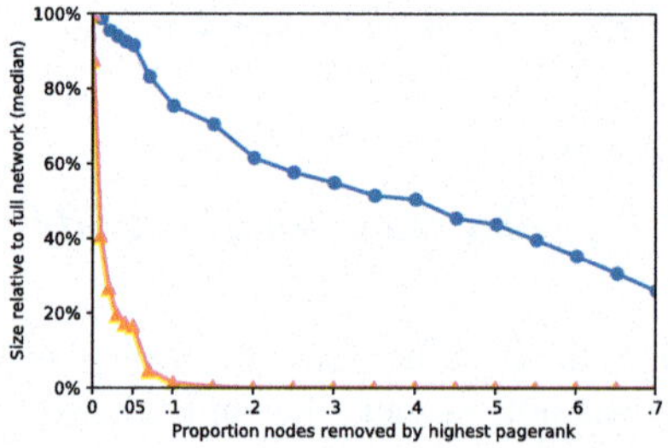

(f) removing nodes by highest pagerank

Fig. 4 Deleting nodes from the darkweb graph and seeing how quickly the WCC and CORE disintegrate. In all plots, we shuffled the order of nodes with the same value until reaching stable statistics, e.g., in 1c, 98% of nodes are tied for the lowest pagerank; so when removing only 10% of the nodes (e.g., Fig. 4e), it's ambiguous which nodes should be deleted first. So in our analysis we shuffled the ordering of the nodes with the same value and recomputed sizes of the WCC/CORE until the median was stable

3.5 Reciprocal Connections

The authors of [23] stress the importance of reciprocal connections in maintaining the www's graph properties. We compute two of their measures. First, we compute [23]'s measure $\frac{\langle k_{in} k_{out} \rangle}{\langle k_{in} \rangle \langle k_{out} \rangle} = \frac{\mathbb{E}[k_{in} k_{out}]}{\mathbb{E}[k_{in}]\mathbb{E}[k_{out}]}$, to quantify in-degree and out-degree's deviation from independence. For the darkweb, we arrive at $\frac{\langle k_{in} k_{out} \rangle}{\langle k_{in} \rangle \langle k_{out} \rangle} = 3.70$. This is in the middle of the road of prior estimates of the www, and means that the out-degree and in-degree are positively correlated. For greater clarity, we also plot the average out-degree as a function of the in-degree, given as,

$$\langle k_{out}(k_{in}) \rangle = \frac{1}{N_{k_{in}}} \sum_{i \in \Upsilon(k_{in})} k_{out,i} \,, \tag{1}$$

which is simply "For all nodes of a given in-degree, what is the mean out-degree?". The results are depicted in Fig. 5. In short, in the darkweb there's no obvious pattern to the relationship between in-degree and out-degree.

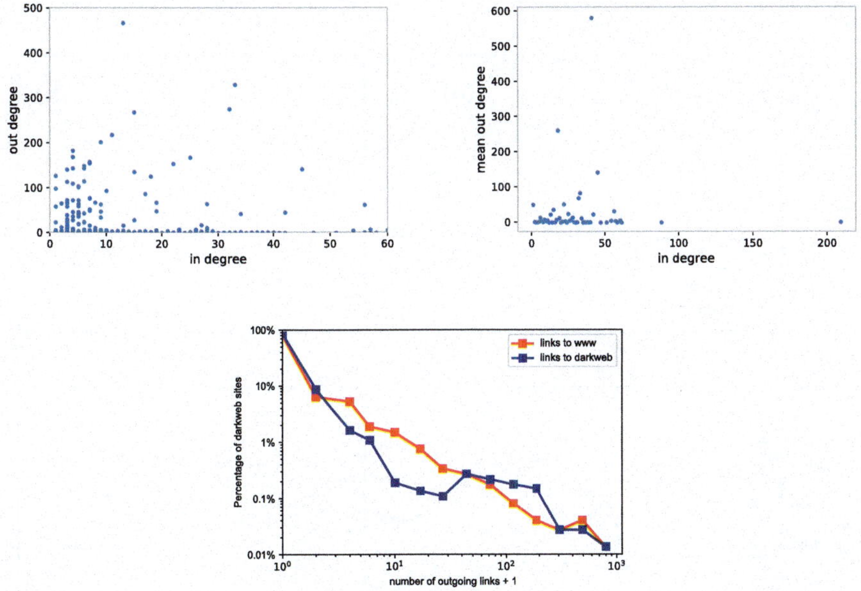

Fig. 5 **a** Scatter plot of k_{in} and k_{out}. **b** Plot of averaged k_{out} versus k_{in}. **c** Comparing the rates of the darkweb sites linking to the www versus linking to other darkweb sites. They are essentially the same

3.6 Network Growth Model

Are standard network growth models able to capture the topology of the darkweb? Here we show a generalized preferential attachment model approximately can. In regular preferential attachment, nodes of only one type are added sequentially to a network. Here, we generalize this procedure to two types of nodes: *pages* and *portals*. Page nodes model the nodes in the darkweb which do not link to anyone (and so have $k_{in} = 0$). Portals model the rest of the nodes, which act as 'portals' to the page nodes. The dynamics of the 'page-portal' model are defined as follows.

At $t = 0$ N_0 portals are arranged in a given network. At each time step after $t > 0$, a new page is added to the network and m different portals are chosen to link to m end nodes (these end nodes can be either pages or portals). The portal which does the linking is chosen according to preferential attachment, that is, chosen with probability $\propto (1 + k_{out}^{\beta})$, where k_{out} is the out degree and β is a free parameter. The end node chosen with probability $\propto (1 + k_{in}^{\beta})$. Notice this scheme means that only portals can link, which means page nodes forever remain with $k_{in} = 0$, as desired. The model has three free parameters m, β, N_0. We chose $N_0 = 397$ so as to match the number of portals in our datasets (defined as all nodes with $k_{in} > 0$) and ran the

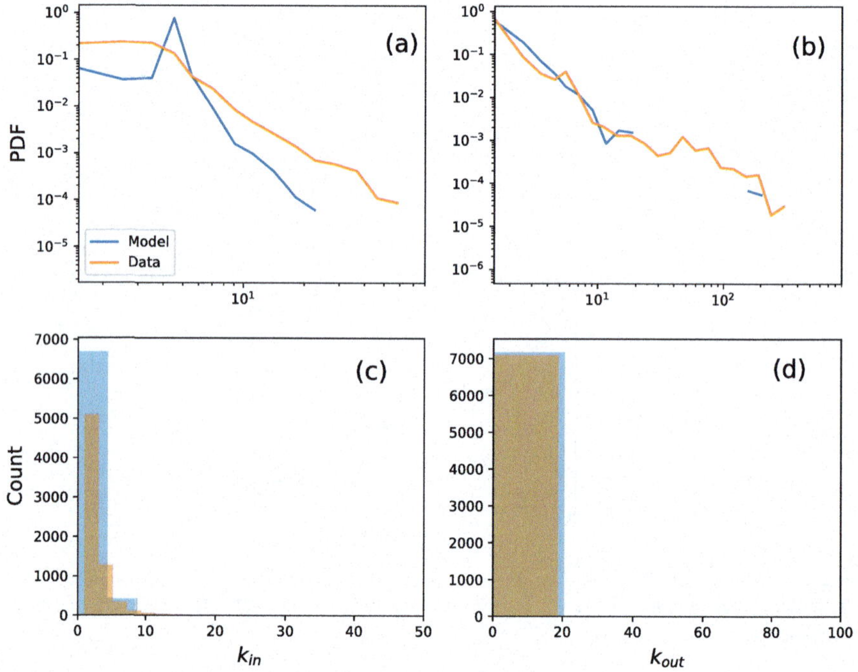

Fig. 6 Comparison of in (k_{in}) and out (k_{out}) degree statistics of darkweb and page-portal model. **a** Probability density functions of k_{in} and k_{out} for page and portal model. **c, d** Histograms of above quantities

model for 6242 timesteps i.e added 6242 page nodes) so that there were 7178 nodes at the end of the process which again matched the dataset.

Figure 6 shows the page-portal model with parameters $(m, \beta) = (4, 2)$ approximates the darkweb, the in and out degree distributions of the model approximately mimicking the data. We report exponents of best fit, found using the powerlaw package python. But keep in mind that as mentioned earlier, there are not enough orders of magnitude in the data for the estimates of the exponents to be reliable; thus the page and portal model is intended as a first step in modeling the darkweb. The values were $(\alpha_{in}, \alpha_{out})_{data} = (3.09, 2.10)$ and $(\alpha_{in}, \alpha_{out})_{model} = (4.3 \pm 1.3, 2.4 \pm 0.2)$, where the model figures report means of 10 realizations and the error is given as half the range. The fitting was performed using the python powerlaw package. α_{in} was much more difficult to measure than α_{out}. As much as half of realizations led to estimates $\alpha_{in} > 50$, which we interpret as the failure of the fitting to converge, and so were discarded.

4 Conclusion

Our primary finding is that the darkweb is a social network anomaly. Its light CORE and massive OUT components distinguishes it from other popular social networks. In fact, calling it a 'web' is a connectivity misnomer; it is more accurate to view it as a set of dark silos—a place of isolation, entirely distinct from the well connected world of the www and other social network. What causes the darkweb to be so isolated? We see two possible explanations:

- **The technological explanation**. In the darkweb, sites go up and go down all the time. Why bother linking if there's little chance that the destination will still exist?
- **The social explanation**. People who create sites on the darkweb are cut from a different social cloth than those who create sites on the www (or at least when using the darkweb, these people behave differently).

To test the technological explanation, we performed a second crawl collecting instances of darkweb sites linking to the www and compared the relative rates of outbound linking in Fig. 5c. There are essentially equal rates of outbound linking to the www as well as the darkweb which tells us (i) the low outbound linking is not due to the impermanence of onion sites and (ii) if onion sites got drastically more stable, we would still see very low rates of linking. Taken together, these indicate the technological explanation is likely untrue. Thus, the social explanation is likely cause of the darkweb's anomalous structure.

Rigorously testing the social hypothesis is however beyond the scope of this work. Although, in a sense we have taken a first step in this direction by generalizing preferential attachment which itself can be viewed as model of social behavior; it is a model of trust: highly linked nodes are perceived as 'trustworthy' sources of information, and so receive a disproportionate number of links; the rich get richer. The isolated silos of the darkweb, however, indicate trust does *not* play a role in the dynamics governing its evolution. Rather, one might say it indicates that *distrust* does. The passive pages of the page and portal model (which recall, do not link to anybody through the dynamics, and are in that sense passive) were a crude way to incorporate this effect. But a more principled behavioral model (i.e. which consistent with known results from psychology) is needed, which were were unable to develop. We hope psychology-fluent researchers will take up this task in future work.

Future work could also study the temporal aspects of the darkweb. Is the topology we have found stationary? For example, in the work most closely related to ours [9], it was found that the resilience of the studied 'darknet' evolved over time (as discussed in the Data Collection section, our darkweb graph is much different to the darknet in [9]). It would be interesting to see if the resilience of our darkweb graph behaves like this too.

Acknowledgements The authors would like to thank all members of the MIT Senseable City Lab consortium.

References

1. Barabási, A.L., Albert, R., Jeong, H.: Scale-free characteristics of random networks: the topology of the world-wide web. Phys. A: Stat. Mech. Appl. **281**, 69–77 (2000)
2. Albert, R., Jeong, H., Barabási, A.-L.: Diameter of the world-wide web. Physics. http://www.physicsweborg/article/world/14/7/09 (1999)
3. Albert, R., Jeong, H., Barabási, A.-L.: Error and attack tolerance of complex networks. Physics. https://www.physicsweborg/article/world/14/7/09 (2000)
4. Wang, L., Jin, H., Dang, J., Jin, Y.: A fault tolerant topology control algorithm for large-scale sensor networks. Physics. http://www.physicsweborg/article/world/14/7/09 (1999)
5. Barabási, A.L., Albert, R.: Emergence of scaling in random networks. Science **286**, 509–512 (1999)
6. Barabási, A.-L., Bonabeau, E.: Scale-free networks. Sci. Am. (1999)
7. Watts, D.J., Strogatz, S.H.: Collective dynamics of small-worldnetworks. Nature (1998)
8. Everton, S.F.: Disrupting Dark Networks, vol. 34. Cambridge University Press (2012)
9. De Domenico, M., Arenas, A.: Modeling structure and resilience of the dark network. Phys. Rev. E **95**, 022313 (2017)
10. Bradley, A., Lawrence, C., Ferguson, E.: Does observability affect prosociality? Proc. Roy. Soc. B: Biol. Sci. (2018)
11. Kristofferson, K., White, K., Peloza, J.: The nature of slacktivism: how the social observability of an initial act of token support affects subsequent prosocial action. J. Consum. Res. (2014)
12. Nowak, M., Highfield, R.: Supercooperators: altruism, evolution, and why we need each other to succeed structure in the web (2011)
13. Appelbaum, J., Muffet, A.: The ".onion" special-use domain name. Internet Eng. Task Force RFC 7686 (2015)
14. Lehmberg, O., Meusel, R., Bizer, C.: Graph structure in the web: aggregated by pay-level domain. In: Proceedings of the 2014 ACM Conference on Web Science. ACM, pp. 119–128 (2014)
15. Project, T.T.: Tormetrics—unique .onion addresses. https://metrics.torproject.org/hidserv-dir-onions-seen.html (2017)
16. Wikipedia (2016). Torchat—wikipedia, the free encyclopedia. https://en.wikipedia.org/w/index.php?title=TorChat&oldid=728783759. Accessed 28 February 2017

17. Wikipedia (2017). Tor (anonymity network)—wikipedia, the free encyclopedia. https://en.wikipedia.org/w/index.php?title=Tor_(anonymity_network)#Tor_Messenger&oldid=767770868. Accessed 28 February 2017
18. Wikipedia (2016). Ricochet (software)—wikipedia, the free encyclopedia. https://en.wikipedia.org/w/index.php?title=Ricochet_(software)&oldid=755044411. Accessed 28 February 2017
19. Clauset, A., Shalizi, C.R., Newman, M.E.: Power-law distributions in empirical data. SIAM Revi. **51**, 661–703 (2009)
20. Meusel, R., Vigna, S., Lehmberg, O., Bizer, C.: The graph structure in the web—analyzed on different aggregation levels. J. Web Sci. **1** (2015)
21. Vigna, S.: Fibonacci Binning. CoRR abs/1312.3749 (2013)
22. Broder, A., Kumar, R., Maghoul, F., Raghavan, P., Rajagopalan, S., et al.: Graph structure in the web. Comput. Netw. **33**, 309–320 (2000)
23. Serrano, M.A., Maguitman, A., Boguñá, M., Fortunato, S., Vespignani, A.: Decoding the structure of the www: a comparative analysis of web crawls. ACM Trans Web **1** (2007)
24. Mislove, A., Marcon, M., Gummadi, K.P., Druschel, P., Bhattacharjee, B.: Measurement and analysis of online social networks (2007)
25. Garcia, D., Mavrodiev, P., Casati, D., Schweitzer, F.: Understanding popularity, reputation, and social influence in the twitter society. Meas. Anal. Online Soc. Netw. (2007)
26. Bollobás, B., Riordan, O.: Robustness and vulnerability of scale-free random graphs. Internet Math. **1**, 1–35 (2004)
27. Borgatti, S.P., Jones, C., Everett, M.G.: Network measures of social capital. Connections **21**, 27–36 (1998)
28. Wu, P., Wen, J.R., Liu, H., Ma, W.Y. (2006). Query selection techniques for efficient crawling of structured web sources. In: Proceedings of the 22nd International Conference on Data Engineering ICDE'06, p. 47. IEEE
29. Adamic, L.A., Huberman, B.A.: Power-law distribution of the world wide web. Science **287**, 2115 (2000)
30. Kumar, R., Raghavan, P., Rajagopalan, S., Sivakumar, D., Tompkins, A., et al.: The web as a graph. In: Proceedings of the Nineteenth ACM SIGMOD-SIGACT-SIGART Symposium on Principles of Database Systems. ACM, pp. 1–10 (2000)
31. Barabási, A.-L.: The physics of the web. Physics (2001). https://www.physicsweborg/article/world/14/7/09
32. Page, L., Brin, S., Motwani, R., Winograd, T.: The pagerank citation ranking: bringing order to the web. Technical Report 1999-66, Stanford InfoLab. http://ilpubs.stanford.edu:8090/422/ (1999). Previous number = SIDL-WP-1999-0120
33. Biryukov, A., Pustogarov, I., Weinmann, R.P.: Trawling for tor hidden services: detection, measurement, deanonymization. In: 2013 IEEE Symposium on Security and Privacy (SP). IEEE, pp. 80–94

On Efficiency and Predictability of the Dynamics of Discrete-Time Boolean Networks

Predrag T. Tošić

Abstract We discuss possible interpretations of "(in)efficient" and "(un)predictable" systems when those concepts are applied to discrete networks and their dynamics. We are interested in *computational notions* of the network dynamics' predictability and efficiency. *Deterministic* discrete network's dynamics is efficient if it settles fairly quickly into an appropriate stationary pattern; that is, efficiency is essentially synonymous with a short transient chain until a stationary system state is reached. An inefficient dynamical system, then, is one that takes a long time to converge to a stationary behavior. The issue of (in)efficiency is related to, but different from, that of the network dynamics' (un)predictability. We call dynamics predictable, if the "ultimate destiny" (i.e., convergence to a stationary behavior) can be efficiently determined or predicted, without performing a step-by-step simulation of the network's dynamics—unless such simulation itself can be done efficiently. For discrete systems with finite configuration spaces, all properties of their dynamics are decidable; however, for non-trivial network sizes, various properties may or may not be possible to predict within reasonable computational resources. In that context, we discuss how computational complexity of reachability problems and enumeration problems about several classes of cellular and network automata relate to whether those networks' dynamics are predictable. We then discuss some examples of cellular automata and other Boolean networks whose dynamics are (i) both predictable and efficient, (ii) efficient but unpredictable, (iii) predictable yet inefficient, and finally (iv) both unpredictable and inefficient. In particular, we briefly overview several classes of Boolean networks that fall into the categories (i), (ii) and (iv).

Keywords Cellular automata · Network automata · Boolean networks · Hopfield networks · Predictability · Efficiency · Computational complexity · Reachability · Fixed point configurations · Asymptotic dynamics

---

P. T. Tošić (✉)
Independent AI and Computer Science Researcher, Spokane, WA, USA
e-mail: predrag.tosic@ieee.org

Department of Mathematics and Statistics (Adjunct), Washington State University, Pullman, WA, USA

D. Braha et al. (eds.), *Unifying Themes in Complex Systems X*, Springer Proceedings in Complexity, https://doi.org/10.1007/978-3-030-67318-5_23

1 Introduction and Motivation

We are interested in discrete networks and their asymptotic dynamics. For simplicity, we focus on deterministic such discrete networks, in which each node can be in one of two states, denoted 0 and 1; a node updates its state according to a single-valued Boolean function; and the nodes update their states in discrete time steps. Such models are called *Boolean Networks* (BNs) or *Boolean Network Automata* (BNA). These models have been extensively studied in physics, theoretical biology, as well as network science and agent-based modeling of a broad variety of physical, cyber-physical, socio-technical and other large-scale decentralized systems. In these models, each node is a simple two-state finite-state machine, where the behavior of each node is some deterministic function of its own current state and those of (some subset of) its neighboring nodes [14, 21, 27]. Classical *Cellular Automata* (CA) are perhaps the best-known example of such discrete networks. A cellular automaton is defined over a regular graph such as a line or ring (1-dimensional cellular spaces) or a higher-dimensional grid or torus (for 2- and higher-dimensional cellular spaces). Each node in a CA updates its state according to the same update rule—some function of the current state of the node itself, and the current states of some of its neighbors [26, 27].

CA generalize to various other network or graph automata models, in which (i) the network topology can be more general than the cellular spaces of CA, and (ii) different nodes are allowed to update according to different update rules [17, 18]. There have been a number of such generalizations studied in various contexts. One of the best-known classes of such models are the (discrete) Hopfield networks [8]. Hopfield nets have been used since the 1980s to model various phenomena both in statistical physics and theoretical biology [8, 9]. More recently, application of Hopfield networks to abstract modeling of computational brains as *associative memories* has gained traction among the research community; see, e.g., [4, 22]. Various discrete network models have been used in other theoretical biology contexts, as well—such as, for mathematical modeling related to protein networks and gene regulatory networks (see, e.g., [6, 10]). Yet other, descriptively fairly similar, discrete network models have been studied in the context of modeling cyber-physical, socio-technical and multi-agent systems [1, 2, 16, 17, 21, 23].

Among many interesting questions about the dynamics of these various discrete network models, their asymptotic dynamics, starting from either an arbitrary or a specific initial configuration, has been of arguably the greatest interest. In that context, characterizing in how many different ways can such a network evolve, and, in particular, enumerating (exactly or at least approximately) all of its stable configurations (or "fixed points") has been studied in the context of Hopfield networks since the 1990s [5, 12, 13]; more recently, this line of research has resulted in tighter theoretical results about possible asymptotic dynamics, as well as interesting interpretations in terms of the storage capacity of associative memories [22].

Enumeration of possible dynamics in general, and stable configurations in particular, has been investigated in the context of discrete network models of large-scale multi-agent and cyber-physical systems, as well; see, e.g., [15, 17, 20]. In particular, attempts have been made to characterize "phase transitions" between relatively simple and predictable network dynamics on one hand, and provably complex and unpredictable dynamics, on the other—as the network parameters such as the underlying "topology" and the local update rules are "tweaked" in different ways [20, 23]. In particular, the present work, focusing on the computational notions of efficiency and predictability, is a continuation of research on phase transitions in network dynamics' complexity found in [23, 24]. We define several discrete network models, previously used in applications as broad as statistical physics, theoretical biology, multi-agent and networked cyber-physical systems, in the next Section, focusing on Boolean networks.

We are interested in efficiency and predictability of the asymptotic dynamics of these discrete (Boolean) networks. Both efficiency and predictability can be defined in different ways; for example, efficiency of a system can be measured in terms of energy flow and dissipation, information flow in and out of the system, etc. However, we are specifically interested in providing intrinsically computational definitions of these two concepts. Since we restrict our discussion to finite, fully deterministic discrete dynamical systems, all properties of such systems' dynamics can be determined "in principle", that is, they are computationally decidable. The problems of efficiency and predictability, then, reduce to which aspects about the system dynamics pertaining to how efficient and how predictable that dynamics is, can be determined computationally efficiently, and which aspects would take prohibitively large computational resources for non-trivial network sizes. We consider the number of nodes of a discrete network to be the parameter that, for our purposes, adequately captures the "network size".

2 Preliminaries and Definitions

In this section, we define the discrete network models of interest; we focus on their Boolean variants, in which each node in the underlying graph or network can be in one of two distinct states, 0 or 1. In this paper, we study only deterministic such Boolean networks. We also formulate some fundamental properties of *asymptotic dynamics* of the general *Boolean Networks* and, as their prominent special cases, classical *Cellular Automata* [14, 26, 27] as well as *Discrete Hopfield Networks* [8, 9].

Definition 1 A *Boolean Network*, also called *Boolean Network Automaton* (BNA for short) is a directed or undirected graph so that each node in the graph has a state, 0 or 1; and each node periodically updates its state, as a function of the current states of (some or all of) its neighboring nodes (possibly, but not necessarily, including itself).

A BNA dynamically evolves in discrete time steps. If the node v_i has k neighbors denoted $v_{i1}, \ldots, v_{ik}$ (where this list may or may not include v_i itself), then the next state of v_i is determined by evaluating a Boolean-valued function $f_i(v_{i1}, \ldots v_{ik})$ of k Boolean variables; f_i is called the *local update function* or *transition rule* for the node v_i. For simplicity, we restrict our attention to the models where each node's next state depends on its own current state (the so-called "automata with memory"), although we point out, all our results herewith hold irrespective of this particular assumption.

Some clarifications are appropriate here. First, in general Boolean Network Automata, different nodes v_i may use different local update functions f_i. This applies to *Discrete Hopfield Nets* [4, 5, 8, 9], as well as many other classes of Boolean Networks, such as those originally introduced by S. Kauffman in the context of systems biology [6, 10], and also several loosely related models proposed in the context of modeling large-scale decentralized cyber-physical and socio-technical systems and infrastructures; see, e.g., [2, 3, 11, 18]. Classical *Cellular Automata* (CA), on the other hand, can be viewed as a special case of BNA, where all the nodes use the same local update rule f_i [14, 20].

The individual node updates can be done either synchronously in parallel, or sequentially, one at a time. While other communication models are worth considering (see, e.g., [19, 21]), the above two possibilities have been studied the most. In this paper, we will focus on the parallel, meaning perfectly synchronous node updates. That is, the next state of the node v_i is determined according to

$$v_i^{t+1} \leftarrow f_i\left(v_{i1}^t, \ldots, v_{ik}^t\right) \tag{1}$$

where the super-script t denotes a discrete time-step.

The tuple of all f_i's put together, $F = (f_1, \ldots, f_n)$, denotes the *global map* that acts on global configurations of a Boolean Network. When all f_i are the same, as is the case with classical cellular automata, the notation is usually slightly abused so that no differentiation is made between the local transition function f_i, acting on a state of a single node, and the global map F, that acts on all the nodes. In most Boolean Network models (other than the classical CA), different nodes in general update according to different local rules. This applies also to Hopfield networks, defined next.

Definition 2 A *Discrete Hopfield Network* (DHN) is made of n binary-valued nodes. Associated to each pair of nodes (v_i, v_j) is (in general, real-valued) their weight, w_{ij}. The weight matrix of a DHN is defined as $W = [w_{ij}]_{i,j=1\ldots n}$. Each node also has a fixed real-valued threshold, hi. A node v_i updates its state x_i from time step t to step t + 1 according to a (binary-valued) *linear threshold function* of the form.

$$x_i^{t+1} \leftarrow \text{sgn}\left(\sum w_{ij} \cdot x_j^t - h_i\right) \tag{2}$$

where the summation is over j = 1, …, n; the term hi is the *threshold* that the weighted sum needs to exceed in order for the node's state to update to + 1; to break ties, we define sgn(0) = + 1. For consistency across all discrete network models, we assume that possible states of a single node are {0, 1} and apply the same convention to Hopfield networks (for which, in much of the literature, the "default states" are labeled {−1, + 1}).

We summarize insights about efficiency and predictability of network models (and, in particular, Boolean Networks as the simplest kind of such discrete networks) based on their asymptotic dynamics. Some of our results were originally formulated in the DHN context [22]; yet other results in our prior work were formulated in the context of two other types of Network Automata (whose nodes' states come from a finite set and, in particular, need not necessarily be Boolean-valued), called *Sequential and Synchronous Dynamical Systems* [2, 15, 16].

Hopfield Networks were originally inspired by biology and especially computational neuroscience; in particular, they were introduced as a model of *associative memory* [8]. Subsequently, in addition to theoretical models in computational biology and neuroscience, Hopfield Networks were used for as a connectionist, self-organizing map model for "learning" and "searching for a solution", i.e., as a powerful tool for various search and optimization problems in computer science, operations research and other areas [9].

We note that some of the earliest Boolean Network models were also originally introduced in the context of systems biology, albeit not specifically neuroscience. Indeed, the very name *Boolean Networks* comes from the seminal work in theoretical biology by Kauffman [10]. In contrast, Sequential and Synchronous Dynamical Systems (SDS and SyDS, respectively) were specifically introduced in the context of agent-based simulation of complex cyber-physical, socio-technical and physical systems [3, 11, 15–17]. We will not formally define these S(y)DS models; instead, we refer the reader to references [2, 3, 11, 15, 17, 18]. We note that the results below, and their implications for the discrete dynamical system's (in)efficiency and (un)predictability, apply to all of the above models (CA, DHNs, SDSs and SyDSs), and indeed other similar discrete-time Boolean (or other finite-domain) Networks found in the literature.

Since Cellular Automata, Boolean Networks and DHN are all deterministic discrete-time dynamical systems, for any given current configuration C^t at time t, there is a unique next-step configuration C^{t+1}. We can therefore define the BN or DHN *configuration spaces*, and also various types of *global configurations* (i.e., tuples capturing the states of all nodes in a network) of interest:

Definition 3 A (global) configuration of a cellular or network automaton or a discrete Hopfield network is a vector $(x_{1, \ldots,} x_n)$ ε $\{0,1\}^n$, where x_i denotes the state of the ith node. A global configuration can also be thought of as a function *F*: V {0,1}, where V denotes the set of nodes in the underlying graph of a CA, BN or DHN.

Definition 4 A *fixed point* (FP) is a configuration such that, once a BN or DHN reaches that configuration, it stays there forever. A *cycle configuration* (CC) is a global state that, once reached, will be revisited infinitely often with a fixed, finite temporal period of 2 or greater. That is, a temporal cycle is made of two or more CCs. A *transient configuration* (TC) is a global configuration that, once reached, is never going to be revisited again. A *Garden-of-Eden* configuration is a TC that has no predecessors.

How do we relate *predictability* of a discrete network to its asymptotic dynamics? We call a discrete dynamical system *locally predictable*, if, starting from a given configuration, we can predict the system's "ultimate destiny", without necessarily having to run step-by-step simulation. What can that ultimate destiny be? For the *deterministic* discrete models such CA, more general Boolean networks (including but not limited to the aforementioned Sequential and Synchronous Dynamical Systems) and the discrete Hopfield networks, eventually the system behavior will become stationary, i.e., one of the following two things will happen: either a FP configuration, or else a temporal cycle (of some fixed finite length), will be reached. The underlying system's dynamics is then predictable, if we can tell *which* FP or temporal cycle will be reached—and if we can do so in a computationally efficient manner, that is, using a computation that takes a number of steps polynomial in the number of network nodes, n.

For certain discrete networks, one can determine that "ultimate destiny" analytically. For others, simulating the system step-by-step might be acceptable, if such convergence to a FP or a temporal cycle takes a number of steps polynomial in the number of nodes (which, in particular, means *polylogarithmic* in the size of the configuration space, 2^n). For yet others, there may not be a computationally feasible way of determining the asymptotic behavior without step-by-step simulation and, moreover, such simulation may take a number of steps that is exponential in n. The first two types of systems, according to our notion of predictability above, would be considered predictable; whereas the third kind of system would be considered unpredictable.

We note that, due to deterministic dynamics and finiteness of configuration spaces involved, it is always possible to determine asymptotic dynamics (or any other aspects of network dynamics) by simulating the system step-by-step, starting from each of the possible 2^n configurations. The *global predictability* question, then, boils down to when can these questions about the asymptotic dynamics be answered (analytically or via simulation), across all possible evolutions from all possible initial configurations, in a number of steps that is polynomial, as opposed to exponential, in the number of network nodes n. Thus, local predictability can be formulated in terms of determining reachability of a particular stationary state (such as, a FP) from a specified initial state; whereas, global predictability relates to both reachability (of particular stationary configurations) and counting or enumeration of different possible dynamical evolutions, across all possible initial configurations.

Definition 5 The *reachability* problem for deterministic discrete dynamical systems is to determine, given two configurations C and C′, whether C′ will eventually be reached starting from C. A variant of this problem, capturing local asymptotic dynamics of the system, is this: starting from a given initial configuration C, can we tell, what will be the ultimate destiny of that system's dynamics, in particular, whether a FP or a temporal cycle will be reached—and which one?

It has been established that, for broad classes of discrete networks, the reachability problem in general tends to be PSPACE-complete [3]; this means that, under the usual assumptions in computational complexity theory, the reachability questions cannot be answered in time polynomial in the size of the network. On the other hand, for certain restricted classes of Boolean Networks, as well as several important (finite-sized) CA and Hopfield networks, the reachability question can be answered efficiently, in polynomial time. In particular, if reachability can be determined in polynomial time for all possible starting configurations, we define the dynamics of such system to be predictable.

This discussion brings us to the second key property of network dynamics that we are interested in: how *efficient* is that dynamics? In our context, we define a computational notion of efficiency to simply mean the following: starting from an arbitrary configuration, how quickly will the system settle down to its asymptotic, stationary behavior—whatever that ultimate behavior may be? That is, we don't care here, whether the ultimate destiny will be a particular point or a temporal cycle, nor which FP or temporal cycle will it end up being. We only care, *how long it will take* to get there. Going back to the fundamental configuration space properties as defined in **Definition 4** above, this boils down to asking, *how long is the transient chain prior to reaching the stationary state.*

While the proposed computational notions of network dynamics' efficiency, local and global predictability are related, they are not synonymous. So, which discrete networks have efficient versus inefficient dynamics, and which networks' dynamics are predictable versus unpredictable? We summarize what is known about a few popular classes of finite CA, Boolean Networks/BNA and Hopfield Networks in the next section.

3 Some Examples of Discrete Dynamical Systems that Are Both Predictable and Efficient

Given the above computationally based notions of dynamics' predictability and efficiency, what would be some examples of interesting classes of CA, Boolean networks and discrete Hopfield networks that are (un)predictable and/or (in)efficient? It turns out, there are interesting classes of such models in each of the four "quadrants" with respect to these two dimensions, viz., predictability and efficiency. In some sense, those systems whose dynamics are predictable yet inefficient, as well as those whose dynamics are efficient yet in general unpredictable, are arguably the most interesting ones. In particular, in this Section, we discuss several examples of discrete dynamical systems that are already known to be, or else can analytically or computationally be shown to be, both efficient and predictable. We then summarize in the next Section a few structurally and description-wise rather simple BN and Hopfield Net classes that are unpredictable; some of those classes turn out to be unpredictable yet efficient, others to be both unpredictable and inefficient, and yet for others, the efficiency status is not known.

We start with the classical CA with some fixed, finite number of nodes *n*. It has been known for over 30 years, based on the seminal work by S. Wolfram [26, 27] and others, that there are certain node update rules that lead to highly complex and in general unpredictable behavior of even 1-dimensional CA, that is, a finite string (or ring) of nodes, where each node updates according to the same local rule $f_{i.}$ However, when the local rule f_i is appropriately restricted, then usually one obtains the resulting dynamics that is both predictable and efficient. What are some specific examples of such update rules?

Consider 1-dimensional CA in which each node updates according to the Boolean AND function; to how many inputs (that is, neighboring nodes) this AND is applied to, is determined by the rule radius; let's say, the input radius is some integer $r > 0$ (typically, this means the input values to the local update rule are the k neighbors to the left, the current state of the node in question itself, and the k neighbors to the right, for the total of $2r + 1$ inputs). It is not hard to convince oneself that, for any starting configuration other than 1^n, eventually the CA will converge to the FP "attractor" 0^n. Further, 1^n is also a FP, however, its entire "basin of attraction" is itself (that is, there are no incoming transients). Likewise, either analytically or via simulation, one can also readily establish that, starting from an arbitrary configuration $C \neq 1^n$, that the convergence to the attractor 0^n will happen fairly fast, certainly in a number of steps polynomial in the number of nodes. (In fact, it can be shown that the speed of convergence is always $O(n)$ and, within a constant multiplicative factor, the bigger the radius k of the local update rule AND, the faster the convergence).

The analysis for the CA in which each node updates according to Boolean OR is basically identical to the analysis for AND, with the roles of states 0 and 1 reversed. In particular, it then immediately follows that cellular automata in which all nodes update either according to Boolean AND or according to Boolean OR, have (asymptotic) dynamics that is both efficient (the convergence is fast) and predictable (we can tell, to which FP will the system converge from a given starting configuration).

Are there update rules more interesting than Boolean AND and OR, that also lead to provably efficient and predictable dynamics? It turns out, there are a number of other such rules, especially in the context of classical CA, in which all the nodes update their states according to the same local update rule. We discuss here just one, perhaps most studied such CA rule across a number of research disciplines, from biology to social sciences to political science to statistical physics: cellular automata all of whose nodes update according to the *Majority* (MAJ) local rule. For simplicity, we focus on the 1-D MAJ CA of radius $r \geq 1$: a node x_i updates its state to 1 if at least $r + 1$ (out of $2r + 1$ input values total, that include the current state of the node x_i itself) of its neighbors are currently in state 1; and the node updates its state to 0, otherwise.

How complex is the dynamics of MAJ CA? In particular, for various underlying cellular spaces and initial configurations, how predictable and efficient is that dynamics? It turns out, some aspects of MAJ CA dynamics are relatively complex;

for example, for many cellular spaces and rule radii r, even in the simplest 1-dimensional case, MAJ CA have *exponentially many* fixed point (FP) configurations [14]. However, it turns out that convergence to each of those FPs happens rather quickly, in time that is linear in the number of CA's nodes, that is, $O(\mathrm{n})$. Moreover, this fast convergence takes place from basically all initial configurations (and, as it turns out, is even faster if the nodes update sequentially one at a time, than if they update in parallel [14, 19]). Therefore, the dynamics of MAJ CA, while non-trivial, is always very efficient.

Insofar as the MAJ CA's predictability, guaranteed fast convergence to a FP (as well as the ability to characterize analytically, under what circumstances and from which initial configurations a temporal cycle might be reached), indicate that these networks are inherently predictable, as well. In particular, given an arbitrary starting configuration and the cellular space ("network topology"), not only can we effectively determine whether the system will evolve to a FP (and how quickly!), but we can also tell, to which among possibly exponentially many FPs will the system converge. While the step-by-step simulation is, as seen in the discussion of efficiency, quite efficient in this particular case, in turns out these questions about the "ultimate destiny" of the system evolving from a given starting state can also be answered analytically; see our prior work [14, 19] for mathematical details.

The AND, OR and MAJ local update rules all belong to a class of arguably the simplest local update functions, called *simple threshold rules* [14]. In 1-D CA case, each simple threshold update rule can be defined as *"at-least-l-out-of-k"* for appropriate values of integers l and k. That is, a node will update its state to 1 if and only if at least l out of its k input values (typically, including its own current state) are 1; and it will update to 0, otherwise. So, for example, for a CA with rule radius r, the MAJ update rule can be described as "at-least-(r + 1)-out-of-(2r + 1)". It turns out, that all these CA with simple threshold (Boolean-valued) functions for their local update rule, have asymptotic dynamics that are both predictable and efficient in our sense [14, 17, 19]. In fact, most of these update rules lead to a small number of different possible dynamics; the only exception is the MAJ rule which, as discussed earlier, results (in general) in a number of fixed points, and therefore a total number of different possible evolutions, that is exponential in the number of CA's nodes n [14, 19].

How complex can the Boolean network dynamics get, once we generalize from the classical CA to more general BN models? If we restrict the local update rules to simple threshold functions, and we require that all the nodes use the same update rule, then making the network topology more complex than the simple cellular spaces such as lines, rings, 2-D rectangular grids or tori, actually won't make much of a difference—the resulting BN dynamics will still remain, in general, both predictable and efficient.

However, once more complex network topologies are combined with not all the BN's nodes having to update their respective states according to the same local rule, the dynamics complexity threshold gets quickly crossed. In particular, once the underlying BN's nodes are allowed to use just two different update rules, even if both rules are restricted to simple threshold functions, the resulting dynamics can become quite complex. This includes the problem of determining the total number of possible evolutions (i.e., different asymptotic behaviors from all possible initial configurations) becoming computationally intractable—formally, **#P**-complete. These remarks, cf. based on our work on BNs only "slightly more general" than the classical finite CA, are a good transition point for us to wrap-up the discussion of those Boolean networks that are both efficient and predictable, and turn to several classes of similar networks that are provably unpredictable—some of which also happen to be provably inefficient in our sense, whereas others, while still unpredictable, turn out to be quite efficient.

4 Some Examples of Discrete Networks Whose Dynamics Are Unpredictable and (In)efficient

That the dynamics of a sufficiently general BN will in general be both unpredictable and inefficient (or at least, inefficient starting from some of the initial configurations), is not at all surprising. By "sufficiently general" here, we mean with respect to any or all of the three key parameters characterizing the structural complexity of a discrete network: (i) the underlying graph (called cellular space in case of CA; and usually referred to as "network topology" for more general BNs); (ii) the local update rules f_i; and (iii) the level of diversity, that is, how many different local update rules (perhaps of a particular restricted kind, as prescribed in ii) are different nodes of a BN allowed to use. We note, that it has been known for a long time, that allowing some complexity with respect to (ii) alone can lead to rather complex—and, in particular, unpredictable and inefficient in our sense—dynamics, as shown for the classical one-dimensional CA with a sufficiently general (single) local update rule [26, 27].

On the other hand, generalizing rather minimally with respect to (i) and (iii), while imposing restrictions on the complexity of local update rules in (ii), has more recently also be shown, to potentially lead to rather complex dynamics. A "minimal generalization" with respect to criterion (iii) simply means, allowing each network node to choose one of two different update rules, rather than every node having to update according to the same update rule [20]. Criterion (i) about the underlying graphs is more complex, and can be generalized (starting with regular cellular spaces of the classical CA) in different ways. The "minimal generalizations" we have investigated, focus on allowing more general underlying graphs that were however still required to be uniformly sparse, that is, where each node has $O(1)$ neighbors. In fact, we have succeeded obtaining highly complex network dynamics, even when the underlying BN's graph is required to be not just uniformly sparse, but also regular—for example, such that $\deg(v) = 3$ for each node v in the graph [20, 23–25].

Limiting our focus to the network dynamics efficiency and predictability, we now state some results about the computational (in)tractability of the reachability-flavored problems, such as: starting from an arbitrary initial configuration, can we (efficiently) tell, whether a FP or a temporal cycle will be eventually reached? In systems that have mostly (or exclusively) fixed points and no temporal cycles, the FP-reachability question becomes, starting from an arbitrary initial configuration, can we tell, to which (among possibly many) FPs will the system's dynamics eventually converge? Examples of systems with no temporal cycles include Sequential CA with simple threshold update rules [14], as well as certain types of discrete Hopfield networks [22].

So, what are some results on the FP (or other types of) reachability, for appropriately restricted sub-classes of BNs and DHNs, that would immediately imply both unpredictability and inefficiency of those systems' asymptotic dynamics? We summarize some of the strongest known to us such results next.

Theorem 1 (based on [3], rephrased)*: The reachability problem for an arbitrary BN whose nodes are restricted to update according to the threshold update rules (and different nodes are allowed to use different such rules), is in general* **PSPACE**-*complete.*

This result immediately implies both general unpredictability and inefficiency of the dynamics of such BNs. To see why such systems are also inefficient in our sense (for at least some initial configurations), notice that, if the system were actually efficient starting from any initial state, that system would converge to its ultimate stationary behavior relatively quickly, and certainly in a number of steps at most polynomial in n, contradicting the established **PSPACE**-completeness of determining what will that ultimate stationary behavior be (a FP vs. a temporal cycle; as well as, which FP or CCs).

For the sufficiently general systems, therefore, it is computationally intractable to determine, starting from an arbitrary initial state, whether a FP or a temporal cycle will be reached. Many systems, however, have a few or none temporal cycles—but may have many FPs. To argue that such systems are predictable, it does not suffice to say, OK, we know the system will converge to an FP—we also want to be able to say, to which (among possibly exponentially many) FP will it converge, given the initial configuration? The related computational problems are those of (i) determining how many FPs a given system has; (ii) determining the size of each FP's basin of attraction; and (iii) given an arbitrary configuration, to which FP's basin of attraction does that configuration belong? It turns out, answering these questions becomes computationally intractable for even much more restricted Boolean networks than those in Theorem 1.

Theorem 2 (based on [20, 23])*: The following problems are all* **#P**-*complete, even when the underlying graphs of a BN or DHN are restricted to star-graphs or wheel-graphs, and all local update rules are monotone Boolean functions:*

- Determining the exact number of all FPs;
- Determining the exact number of all TCs;
- Determining the exact number of only those TCs that are Gardens-of-Eden;
- Determining the total number of predecessors of an arbitrary configuration.

Note that the established complexity of certain properties of the system dynamics, as summarized in Theorem 2, still do not imply that those systems are necessarily inefficient. Namely, it is possible that, for an arbitrary starting configuration, the convergence to the eventual stationary pattern (typically, to an FP), actually takes place fast, implying dynamics' efficiency. For sufficiently general BNs, instances can be found where this convergence is not fast, implying inefficiency. However, with some further restrictions on the update rules, such as that they all be monotone or of simple threshold variety, the convergence to a FP can be shown to actually always be fast, thus resulting in systems with many possible dynamics and, in particular, where determining that number of different evolutions is computationally intractable; hence, these systems are globally unpredictable, yet uniformly efficient. Another similar example of unpredictable yet efficient systems is discussed in [22] in the context of Hopfield networks.

Our last theoretical results, establishing complex dynamics of descriptively simple and structurally rather constrained sparse networks, and in particular showing the unpredictability of that dynamics (with the question of efficiency being, as far as we are aware, as yet unanswered), is summarized in the following theorem (based on [23]):

Theorem 3 *Let a Boolean Network be defined over a uniformly sparse graph; and let all BN's local update rules be simple threshold functions. Enumerating all possible dynamics (in particular, all FPs) of such BN is* **#P***-complete and therefore computationally intractable, even when these restrictions simultaneously hold: (i) each node in the underlying graph has at most (alternatively, exactly 3) neighbors; and (ii) each node gets to choose from just two given simple threshold functions, that is, the "diversity" of the local interactions across the entire BN includes just two different update rules.*

5 Summary

We study discrete Boolean networks from a computational standpoint; in particular, we focus on those networks' dynamics predictability and efficiency. While those two concepts can be defined in different ways, we choose computation-centric definitions. We re-interpret a number of results about CA and BN dynamics in terms of our computational notions of efficiency and predictability. Lastly, we show several classes of such networks whose dynamics are both efficient and predictable, those that are unpredictable yet efficient, and those that are both unpredictable and inefficient.

References

1. Bandini, S., Manzoni, S., Simone, C.: Heterogeneous Agents Situated in Heterogeneous Spaces. Appl. Artif. Intell. Int. J. **16**(9–10), 831–852 (2002)
2. Barrett, C., et al.: Gardens of Eden and fixed points in sequential dynamical systems. Discrete Math. Theor. Comp. Sci. (DMTCS) AA, 95–110 (2001)
3. Barrett, C., et al.: Reachability problems for sequential dynamical systems with threshold functions. Theor. Comput. Sci. **295**(1–3), 41–64 (2003)
4. Davey, N., Calcraft, L., Adams, R.: High capacity, small world associative memory models. Connect. Sci. **18**(3), 247–264 (2006)
5. Floreen, P., Orponen, P.: On the computational complexity of analyzing hopfield nets. Complex Syst. **3**, 577–587 (1989)
6. Graudenzi, A. et al.: Dynamical properties of a Boolean model of a gene regulatory network with memory. J. Comp. Biol. **18**(10), (2011)
7. Helbing, D.: Social Self-Organization. Springer-Verlag, Berlin, Understanding Complex Systems series (2012)
8. Hopfield, J.: Neural networks and physical systems with emergent\collective computational abilities. Proc. Natl. Acad. Sci. **79**, 2554–2558 (1982)
9. Hopfield, J., Tank, D.: Neural computation of decisions in optimization problems. Biol. Cybern. **52**, 141–152 (1985)
10. Kauffman, S.A.: Emergent properties in random complex automata. Physica D **10**(1–2), 145–156 (1984)
11. Mortveit, H., Reidys, C.: Discrete sequential dynamical systems. Discrete Math. **226**(1–3), 281–295 (2001)
12. Orponen, P.: Computational complexity of neural networks: a survey. Nordic J. Comput. (1996)
13. Sima, J., Orponen, P.: General-purpose computation with neural networks: a survey of complexity theoretic results. Neural Comput. **15**(12), 2727–2778 (2003)
14. Tosic, P., Agha, G.: Characterizing configuration spaces of simple threshold cellular automata. In: Proceedings of the 6th International Conference on Cellular Automata for Research & Industry (ACRI'04), Amsterdam, The Netherlands, pp. 861–870. Springer (2004)
15. Tosic, P., Agha, G.: On computational complexity of counting fixed points in symmetric Boolean graph Automata. In: Proceedings of the 4th International Conference on Unconventional Computing (UC'05), Seville, Spain, LNCS vol. 3699, pp. 191–205. Springer (2005)
16. Tosic, P.: Cellular automata for distributed computing: models of agent interaction and their implications. In: Proceedings of the International Conference Systems, Man & Cybernetics (SMC'05), pp. 3204–3209. IEEE (2005)
17. Tosic, P.: On modeling and analyzing sparsely networked large-scale multi-agent systems with cellular and graph automata. In: V. Alexandrov et al. (eds.) Computer Science ICCS-2006: 6th International Conference (Proc. Part III), LNCS vol. 3993, pp. 272–280. Springer (2006)
18. Tosic, P.: On the complexity of counting fixed points and gardens of Eden in sequential & synchronous dynamical systems. Int. J. Found. Comput. Sci. (IJFCS) **17**(5), 1179–1203. World Scientific (2006)
19. Tosic, P.: Cellular automata communication models: comparative analysis of parallel, sequential and asynchronous CA with simple threshold update rules. Int. J. Nat. Comput. Res. (IJNCR) **1**(3), 66–84 (2010)
20. Tosic, P.: On the complexity of enumerating possible dynamics of sparsely connected Boolean network automata with simple update rules. Discrete Math. Theor. Comput. Sci. (DMTCS), pp. 125–144 (2010)
21. Tosic, P.: Modeling large-scale multi-agent systems with sequential and genuinely asynchronous cellular automata. Acta Physica Polonica B (Proc. Supplement) **4**(2), 217–236. Polish Academy of Sciences (2011)

22. Tosic, P.: On simple models of associative memory: network density is not required for provably complex behavior. In: Proceedings of the Brain Informatics and Health, Lecture Notes in Computer Science (LNCS) series, vol. 9919, pp. 61–71. Springer (2016)
23. Tosic, P.: On phase transitions in dynamics of cellular and graph automata models of sparsely interconnected multi-agent systems. In: ACM Proceedings of the Autonomous Agents & Multi-Agent Systems (AAMAS'17), pp. 474–483. Sao Paulo, Brazil (2017)
24. Tosic, P., Ordonez, C.: Boolean network models of collective dynamics of open and closed large-scale multi-agent systems. In: Proceedings of the 8th International Conference on Industrial Applications of Holonic & Multi-Agent Systems (HoloMAS'17), pp. 95–100. Lyon, France (2017)
25. Tosic, P.: Examples of NP-complete and CoNP-complete problems about determining certain configuration space properties of Sequential and Synchronous Dynamical Systems. Urbana, IL, USA (2002).(unpublished)
26. Wolfram, S.: Theory and Applications of CA. World Scientific, Singapore (1986)
27. Wolfram, S.: Cellular Automata and Complexity (collected papers). Addison-Wesley, Reading, UK (1994)

Joint Lattice of Reconstructability Analysis and Bayesian Network General Graphs

Marcus Harris and Martin Zwick

Abstract This paper integrates the structures considered in Reconstructability Analysis (RA) and those considered in Bayesian Networks (BN) into a joint lattice of probabilistic graphical models. This integration and associated lattice visualizations are done in this paper for four variables, but the approach can easily be expanded to more variables. The work builds on the RA work of Klir [22], Krippendorff [26], and Zwick [44], and the BN work of Pearl [33–36], Verma and Pearl [39], Heckerman et al. [20], Chickering [9], Andersson et al. [1], and others. The RA four variable lattice and the BN four variable lattice partially overlap: there are ten unique RA general graphs, ten unique BN general graphs, and ten general graphs common to both RA and BN. For example, the specific graph having probability distribution p(A)p(C)p(B | AC) is unique to BN, the RA specific graph AB:AC:BC, which contains a loop, is unique to RA, and the specific graph ACD:BCD with probability distribution p(A | CD)p(B | CD)p(D | C)p(C) is common to both RA and BN. The joint RA-BN lattice of general graphs presented in this paper expands the set of general graphs with unique independence structures beyond what was previously available by either RA alone or BN alone, thus allowing for representations of complex systems which are (i) more accurate relative to data and/or (ii) simpler and thus more comprehensible and more generalizable than would be possible by modeling only with RA or only with BN.

1 Introduction

Reconstructability Analysis (RA) is a data modeling approach developed in the systems community ([2, 4, 7, 14, 15, 21–23, 25, 26, 43–46]; and others) that combines graph theory and information theory. Its applications are diverse, including

---

M. Harris (✉) · M. Zwick
Portland State University, Portland, OR, USA
e-mail: maharris@pdx.edu

M. Zwick
e-mail: zwick@pdx.edu

D. Braha et al. (eds.), *Unifying Themes in Complex Systems X*, Springer Proceedings in Complexity, https://doi.org/10.1007/978-3-030-67318-5_24

time-series analysis, classification, decomposition, compression, pattern recognition, prediction, control, and decision analysis [45]. It is designed especially for nominal variables, but continuous variables can be accommodated if their values are discretized. Graph theory specifies the structure of the model (the set of relations between the variables), which is independent of actual data (except for specification of variable cardinalities); information theory uses the data to characterize the strength and the precise nature of the relations. Data applied to a graph structure yields a probabilistic graphical model of the data.

In RA, graphs are undirected, although directions are implicit if one variable is designated as the response variable (dependent variable), while all other variables are designated as explanatory variables (independent variables). In this paper, such IV-DV designations are not made, so we are concerned with only what are called "neutral systems." Graphs are either general or specific. A general graph identifies relations among variables that are unlabeled; a specific graph labels the variables. For example, for a system consisting of variables A, B, and C, AB:BC is a specific graph where nodes A and B are linked and B and C are also linked. In this notation, the order of variables in any relation is arbitrary, as is the order of the relations. Relations include all of their embedded relations. For example, the relation ABC includes embedded relations AB, AC and BC and the univariate margins A, B, and C. Specific graphs AB:BC, BA:AC and AC:CB are all instances of the same general graph that contains a unique independence structure regardless of variable labels.

For a three variable neutral system with loops there are 5 general graphs and 9 specific graphs; for four variables there are 20 general graphs and 114 specific graphs. The number of graphs increases hyper-exponentially with the number of variables.

A Bayesian Network (BN) is another graphical modeling approach for data modeling that is closely related to RA; indeed where it overlaps RA the two methods are equivalent, but RA and BN each has distinctive features absent in the other methodology. BNs have origins in the type of path model originally described by Wright [40, 41], but it was not until the 1980s that BNs were more formally established [32–34, 36].

As does RA, BN combines graph theory and probability theory; graph theory provides the structure and probability theory characterizes the nature of relationships between variables. BNs are represented by a single type of graph structure; a directed acyclic graph, which is a subset of chain graphs [28]. Graphs in RA can be hypergraphs, i.e., can include relations between more than two variables. BN graphs have only pairwise links, but V-structures (explained below) are interpreted as higher ordinatlity relations. For a three variable BN lattice, there are 5 general graphs and 11 specific graphs; for four variables there are 20 general graphs and 185 specific graphs.

This paper integrates RA and BN and visualizes their joint lattice of general graphs for four variables.

2 RA Lattice

The lattice of four variable RA general graphs of Fig. 1 (adapted from [22, 26]) represents all four variable RA graphs with unique independence structures. In these graphs, lines (which can branch) are variables; boxes are relations between variables. Where only two lines extend from a box, the relation is dyadic. If more than two lines extend from a box, the graph is a hypergraph. Bolded general graphs in Fig. 1 do not have loops whereas non-bolded general graphs have loops. The figure shows all 20 general graphs for four variables. G1 is the most complex general graph, in which all four variables are connected in a tetradic relation. Graphs below G1 are increasingly less complex decompositions of G1, ending with G20, the least complex graph, representing complete independence among all four variables.

A general graph represents a unique independence structure which disregards all possible ways that variables could be labeled. For example, the left-most and right-most variables in G7 in Fig. 1 are independent of one another given the two central variables that connect both relations, where these four variables have not been assigned any specific identities.

When, in RA or BN, the variables of a general graph are labeled, it is called a specific graph equivalence class or specific graph. A specific graph, given data, produces a unique probability distribution. (This is the case for RA; as explained below, two different directed specific BN graphs can have the same probability distribution.) Since the number of graphs increases hyper-exponentially with the number of variables, searching the entire lattice for the best model can be very computationally expensive as the number of variables increases.

The colon notation for RA specific graphs represents marginal or conditional independence between variables. For example, G20 from Fig. 1, having notation A:B:C:D, has the independence structure (A _I_ B, C, D), (B _I_ C, D), (C _I_ D); A, B, C, and D are all marginally independent of each other. G17, having notation AB:BC:D, has the independence structure (D _I_ A, B, C), (A _I_ C | B); A, B, and C are all marginally independent of D, and A is conditionally independent of C given B. The overlap of B in the AB and BC relations encodes the conditional independence of A and C given B.

3 BN Lattice

The primary differences between RA and BN are two-fold: (1) BNs are directed and acyclic whereas RA graphs are undirected and can have loops or not have loops, and (2) some BN graphs contain converging edges, called a V-structure, which encodes a unique independence structure not found in RA general graphs.

As in RA, there are BN general graphs and BN specific graphs, in the BN literature referred to as essential graphs and labeled essential graphs respectively [1]. BN general graphs of Fig. 2 represent unique independence structures of variables, edges,

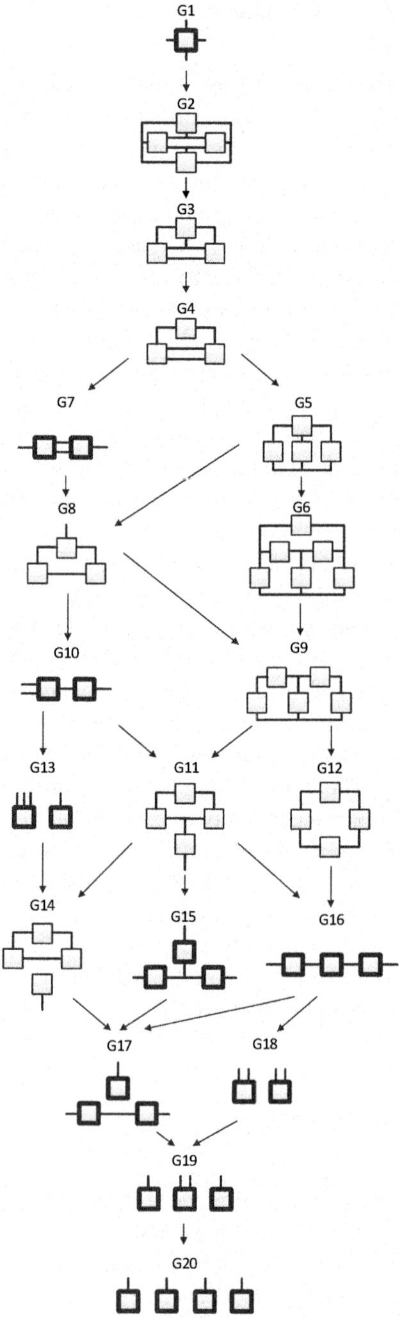

Fig. 1 RA general graph lattice

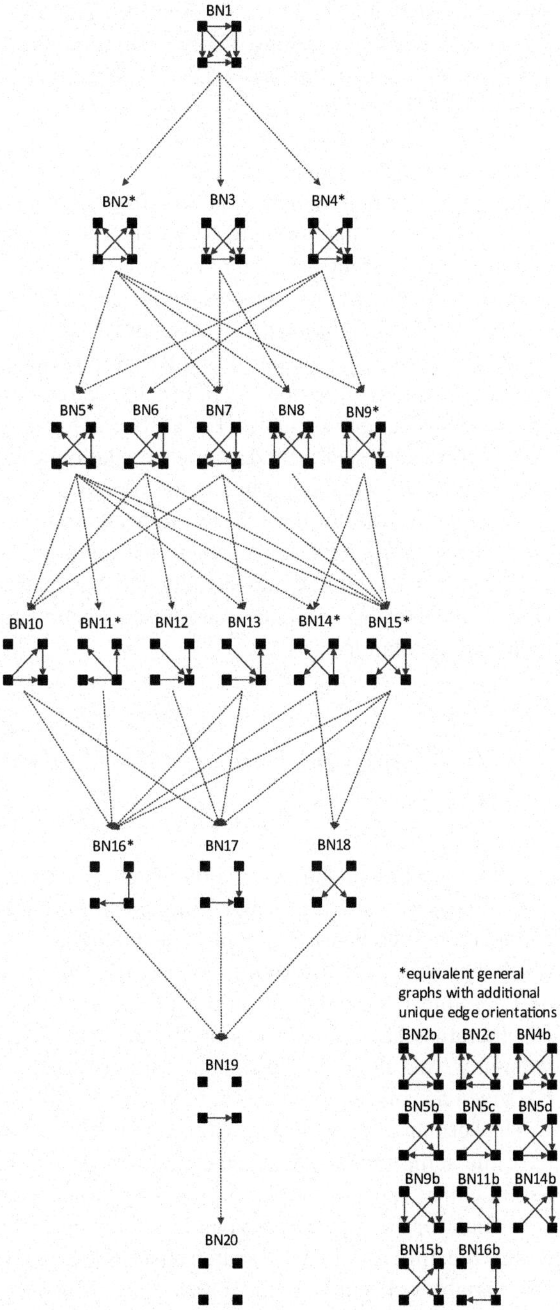

Fig. 2 BN general graph lattice

and edge orientations, where specific identities are not assigned to the variables. Figure 2 shows the hierarchy of BN general graphs for four variables. There are 20 BN general graphs in the lattice, i.e., 20 unique general independence structures for four variable BNs.

In Fig. 2, general graphs are labeled BN1, BN2…BN19, BN20. Solid squares represent variables, edges are represented by directed arrows from one square to another. The dashed lines with arrows from one general graph to another represent the hierarchy of general graphs, with parent graphs being above child graphs. Child graphs result from the deletion of one edge from the parent graph. The bottom-right addition to the lattice that tabulates equivalent general graphs is discussed in Sect. 3.2.

The alphabetic notation that we use for BN specific graphs is derived from the RA notation described previously. For a BN graph without a V-structure, the notation is identical to the RA notation. As in RA, the colon represents marginal or conditional independence among variables. For a BN graph with a V-structure, the notation adds subscripts to represent the independence relations encoded by the V-structure, which are unique to BNs and not found in RA. For example, BN17 from Fig. 2, with labels A, B, C, D in order of top left, top right, bottom left, bottom right respectively, is given the notation $BCD_{B:C}$:A, where the colon between $BCD_{B:C}$ and A states the independency (A _I_ B, C, D), namely that A is marginally independent of B, C, and D. The subscript $_{B:C}$ encodes marginal independence between B and C within the triadic BCD relation.

3.1 *Generating the Lattice of BN General and Specific Graphs*

The BN literature on lattices predominately focuses on search algorithms to find the best BN given a scoring metric. Implicit in these search algorithms is a lattice of candidate graphs being explored in search of the best model. Chickering [13] and others have shown the search problem to be NP-hard, with four variables there are 543 possible BNs, with 10 variables there are O(10^18) [31]. Because of this, the BN literature has focused less on the description of the exhaustive lattice of BN structures, and more on advancing search heuristics to efficiently identify the best BN given a scoring metric ([3, 5, 6, 9, 11, 12, 16–18, 20, 24, 27, 29]; and others).

Heckerman et al. [20] first showed that BNs with differing edge topologies can have the same independence structure and thus the same probability distribution, herein called specific graphs. In contrast to heuristics that search all BNs, search heuristics for BN specific graphs have proven to be more efficient because they reduce the dimensionality of search space ([1, 8–10, 19, 37, 38, 42]; and others). For four variables, this reduces the search space from 543 BNs to 185 specific graphs [1]. Removing variable labels, these specific graphs can be summarized by 20 general graphs.

Building from the RA work of Klir [23] and Zwick [44], and the BN work of Pearl [33–36], Verma and Pearl [39], Heckerman et al. [20], Chickering [9], Andersson et al. [1], and others, the following procedure was used to generate the four variable BN general and specific graph lattice in a way that can be integrated with the four variable RA general graph lattice. While this procedure is applied in this paper to four variables, it could be used for any number of variables.

Procedure to generate the BN general and specific graph lattice for any number of variables

1. Generate all graphs for n variables by permuting all possible variable labels, edge connections and edge orientations. Eliminate any graphs with loops. The result is the set of all directed acyclic graphs for n variables.
2. For each directed acyclic graph, evaluate the specific independence structure following the d-separation procedure [30] to generate the exhaustive list of independencies for each graph.
3. Organize graphs with the same labeled independencies into specific graph equivalence classes.
4. From each specific graph equivalence class, select a single edge topology to represent the general graph equivalence class. List any additional equivalent general graphs with unique edge topologies separately, as done in Fig. 2.
5. Organize general graphs into levels based upon the number of edges in each general graph and link nested general graphs in the lattice to reflect parent-child general graphs.

3.2 Additional Representations of BN General Graph Equivalence Classes

There are 20 general graphs in the four variable BN lattice. However eight of these, marked with asterisks in Fig. 2, can be represented by additional unique edge topologies which, however, result in identical probability distributions when applied to data. These are BN2*, BN4*, BN5*, BN9*, BN11*, BN14*, BN15*, and BN16*. Thus, for example, BN2b has edge orientations that are different from (and cannot be mapped onto) those of BN2*, but when variables are labeled in BN2* and BN2b, identical probability distributions result. This property is unique to BN and is not found in RA, in which there is a single unique representation of each RA general graph.

4 Integrating the Rho, RA and BN Lattices

This section integrates the RA and BN general graph lattices using the four variable Rho lattice [22]. The joint RA-BN lattice of general graphs presented in this paper expands the set of general graphs with unique independence structures beyond what was previously available by either RA alone or BN alone. The lattice identifies general independence structures unique to RA, general independence structures unique to BNs, and general independence structures that are equivalent across RA and BNs. Where two or more RA or BN graphs have the same general independence structure regardless of variable labels, they are equivalent.

4.1 Lattice of Four Variable Rho Graphs

The four variable Rho (ρ) lattice of Fig. 3 (adapted from [22, p. 237]) is a simplification of the RA lattice of general graphs of Fig. 1. The Rho lattice represents all possible undirected relations between four variables, an even more general lattice than that of the RA general graph lattice and general enough to map both RA and BN four variable general graphs to one of the eleven represented structures. Solid dots represent variables; lines connecting dots represent relations between variables. In terms of the RA general graph lattice of Fig. 1, if two variables are directly connected by any box (a relation of arbitrary ordinality), a line connects them in the Rho lattice. Arrows from one Rho graph to another represent hierarchy, i.e., the generation of a child graph from a parent graph. The graph ρ1 represents maximal connectedness, or interdependence, between variables, and the graph ρ11 represents independence among all variables. Graphs in-between ρ1 and ρ11 represent a mix of dependence and independence among variables. Each RA or BN graph corresponds to one, and only one, of the eleven Rho graphs.

4.2 Equivalent RA and BN General Graphs

Out of 20 RA general graphs and 20 BN general graphs, there are 10 RA general graphs, comprising all of the graphs without loops in the RA lattice that are equivalent to BN general graphs. Each of these RA-BN equivalent pairs corresponds to one of the 11 Rho graphs from Fig. 3, with the exception of ρ4. ρ4 has corresponding RA and BN general graphs, but these do not have equivalent independence structures, and are discussed in the following section on non-equivalent RA and BN general graphs.

Figure 4 shows an example of equivalent RA and BN graphs, namely G7 and BN2*, respectively. Labeled variables in G7 results in independencies (A _|_ B | C, D) and thus the RA specific graph notation ACD:BCD. Assigning labels to BN2* yields the same independencies and thus the same specific graph ACD:BCD.

Table 1 shows the list of all equivalent Rho, RA and BN four variable general graphs, an example of their specific graph notation, and their independences. These specific graph examples align with the BN general graphs of Fig. 2 assuming labeling of nodes A, B, C, D in the order of top left, top right, bottom left, bottom right.

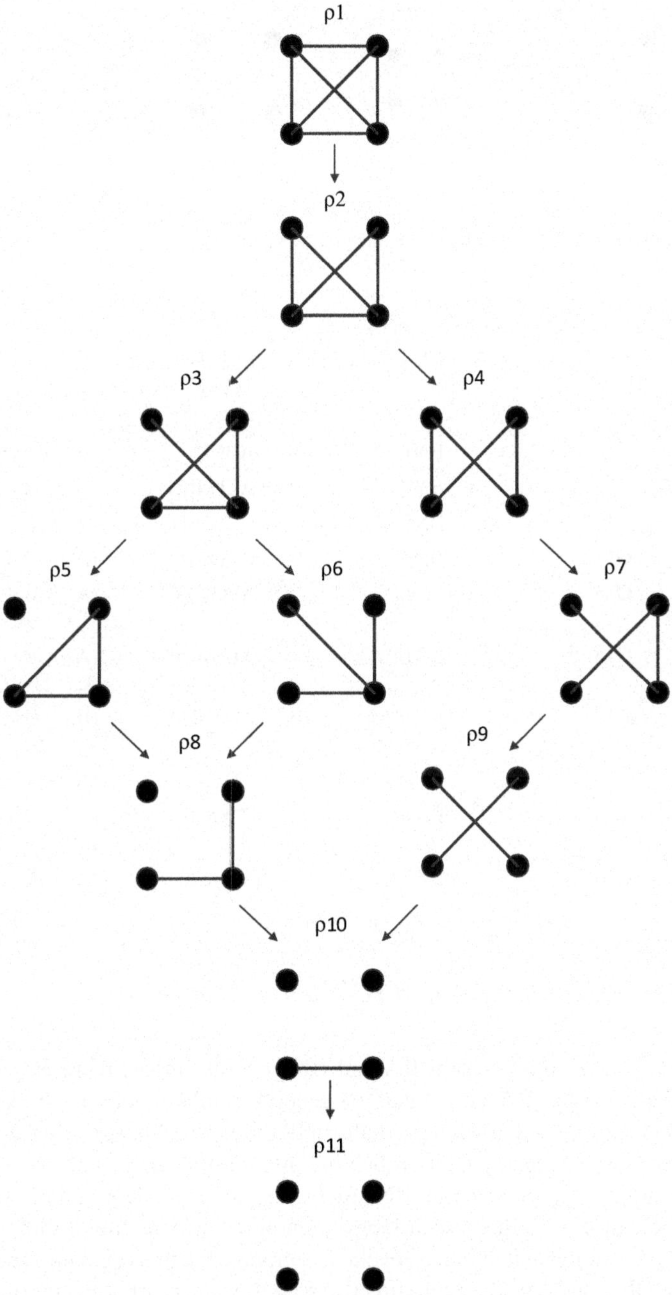

Fig. 3 Rho lattice

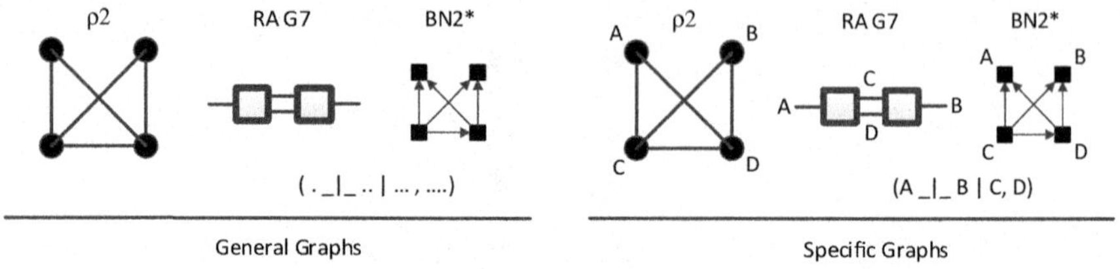

Fig. 4 RA G7 and BN2* general and specific graph example

Table 1 Rho, RA and BN equivalent graphs

Rho graph	RA general graph	BN general graph	Specific graph example (RA notation)	Independencies
ρ1	G1	BN1	ABCD	No independencies
ρ2	G7	BN2*	ACD:BCD	(A _I_ B I C, D)
ρ3	G10	BN5*	BCD:AD	(A _I_ B, C I D)
ρ5	G13	BN10	BCD:A	(A _I_ B, C, D)
ρ6	G15	BN11*	AD:BD:CD	(A _I_ B, C I D), (B _I_ C I D)
ρ7	G16	BN14*	AD:BC:BD	(A _I_ B I D), (C _I_ A, D I B)
ρ8	G17	BN16*	BD:CD:A	(B _I_ C I D), (A _I_ B, C, D)
ρ9	G18	BN18	AD:BC	(A, D _I_ B, C)
ρ10	G19	BN19	CD:A:B	(B _I_ C, D), (A _I_ B, C, D)
ρ11	G20	BN20	A:B:C:D	(A _I_ B, C, D), (B _I_ C, D), (C _I_ D)

4.3 Non-equivalent RA and BN General Graphs

In addition to the 10 equivalent RA and BN general graphs, there are 10 general graphs unique to the RA lattice and 10 general graphs unique to the BN lattice. All 10 non-equivalent RA general graphs in the four variable lattice have loops and require iteration to generate their probability distributions. BNs are acyclic and have analytic solutions, so there are no BN graphs that are equivalent to these RA graphs with loops. Since RA graphs with loops are undirected, one might think that there could be some equivalent directed acyclic BN graphs, but this is not the case, because BN graphs that are acyclic when directions are considered but cyclic if directions are ignored have V-structure interpretations, as described previously. All 10 non-equivalent BN graphs have such V-structures, which encode independence relations unique to BNs.

To illustrate: the structure A → B, B → C, C → D, D → A is cyclic and not a legitimate BN structure, but the directed structure of A → B, B → C, C → D, A → D, which has the same undirected links, is not cyclic, and is a legitimate BN structure, namely BN9b. However, this latter structure is not interpreted as a set of dyadic relations, which would be written in RA notation as AB:BC:CD:AD. Rather, the V-structure consisting of C → D and A → D is interpreted as a triadic relation, which contributes a p(D|AC) to the probability expression, p(A) p(B|A)p(C|B) p(D|AC), which does not correspond to any RA structure.

4.4 *Joint Lattice of RA and BN General Graphs*

The joint lattice of RA and BN general graphs is organized using the Rho lattice [22] of Fig. 3, the RA general lattice of Fig. 1 [22, 26] and the BN general lattice of Fig. 2. All 20 RA general graphs and all 20 BN general graphs for each Rho graph are shown in the joint lattice of RA and BN general graphs of Fig. 5.

In Fig. 5, within each Rho graph, where RA and BN general graphs are equivalent, that is, their independence structures are identical, the BN graph is placed underneath the RA equivalent graph. Where RA or BN graphs are not equivalent, representing an independence structure unique to RA or to BN, they stand alone side-by-side. Arrows from one graph to another in the joint lattice represent the hierarchy of the RA lattice only.

5 Conclusions

The joint lattice of RA and BN general graphs for four variables increases the number of general graphs with unique independence structures from 20 in the four variable RA lattice and 20 in the four variable BN lattice to 30 in the joint RA-BN lattice, and when variable labels are added, increases the number of unique specific graphs from 114 in the RA lattice and 185 in the BN lattice to 238 in the joint lattice.

The integration of the two lattices offers a richer and more expansive way to model complex systems leveraging the V-structure unique to BN graphs and the allowability of loops in RA graphs. The joint RA-BN lattice of general graphs presented in this paper expands the set of general graphs with unique independence structures (or, equivalently, with unique interdependence structures) beyond what was previously available by either RA alone or BN alone, thus allowing for representations of complex systems which are (i) more accurate relative to data and/or (ii) simpler and thus more comprehensible and more generalizable than would be possible by modeling only with RA or only with BN. This joint lattice thus demonstrates how these two related frameworks—RA and BN—both members of the family of probabilistic graphical modeling methodologies, can be integrated into a unified framework. Extension of this work will include designing algorithms to search this

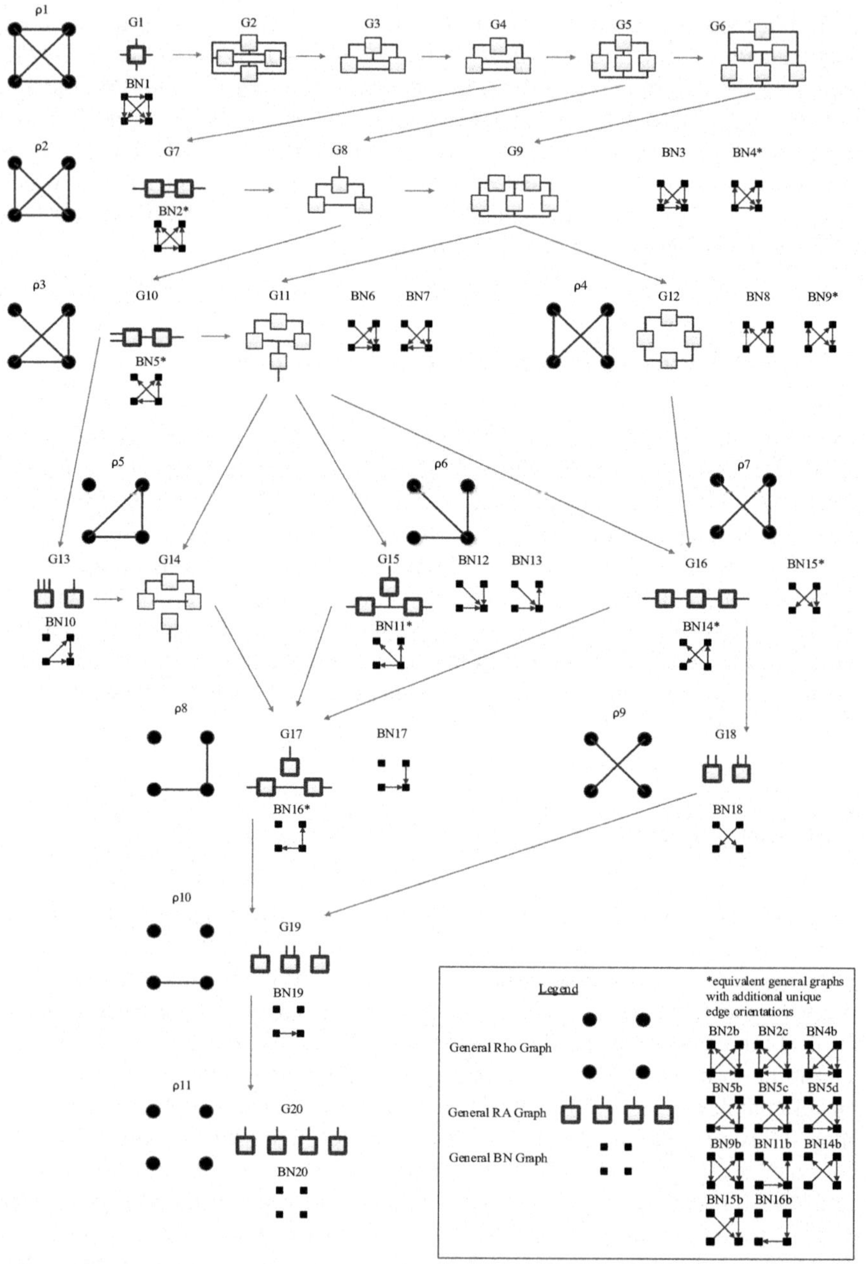

Fig. 5 Joint lattice of RA and BN general graphs

joint RA-BN lattice, analysis of RA and BN predictive models in which the IV-DV distinction is made, consideration of "hybrid" RA-BN models, and other topics.

References

1. Andersson, S., Madigan, D., Perlman, M.: A characterization of Markov equivalence classes for acyclic digraphs. Ann. Stat. (1997)
2. Ashby, W.R.: Constraint analysis of many-dimensional relations. Gen. Syst. Yearb. **9**, 99–105 (1964)
3. Bouckaert, R.: Properties of Bayesian belief network learning algorithms. In: Uncertainty Proceedings, pp. 102–109 (1994)
4. Broekstra, G.: Nonprobabilistic constraint analysis and a two stage approximation method of structure identification. In: Proceedings of the 23rd Annual SGSR Meeting (1979)
5. Buntine, W.: Classifiers: a theoretical and empirical study. In: Proceedings IJCAI, pp. 638–644. Morgan Kaufmann, Sydney (1991)
6. Buntine, W.: Theory refinement on Bayesian networks. In: Proceedings of the Seventh Conference on Uncertainty in Artificial Intelligence, pp. 52–60, Los Angeles, CA (1991)
7. Cavallo, R.: The Role of System Science Methodology in Social Science Research. Martinus Nijhoff Publishing (1979)
8. Chen, Y.: Structure Discovery in Bayesian Networks: Algorithms and Applications. Iowa State University graduate thesis (2016)
9. Chickering, D.: Transformational characterization of equivalent Bayesian network structures. In: Proceedings of the Eleventh Conference on Uncertainty in Artificial Intelligence, pp. 87–98 (1995)
10. Chickering, D.: Learning equivalence classes of Bayesian-network structures. J. Mach. Learn. Res. (2002)
11. Chickering, D., Geiger, D., Heckerman, D.: Learning Bayesian networks: search methods and experimental results. In: Proceedings of the Fifth Conference on Artificial Intelligence and Statistics, pp. 112–128. Society for Artificial Intelligence and Statistics, Ft. Lauderdale, FL (1995)
12. Chickering, D., Heckerman, D., Meek, C.: A Bayesian approach to learning Bayesian networks with local structure. In: Geiger, D. (ed.) Proceedings of the Thirteenth Conference on Uncertainty in Artificial Intelligence, pp. 80–90. Morgan Kaufmann, Providence, RI (1997)
13. Chickering, D., Heckerman, D. Meek, C.: Large-sample learning of Bayesian networks is NP-hard. J. Mach. Learn. Res. 1287–1330 (2004)
14. Conant, R.: Mechanisms of Intelligence: Ashby's Writings on Cybernetics. Intersystems Publications (1981)
15. Conant, R.: Extended dependency analysis of large systems. Int. J. Gen. Syst. (1988)
16. Cooper, D.A.: A Bayesian method for the induction of probabilistic networks from data. Mach. Learn. **9**, 309–347 (1992)
17. Friedman, N., Goldszmidt, M.: Building classifiers using Bayesian Networks. AAAI-96 Proceedings (1996)
18. Friedman, N., Koller, D.: Being Bayesian about network structure: a Bayesian approach to structure discovery in Bayesian networks. Mach. Learn. **50**(1–2), 95–125 (2003)
19. Gillispie, S., Perlman, M.: Enumerating Markov equivalence classes of acyclic digraph models. In: Proceedings of the Seventeenth Conference on Uncertainty in Artificial Intelligence, pp. 171–177 (2001)
20. Heckerman, D., Geiger D., Chickering, D.: Learning Bayesian networks: the combination of knowledge and statistical data. In: Proceedings of Tenth Conference on Uncertainty in Artificial Intelligence, pp. 293–301. Morgan Kaufmann, Seattle, WA (1994)

21. Klir, G.: Identification of generative structures in empirical data. Int. J. Gen. Syst. 89–104 (1976)
22. Klir, G.: The Architecture of Systems Problem Solving. Plenum Press, New York (1985)
23. Klir, G.: Reconstructability analysis: an offspring of Ashby's constraint theory. Syst. Res. 267–271 (1986)
24. Koivisto, M., Sood, K.: Exact Bayesian structure discovery in Bayesian networks. J. Mach. Learn. Res. 549–573 (2004)
25. Krippendorff, K.: An algorithm for identifying structural models of multivariate data. Int. J. Gen. Syst. 63–79 (1981)
26. Krippendorff, K.: Information theory: structural models for qualitative data. In: Quantitative Applications in the Social Sciences, vol. 62. Sage, Beverly Hills (1986)
27. Larranaga, P., Kuijpers, C., Murga, R., Yurramendi, Y.: Learning Bayesian network structures by searching for the best ordering with genetic algorithms. IEEE Trans. Syst. Man Cybern. **26**(4), 487–493 (1996)
28. Lauritzen, S.: Graphical models. In: Oxford Statistical Science Series. Oxford University Press, New York (1996)
29. Malone, B., Yuan, C., Hanse, E.: Memory-efficient dynamic programming for learning optimal Bayesian networks. In: Proceedings of the Twenty-Fifth AAAI Conference on Artificial Intelligence, pp. 1057–1062 (2011)
30. MIT: D-Separation (2015). http://web.mit.edu/jmn/www/6.034/d-separation.pdf
31. Murphy, K.: A Brief Introduction to Graphical Models and Bayesian Networks (1998). https://www.cs.ubc.ca/~murphyk/Bayes/bayes_tutorial.pdf
32. Neapolitan, R.: Probabilistic Reasoning in Expert Systems: Theory and Algorithms. Wiley, New York (1989)
33. Pearl, J.: Bayesian networks: a model of self-activated memory for evidential reasoning. In: Proceedings of the 7th Conference of the Cognitive Science Society, pp. 329–334, Irvine, CA (1985)
34. Pearl, J.: Probabilistic Reasoning in Intelligent Systems: Networks of Plausible Inference. Morgan Kaufmann Publishers Inc., San Francisco (1988)
35. Pearl, J.: Causality: Models, Reasoning, and Inference. Cambridge University Press (2000)
36. Pearl, J., Verma, T.: The logic of representing dependencies by directed graphs. In: Proceedings of the 6th National Conference on Artificial Intelligence, Seattle, WA (1987)
37. Studený, M., Vomlel, J.: Geometric view on learning Bayesian network structures. Int. J. Approx. Reason. (2010)
38. Tian, J., He, R., Ram, L.: Bayesian model averaging using the k-best Bayesian network structures. In: Proceedings of the Twenty-Sixth Conference on Uncertainty in Artificial Intelligence (2010)
39. Verma, T., Pearl, J.: Equivalence and synthesis of causal models. In: Proceedings of the Sixth Annual Conference on Uncertainty in Artificial Intelligence, pp. 255–270 (1990)
40. Wright, S.: Correlation and causation. J. Agric. Res. 557–585 (1921)
41. Wright, S.: The method of path coefficients. Ann. Math. Stat. **5**(3), 161–215 (1934)
42. Zhang, H. (2004). The optimality of Naive Bayes. In: FLAIRS Conference
43. Zwick, M.: Reconstructability analysis of epistasis. Ann. Hum. Genet. 157–171 (2010). https://doi.org/10.1111/j.1469-1809.2010.00628.x, https://works.bepress.com/martin_zwick/3/
44. Zwick, M.: Wholes and parts in general systems methodology. In: Wagner, G. (ed.) The character concept in evolutionary biology. Academic Press, New York (2001). https://works.bepress.com/martin_zwick/52/
45. Zwick, M.: An overview of reconstructability analysis. Kybernetes **33**, 887–905 (2004). https://works.bepress.com/martin_zwick/57/
46. Zwick, M., Carney, N., Nettleton, R.: Exploratory reconstructability analysis of accident TBI data. Int. J. Gen. Syst. **47**(2), 174–191 (2018). https://works.bepress.com/martin_zwick/80/

Hypernetwork Science: From Multidimensional Networks to Computational Topology

Cliff A. Joslyn, Sinan G. Aksoy, Tiffany J. Callahan, Lawrence E. Hunter, Brett Jefferson, Brenda Praggastis, Emilie Purvine, and Ignacio J. Tripodi

Abstract As data structures and mathematical objects used for complex systems modeling, hypergraphs sit nicely poised between on the one hand the world of network models, and on the other that of higher-order mathematical abstractions from algebra, lattice theory, and topology. They are able to represent complex systems interactions more faithfully than graphs and networks, while also being some of the simplest classes of systems representing topological structures as collections of multidimensional objects connected in a particular pattern. In this paper we discuss the role of (undirected) hypergraphs in the science of complex networks, and provide a mathematical overview of the core concepts needed for hypernetwork modeling, including duality and the relationship to bicolored graphs, quantitative adjacency and incidence, the nature of walks in hypergraphs, and available topological relationships and properties. We close with a brief discussion of two example applications: biomedical databases for disease analysis, and domain-name system (DNS) analysis of cyber data.

1 Hypergraphs for Complex Systems Modeling

In the study of complex systems, graph theory has been the mathematical scaffold underlying network science [4]. A graph $\mathcal{G} = \langle V, E \rangle$ comprises a set V of vertices connected in a set $E \subseteq \binom{V}{2}$ of edges (where $\binom{V}{2}$ here means all unordered pairs of $v \in V$), where each edge $e \in E$ is a pair of distinct vertices. Systems studied in biology, sociology, telecommunications, and physical infrastructure often afford a representation as such a set of entities with binary relationships, and hence may be analyzed utilizing graph theoretical methods.

Graph models benefit from simplicity and a degree of universality. But as abstract mathematical objects, graphs are limited to representing *pairwise* relationships

C. A. Joslyn (✉) · S. G. Aksoy · B. Jefferson · B. Praggastis · E. Purvine
Pacific Northwest National Laboratory, Seattle, WA, USA
e-mail: Cliff.Joslyn@pnnl.gov

T. J. Callahan · L. E. Hunter · I. J. Tripodi
University of Colorado Anschutz Medical Campus, Denver, CO, USA

D. Braha et al. (eds.), *Unifying Themes in Complex Systems X*, Springer Proceedings in Complexity, https://doi.org/10.1007/978-3-030-67318-5_25

between entities, while real-world phenomena in these systems can be rich in *multi-way* relationships involving interactions among more than two entities, dependencies between more than two variables, or properties of collections of more than two objects. Representing *group* interactions is not possible in graphs natively, but rather requires either more complex mathematical objects, or coding schemes like "reification" or semantic labeling in bipartite graphs. Lacking multi-dimensional relations, it is hard to address questions of "community interaction" in graphs: e.g., how is a collection of entities A connected to another collection B through chains of other communities?; where does a particular community stand in relation to other communities in its neighborhood?

The mathematical object which *natively* represents multi-way interactions in networks is called a "hypergraph" [6].[1] In contrast to a graph, in a hypergraph $\mathcal{H} = \langle V, E \rangle$ those same vertices are now connected generally in a family E of hyperedges, where now a hyperedge $e \in E$ is an arbitrary subset $e \subseteq V$ of k vertices, thereby representing a k-way relationship for any integer $k > 0$. Hypergraphs are thus the natural representation of a broad range of systems, including those with the kinds of multi-way relationships mentioned above. Indeed, hypergraph-structured data (i.e. hypernetworks) are ubiquitous, occurring whenever information presents naturally as set-valued, tabular, or bipartite data.

Hypergraph models are definitely more complicated than graphs, but the price paid allows for higher fidelity representation of data which may contain multi-way relationships. An example from bibliometrics is shown in Fig. 1.[2] On the upper left is a table showing a selection of five papers co-authored by different collections of four authors. Its hypergraph is shown in the lower left in the form of an "Euler diagram", with hyperedges as colored bounds around groups of vertices. A typical approach to these same data would be to reduce the collaborative structure to its so-called "2-section": a graph of all and only the two-way interactions present, whether explicitly listed (like paper 1) or implied in virtue of larger collaborations (as for paper 2). That co-authorship graph is shown in the lower right, represented by its adjacency matrix in the upper right. It can be seen that the reduced graph form necessarily loses a great deal of information, ignoring information about the single-authored paper (3) and the three-authored paper (2).

Hypergraphs are closely related to important objects in discrete mathematics used in data science such as bipartite graphs, set systems, partial orders, finite topologies, abstract simplicial complexes, and especially graphs proper, which they explicitly generalize: every graph is a 2-uniform hyergraph, so that $|e| = k = 2$ for all hyperedges $e \in E$. Thus they support a wider range of mathematical methods, such as those from

[1]Throughout this paper we will deal only with "basic" hypergraphs in the sense of being undirected, unordered, and unlabeled. All of these forms are available and important [3, 16].

[2]Hypergraph calculations shown in this paper were produced using PNNL's open source hypergraph analytical capabilities HyperNetX (HNX, https://github.com/pnnl/HyperNetX) and the Chapel Hypergraph Library (CHGL, https://github.com/pnnl/chgl); and additionally diagrams were produced in HNX.

Paper #	Authors
1	Andrews, Davis
2	Andrews, Carter, Davis
3	Davis
4	Andrews, Bailey
5	Bailey, Carter

	Andrews	Bailey	Carter	Davis
Andrews		Y	Y	Y
Bailey	Y		Y	
Carter	Y	Y		Y
Davis	Y		Y	

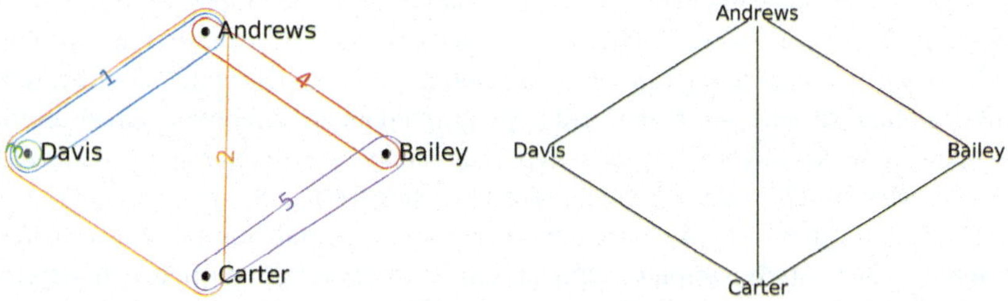

Fig. 1 Bibliometrics example comparing graphs and hypergraphs. (Upper left) Collaborative authorship structure of a set of papers. (Lower left) Euler diagram of its hypergraph. (Upper right) Adjacency matrix of the 2-section. (Lower right) Co-authorship graph resulting from the 2-section

computational topology, to identify features specific to the high-dimensional complexity in hypernetworks, but not available using graphs.

Hypergraph methods are well known in discrete mathematics and computer science, where, for example, hypergraph partitioning methods help enable parallel matrix computations [9], and have applications in VLSI [22]. In the network science literature, researchers have devised several path and motif-based hypergraph data analytics (albeit fewer than their graph counterparts), such as in clustering coefficients [31] and centrality metrics [14]. Although an expanding body of research attests to the increased utility of hypergraph-based analyses [17, 20, 24, 30], and are seeing increasingly wide adoption [19, 28, 29], many network science methods have been historically developed explicitly (and often, exclusively) for graph-based analyses. Moreover, it is common for analysts to reduce data arising from hypernetworks to graphs, thereby losing critical information.

As explicit generalizations of graphs, we must take care with axiomatization, as there are many, sometimes mutually inconsistent, sets of possible definitions of hypergraph concepts which can yield the same results (all consistent with graph theory) when instantiated to the 2-uniform case. And some graph concepts have difficulty extending naturally at all. For example, extending the spectral theory of graph adjacency matrices to hypergraphs is unclear: hyperedges may contain more than two vertices, so the usual (two-dimensional) adjacency matrix cannot code adjacency relations. In other cases, hypergraph extensions of graph theoretical concepts may be natural, but trivial, and risk ignoring structural properties only in hypergraphs. For example, while edge incidence and vertex adjacency can occur in at most one vertex or edge for graphs, these notions are set-valued and hence *quantitative* for hypergraphs. So while subsequent graph walk concepts like connectivity are applicable to hypergraphs, if applied simply, they ignore high-order structure in not accounting for the "widths" of hypergraph walks.

Researchers have handled the complexity and ambiguity of hypergraphs in different ways. A very common approach is to limit attention to k-uniform hypergraphs, where all edges have precisely k vertices (indeed, one can consider graph theory itself as actually the theory of 2-uniform hypergraphs). This is the case with much of the hypergraph mathematics literature, including hypergraph coloring [11, 25], hypergraph spectral theory [7, 8], hypergraph transversals [2], and extremal problems [32]. k-uniformity is a very strong assumption, which supports the identification of mathematical results. But real-world hypergraph data are effectively *never* uniform; or rather, if presented with uniform hypergraph data, a wise data scientists would be led to consider other mathematical representations for parsimony.

Another prominent approach to handling real-world, and thus non-uniform, hypergraph data is to study simpler graph structures implied by a particular hypergraph. Known by many names, including line graph, 2-section, clique expansion, and one-mode projection, such reductions allow application of standard graph-theoretic tools. Yet, unsurprisingly, such hypergraph-to-graph reductions are inevitably and strikingly lossy [10, 23]. Hence, although affording simplicity, such approaches are of limited utility in uncovering hypergraph structure.

Our research group is dedicated to facing the challenge of the complexity of hypergraphs in order to gain the formal clarity and support for analysis of complex data they provide. We recognize that to enable analyses of hypernetwork data to better reflect their complexity but remain tractable and applicable, striking a balance in this faithfulness-simplicity tradeoff is essential. Placing hypergraph methods in the context of the range of both network science methods on the one hand, and higher-order topological methods on the other, can point the way to such a synthesis.

The purpose of this paper is to communicate the *breadth* of hypergraph methods (which we are exploring in depth elsewhere) to the complex systems community. In the next section we survey the range of mathematical methods and data structures we use. Following that we will illustrate some uses by showing examples in two different contexts: gene set annotations for disease analysis and drug discovery, and cyber analytics of domain-name system relations.

2 Hypergraph Analytics

A **hypergraph** is a structure $\mathcal{H} = \langle V, E \rangle$, with $V = \{v_j\}_{j=1}^n$ a set of vertices, and $E = \{e_i\}_{i=1}^m$ an indexable family of hyperedges $e_i \subseteq V$. Hyperedges come in different sizes $|e_i|$ possibly ranging from the singleton $\{v\} \subseteq V$ (distinct from the element $v \in V$) to the entire vertex set V.

Table 1 Incidence matrix of example hypergraph Fig. 1

	1	2	3	4	5
a	1	1	0	1	0
b	0	0	0	1	1
c	0	1	0	0	1
d	1	1	1	0	0

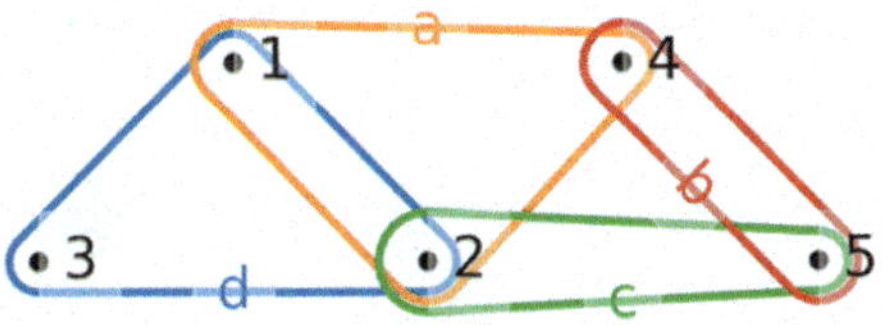

Fig. 2 Dual hypergraph $\mathcal{H}^*$ of our example

A hyperedge $e = \{u, v\}$ with $|e| = 2$ is the same as a graph edge. Indeed, all graphs $\mathcal{G} = \langle V, E \rangle$ are hypergraphs: in particular, graphs are "2-uniform" hypergraphs, so that now $E \subseteq \binom{V}{2}$ and all $e \in E$ are unordered pairs with $|e| = 2$. It follows that concepts and methods in hypergraph theory should explicitly specialize to those in graph theory for the 2-uniform case. But conversely, starting from *graph theory* concepts, there can be many ways of consistently extending them to hypergraphs. The reality of this will be seen in a number of instances below.

Hypergraphs can be represented in many forms. In our example in Fig. 1 above, letting $V = \{a, b, c, d\}$ for the authors and $E = \{1, 2, 3, 4, 5\}$ for the papers, we first represent it as a set system

$$\mathcal{H} = \{\{a, d\}, \{a, c, d\}, \{d\}, \{a, b\}, \{b, c\}\}.$$

We commonly compress set notation for convenience, and when including edge names as well, this yields the compact set system form $\mathcal{H} = \{1{:}ad, 2{:}acd, 3{:}d, 4{:}ab, 5{:}bc\}$. This representation in turn points to the fact that a hypergraph $\mathcal{H}$ is determined uniquely by its Boolean **incidence matrix** $B_{n \times m}$, where $B_{ji} = 1$ iff $v_j \in e_i$, and 0 otherwise. The incidence matrix for the example from Fig. 1 is shown in Table 1. Note that hypergraph incidence matrices are general Boolean matrices, rectangular and non-symmetric, unlike the adjacency matrices typically used to define graphs; while the incidence matrices of graphs are restricted to having precisely two 1's in each column. Also, adjacency structures for hypergraphs are substantially more complicated than for graphs, and in fact are not matrices at all.

The **dual hypergraph** $\mathcal{H}^* = \langle E^*, V^* \rangle$ of $\mathcal{H}$ has vertex set $E^* = \{e_i^*\}_{i=1}^m$ and family of hyperedges $V^* = \{v_j^*\}_{j=1}^n$, where $v_j^* := \{e_i^* : v_j \in e_i\}$. $\mathcal{H}^*$ is just the hypergraph with the transposed incidence matrix B^T, and $(\mathcal{H}^*)^* = \mathcal{H}$. We thus consider that hypergraphs always present as dual *pairs*, which we call the "forward" and the "dual" somewhat arbitrarily, depending on how the data are naturally presented. But this is not true for graphs: the dual $\mathcal{G}^*$ of a graph $\mathcal{G}$ is 2-uniform (and hence still a graph) if and only if $\mathcal{G}$ is 2-regular (all vertices have degree 2), in which case $\mathcal{G}$ is a cycle or disjoint union of cycles. The dual of our example is shown in Fig. 2.

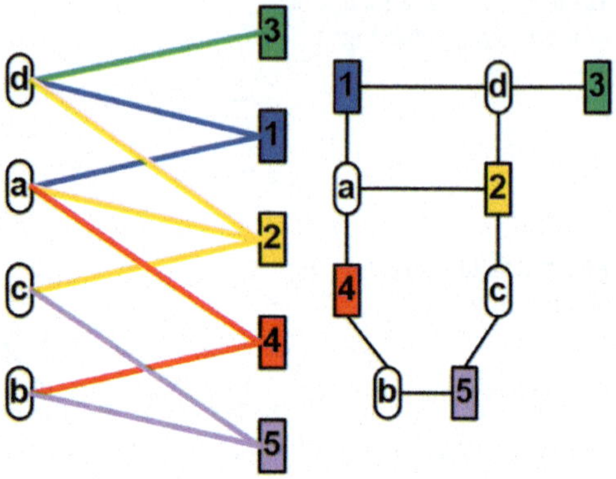

Fig. 3 Two different layouts of the bicolored graph representation of the hypergraph $\mathcal{H}$

There is a bijection between the class of hypergraphs and that of **bicolored graphs**[3] $\mathcal{G} = \langle V, E, \phi \rangle$, where now $E \subseteq \binom{V}{2}$ is a set of unordered pairs of vertices (graph edges), and $\phi \colon V \to \{0, 1\}$ with $\{v_i, v_j\} \in E$ iff $\phi(v_i) \neq \phi(v_j)$. Conversely, any bicolored graph $\mathcal{G}$ determines a hypergraph $\mathcal{H}$ by associating the vertices and hyperedges of $\mathcal{H}$ with the two colors respectively, and then defining $B_{j,i} = 1$ if and only if $\{v_i, v_j\} \in E$. Figure 3 shows two different layouts of the bicolored graph form of our example hypergraph $\mathcal{H}$.

Moving from the bijection between bicolored graphs and hypergraphs to establishing canonical isomorphisms still requires careful consideration, ideally in a categorical context [12, 15, 33]. While a number of complex network analytics for bipartite graph data can be applied naturally to hypergraphs, and *vice versa*, depending on the semantics of the data being modeled, questions and methods for data with this common structure may be better addressed in one or another form, if only for algorithmic or cognitive reasons. Nor does it mean that graph theoretic methods suffice for studying hypergraphs. Whether interpreted as bicolored graphs or hypergraphs, data with this structure often require entirely different network science methods than (general) graphs. An obvious example is triadic measures like the graph clustering coefficient: these cannot be applied to bicolored graphs since (by definition) bicolored graphs have no triangles. Detailed work developing bipartite analogs of modularity [5], community structure inference techniques [26], and other graph-based network science topics [27] further attests that bipartite graphs (and hypergraphs) require a different network science toolset than for graphs.

In graphs, the structural relationship between two distinct vertices u and v can *only* be whether they are adjacent in a *single* edge ($\{u, v\} \in E$) or not ($\{u, v\} \notin E$); and dually, that between two distinct edges e and f can *only* be whether they are incident at a single vertex ($e \cap f = \{v\} \neq \emptyset$) or not ($e \cap f = \emptyset$). In hypergraphs, both of these concepts are applicable to *sets* of vertices and edges, and additionally become *quantitative*. Define $\text{adj} \colon 2^V \to \mathbb{Z}_{\geq 0}$ and $\text{inc} \colon 2^E \to \mathbb{Z}_{\geq 0}$, in both set notation and

[3] Typically the concept of a **bipartite** graph is used here, which is a graph that admits to at least one bicoloring function. The resulting differences are interesting, but not significant for this paper.

(polymorphically) pairwise:

$$\text{adj}(U) = |\{e \supseteq U\}|, \quad \text{adj}(u, v) = |\{e \supseteq \{u, v\}\}|$$

$$\text{inc}(F) = |\cap_{e \in F} e|, \quad \text{inc}(e, f) = |e \cap f|$$

for $U \subseteq V, u, v \in V, F \subseteq E, e, f \in E$. In our example, we have e.g. $\text{adj}(a, d) = 3$, $\text{adj}(\{a, c, d\}) = 1$, $\text{inc}(1, 2) = 2$, $\text{inc}(\{1, 2, 3\}) = 1$. These concepts are dual, in that adj on vertices in $\mathcal{H}$ maps to inc on edges in $\mathcal{H}^*$, and *vice versa.* And for singletons, $\text{adj}(\{v\}) = \deg(v) = |e \ni v|$ is the degree of the vertex v, while $\text{inc}(\{e\}) = |e|$ is the size of the edge e.

This establishes the basis for extending the central concept of graph theory, a **walk** as a sequential visitation of connected nodes, to hypergraphs. Consider a (graph) walk of length ℓ as a sequence $W = v_0, e_0, v_1, e_1, \ldots, e_\ell, v_{\ell+1}$ where v_i, v_{i+1} are adjacent in $e_i, 0 \leq i \leq \ell$, and (dually!) e_i, e_{i+1} are incident on $v_{i+1}, 0 \leq i \leq \ell - 1$. Then W can be equally determined by either the vertex sequence $v_0, \ldots, v_{\ell+1}$, or the edge sequence $e_0, \ldots, e_\ell$. In contrast, with quantitative adjacency and incidence in hypergraphs, sequences of vertices can be adjacent, and sequences of hyperedges incident, in quantitatively different ways, and need not determine each other. Indeed, vertex sequences become hyperedge sequences in the dual, and *vice versa.* For parsimony we work with edgewise walks, and define [1] an ***s*-walk** as a sequence of edges $e_0, e_1, \ldots, e_\ell$ such that $s \leq \text{inc}(e_i, e_{i+1})$ for all $0 \leq i \leq \ell - 1$. Thus walks in hypergraphs are characterized not only by length ℓ, indicating the distance of interaction, but also by "width" s, indicating a *strength* of interaction (see Fig. 4).

For a fixed $s > 0$, we define the ***s*-distance** $d_s(e, f)$ between two edges $e, f \in E$ as the length of the shortest s-walk between them, or infinite if there is none. Note that a graph walk is a 1-walk. We have proved [1] that s-distance is a metric, and can thus define the ***s*-diameter** as the maximum s-distance between any two edges, and an ***s*-component** as a set of edges all connected pairwise by an s-walk. Connected components in graphs are simply 1-components, and our example graph is "connected" in that sense, having a single 1-connected component. But it has *three* 2-components, the hyperedge sets {1, 2}, {4}, and {5}. Other network science methods generalize from graphs to hypergraphs [1], including vertex ***s*-degree**, ***s*clustering coefficients**, and both ***s*-closeness** and ***s*-betweenness centralities**.

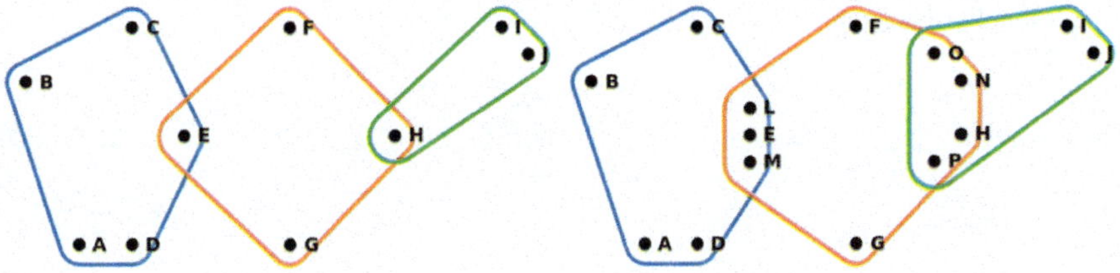

Fig. 4 Two s-walks of length $\ell = 2$. (Left) Lower width $s = 1$. (Right) Higher width $s = 3$

In graphs, two edges e, f are only incident or not, but in hypergraphs, there could additionally be an **inclusion** relationship $e \subseteq f$ or $f \subseteq e$. Indeed, define a **toplex** or **facet** as a maximal edge e such that $\nexists f \supseteq e$, and let $\check{E} \subseteq E$ be the set of all toplexes. Then the **inclusiveness** $I(\mathcal{H}) \in [0, 1)$ of a hypergraph $\mathcal{H}$ is the proportion of included edges, that is, the ratio of non-facets to all edges: $I(\mathcal{H}) := |E \setminus \check{E}|/|E|$. For a hypergraph $\mathcal{H}$, let $\check{\mathcal{H}} = \left\langle V, \check{E} \right\rangle$ be the **simplification** of $\mathcal{H}$, and we call $\mathcal{H}$ **simple** when $\mathcal{H} = \check{\mathcal{H}}$. $I(\mathcal{H}) = 0$ iff $\mathcal{H} = \check{\mathcal{H}}$ is simple, so that all edges are toplexes. In our example, there are three toplexes $\check{E} = \{acd, ab, bc\}$, so that $I(\mathcal{H}) = 2/5$.

Maximal $I(\mathcal{H})$, on the other hand, is more complicated, and the case when all possible sub-edges are present, so that E is closed by subset. This yields $\mathcal{H}$ as an **abstract simplicial complex (ASC)**, so that if $e \in E$ and $f \subseteq e$ then $f \in E$. Let $\widehat{\mathcal{H}} = \langle V, \widehat{E} \rangle$ be the ASC generated by $\mathcal{H}$, so that $\widehat{E} = \{g \subseteq e\}_{e \in E}$ is the closure of the hyperedges by subset. Each hypergraph $\mathcal{H}$ then maps to a class of hypergraphs we call a **hyperblock** $[\mathcal{H}]$, so that each pair of hypergraphs $\mathcal{H}', \mathcal{H}'' \in [\mathcal{H}]$ have the same ASC: $\widehat{\mathcal{H}'} = \widehat{\mathcal{H}''}$. It follows that they also have the same toplexes: $\widecheck{\mathcal{H}'} = \widecheck{\mathcal{H}''}$.

This results in another representation we call a **simplicial diagram**, shown for our example in Fig. 5. The toplexes $\check{E}$ of $\mathcal{H}$ are shown as a collection of hyper-tetrahedrons joined where they intersect. This is also sufficient to indicate the ASC $\widehat{\mathcal{H}}$, and, indeed, all the hypergraphs $\mathcal{H}' \in [\mathcal{H}]$ in the hyperblock of $\mathcal{H}$ are included in the diagram. They are distinguished by additionally labeling the hyperedges (and circling singletons) actually included in a particular hypergraph $\mathcal{H}' \in [\mathcal{H}]$, including both their toplexes and their included edges. In our example, these are $3 = \{d\} \subseteq 1 = \{a, d\} \subseteq 2 = \{a, c, d\}$. Contrast with the singleton $\{b\}$ or graph edge $\{a, c\}$, which are only in the ASC $\hat{\mathcal{H}}$, and not edges in $\mathcal{H}$ itself.

Given a hypergraph $\mathcal{H}$, we can define its ***k*-skeleton** $k\text{-skel}(\mathcal{H}) = \{e \in E : |e| = k\}$ as the set of hyperedges of size k. Each k-skel$(\mathcal{H})$ is thus a k-uniform sub-hypergraph of $\mathcal{H}$, and we can conceive of $\mathcal{H}$ as the disjoint union of its k-uniform skeletons: $\mathcal{H} = \bigsqcup_k k\text{-skel}(\mathcal{H})$. Where the k-skeleton is the set of all edges of size k *present* in a hypergraph $\mathcal{H}$, in contrast the ***k*-section** is the set of all edges of size k *implied* by $\mathcal{H}$, that is, all the vertex sets which are sub-edges of some hyperedge. Formally, $\mathcal{H}_k = k\text{-skel}\left(\widehat{\mathcal{H}}\right)$, so that the k-section is the k-skeleton of the ASC of $\mathcal{H}$, and the ASC is the disjoint union of the sections: $\widehat{\mathcal{H}} = \bigsqcup_k \mathcal{H}_k$.

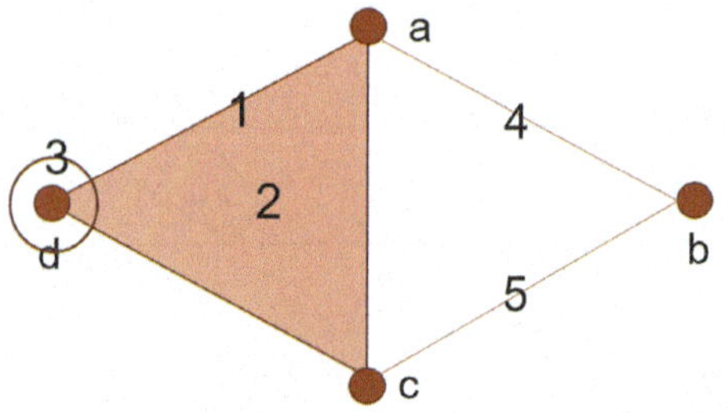

Fig. 5 Example hypergraph $\mathcal{H}$ as a simplicial diagram

Since the k-skeletons are all uniform, and any k-section or union of k-sections is smaller than the entire hypergraph $\mathcal{H}$, there is substantial interest in understanding how much information about a hypergraph is available using only them. The 2-section in particular, which is a graph with adjacency matrix BB^T, can be thought of as a kind of "underlying graph" of a hypergraph $\mathcal{H}$. Also of key interest is the 2-section of the dual hypergraph $\mathcal{H}^*$, called the **line graph** $L(\mathcal{H}) = (\mathcal{H}^*)_2$, which dually is a graph with adjacency matrix $B^T B$. As noted above in the discussion of Fig. 1, these are particularly widely used in studies when confronted with complex data naturally presenting as a hypergraph. The limitations of this are evident in the example in Fig. 6. On the left are our example hypergraph and its dual, and in the center the 2-section $\mathcal{H}_2$ and the line graph $L(\mathcal{H}) = (\mathcal{H}^*)_2$. On the right are the results of taking the maximal cliques of the 2-sections as hyperedges in an attempt to "reconstruct" the original hypergraph $\mathcal{H}$. It is clear how much information is lost.

The ASC $\widehat{\mathcal{H}}$ is additionally a topological complex, that is, a collection of different k-dimensional structures attached together in a particular configuration or pattern. Indeed, the hyperblock $[\mathcal{H}]$ of a (finite) hypergraph $\mathcal{H}$ generates a number of (finite) topological spaces of interest [13]. The most cogent of these is the Alexandrov topology with a sub-base consisting of m open sets $T_j = \{e \supseteq e_j\}_{e \in E}$; that is, for each hyperedge $e_j \in E$, the sub-base element T_j is constructed by collecting all its superedges. The full topology $\mathcal{T}(\mathcal{H})$ is generated by taking all unions of all intersections of these sub-base elements T_j.

The topological space $\mathcal{T}(\mathcal{H})$ will reflect the inherent complexity of the overall "shape" of the hypergraph $\mathcal{H}$. This includes those portions which are connected enough to be contracted, and also the presence of open loops, "holes" or "voids" of different dimension, which can obstruct such contractions. This is

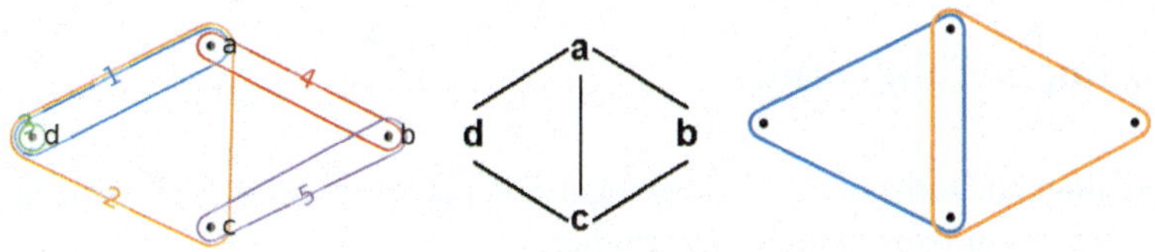

(a) Forward hypergraph $\mathcal{H}$. (b) Forward 2-section $\mathcal{H}_2$. (c) Forward clique reconstruction.

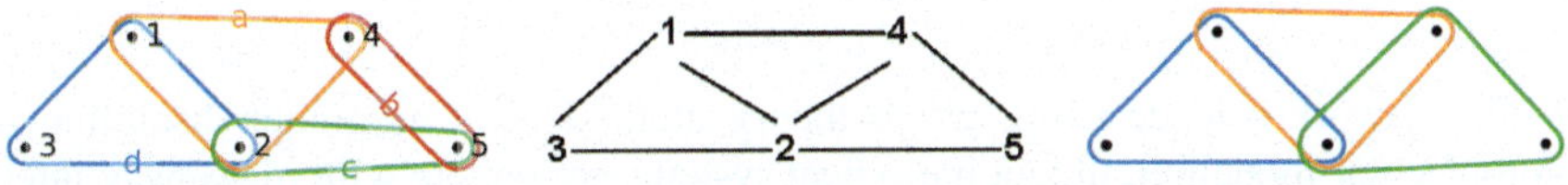

(d) Dual hypergraph $\mathcal{H}^*$. (e) Line graph $L(\mathcal{H}) = (\mathcal{H}^*)_2$. (f) Dual clique reconstruction.

Fig. 6 2-sections and their clique reconstructions

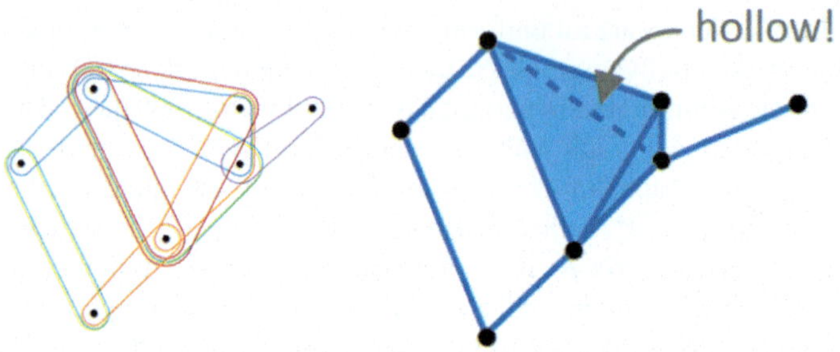

Fig. 7 An example simple hypergraph $\check{\mathcal{H}}$ with $\beta = \langle 1, 1, 1 \rangle$. (Left) Euler diagram of $\check{\mathcal{H}}$. (Right) Simplicial diagram of $\check{\mathcal{H}}$

called the **homology** of the space $\mathcal{T}(\mathcal{H})$, and is characterized by its **Betti numbers** $\beta_k, 0 \leq k \leq \max |e_j| - 1$, of $\mathcal{H}$, indicating the humber of holes of dimension k present in $\mathcal{H}$. We collect the Betti nubmers to create a **Betti sequence** $\beta = \langle \beta_k \rangle_{k=0}^{\max |e_j|-1}$.

The presence of such gaps may invoke questions or hypotheses: what is stopping the connectivity of these holes, of filling them in? In our example, $\beta_1 = 1$ because of the single open 1-cycle indicated by edges ab, bc, and ca. Contrast this with the similar cycle acd, which is closed in virtue of the hyperedge 2. $\beta_0 = 1$, indicating that $\mathcal{H}$ is 1-connected; while $\beta_2 = 0$, so that $\beta = \langle 1, 1, 0 \rangle$. By comparison, the simplicial diagram of the simple hypergraph $\check{\mathcal{H}}$ shown in Fig. 7 contains a hollow tetrahedron (four triangles surrounding a void), in addition to the open cycle of graph edges on the left. Thus its Betti sequence is $\beta = \langle 1, 1, 1 \rangle$.

3 Example Applications

Here we illustrate some of the mathematical structures and methods introduced above in brief reports of two example case studies.

3.1 *Human Gene Set Example*

While network science using graph theory methods is a dominant discipline in biomolecular modeling, in fact biological systems are replete with many-way interactions likely better represented as hypergraphs. Genes interact in complex combinations, as recorded in a panoply of biomedical databases. We have begun an exploratory examination of the usefulness of hypergraphs in elucidating the relationships between human genes, *via* their annotations to semantic categories of human

Table 2 Distribution of annotations across biological databases

Database	Annotations
GO biological process	12,305
Reactome pathways	2,291
Chemicals	3,289
Diseases	2,683
Total	**20,568**

biological processes in the Gene Ontology[4] and the Reactome pathway database,[5] chemicals from the Chemical Entities of Biological Interest ontology,[6] and diseases from the Human Disease Ontology.[7] These data were selected to help us better understand potential overlaps between gene sets with causative relations in metabolic rare diseases, like phenylketonuria and Alpha-1 antitrypsin deficiency, and their known biological processes and chemical interactions. We also seek to explore potential overlaps in pathway, biological process, and chemical gene sets as a means to elucidate novel gene targets for drug repurposing, which can then be evaluated in the lab.

Data for this analysis were obtained from PheKnowLator v2.0.0.[8] We compiled a hypergraph $\mathcal{H} = \langle V, E \rangle$ with $|V| = 17{,}806$ human genes as vertices against $|E| = 20{,}568$ annotations as hyperedges, distributed across the source databases as shown in Table 2 (noting that there is substantial overlap among these sources). Of these edges, 8,006 are toplexes, yielding an inclusivity of $I(\mathcal{H}) = 61.1\%$, and the density of the incidence matrix B is 0.000926. Figure 8 shows the distribution of vertex degree $\deg(v) = \text{adj}(\{v\})$ and edge size $|e| = \text{inc}(\{e\})$, with the expected expoential distribution.

Figure 8 shows only the lowest, "first order" distribution of the hypergraph structure, the adjacency and incidence of singletons. Consideration of higher-order interactions would require expensive combinatorial calculations of, for example, k-way intersections and hyperedge inclusions of arbitrary sets of vertices. A modest step towards that goal in our methodology is first to focus on toplexes, which determine the topological structure, and then their pairwise intersections: $\text{inc}(e, f)$ for $e, f \in \check{E}$. This is shown in the left of Fig. 9, which reveals a long tail, indicating a significant number of pairs of annotations with large intersections of genes. Attending to incidences of even higher order would reveal the increasingly rich complex interactions of gene sets.

[4]http://geneontology.org/docs/ontology-documentation.

[5]https://reactome.org.

[6]https://www.ebi.ac.uk/chebi/.

[7]https://disease-ontology.org/.

[8]https://github.com/callahantiff/PheKnowLator, downloaded on 03/05/20, see also https://github.com/callahantiff/PheKnowLator/wiki/v2-Data-Sources.

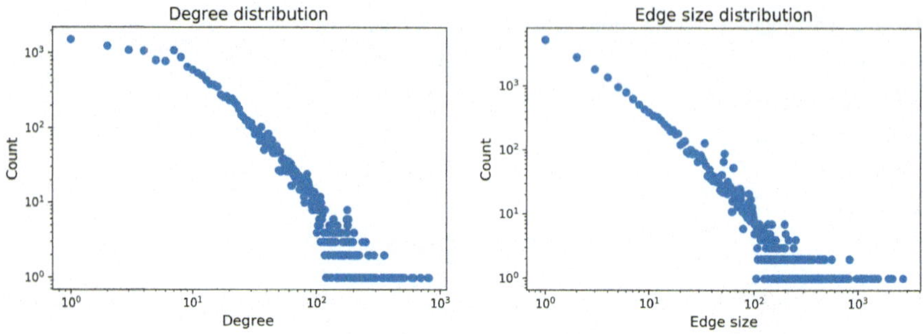

Fig. 8 Distributions of: (Left) Vertex degree $\deg(v) = \text{adj}(\{v\})$; (Right) Edge size $|e| = \text{inc}(\{e\})$

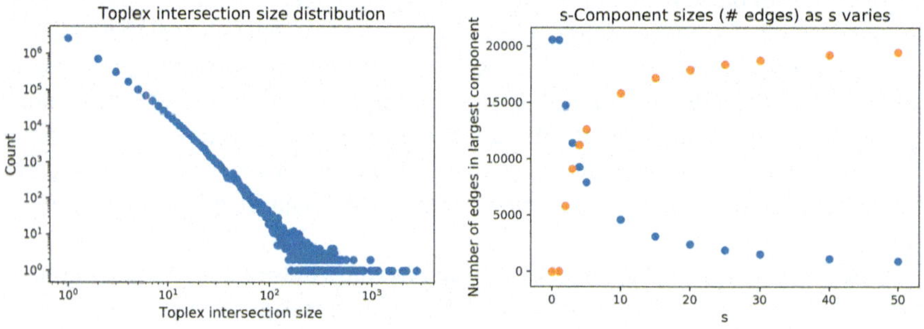

Fig. 9 (Left) Distribution of the size of toplex intersections: $\text{inc}(e, f)$ for $e, f \in \check{E}$. (Right) # components (orange) and size of largest component (blue)

Of even more interest is the right of Fig. 9, which shows distribution information about the connected s-components for different intersection levels s. On the top the number of s-components is shown in orange, and the size of the largest component in blue, all as a function of increasing s. Expectedly these appear monotonic increasing and decreasing respectively, but it's notable that even for large s there persist some very large components, again demonstrating the large multi-way interactions amongst these gene sets.

3.2 DNS Cyber Example

The Domain Name System (DNS) provides a decentralized service to map from domain names (e.g., www.google.com) to IP addresses. Perhaps somewhat counter-intuitively, DNS data present naturally as a hypergraph, in being a many-many relationship between domains and IPs. While typically this relationship is one-to-one, with each domain uniquely identifying a single IP address and *vice versa*, there are a number of circumstances which can violate this, for example domain aliasing,

	2wzl79.com	37b82p.com	5bewa2.com	8sa5zz.com	8zjcn3.com	9mt5sd.com	dyewir.com	g53fag.com	j4bpk1.com	kbeuu8.com	sgbxhl.com	ue594m.com		
103.86.122.130							X			X	X			
103.86.122.148		X	X	X				X				X		
103.86.122.149			X					X				X		
103.86.122.152	X	X		X										
103.86.122.154					X	X	X			X	X			
103.86.122.160							X				X	X		
103.86.122.169			X											
103.86.122.173	X													
103.86.122.181	X			X				X						
103.86.122.192					X									
103.86.122.195						X			X					
103.86.122.220			X											
103.86.122.222		X					X					X		
103.86.122.223					X	X			X	X	X			
103.86.122.225	X	X		X				X						
103.86.122.238									X					
103.86.122.242										X				

Fig. 10 Portion of the incidence matrix for ADNS data

hosting services where one IP serves multiple websites, or duplicated IPs to manage loads against popular domains.

ActiveDNS (ADNS) is a data set maintained by the Astrolavos Lab at the Georgia Institute of Technology.[9] It submits daily DNS lookups for popular zones (e.g., .com, .net, .org) and lists of domain names. Using data from April 26, 2018 as an example, this day consists of 1,200 Avro files with each file containing on average 900K records. Our hypergraph representation coded each domain (hyperedges $e \in E$) as a collection of its IPs (vertices $v \in V$). A small portion of the incidence matrix B is shown in Fig. 10.

To identify some of the simplest hypergraph-specific properties, we looked [21] specifically at the 2-components, and identified the one with maximum 2-diameter (6), which is shown in Fig. 11. The IP addresses in this component all belong to the IP range 103.86.122.0/24 and the domains are registered to GMO INTERNET, INC according to WHOIS records. Moreover, current DNS queries for most of these domains at a later date resolve to IPs in the range 103.86.123.0/24 and have a "time to live" of only 120 s. This pattern of quickly changing of IP address is consistent with the "fast flux" DNS technique which can be used by botnets to hide malicious content delivery sites and make networks of malware more difficult to discover [18].

---

[9]https://activednsproject.org.

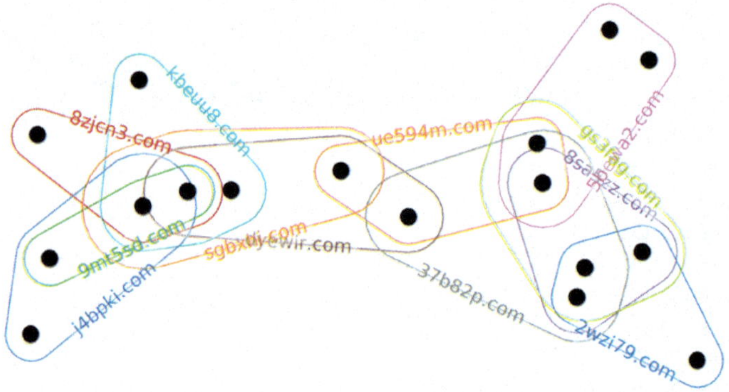

Fig. 11 The 2-component with largest 2-diameter, possibly indicating fast flux behavior

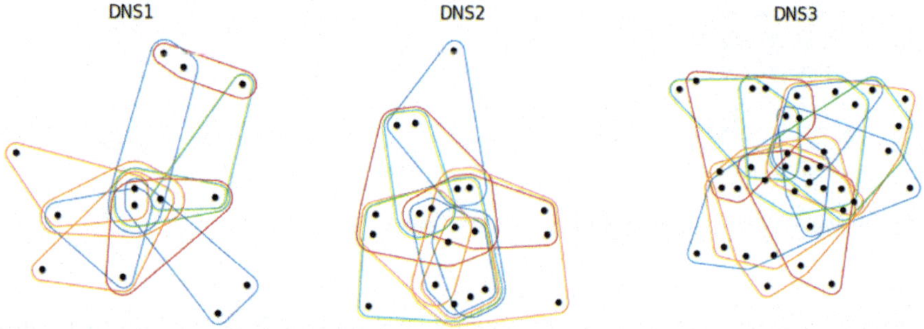

Fig. 12 Three 2-components with non-trivial homologies

Other 2-components reveal non-trivial homologies, three of which are shown in Fig. 12. DNS1 has $\beta = \langle 1, 1, 0, 0, \ldots \rangle$, with a visible hole surrounded by a 1 dimensional loop on the top. DNS2 has $\beta = \langle 1, 1, 2, 0, \ldots \rangle$, and DNS3 has $\beta = \langle 1, 3, 1, 0, \ldots \rangle$, indicating one and three 1-dimensional holes, and two and one 2-dimensional voids, respectively. The open loops in DNS2 and DNS3 are harder to visualize, so Fig. 13 shows a simplicial diagram of one of the two 2-dimensional voids in DNS2. There are two solid tetrahedrons for the domains potterybarn.com and pbteen.com, each with four IPs, three of which (those ending in .160, .9, and .105) they share. Then wshome.com is a triangle in the foreground, and westelm.com a triangle behind (see caption for details). These are effectively "transparent window panes" surrounding a hollow tetrahedral space. Identification of such multi-dimensional open-loop structures affords the opportunity to consider these as hypotheses: on what basis is the Pottery Barn company structuring its multiple domains over their multiple IPs in this complex pattern?

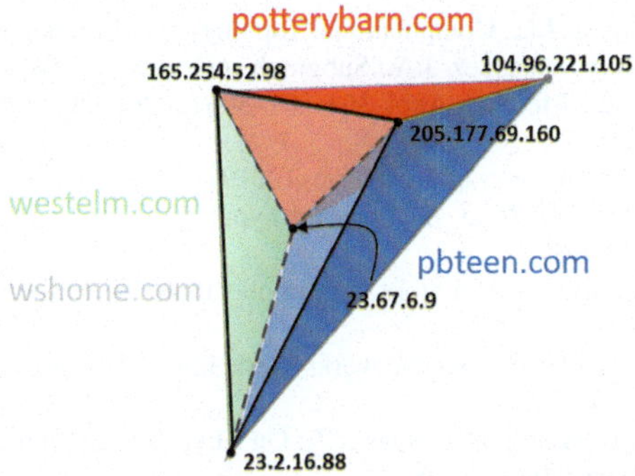

Fig. 13 Simplicial diagram of the complex pattern of IPs shared by some Pottery Barn domains around one of the two 2-dimensional voids in DNS2 of Fig. 12. wshome.com is the transparent foreground triangle with IPs ending in .160, .88 and .98; and westelm.com the background triangle with IPs ending in .98, .88, and .9. Potterybarn.com and pbteen.com are solid tetrahedrons with four IPs each

Acknowledgements PNNL-SA-152208. This work was also partially funded under the High Performance Data Analytics (HPDA) program at the Department of Energy's Pacific Northwest National Laboratory. Pacific Northwest National Laboratory is operated by Battelle Memorial Institute under Contract DE-ACO6-76RL01830.

References

1. Aksoy, S.G., Joslyn, C.A., Marrero, C.O., Praggastis, B., Purvine, E.A.: Hypernetwork science via high-order hypergraph walks. EPJ Data Sci. **9**, 16 (2020)
2. Alon, N.: Transversal numbers of uniform hypergraphs. Graphs Comb. **6**(1), 1–4 (1990)
3. Ausiello, G., Fanciosa, P.G., Frigioni, D.: Directed hypergraphs: Problems, algorithmic results, and a novel decremental approach. In: ICTCS 2001, LNCS, vol. 2202, pp. 312–328 (2001)
4. Barabási, A.L.: Network Science. Cambridge University Press, Cambridge (2016)
5. Barber, M.J.: Modularity and community detection in bipartite networks. Phys. Rev. E **76**(6) (2007). https://doi.org/10.1103/physreve.76.066102
6. Berge, C., Minieka, E.: Graphs and Hypergraphs. North-Holland (1973)
7. Chung, F.: The laplacian of a hypergraph. Expanding graphs (DIMACS series), pp. 21–36 (1993)
8. Cooper, J., Dutle, A.: Spectra of uniform hypergraphs. Linear Algebr. Its Appl. **436**(9), 3268–3292 (2012)
9. Devine, K., Boman, E., Heaphy, R., Bisseling, R., Catalyurek, U.: Parallel hypergraph partitioning for scientific computing. In: Proceedings 20th IEEE International Parallel & Distributed Processing Symposium. IEEE (2006). https://doi.org/10.1109/ipdps.2006.1639359
10. Dewar, M., Healy, J., Pérez-Giménez, X., Prałat, P., Proos, J., Reiniger, B., Ternovsky, K.: Subhypergraphs in non-uniform random hypergraphs. Internet Math. (2018). https://doi.org/10.24166/im.03.2018
11. Dinur, I., Regev, O., Smyth, C.: The hardness of 3-uniform hypergraph coloring. Combinatorica **25**(5), 519–535 (2005)
12. Dörfler, W., Waller, D.A.: A category-theoretical approach to hypergraphs. Archiv der Mathematik **34**(1), 185–192 (1980). https://doi.org/10.1007/bf01224952

13. Edelsbrunner, H., Harer, J.L.: Computational Topology: An Introduction. AMS (2000)
14. Estrada, E., Rodríguez-Velázquez, J.A.: Subgraph centrality and clustering in complex hyper-networks. Phys. A: Stat. Mech. Its Appl. **364**, 581–594 (2006). https://doi.org/10.1016/j.physa.2005.12.002
15. Fong, B., Spivak, D.I.: Hypergraph categories (2019)
16. Gallo, G., Longo, G., Pallottino, S.: Directed hypergraphs and applications. Discret. Appl. Math. **42**, 177–201 (1993)
17. Iacopini, I., Petri, G., Barrat, A., Latora, V.: Simplicial models of social contagion. Nat. Commun. **10**, 2485 (2019)
18. jamie.riden: How Fast-Flux Service Networks Work. http://www.honeynet.org/node/132. Last accessed 26 Nov 2018
19. Javidian, M.A., Lu, L., Valtorta, M., Qang, Z.: On a hypergraph probabilistic graphical model (2018). https://arxiv.org/abs/1811.08372
20. Johnson, J.: Hypernetworks in the Science of Complex Systems. Imperial College Press, London (2013)
21. Joslyn, C.A., Aksoy, S., Arendt, D., Firoz, J., Jenkins, L., Praggastis, B., Purvine, E.A., Zalewski, M.: Hypergraph analytics of domain name system relationships. In: Kaminski, B., et al. (eds.) 17th Workshop on Algorithms and Models for the Web Graph (WAW 2020). Lecture Notes Comput. Sci. **12901**, 1–15. Springer (2020)
22. Karypis, G., Kumar, V.: Multilevel k-way hypergraph partitioning. VLSI Des. **11**(3), 285–300 (2000). https://doi.org/10.1155/2000/19436
23. Kirkland, S.: Two-mode networks exhibiting data loss. J. Complex Netw. **6**(2), 297–316 (2017). https://doi.org/10.1093/comnet/cnx039
24. Klamt, S., Haus, U.U., Theis, F.: Hypergraphs and cellular networks. PLoS Comput. Biol. **5**(5), e1000,385 (2009)
25. Krivelevich, M., Sudakov, B.: Approximate coloring of uniform hypergraphs. J. Algorithms **49**(1), 2–12 (2003)
26. Larremore, D.B., Clauset, A., Jacobs, A.Z.: Efficiently inferring community structure in bipartite networks. Phys. Rev. E **90**(1) (2014). https://doi.org/10.1103/physreve.90.012805
27. Latapy, M., Magnien, C., Vecchio, N.D.: Basic notions for the analysis of large two-mode networks. Soc. Netw. **30**(1), 31–48 (2008). https://doi.org/10.1016/j.socnet.2007.04.006
28. Leal, W., Restrepo, G.: Formal structure of periodic system of elements. Proc. R. Soc. A. **475** (2019). https://doi.org/10.1098/rspa.2018.0581
29. Minas, M.: Hypergraphs as a unifrom diagram representation model. In: Proceedings of the 6th International Workshop on Theory and Applications of Graph Transformations (2000). ftp://ftp.informatik.uni-erlangen.de/local/inf2/Papers/tagt98.pdf
30. Patania, A., Petri, G., Vaccarino, F.: Shape of collaborations. EPJ Data Sci. **6**, 18 (2017)
31. Robins, G., Alexander, M.: Small worlds among interlocking directors: network structure and distance in bipartite graphs. Comput. Math. Organ. Theory **10**(1), 69–94 (2004). https://doi.org/10.1023/b:cmot.0000032580.12184.c0
32. Rödl, V., Skokan, J.: Regularity lemma for k-uniform hypergraphs. Random Struct. Algorithms **25**(1), 1–42 (2004)
33. Schmidt, M.: Functorial approach to graph and hypergraph theory (2019)

Multi-level Co-authorship Network Analysis on Interdisciplinarity: A Case Study on the Complexity Science Community

Robin Wooyeong Na and Bongwon Suh

Abstract Determining interdisciplinarity has become crucial as the importance of such collaboration is increasingly appreciated. We aim to demonstrate the interdisciplinarity of the complexity science community through a multi-level co-authorship network analysis. We conduct a multi-level analysis to avoid the pitfalls between multidisciplinarity and interdisciplinarity. We build a weighted co-authorship network and label each author's disciplines based on their self-declared interests available in the Google Scholar account. Our work uses an applied form of Shannon entropy to measure the diversity of disciplines. While showing the multidisciplinarity through measuring the diversity at the global level, we utilize a community detection algorithm and show the interdisciplinarity via a group level analysis. We also conduct an individual level analysis by measuring the *neighbor diversity* for each node and comparing it with the author's degree centrality. Our research shows that the diversity of disciplines in the complexity science community comes from the interdisciplinary characteristics at the group-level and individual-level analysis. The model can be applied to other heterogeneous communities to determine how truly interdisciplinary they are.

Keywords Interdisciplinary studies · Diversity · Complexity science · Co-authorship network · Network communities · Multi-level analysis

1 Introduction

As problems encountered by scientists get increasingly complex, collaboration between experts in different fields have become more common throughout the years [1, 2]. This has led to the emergence of many interdisciplinary fields and programs.

R. W. Na
College of Liberal Studies, Seoul National University, Seoul, Republic of Korea

B. Suh (✉)
Graduate School of Convergence Science and Technology, Seoul National University, Seoul, Republic of Korea
e-mail: bongwon@snu.ac.kr

D. Braha et al. (eds.), *Unifying Themes in Complex Systems X*, Springer Proceedings in Complexity, https://doi.org/10.1007/978-3-030-67318-5_26

The intangible aspects of interdisciplinarity however, has risen questions about its concept. Does heterogeneity of fields in a research group imply solid collaboration between them? To answer this question, a lot of earlier researches have worked on measuring the interdisciplinarity of research groups [3].

Many researches on interdisciplinarity use complex networks approach to map the characteristics of science communities, mostly through bibliographic methods. A literature review summarizes how different units such as authors, documents, journals, and words can be used to assess interdisciplinarity [3]. Among these, author-based analysis is not preferred in most earlier researches because information about their disciplines are not readily available in most datasets. It has also been suggested by an author-based study that reviewing the affiliations of authors by going through their manuscripts can be tedious [4]. Also, departmental affiliation in manuscripts may not successfully embed authors to disciplinary fields [5]. Given these drawbacks, many researches for measuring interdisciplinarity adopt bibliographic methods through journal citation data [6–17].

Nonetheless, some researchers acknowledge that author-based analysis has a great potential. Not only has author-based analysis been suggested as one of the simplest ways to collaborate between different fields [13], it is also considered as an effective way of examining the social and intellectual structure [18]. Therefore, the approach of measuring interdisciplinarity is shifting from semantic analysis to author-based network analysis [5]. Co-authorship analysis is widely used in solving questions related to authors' collaboration [19].

While a few studies on interdisciplinarity provide effective methods of author-based analysis, they do not use powerful co-authorship network analysis tools such as community detection algorithms. Furthermore, many researches concentrate exclusively on low-level analysis such as individual author or article.

We aim to build a co-authorship network of a well-known interdisciplinary field, complexity science, to illustrate the field's interdisciplinarity in multiple levels. We collect the data from the Google Scholar database by searching for every author who has labeled their interest in complexity science. For the classification of disciplines, we analyze the self-declared interests of each researcher in addition to their affiliation enlisted in the Google Scholar profile. We measure the diversity using an applied form of Shannon entropy for each level.

We start with demonstrating the multidisciplinarity of the complexity science community by measuring the diversity of disciplines on the global level. At the group level, we divide the co-authorship network using a community detection algorithm and show that the diversity of disciplines for each divided community remains high. Compared with the global level diversity, this implies that authors do not form their groups based on their traditional disciplines. At the individual level, we measure the interdisciplinarity of each author by measuring the diversity of disciplines in their neighbors called *neighbor diversity* and compare them with the degree centrality. The high correlation between these demonstrate that researchers are very open to collaborating with scholars in diverse disciplines.

Our research has three main contributions. First, we conduct a multi-level analysis on the interdisciplinarity using a social network approach, contrast to many single level analyses relying on bibliographic methods. Secondly, we demonstrate the relationship between neighbor diversity and degree centrality and discuss how the relationship can show the interdisciplinarity of a community. Lastly, we propose a new way of collecting author data and labeling them to disciplines by the self-declaration each scholar has made. Our research design can be applied for determining interdisciplinarity in many other heterogeneous communities such as multidisciplinary fields, research institutes, or liberal arts programs in universities.

2 Concepts of Interdisciplinarity and Diversity

2.1 Multidisciplinarity Versus Interdisciplinarity

Interdisciplinarity is defined as approaches to integrate separate disciplinary data, methods, tools, concepts, or theories [3]. Unlike interdisciplinarity, multidisciplinarity is no more than a simple sum of separate disciplines where each discipline retains its original identity. Hence, the process of integration should be considered while assessing interdisciplinarity. Literature review observes bibliographic analysis and citation-based measurements to be the most common in capturing the integrating properties [3]. Meanwhile, teamwork is suggested as a crucial part of interdisciplinarity which is difficult to capture through these bibliographic analyses [20]. This strengthens our pursuit for co-authorship networks.

The difference between multidisciplinarity and interdisciplinarity is a caveat for this study, which is why interdisciplinarity should be assessed through both high-level and low-level approaches. An example of this pitfall is demonstrated in the interdisciplinary analysis in the field of nanoscience [4]. Although nanoscience can be considered multidisciplinary given its heterogeneity in disciplines, it cannot be considered interdisciplinary since not many scholars show cross-disciplinary co-authoring process and merely share the prefix "nano" in their publication titles or journals. Therefore, there is no effective interaction among disciplines. To demonstrate the presence of heterogeneous teamwork, we measure the diversity of complexity scientists in three different levels.

2.2 Diversity Measures

Earlier studies propose diversity measures as methods of assessing interdisciplinarity [12, 15, 21]. High diversity implies heterogeneity of research fields in our observation. It should be handled with care, however, since heterogeneity may imply multidisciplinarity but not necessarily interdisciplinarity. Integration must be considered, which is why we do not stop at measuring diversity but compare them using community detection algorithms in the group level and degree centralities in the individual level.

In this research, the Shannon entropy from information theory is used to measure the diversity. In n types of disciplines with each of it holding the ratio of p_i, the entropy is measured through logarithmic function of a base b in the following form. The entropy is higher when we have a larger n and when p_i are equally distributed. Therefore, even if there are many fields in our system, entropy will not be high if it is highly biased towards a few disciplines.

$$S = -\sum_{i}^{n} p_i \log_b p_i \quad (1)$$

In measuring interdisciplinarity, the Shannon entropy is suggested to be more effective compared to the diversity index of the Gini coefficient where it is usually used to measure inequality [8]. The entropy is effective for measuring the variety and balance of an interdisciplinary system [21]. Recent studies also use the Shannon entropy to measure multidisciplinarity or interdisciplinarity [22, 23].

In our research, we apply an exponential transformation to this entropy to show diversity. The Shannon entropy can be misleading in terms of scales because a two times larger Shannon entropy does not imply a two times larger diversity [24]. Therefore, the concept of "true diversity" is suggested where the base b is taken to the power of entropy S. True diversity is intuitive in describing the heterogeneity. For example, if there is only one discipline in a group, the entropy would be 0 and thus making the true diversity 1. If there are two disciplines equally composed in a group, the true diversity would be 2. In this case, the latter system is two times more diverse than the former. To sum up, true diversity refers to the number of disciplines equally contributing in our system. We use this true diversity to intuitively estimate how diverse each group or individual is in terms of disciplines.

3 Research Design

3.1 Analytical Framework

To demonstrate the interdisciplinarity of a co-authorship network of complexity scientists, we conduct a three-phase research. As depicted in Fig. 1, this involves data acquisition, classification and network construction, and interdisciplinarity analysis. We first collect author information about complexity scientists by searching through the Google Scholar pages. In the next phase, we classify the disciplines of scholars based on their interests and affiliation while constructing a weighted co-authorship network involving all of them. Lastly, we measure the diversity at three levels. At the global level, co-authorship network is not used but only the true diversity of

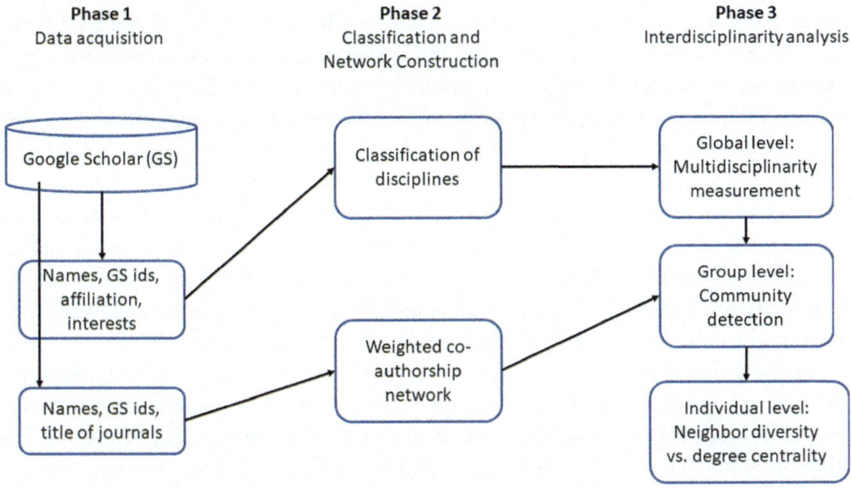

Fig. 1 Research flow chart

disciplines is measured. On this level, the multidisciplinarity is demonstrated but not the interdisciplinarity. It is in the group level and the individual level that we use co-authorship networks to analyze interactions that underlie in the multidisciplinarity at the global level. For the group level, we use the community detection algorithm and measure the diversity of each divided community. For the individual level, we define a diversity index called *neighbor diversity* for each node and compare them with the degree centrality.

3.2 Collecting and Preprocessing the Data

In our data acquisition phase, we aim to collect author data which include information that can be used to label their disciplines. We collect the names, user ids, affiliations, lists of research interests, and lists of publication titles for 4,763 complexity scientists from Google Scholar at May 15, 2019. Many studies collect author data either by searching for publications in journals [18, 25] or specific institutions such as universities [5, 7, 14, 26, 27]. However, they are not easily applicable in the case of complexity science. First, due to the nature of interdisciplinarity, topics related to complexity science is not exclusively published in certain domains or types of journals. Also, many complexity scientists are affiliated in institutions where complexity science division is not explicitly organized.

Out of other databases, three significant reasons make the Google Scholar data suitable for our research. First, Google Scholar provides a list of research interests for each author. Since it is difficult to determine whether a scholar is a complexity scientist, we rely on their self-declared interest which is the list of research interests from the Google Scholar account. We collector authors who possess a Google Scholar account and has included "complexity science" or other synonyms in their research interest. Synonyms include "complex system(s)," "complexity(ies)," "complexity sciences," "complex system(s) science", "complex network(s)," "complex dynamical network(s)," "complex adaptive system(s)," "complexity theory(ies)," "network and complexity science" and "complex dynamics." Subfields of complexity science such as "networks" or "nonlinear dynamics" are not included since not all researchers of those fields are likely to declare themselves working in the field of complexity science. For example, scholars with interest in networks may have interest in computer networks but little interest in complex networks.

Secondly, while studies on co-authorship network suffer the issue of homonomy [28], Google Scholar helps us circumvent it. Users with the same name but different Google Scholar ID can reasonably be assumed as different people. There are few situations where an author is suspected to run two different Google Scholar accounts, specifically when their name, affiliation, research interest, and list of publications match significantly. These cases are easily removed through inspection.

Lastly, collecting author information from journals related to complexity science may include scholars who are doubtful to be considered as complexity scientists. It is possible to collect complexity science related articles through keywords [29] or publication data provided by complex systems science institutes such as the Santa Fe Institute. Nevertheless, making an author list from publications can be problematic. Not every author in those publications co-author for the sake of being a complexity scientist, especially in fields where it is common for tens of authors to collaborate on a paper. For these reasons, we rely on interests listed in the Google Scholar profile for each author and apply this method in classifying their traditional disciplines as well.

3.3 Labeling of Disciplines

In this process, we label the disciplines of each author also based on their self-declared interest. Identifying the traditional disciplinary background of complexity scientists is difficult as it requires an expert judgement [3]. Though some researches categorize them based on their affiliated department [4, 5, 30], they acknowledge the possible inaccuracy that may be caused through inconsistent naming. Working in an interdisciplinary field, complexity scientists are assumed to have higher possibility of falling into situations where their departmental affiliation does not fully represent their disciplines. For example, a researcher with a doctorate degree in physics may work in an economics department for postdoctoral position and become a faculty at an information science department. Such cases are not uncommon among complexity scientists.

Therefore, it is very difficult to label the disciplines of scholars unless each one is closely examined with expertise. However, a solution can be reached by relying on their own declaration of interest. More precisely, we use the list of research interests for each author in their Google Scholar account once again. We also use the affiliation information available for some Google Scholar users. For instance, if an author has included subfields of physics in their interest or is affiliated to a physics department, we label that person's disciplines as physics.

Google Scholar users enter their research interest without strict guidelines or restrictions, contrary to choosing several topics from a drop-down menu. Hence, classification for each research interest label is not officially provided. Thus, we develop our own method of mapping each interest into traditional disciplines. To begin with, we make a list of disciplines to label based on the list of departments of Seoul National University, a well-established research university where the two authors are affiliated.

Next, we design a model to map each author's research interest to these disciplines. Examining the entire research interests in our dataset, we make a list of expressions that can be categorized into specific disciplines. For example, a researcher including "medic" or "physio" in their research interest is labeled as a medical scientist. If a researcher has an interest in "econophysics" and "ecology," we label their disciplines to physics, economics and ecology without any weight. Thus, one author can be labeled into more than one discipline and on average, there are 1.35 disciplines assigned for each author. To show the validity of this criteria, the interrater agreement has been measured between the two authors of this work. The Kippendorff's alpha value is 0.80.

While designing the model, we were careful not to include interdisciplinary topics such as "artificial intelligence" to the list since that will increase the risk of misclassification. Also, we were strict on distinguishing "computer science" from "computational science." Computational skills are mandatory in most cases to conduct research on complex systems, and computer scientists concentrate on different issues compared to computational social scientists.

3.4 Building the Co-authorship Network

We construct a co-authorship network of 4,763 nodes and 12,663 edges using about 274 thousand publications. Authors are set as nodes and undirected links between them are added with weight equal to the number of papers co-authored by each pair of nodes. This way, authors who have co-authored many times have stronger ties and it is considered during the community detection.

In our network, the true diversity of disciplines is measured in the following manner. Author A with disciplines labeled "mathematics" and "physics" have their own diversity of 2 since the two disciplines are unweighted. If Author B is labeled with "social science" and "computer science," the group of Authors A and B carry the true diversity of 4. If Author C is labeled with "physics," the group of Authors A, B and C carry the true diversity of 3.79. This is smaller than 4 since there are 4 distinct disciplines but they are not equally frequent.

3.5 *Multidisciplinarity at the Global Level and Interdisciplinarity at the Group Level*

We first calculate the diversity of discipline in our whole network of complexity scientists using the true diversity measurement. As discussed above, the high diversity at the global level may prove multidisciplinarity but not interdisciplinarity. To assert interdisciplinarity, we need to show that there are essential interactions between distinct disciplines. For this purpose, we proceed a group level analysis by using the modularity-based community detection algorithm [31]. We measure the diversity of 21 subcommunities with the number of nodes larger than 40.

3.6 *Interdisciplinarity at the Individual Level*

For the individual level analysis, we assess if authors are capable of co-authoring with scholars in diverse disciplines. We measure the *neighbor diversity* for each node. Neighbor diversity of an author is the diversity of a group that includes the author and their first-degree neighbors. Degree centrality is the number of first-degree neighbors for each node. We compare these two values for analysis. Since higher degree centrality increases the chance for connecting with people in diverse disciplines, it is reasonable to predict that neighbor diversity and degree centrality will have positive correlation. To determine interdisciplinarity however, it is important to examine how strongly correlated they are. For example, if a scholar with discipline placed as physics and mathematics have many neighbors but most of them from the same disciplines, their neighbor diversity should stay low regardless of their high degree centrality. Hence, we provide an exponential variogram fit to discuss how this relationship could be used to assess interdisciplinarity.

4 Results and Discussions

4.1 *Multidisciplinarity at the Global Level*

According to the global level analysis, the complexity science community is highly multidisciplinary to say the least. Figure 2 shows that there are 28 disciplines contributing to the field of complexity science according to our labeling criteria. The red bar at Fig. 3 shows that the true diversity at the global level is about 17.19. This means that about 17 out of 28 disciplines can be thought to fairly compose the science community. Thus, from a global-level analysis, complexity science is highly multidisciplinary to say the least.

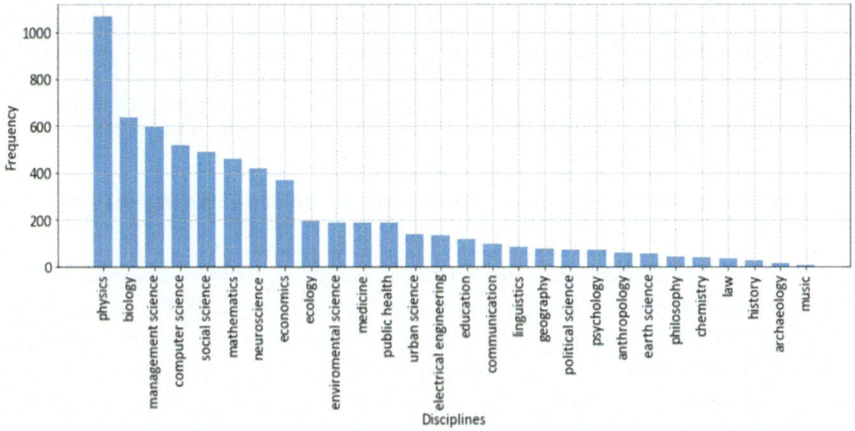

Fig. 2 Number of complexity scientists in each field. Each scholar may be affiliated to more than one discipline

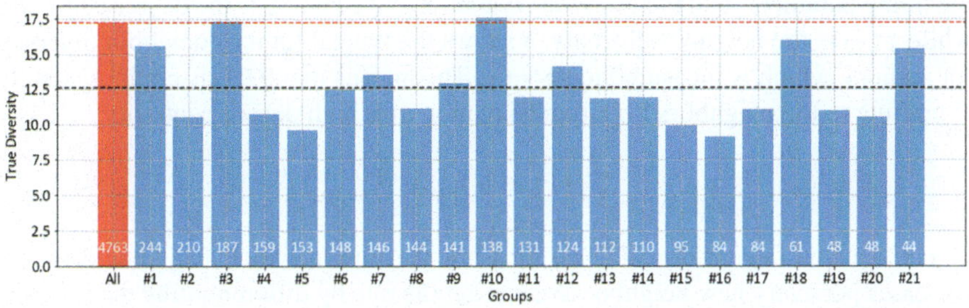

Fig. 3 Comparison of true diversities at the global level and at the group level. Groups are ordered by the size of each subcommunity with the white number for each bar representing the size. The red horizontal line indicates the global diversity while the black horizontal line represents the average of diversity for the 21 communities

4.2 Demonstrating Interdisciplinarity at the Group Level

The diversity for divided communities are mostly low compared to the whole network, but not extremely low since the 21 communities show the mean diversity of 12.58 and minimum of 9.17 with standard deviation of 2.44. Given that communities are divided according to the weighted edges between authors, each community shows the divided social groups based on network properties [26]. Measuring the diversities of these social groups, we can discuss whether authors tend to co-author with researchers from different disciplines. Since the minimum value is 9.17, more than 9 disciplines are equally composed in these social groups. On average, more than 12 disciplines are fairly contributing in a social group of size more than 40. Note that the largest community is less than 6% of the total size and the smallest community is less than 1%. Even in the small-sized social groups, the diversity of disciplines is still maintained, strengthening our argument that the complexity science community is truly interdisciplinary.

4.3 Demonstrating Interdisciplinarity at the Individual Level

We examine how degree centrality relates to neighbor diversity with the goal of showing interdisciplinary interactions at the individual level. In Fig. 4, we fit the relationship with the following exponential variogram function with the r-squared value of 0.6980. At the low degree centrality domain, linear relationship is shown through the orange line. At the high degree centrality domain however, saturation occurs. This shows that the rate of diversity growth per new neighbor decreases exponentially. The two domains show different characteristics, and we can analyze both domains to assert interdisciplinarity.

For the high degree domain, we measure the value of saturation point which is the fitting parameter b in the following equation. Given that x is the degree centrality while $f(x)$ is the neighbor diversity, b shows the typical limit of neighbor diversity for authors, which is measured to be 9.61. This means that 9.61 disciplines equally contribute to the neighbors for a typical author with high degree centrality.

$$f(x) = b - ae^{-cx} \tag{2}$$

For the low degree domain, we measure the rate at which neighbor diversity increases per every new neighbor. We can get the rate by differentiating the function near the origin. In this case, $f'(0) = ac$ should be the rate that neighbor diversity

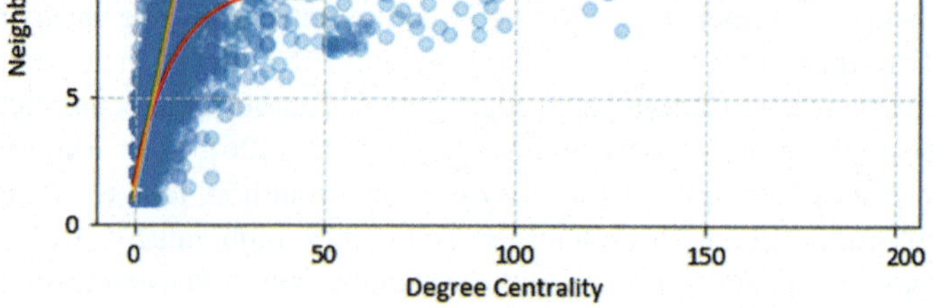

Fig. 4 Relationship between degree centrality and neighbor diversity of each node. The R-squared value for variogram fit is 0.6980 with $b = 9.61$ and $ac = 0.73$ in $f(x) = b - ae^{-cx}$

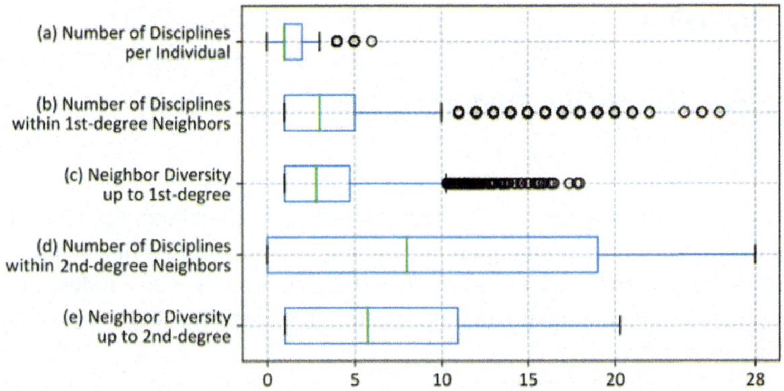

Fig. 5 Distribution of the five measurements for every author in our data

increases before it gets close to the saturation point. This is also the slope value for the orange line at Fig. 4, which is measured to be 0.73. This indicates that authors with low degree centrality or neighbor diversity less than 5 typically experience an 0.73 increase in neighbor diversity per every new neighbor. This means that scholars are comfortable with making connections and co-authoring with colleagues from different disciplines. Also, a few authors follow the orange line even in high neighbor diversity level, proving that there exist authors not experiencing the saturating effect suggested in the variogram fit.

The boxplots in Fig. 5 strengthens our assertion that the complexity science network is interdisciplinary in the individual level. If (c) showed much lower distribution compared to (b), it means that authors have biased distributions in their neighbors' disciplines. Therefore, the very similar boxplots of (b) and (c) show that many authors co-author with scholars in various disciplines, not being significantly biased to some fields.

Furthermore, comparing (a), (b) and (d), we can see that the number of disciplines an author gets to connect with increases significantly when they get to know their first-degree and second-degree neighbors. (d) Shows the whisker at 28, showing that it is not highly unlikely for an author to be connected to all 28 fields from our list. This resembles the small-world network property where it does not take too many steps in the network to connect with a random node [32]. In our case, it is suggested that it does not take many steps to get to know every discipline in complexity science.

5 Conclusion

5.1 Summary of Our Findings

Through a multi-level analysis of the co-authorship network, we demonstrate that the field of complexity science is truly interdisciplinary. Co-authorship network helps us analyze interdisciplinary interactions in a social context rather than a bibliometric context. By connecting network properties with the true diversity index, this study shows how complexity scientists tend to co-author significantly with scholars of different disciplines.

The global level analysis in multidisciplinarity states that out of 28 distinct fields in complexity science, the author network can be thought as a community with about 17 equally contributing disciplines. Our group level and individual level analysis suggests that this multidisciplinarity comes from interdisciplinary interaction between authors.

The group level analysis implies that in every subcommunity with size larger than 40, at least 9 disciplines equally contribute to the subcommunity (Fig. 3). This implies that complexity scientists do not tend to form discipline-based groups such as "complexity physicists," or "complexity economists" that exclude scholars from other disciplines. For the best of our knowledge, this is one of the few co-author-based analyses on measuring interdisciplinarity and the first research in diligently applying community detection methods with their social implication.

For the individual-level analysis, instead of relying on linear fitting or Pearson correlation measurement to explain the relationship between degree centrality and neighbor diversity, we take the saturation point into account and fit the relation to an exponential variogram function (Fig. 4). This shows that scholars with high degree centrality can expect their neighbor diversity almost up to the value of 9.61 while authors with low degree centrality can expect their neighbor diversity to increase by 0.73 for every new connection. This shows how comfortable complexity scientists are in co-authoring with scientists outside of their disciplines. Also, no significant difference in the neighbor diversity (Fig. 5b) and number of fields of neighbors (Fig. 5c) shows how unbiased complexity scientists generally are in terms of collaborating with scholars of various disciplines.

We also introduce new ways of collecting data, labeling authors in disciplines, and measuring interdisciplinarity through network analysis. Unlike the worries that using Google Scholar and author-based analysis is undesirable due to the unavailability of data, we overcome this problem by using the Google Scholar profile information and relying on each author's declared research interests.

Our research can be applied to measuring interdisciplinarity of other science communities. While many studies on interdisciplinarity measurement concentrate on measuring the interdisciplinarity of disciplines, journals or individuals, we measure the interdisciplinarity of the interdisciplinary community itself. This can be combined with other co-authorship network studies that analyzes interdisciplinary research or claim the presence of interdisciplinarity but do not provide quantitative measurement in their network analysis [1, 28]. Furthermore, our methods can be used in social network studies in measuring how diverse a community is in terms of disciplines. For example, social network from a liberal arts institute may be analyzed to demonstrate whether students are familiar with making connections to other students with different majors.

5.2 Limitations and Future Work

Since we collect author data solely from Google Scholar, our analysis may not reflect the characteristic of the whole population of complexity scientists. Given that we rely on each author's own declaration of their interests, the information extracted from our data may lack consistency. Some scholars may declare themselves as complexity scientist or be affiliated at complexity science institutes but do not bother listing the specific word in their Google Scholar profile. Some scholars may want to list as many interests that comes to their mind while others may list only a few subjects they find themselves most devoted into. Also, not weighting the disciplines for authors classified into more than two fields may inaccurately describe their expertise.

To overcome this issue, we can perform text analysis on abstracts of the publication data. By extracting relevant keywords that can be classified into certain disciplines, we can assign a vector to each author representing how close they are to each of the disciplines. Also, this vectorization can be applied to measure distances between individual authors or network communities, giving idea on how a pair of disciplines is close to one another in the field of complexity science.

References

1. Luzar, B., Levnajic, Z., Povh, J., Perc, M.: Community structure and the evolution of interdisciplinarity in Slovenia's scientific collaboration network. PLoS ONE **9**(4), e94429 (2014). https://doi.org/10.1371/journal.pone.0094429
2. Van Noorden, R.: Interdisciplinary research by the numbers. Nature **525**(7569), 306–307 (2015). https://doi.org/10.1038/525306a
3. Wagner, C.S., Roessner, J.D., Bobb, K., Klein, J.T., Boyack, K.W., Keyton, J., Rafols, I., Börner, K.: Approaches to understanding and measuring interdisciplinary scientific research (IDR): a review of the literature. J. Informetrics **5**(1), 14–26 (2011). https://doi.org/10.1016/j.joi.2010.06.004
4. Schummer, J.: Multidisciplinarity, interdisciplinarity, and patterns of research collaboration in nanoscience and nanotechnology. Scientometrics **59**(3), 425–465 (2004)
5. Abramo, G., D'Angelo, C.A., Di Costa, F.: Identifying interdisciplinarity through the disciplinary classification of coauthors of scientific publications. J. Am. Soc. Inform. Sci. Technol. **63**(11), 2206–2222 (2012). https://doi.org/10.1002/asi.22647
6. Carley, S., Porter, A.L.: A forward diversity index. Scientometrics **90**(2), 407–427 (2011). https://doi.org/10.1007/s11192-011-0528-1

7. Cassi, L., Mescheba, W., de Turckheim, É.: How to evaluate the degree of interdisciplinarity of an institution? Scientometrics **101**(3), 1871–1895 (2014). https://doi.org/10.1007/s11192-014-1280-0
8. Leydesdorff, L.: Betweenness centrality as an indicator of the interdisciplinarity of scientific journals. J. Am. Soc. Inform. Sci. Technol. **58**(9), 1303–1319 (2007). https://doi.org/10.1002/asi.20614
9. Leydesdorff, L., Rafols, I.: Indicators of the interdisciplinarity of journals: diversity, centrality, and citations. J. Informetrics **5**(1), 87–100 (2011). https://doi.org/10.1016/j.joi.2010.09.002
10. Leydesdorff, L., Wagner, C.S., Bornmann, L.: Betweenness and diversity in journal citation networks as measures of interdisciplinarity: a tribute to Eugene Garfield. Scientometrics **114**(2), 567–592 (2018). https://doi.org/10.1007/s11192-017-2528-2
11. Porter, A.L., Cohen, A.S., David Roessner, J., Perreault, M.: Measuring researcher interdisciplinarity. Scientometrics **72**(1), 117–147 (2007). https://doi.org/10.1007/s11192-007-1700-5
12. Porter, A.L., Rafols, I.: Is science becoming more interdisciplinary? Measuring and mapping six research fields over time. Scientometrics **81**(3), 719–745 (2009). https://doi.org/10.1007/s11192-008-2197-2
13. Porter, A.L., Roessner, D.J., Heberger, A.E.: How interdisciplinary is a given body of research? Res. Eval. **17**(4), 273–282 (2008). https://doi.org/10.3152/095820208x364553
14. Rafols, I., Leydesdorff, L., O'Hare, A., Nightingale, P., Stirling, A.: How journal rankings can suppress interdisciplinary research: a comparison between innovation studies and business & management. Res. Policy **41**(7), 1262–1282 (2012). https://doi.org/10.1016/j.respol.2012.03.015
15. Rafols, I., Meyer, M.: Diversity and network coherence as indicators of interdisciplinarity: case studies in bionanoscience. Scientometrics **82**(2), 263–287 (2009). https://doi.org/10.1007/s11192-009-0041-y
16. Silva, F.N., Rodrigues, F.A., Oliveira, O.N., and da F. Costa, L.: Quantifying the interdisciplinarity of scientific journals and fields. J. Informetrics, **7**(2), 469–477 (2013). https://doi.org/10.1016/j.joi.2013.01.007
17. Wang, X., Wang, Z., Huang, Y., Chen, Y., Zhang, Y., Ren, H., Li, R., Pang, J.: Measuring interdisciplinarity of a research system: detecting distinction between publication categories and citation categories. Scientometrics **111**(3), 2023–2039 (2017). https://doi.org/10.1007/s11192-017-2348-4
18. Boyack, K.W., Klavans, R., Börner, K.: Mapping the backbone of science. Scientometrics **64**(3), 351–374 (2005). https://doi.org/10.1007/s11192-005-0255-6
19. Glänzel, W., Schubert, A.: Analyzing scientific networks through co-authorship, pp. 99–109. Proc. International Workshop on Webometrics, Informetrics and Scientometrics (2004)
20. Fiore, S.M.: Interdisciplinarity as teamwork. Small Group Res. **39**(3), 251–277 (2008). https://doi.org/10.1177/1046496408317797
21. Stirling, A.: A general framework for analysing diversity in science, technology and society. J. R. Soc. Interface **4**(15), 707–719 (2007). https://doi.org/10.1098/rsif.2007.0213
22. Sayama, H., Akaishi, J.: Characterizing interdisciplinarity of researchers and research topics using web search engines. PLoS ONE **7**(6), e38747 (2012). https://doi.org/10.1371/journal.pone.0038747
23. Zuo, Z., Zhao, K.: The more multidisciplinary the better? The prevalence and interdisciplinarity of research collaborations in multidisciplinary institutions. J. Informetrics **12**(3), 736–756 (2018). https://doi.org/10.1016/j.joi.2018.06.006
24. Jost, L.: Entropy and diversity. Oikos **113**(2), 363–375 (2006). https://doi.org/10.1111/j.2006.0030-1299.14714.x
25. Morillo, F., Bordons, M., Gómez, I.: Interdisciplinarity in science: A tentative typology of disciplines and research areas. J. Am. Soc. Inform. Sci. Technol. **54**(13), 1237–1249 (2003). https://doi.org/10.1002/asi.10326
26. Girvan, M., Newman, M.E.: Community structure in social and biological networks. Proc. Natl. Acad. Sci. U S A **99**(12), 7821–7826 (2002). https://doi.org/10.1073/pnas.122653799

27. Nagpaul, P.S.: Visualizing cooperation networks of elite institutions in India. Scientometrics **54**(2), 213–228 (2002). https://doi.org/10.1023/a:1016036711279
28. Liu, P., Xia, H.: Structure and evolution of co-authorship network in an interdisciplinary research field. Scientometrics **103**(1), 101–134 (2015). https://doi.org/10.1007/s11192-014-1525-y
29. Grauwin, S., Beslon, G., Fleury, É., Franceschelli, S., Robardet, C., Rouquier, J.B., Jensen, P.: Complex systems science: Dreams of universality, interdisciplinarity reality. J. Am. Soc. Inform. Sci. Technol. **63**(7), 1327–1338 (2012). https://doi.org/10.1002/asi.22644
30. Zhang, L., Sun, B., Chinchilla-Rodríguez, Z., Chen, L., Huang, Y.: Interdisciplinarity and collaboration: on the relationship between disciplinary diversity in departmental affiliations and reference lists. Scientometrics **117**(1), 271–291 (2018). https://doi.org/10.1007/s11192-018-2853-0
31. Blondel, V.D., Guillaume, J.-L., Lambiotte, R., and Lefebvre, E.: Fast unfolding of communities in large networks. J. Stat. Mech. Theory Exp. **2008**(10), (2008). https://doi.org/10.1088/1742-5468/2008/10/p10008
32. Milgram, S.: The mind-body problem: the perspective of psychology. Open J. Philosophy **08**(01), 60–75 (1967). https://doi.org/10.1037/e400002009-005

Complexity and Corruption Risk in Human Systems

James K. Hazy

Abstract This paper applies concepts from complexity science to describe corruption risk in the context of various types of uncertainty intrinsic to social interaction systems [12]. This is in contrast to typical approaches that focus on individual ethics and agency issues that are observed during human organizing (cf. [23]). More specifically, this article argues that disequilibrium conditions in organizations [15, 16] emerge as opportunity tension [7] creates value potential which perturbs the alignment of attention among the agent population related to their expectations about the benefits of participating in the organizational system.

As new value potential captures the attention of some agents, disequilibrium conditions increase the complexity of the decision space encountered by each agent. Rather than just going along with others and following the momentum of internal processes, they have alternatives: Does one continue going along with what the organization is doing or take a chance on this new opportunity? On the margin, this dilemma reduces the level of participation in ongoing activities further weakening the system's internal process momentum. This further increases the perceived relative value of the new opportunity. As a consequence of this intensifying value-identifying positive feedback loop [8, 9], more agents consider acting outside the norm in an effort to realize this new value. In contrast, in a healthy organization, information about any new value would be fully integrated into the system.

Corruption enters the calculation when local decisions involve contingent pathways that might disproportionally advantage decision-makers while harming others in the organization [2, 19]. Decisions along potentially harmful pathways support 'value sinks' which, like heat sinks in thermodynamics, drain value from the system. Thus, corruption risk is systemic and is present as networks absorb new information from more complex networks in the ecosystem [24].

J. K. Hazy (✉)
Adelphi University, New York, USA
e-mail: hazy@adelphi.edu

D. Braha et al. (eds.), *Unifying Themes in Complex Systems X*, Springer Proceedings in Complexity, https://doi.org/10.1007/978-3-030-67318-5_27

1 Three Pillars of Structure in the Literature

This article explores the notion of 'corruption risk' in the context of three specific challenges that agents face as they are driven to use organizing as a means to maintain or improve predictability across scale. Recent research, primarily from the computational sciences, has uncovered conditions that enable information exchange among complex networks in the context of scale-free relationships evident in the probability density functions (PDFs) of certain agent states in the population [14, 24, 25]. Hazy [11] argues that agents intentionally act to further these dynamics in the population by constructing what are called structural attractors [1]. These are recognizable physical, social or conceptual objects that increase local predictability for agents who participate in them. One might consider a road sign that identifies one's position, for example.

Hazy [11] identifies three distinct types: spatial (including relative position in an abstract as well as physical space), temporal, or social, according to the type of complexity that each type is intended to navigate. These sources of attraction 'draw' participation by enhancing predictability—and thus value—in the face of three aspects of complexity: uncertainty about 'where,' that is, the spatial complexity regarding where opportunities and threats will be found, uncertainty about 'when' opportunities and threats will occur, that is, the temporal complexity of unfolding events; and uncertainty about 'who,' that is, who know what, who will do what, who can decide what, who will cooperate, and who will compete with the agent. This paper argues that the contingent nature of intentional construction of these structural attractors around newly discovered sources of value offers a new lens to study corruption.

To explore this idea, I propose a conceptual framework that reimagines the definition of value creation. In particular, it defines 'value' as a type of discovered or constructed predictability with respect to access to a relevant object—be it physical, temporal, social, or even abstract—that persists whether it does so in some physical or abstract space, over time, or though social exchange. Notably, as can be seen in Table 1, the tri-partite nature of 'value' in the proposed framework, aligns roughly with the three analogous macro-level structuration processes of Giddens [6], the three analogous micro-level factors of Vroom's [22] expectancy theory, and three of the dynamic states identified by Tunstrøm et al. [21] to define cognition during organizing. The framework also aligns with the mutually reinforcing normalizing processes that Ashforth and Anand [3] suggest operate to embed corruption in organizations.

Using Vroom's language, but framed in the context of human organizing, each of the three types of structural attraction exerts its own putative force called respectively, spatial, temporal or social 'value search'. Each draws agents to participate in the social dynamics that these structures promote enable to increase local predictability during their search for value along spatial dimensions, with respect to the timing of unfolding events, and through social exchange within the larger ecosystem.

Table 1 Structural attraction as constructed predictability creates value

Name	Complexity	Macro	Micro	Predictability and value
Expectancy alignment	Spatial	Signification	Expectancy	Resolves opportunity tension through goals and objectives by focusing attention where predictable opportunities and treats are located
Instrumental momentum	Temporal	Domination	Instrumentality	Identifies value in temporal patterns (e.g., periodicity) that predict events, resource flows, rewards and incentives, or coercion from others
Following valence	Social	Legitimation	Valence	Reduce risk by clarifying who is trusted to share knowledge, or with authority to act as 'us' that includes 'me' when sharing resources

2 Three Structural Pillars of Human Organization

For the purpose of this analysis, I define an 'organization' as an observable potentially relevant conceptual object shared by member agents in the context of predictable interacting dynamically stable scale free patterns of human interaction. Further, I posit that three dynamically stable patterns can be objectively quantified in a population. These are: Expectancy Alignment (EA), Instrumental Momentum (IM), and Following Valence (FV). These patterns emerge as metrics during interactions among subsets of agents as they collectively identify 'value' opportunities for participating agents by exploiting predictable coarse-grain organizational properties [5]. As shown in Fig. 1, these patterns are observed in the context of inverse power law (IPL) relationships within distinct probability distribution functions (PDFs) of identifiable shared dimensions of the agents' decision space.

Each metric emerges as a defined quantity when IPL conditions are such that the statistical mean and variance in the relevant PDF is defined, that is, when the $c \geq 3$ for the critical exponent, $-c$, or when, although the mean is defined, the variance is not finite, that is, when $2 < c < 3$. For example, in cases where $c \geq 3$ for EA in the population, each agent can assume that on average, 'we' are aligned with a particular 'opportunity' in the ecosystem. This emergent predictability related to "like others" along the three relevant dimensions, EA, IM, and FV, each of which crosses scale, facilitates the identification and gathering of resources [8]. These, I argue, are the tri-partite genesis of structural attraction, the social-force that enables human organizing to add predictive value for participating agents. Structural attraction guides the interaction dynamics of value creation observed during interactions among agents.

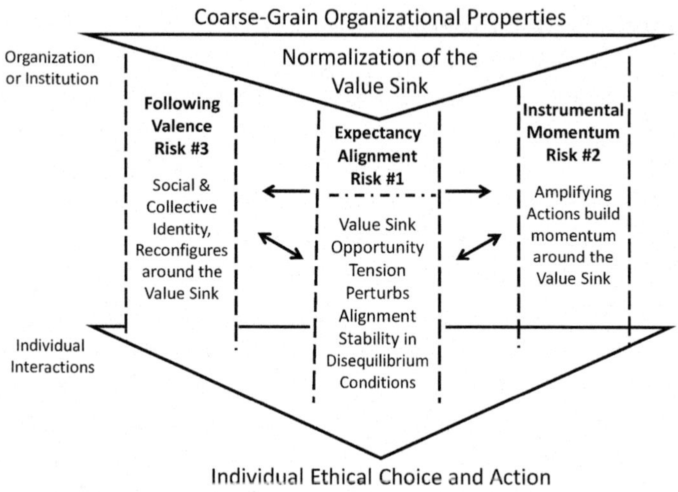

Fig. 1 Corruption risk exists when a stable sub-organization might form around a value sink

To sustain baseline predictability in a population and realize the benefits of social ordering, agents are therefore driven to maintain the characteristic exponent of $-c$, where $c \geq 3$ in the IPL for PDFs of one or more of their EA, IM, or VL states. This is the human equivalent of a 'defensible nest' in evolutionary biology [20]. Thus, for agents, an 'organization' is a zone of higher predictability (and thus less perceived complexity) and is experienced simultaneously by a bounded collection of agents within a complex ecosystem. This analysis suggests three propositions:

Proposition 1: *In a stable organization, the PDF of a metric for 'Expectancy Alignment' of states among subsets of agents in the face of 'opportunity tension' reflects an IPL with critical exponent* $-c$ *such that* $c \geq 3$.

Proposition 2: *In a stable organization, the PDF of 'Instrumental Momentum' of states among subsets of agents in the face of 'informational differences' reflects an IPL with critical exponent* $-c$ *such that* $c \geq 3$.

Proposition 3: *In a stable organization, the PDF of 'Following Valence' of states among subsets of agents who viscerally 'trust' one another reflects an IPL with critical exponent* $-c$ *such that* $c \geq 3$.

3 Corruption Risk in Organizations

Corruption enters the picture when agents build structural attractors that allows locally realized value to be exploited in a way that the new value is lost to the broader reference system. In particular, a value sink is defined as a self-contained 'organization' nested within another organization such that the value created from an opportunity is preserved for 'insiders' of the value sink, and is done in a way that harms others or denies 'outsider' members of the reference organization access to that value. To do this, a renegade subset of organized agents destabilizes one or more of the pillars of the reference organization by increasing the critical exponent, $-c$, beyond the threshold where $c < 2$ making choices and rewards along that dimension less predictable. This creates conditions such that for a rational agent, there is a reasonable loss of confidence in the predictability of existing internal interactions because both mean and variance are no longer defined [24]. As described in more detail in the next paragraphs, risk of corruption, that is, risk that a dissipative value sink structure will emerge to replace lost predictability, occurs as any of these pillars becomes unstable in the variance, i.e. where $c < 3$.

3.1 Risk 1: Expectancy Alignment Instability Increases the Probability that a Value Sink Will Emerge

The first stabilizing pillar that crosses levels of scale relates to mechanisms which help to resolve opportunity tension by identifying and sharing broadly among agents the 'what' and 'where' of an opportunity or threat around which individuals might align. For example, individuals might align toward a water source, marketplace, or warehouse. In stable organizations, agents align with operationally relevant opportunities and goals and form projects that are focused on specific opportunities or threats [10]. Signification structures such as project plans and priority lists, attract individuals to their use, and in their use bias the choices that agents take in ways that to align their attention and expectations with like-others—what would be called 'us'. These activities promote and maintain a significant, observable level of 'expectancy alignment' (EA) in the population.

I posit that when the PDF that measures the level of EA across agent states reflects an IPL pattern with a critical exponent $-c$ such that $c \geq 3$ the pattern enables a level of predictability for each agent with respect to the expectations of other agents in the population. This emergent predictability adds value to the collective by providing relevant information about opportunities in the ecosystem. For example, activities that 'inspire a vision' of the future, or 'set clear objectives' would be examples of individuals using signification and the expectancy alignment pillar to attract agents to the organization's goals as they evaluate their own expectations and make decisions. The emergence of sudden collective order as reflected in covariant choices and action across a population's members has been called "collective cognition" [4].

These activities along with social contagion dynamics [13] align the polarity of fine-grain attention among agents and this literally stores an information bit about the opportunity in its structure. When this bit of information is erased, rational processing by agents is less efficacious which increases the risk that a value sink will emerge. This suggests a proposition:

Proposition 4: *If the IPL of the PDF of Expectancy Alignment among subsets of agents of the organization has a critical exponent* $-c$ *such that* $c < 3$*, then there is risk of corruption.*

3.2 Risk 2: Instrumental Momentum Instability Signals a Nascent Value Sink

The second stabilizing pillar reflects the 'when' of access to existing internal resources and information. For example, this includes processes such as promotions and pay treatment, bureaucratic process structures that distribute resources, and services such as payroll and incentive payments. These scale-free structural mechanisms sustain the momentum of the nested internal dynamics of resource acquisition and distribution as they operate predictably within a population. These domination structures include mechanisms like periodic forcing actions that use, for example, incentive payments and bonuses or the threat of punishment, to sustain and increase temporal oscillations and repeated patterns in the organization. These predictable domination structures attract agents to belong to organizations due to the predictability of trusted institutional events, for example meals, work hours, payroll cycles etc.

Activities that build and sustain 'Instrumental Momentum' promote and sustain an observable level of predictability over time and across scale while leveraging the relative levels of political, economic and positional power of certain individuals across the social network. These structures support agents' choices and actions by storing information about and thus stabilizing the relative predictability of the stocks and flows of requisite resources, like bi-weekly pay checks. These activities build and sustain the instrumentality that engenders collective efficacy. The presence of internal momentum as reflected in changing relative patterns among agents, are elements of "collective cognition" [4]. If this predictability dissipates, corruption risk increases. This analysis suggests a proposition:

Proposition 5: *If the IPL of the PDF of Instrumental Momentum states among subsets of agents of the organization has a critical exponent* $-c$ *such that* $c < 3$*, then there is risk of corruption.*

3.3 Risk 3: Following Valence Legitimizing a Value Sink Signals Structural Reconfiguration

The third stabilizing pillar reflects the 'who' of legitimate access to the benefits and value created by the 'organization' as a distinct ontological entity. For example, the PDF structure of social identity differences among 'us', as well as between 'us' and 'them', and how this distribution, perhaps displayed during rituals, determines claims to the organization's resources, past, present and future, as well as guarantees with respect to the agent's health and safety with respect to various threats internal and external to the organization. Thus, the third stabilizing pillar consists of interactions within a collective that, due to generalized trust or 'valence level' among individual interactions, creates a level of viscosity during social interactions that resists an agent's tendency to go its own way.

This stored information allows agents to form cognitive models that predict persistent levels of covariance in localized choice and action within subgroups. This nested conformity among agents in groups form communities of otherwise independent actors as they imitate or follow one another's choices in complex decision spaces [17, 18]. For example, this would involve knowing who has legitimate authority with respect to resource distribution and thus deserves loyalty and imitation. The "valence" of "us" has the enabling effect of identifying for individuals, the subset of others who "we" can trust, and therefore whom "I" should follow instinctively absent other information. When this structural predictability breaks down, corruption risk increases. This suggests a final proposition:

Proposition 6: *If the IPL of the PDF of Following Valence among subsets of agents of the organization has a critical exponent $-c$ such that $c < 3$, then there is risk of corruption.*

References

1. Allen, P.M., Strathern, M.: Evolution, emergence, and learning in complex systems. Emergence **5**(4), 8–33 (2006)
2. Appiah, K.A.: The Ethics of Identity. Princeton University Press (2005)
3. Ashforth, B.E., Anand, V.: The normalization of corruption in organizations. Res. Org. Beh. (2003). https://doi.org/10.1016/S0191-3085(03)25001-2
4. Couzin, I.D.: Collective minds. Nature **445**(14), 715 (2007)
5. Gell-Mann, M.: What is complexity? In: Curzio, A.Q., Fortis, M. (eds.) Complexity and Industrial Clusters: Dynamics and Models in Theory and Practice. Physica-Verlag, Berlin (2002)
6. Giddens, A.: The Constitution of Society. University of California Press, Berkeley, CA (1984)
7. Goldstein, J., Hazy, J.K., Lichtenstein, B.: Complexity and the Nexus of Leadership: Leveraging Nonlinear Science to Create Ecologies of Innovation. Palgrave Macmillan, Englewood Cliffs (2010)

8. Hazy, J.K.: Leadership in complex systems: a meta-level information processing capabilities that bias exploration and exploitation. In: Proceedings of the 2004 NAACSOS Conference, Carnegie Mellon University, Pittsburgh, PA (2004)
9. Hazy, J.K.: A leadership and capabilities framework for organizational change: simulating the emergence of leadership as an organizational meta-capability. Unpublished Dissertation, The George Washington University, Washington, D.C. (2005)
10. Hazy, J.K.: Leading large: emergent learning and adaptation in complex social networks. Int. J. Complex. Leader. Manag. **14**(1/2), 52–73 (2012)
11. Hazy, J.K.: How and why effective leaders construct and evolve structural attractors to overcome spatial, temporal, and social complexity. In: Sturmberg, J.P. (ed.) Embracing Complexity in Health, pp. 223–235. Springer, Basel, Switzerland (2019). https://doi.org/10.1007/978-3-030-10940-0_13
12. Hazy, J.K., Backström, T.: Human interaction dynamics (HID): foundations, definitions, and directions. Emerg. Complex. Org. **15**(4), 91–111 (2013)
13. Hazy, J.K., Boyatzis, R.E.: Emotional contagion and proto-organizing in human interaction dynamics. Front. Psychol. **6**(806), 1–16 (2015)
14. Hidalgo, C.: Why Information Grows. Basic Books, New York (2015)
15. Leifer, R.: Understanding organizational transformation using a dissipative structure model. Hum. Relat. (1989). https://doi.org/10.1177/001872678904201003
16. Lichtenstein, B.: Generative Emergence: A New Discipline of Organizational, Entrepreneurial, and Social Innovation. Oxford University Press, New York City (2014)
17. Mahmoodi, K., Grigolini, P., West, B.J.: On social sensitivity to either zealot or independent minorities. Chaos Solitons Fractals **110**, 185–190 (2018)
18. Mahmoodi, K., West, B.J., Grigolini, P.: Self-organizing complex networks: individual versus global rules. Front Physiol. **8**, 478 (2017)
19. Mill, J.S.: On Liberty and Other Writings. Cambridge University Press (1859/1989)
20. Nowak, M., Tarnita, C.E., Wilson, E.O.: The evolution of eusociality. Nature **466**, 1057 (2010). https://doi.org/10.1038/nature0920
21. Tunstrøm, K., Katz, Y., Ioannou, C.C., Cristian, C., Lutz, M.J., Couzin, I.D.: Collective states, multistability and transitional behavior in schooling fish. PLOS Comp. Bio. **9**(2), e1002915 (2013)
22. Vroom, V.H.: Work and Motivation. Wiley, New York (1964)
23. Warren, D.E.: Constructive and destructive deviance in organizations. Acad. Manag. Rev. **28**(4), 622–632 (2003)
24. West, B.J.: Fractional Calculus View of Complexity: Tomorrow's Science. CRC Press, Boca Raton, FL (2016)
25. West, G.: Scale: The Universal Laws of Growth, Innovation, Sustainability, and the Pace of Life in Organizations, and Companies. Penguin Press, New York (2017)

A Novel Viewpoint on Social Complexity and the Evolution Model of Social Systems Based on Internal Mechanism Analysis

Wei Wang

Abstract Social systems are composed by people and the environments they lived in. The systems reflect the relationships among people, and the relationships between human beings and environments. Because individuals are with different beliefs, wishes or demands, and the environments are with a series of uncertainties, the complexity of social systems is inevitable. To analyze and understand the complexity is very important. It is helpful or useful for us to model and predict the systems and to reconstruct a better society. In this chapter, some of the definitions and methods, such as, the definitions of the complexity, the components of social systems, hierarchical and multiscale analysis of social systems, networks methods of the social factors, statistical methods, index methods, and so on, are summarized. From the viewpoint of original motivations of the social development, the analysis on the complexity of social systems is considered. For modelling the evolution of the main social indexes, a novel spatial-temporal model is proposed, in which, the spatial part is a fractal structure model which is iterated by the self-interactive process with the external factors, and the temporal part is a diffusion process which is determined by the temporal natural, social, and political conditions.

Keywords Social system · Complexity · Internal mechanism · Spatial- temporal model

1 Introduction

During the long history of social developments, people have been struggling for improving their living conditions and pursuing for their rights of freedom. In such periods, science and technology provide convenient and efficient thoughts, tools, and means for people's working and living. Science and technology also provide more chances or opportunities for people to develop their skills and careers. On the

W. Wang (✉)
School of Mathematics, Renmin University of China, No. 59, Zhongguancun, Haidian, Beijing 100872, China
e-mail: wwei@ruc.edu.cn

D. Braha et al. (eds.), *Unifying Themes in Complex Systems X*, Springer Proceedings in Complexity, https://doi.org/10.1007/978-3-030-67318-5_28

other hand, because of the differences in the possession of natural resources and the abilities to adapt the societies, the conflicts or the fights for the equalities or rights have been going on and on. That is a part of the main reasons causing social complexity. Complexity analysis has been one of the recent emerging topics in the study of animal and human societies [1]. It is the eager for better understanding and describing the social complexity that raises new requirements for us to get to know the social phenomenon and to reveal the internal mechanism of complexity in the procedures of social developments.

In fact, there have been many methods proposed for analyzing the complexity of social systems, for example, the components in social systems (see [1] and the references therein), human cooperations [2], and regression models of social factors (see [3] and the references therein), and the index methods for the comprehensive analysis of the social developments [4], and so on.

Because of inconsistent definitions and operationalization on key social traits of the fundamental components of social complexity, to compare the across studies and species, as distinct components of a social system, the definition and characterization of social organization, social structure, mating system, and care system were discussed [1].

Existing theories are usually formulated as verbal models and, as a result, do not yield sharply defined, quantitative predictions that could be unambiguously tested with data [5]. In [5], a cultural evolutionary model, that predicts where and when the largest-scale complex societies arose in human history, was developed.

A systematic review of statistical physics research was provided to advance our understanding of human cooperations, in particular focusing on the application of Monte Carlo methods, the theory of phase transitions to understand pattern formation, the spatiotemporal dynamics of the systems, and the principles of self-organization that may lead to socially favorable evolutionary outcomes [2].

However, the concept remains both poorly defined and understood [1]. Although the emergence of complex societies has been identified as one of the major transitions in evolutionary biology [6], it is frequently unclear what is meant by 'social complexity' and definitions vary [1].

At the same time, many of the methods are based on the regression analysis on the relevant factors of the social phenomena. That is to say, they are about the statistical analysis or the analysis on the relevance of different factors. There is little concerns about the complexity from the internal mechanism of social phenomena or social systems [7].

In this chapter, firstly, we will give a very brief summary on the analysis methods for social complexity, such as the correlation analysis of factors, the components analysis and the regression models, index methods, and cultural models, etc. and we point out the limitation and challenge of the analysis methods. Secondly, we intend to analyze the complexity from another point of view, i.e., the origin motivations of social systems. Thirdly, we establish the primary model, a novel spatial-temporal model, on the social developments.

The arrangement of this chapter is as follows. In Sect. 2, some of the methods proposed for the analysis of social systems are reviewed briefly. In Sect. 3, from the viewpoint of the original motivations of the social developments, the analysis on the complexity of social systems is considered. In Sect. 4, for modelling the evolution of the social processes derived from the internal mechanism, a novel spatial-temporal model is proposed, in which, the spatial part is a fractal structure model, which is iterated by the self-interactive process with the external factors, and the temporal part is a diffusion process, followed by some conclusions.

2 Summary on Definitions and Methods for the Complexity of Social Systems

As mentioned above, there have been some definitions and methods proposed for the analysis on social systems or social complexity, such as, the definitions of social complexity, the components of social systems, the structure models, cultural development models, regressive models, index methods, and qualifying models, etc. Here, we will give a very brief summary on those methods.

2.1 *Definitions on Social Complexity*

Generally speaking, complexity and simplicity are oppositive. A system is said to be complex if the system is with the following features: (1) the diversity of its components; (2) the hierarchical and multiscale of its structure; (3) the diversity, overlap, dynamic variable of their interactive, and its parts are all neither arranged completely randomly nor completely ordered [1, 6, 8].

According to the very general definition, which also applies to phenomena in physics or molecular biology, complex systems are made up of multiple parts that interact in multiple ways, generating non-linear emergent effects that are greater than the sum of their parts [9, 10].

Social interactions and relationships contribute to more social complexity when interactions take multiple forms and occur in different contexts, when individuals recognize each other, memorize past interactions, have many opportunities to interact with the same individuals again in the future, and if they modify their behavior in response to previous interactions [11].

Such kind of definitions gives us a more intuitive description to the relationship among the different factors or integrants. It is helpful for the analysis of social systems.

Based on the definitions of social complexity, the mode and the potential process of social systems, the definitions and characterization on social organization, social structure, mating system, and care system as distinct components of a social system were discussed [1].

Moreover, aspects of reproduction and parental care have also been invoked to characterize levels of social complexity, so that no single comprehensive measure is readily available [2].

2.2 *The Components of Social Systems and Physical Approach to Complex Systems*

Typically, natural or social systems consist of a large number of nonlinearly interacting elements. These systems are open systems, they interchange information or mass with environment and constantly modify their internal structure and patterns of activity in the process of self-organization. As a result, they are flexible and easily adapt to variable external conditions. However, the most striking property of such systems is the existence of emergent phenomena which cannot be simply derived or predicted solely from the knowledge of the systems structure and the interactions among their individual elements.

It is just the social complexity that raises a demand on the analysis about it. The property related to the complexity points to the holistic approaches which require giving parallel descriptions of the same system on different levels of its organization. There is strong evidence—consolidated also in the present review—that different, even apparently disparate complex systems can have astonishingly similar characteristics both in their structures and in their behaviours. One can thus expect the existence of some common, universal laws that govern their properties.

Physics methodology proves to be helpful in addressing many of the related issues. In the reviews [2, 12], the authors advocate some of the computational methods which in their opinions are especially fruitful in extracting information on selected—but at the same time most representative—complex systems like human brain, financial markets and natural language, from the time series representing the observables associated with these systems. The properties focus on comprising the collective effects and their coexistence with noise, long-range interactions, the interplay between determinism and flexibility in evolution, scale invariance, criticality, multifractality and hierarchical structure [13]. The methods described either originate from 'hard' physics—like the random matrix theory—and then were transmitted to other fields of science via the field of complex systems research, or they originated elsewhere but turned out to be very useful also in physics—like, for example, fractal geometry [14].

2.3 The Hierarchical, the Multiscale Complexity in Networks and Dynamical Networks for Social Systems

In order to reveal the hierarchical structure and the multiscale complexity, in [15], the authors provided a comprehensive review on the features of social networks.

Networks have become a key approach to understanding systems of interacting objects. Meanwhile, many networks are known to possess hierarchical organization, where communities are recursively grouped into a hierarchical structure, see the references in [8, 15].

In [15], the authors reinvented communities as groups of links rather than nodes and shown that this unorthodox approach successfully reconciles the antagonistic organizing principles of overlapping communities and hierarchy. In contrast to the existing literature, which has entirely focused on grouping nodes, link communities naturally incorporate overlap while revealing hierarchical organization. Their results imply that link communities are fundamental building blocks that reveal overlap and hierarchical organization in networks to be two aspects of the same phenomenon.

These networks possess rich meta data that allow us to describe the structural and functional roles of each node. By calculating meta data-based similarity measures between nodes, we can determine the quality of communities by the similarity of the nodes they contain (community quality). Likewise, we can use meta data to estimate the expected amount of overlap around a node, testing the quality of the discovered overlap according to the meta data (overlap quality).

Furthermore, dynamical networks are powerful tools for modelling a broad range of complex systems, including financial markets, brains, and ecosystems. They encode how the basic elements (nodes) of these systems interact altogether (via links) and evolve (nodes dynamics). Despite substantial progress, little is known about why some subtle changes in the network structure, at the so-called critical points, can provoke drastic shifts in its dynamics. This challenging problem can be tackled by introducing a method that reduces any network to a simplified low-dimensional version. It can then be used to describe the collective dynamics of the original system [16].

This dimension reduction method relies on spectral graph theory and, more specifically, on the dominant eigenvalues and eigenvectors of the network adjacency matrix. Contrary to previous approaches, the method is able to predict the multiple activation of modular networks as well as the critical points of random networks with arbitrary degree distributions.

2.4 The Cultural Evolutionary Model

How did human societies evolve from small groups, integrated by face-to-face cooperation, to huge anonymous societies of today, typically organized as states? Why is there so much variation in the ability of different human populations to construct viable states? In [5], a cultural evolutionary model was developed, it can predicts where and when the largest-scale complex societies arose in human history. The central premise of the model is that costly institutions that enabled large human groups to function without splitting up evolved as a result of intense competition between societies—primarily warfare. Warfare intensity, in turn, depended on the spread of historically attested military technologies (e.g.., chariots and cavalry) and on geographic factors (e.g., rugged landscape) [5].

2.5 *Quantifying Social Complexity—Index Methods, Regression Analysis*

Apart from the analysis of structure and components mentioned above, there are some quantitative methods for social systems.

2.5.1 Some Index Models

For a specific question to compare levels of social complexity across species, it is required a scale that ideally integrates measures of both group size and relationship diversity. Most previous studies have focused on either one of these measures, for example, in [17], the authors proposed an index of social complexity that integrated information on interspecific variation in the representation of different age and sex classes. In [18], the authors proposed a continuous index of sociality based upon measures of philopatry, grouping tendencies, and the tendency of individuals to contribute to reproductive altruism.

In [4], the authors proposed the index method for measuring the levels of social developments, especially for that of China.

Other measures of social complexity focused on social relationships. In [19], the authors recommended that social complexity should be measured as the number of differentiated relationships that individuals have, where differentiated relationships are those that can be distinguished by an observer. In [20], the authors proposed a more specific procedure to quantify the diversity of differentiated relationships among group members, but it did not incorporate information on group size either.

In the past, scholars in history, archeology, sociology, anthropology, and other related disciplines have used very different and mostly single measures of social complexity. A recent multi-disciplinary collaborative project that also included evolutionary biologists used a list of 51 variables that reflect the size and hierarchical complexity of social units along with some human-specific features, such as the information and monetary system [21].

2.5.2 Some Regression Models

Classical regressions, such as least absolute shrinkage and selection operator techniques, provide models with linearity, convexity and differentiability attributes, which are highly appreciated for describing systems performance. However, the automatic generation of accurate models for complex systems is a difficult challenge that designers have not yet fulfilled by using analytical approaches [22].

For comparative biology, such as, phylogenies, taxonomies and multi-trait models for continuous and categorical characters, the general quantitative genetic methods were proposed. And regressions were performed while accounting for phylogenetic relatedness among species using MCMC generalized linear mixed models [23].

To understand the relationships among the traits associated with social complexity, and the traits associated with the complexity of social relationships, the following regression models have been built: four models with average levels of kinship as a predictor of each of the traits of organisational complexity; five models with average levels of kinship as a predictor of each of the traits associated with relational complexity; models with absolute brain mass as the response variable and the combined measure of relational complexity as predictor variable on its own, the combined measure of relational complexity together with body mass, and together with body mass and diet/arboreality; and models with absolute brain mass as the response variable and average levels of kinship as predictor variable on its own, average levels of kinship together with body mass, and together with body mass and diet/arboreality [3].

2.6 Statistical Physics Methods of Human Cooperations

Human cooperations are the results of their struggling for survival. In a reality world, extensive cooperation among unrelated individuals is unique to humans, who often sacrifice personal benefits for the common good and work together to achieve what they are unable to execute alone. The evolutionary success of our species is indeed due, to a large degree, to our unparalleled other-regarding abilities. Yet, a comprehensive understanding of human cooperation remains a formidable challenge. Recent research in the social sciences indicates that it is important to focus on the collective behavior that emerges as the result of the interactions among individuals, groups, and even societies. Non-equilibrium statistical physics has proven to be very valuable for understanding counterintuitive evolutionary outcomes. The related references can be found in [2].

The analytical methods mentioned above are based on the statistical analytical methods. They depend on the sample obtained. At the same time, they cannot reflect the master-slave relationship among the factors, and they cannot reflect the motivations for the evolution of social systems. It means that the methods don't reflect the causality of social systems.

2.7 The Challenges of the Analysis on Social Systems

As to the challenges on the analysis of social systems, first of all, the problem is that there is no proper model to reflect the essential characteristics from the internal mechanism of social systems [7].

The second one is that the data in social computation are all with serious uncertainties. Models based on traditional methods of time series or stochastic methods are often depending on the obtained samples, which will seriously influence on the adaptability of the model to the drastic changing of evolutionary processes, and they will also influence on the accuracy of estimation or prediction [24].

At the same time, with the development of science and technology, especially the information technology, the appearance of new technology, such as, big data, machine learning, artificial intelligence, and block chains, etc. the advances provide new possibility for social activities, they also bring about new problems for the innovation of analysis on social systems.

It is based on the statement above, we intend to provide a novel viewpoint on social complexity and a new spatial-temporal model for the evolution of social systems based on the the theory related to the evolution processes of polymer [25] and the self-interactive systems with external factors [24], which with the feature of self-interactive processes that can reflect the behavioral characteristics of humans.

3 The Novel Viewpoint on Complexity Analysis of Social Systems

As mentioned above, there have been many aspects about the analysis on social complexity. However, many of the analysis concern about the phenomenon, the performance of the complexity. There has been little concerns about the causes or the motivations of complexity.

3.1 The Analysis on the Causes of the Social Developments

Generally speaking, the evolutionary history of human societies is the history of human beings fighting against the natural environments and struggling with themselves. It is the strong aspiration for improving the natural environments and extending the living space that makes great progress of science and technology. The designing and manufacturing of tools make the people having more ability to provide more materials they want, and having the ability to go to a longer distance or to reach a wider space. The pursuits for more ability and more resources make the society moving forward. In fact, to pursue the more ability of developments, the progress of science and technology has been pushing forward forever. Science and technology provide the way for developments.

At the same time, the development of technology has expedited the development of business. The companies provide the means of the materials and facilities we need. Human beings are not only the providers bus also the consumers. The interaction of consumers and providers is also one of the main reasons for cooperations and social developments. It is the social developments that lead to the emergence of the economic, commercial, and financial sectors, and even to the promoting of human civilizations.

On the other hand, owing to the weakness of humanity, humans' desires, greed, and selfishness have caused many tragedies and regrets. It is believed that there is only permanent interests but no friends forever. For interests (resources, or money), someone may give up family and friendship. Even if they are members of family or friends, owing to the conflict of gains, they may go to court. Because of the limited resources, in order to occupy the more resources, the conflicts and fights between groups may be inevitable, the countries may get into war.

In fact, people also seek for the methods to fight against the weakness of human beings, in the early periods of the history, to eliminate the evil, the Jesus Christ and the God were created in the western, the Heaven and the Jade Emperor were created in the east countries. In a reality world, to avoid the confuse, people established the social systems and political rules to regulate the society and to ensure the developments and progresses.

Under the conditions of having common beliefs or interests, individuals may come together to form groups, to cooperate to achieve their targets. The pursuit of interests and the difference among of individuals or groups are the original motivations of competition. On the contrary, it is the competition that may also improve the social developments.

3.2 *Hierarchical Structure of the Social Components*

It is just because of the limitations of a single person that people must join together to form a joint force, to overcome greater troubles or obstacles they meet. Human beings have the feature of sociality. The cooperations of people can make them the best of both sides, promote the production, resist foreign aggression powerfully, and so on. It is just the proper organization that we can obtain the effect, i.e., one plus one is greater than two. Therefore, people have formed a series of organizations.

Usually, social organizations are in the form of hierarchical structure. The structure has more scales, from a single person, to a small family, to a larger family or family in law, to a friendship circle, to a community, to a district, to a city, to a sate, to a country, to a continent, to the globe. It is called the pyramid of societies. However, each scale will have its complexity.

With the development of information technology, there appear more and more social contacts or social circles. From the viewpoint of scientific analysis, the relationships can be expressed by graphs or networks, and such kinds of graphs are dynamical ones.

The emergence of some high-tech companies, such as, Google, Facebook, Tencent, etc. has made people's communication and contact in a more and more convenient and efficient way. Such kinds of technical development also prompt the social development in a new form. That also makes social analysis in a new stage, i.e., social computation.

3.3 The Viewpoints on Social Complexity from the Origins of the Systems

In fact, any social system is a complex system. The main reason is that they involve many different types of people, groups, organizations, and even countries. It is a complicated system because different individuals have different backgrounds, experiences, cognitive levels, and different persons have different views on world's view, values, and outlooks on life. Those factors lead to great differences in attitudes, material needs, contributions to the societies, the senses of identity of the communities, and even the identity of countries. It is because of the differences of those factors that promote individuals' growths and progresses at different levels. At the same time, different individuals will be restricted by different families, regions, countries and social rules. In short, these differences will also restrict the levels and degrees of developments of persons, groups, organizations, and even countries.

The complexities of higher scales depend also on the complexities of their lower scales. That is to say, as the groups, the states, or even countries, the complexities come from the forms of the organizations, political rules, the levels of developments, and even the natural resources.

These factors are the main components of causing the complexities of the societies. So to model the social system, the original reasons should be considered. For a single person, he or she can be regarded as a feedback system, the beliefs, pursuits and experiences are the input variables, and natural environments, social rules, ethics are the restricted conditions, the index function is the level of growth. So each individual is a closed-loop system. Any scale of the groups is also a closed-loop system. The integrated society is, of course, a complex feedback system. The differences of those systems are the levels and degrees of complexities.

4 The New Strategy for Modelling the Evolution of Social Systems

To solve the problem of modelling the evolution of social systems from the internal mechanism of the systems, we intend to build the model from two aspects, such as, the spatial one and the temporal one.

4.1 The Basic Consideration About the Model

Social systems are composed by all kinds of individuals, groups and the environments they lived in. How can we build a proper model for the evolution of the systems?

Here, we solve the problem from another point of view. We imagine the scenario that a group of ships sail through the flowing river filled with dangerous shoals and rapids. From the viewpoint of the spatial aspect, a lot of people gather in a series of huge cabins. They will cooperate in certain structure to perform the restrictive tasks. From the viewpoint of the temporal one, the ships are being organized tightly, going or surfing forward. That is to say, on one hand, they form certain social structure. On the other hand, they move in the flowing river of the social history. It is such kind of consideration that we can model the society system as a complex structure moving in a changing environment. So, the evolution of social systems can be modeled by certain kind of diffusion processes [26].

From the viewpoint of us, the spatial structure of social systems will be the fractal structure [7]. It is modeled by some multifractal structure, such as, Brownian motion iterated by a lot of Brownian motions [14]. That is to say, in the lower scale, the model is the one of self-interactive process with certain external factors [24], and in the more higher scale, the model is still the self-interactive process with certain external factors, which is iterated by the lower scale one. Meanwhile, the temporal part of the social systems is determined by the macro conditions, such as, the natural environments, development levels, economic strengths, etc.

Owing to the essential characteristics of individuals are the self-interactive with himself/herself, to have a better understand for the novel model, we would like to have an interpretation to the self-interactive processes at first.

4.2 The Self-interactive Diffusion Process

More precisely, self-interacting diffusions have been first introduced by Norris, Rogers, and Williams under the name of Brownian polymers [27]. They proposed a model for the shape of a growing polymer. This process has been studied by different authors, such as [25]. They shown, in particular in the self-attracting case, that x converges almost surely. Later, another model of polymers has been proposed by Benaïm et al. [28]. The model was proposed as the following stochastic differential equation:

$$dx_t = -\left(\int_0^t f(x_t - x_s)ds\right)dt + dB_t + du_t \tag{1}$$

where x_t refers to the state of the main index, B_t refers to the Brownian motion, f is a proper function, and u_t is an interactive factor.

In fact, for the study of processes which are with path-interaction since the seminal work of Norris, Rogers, and Williams [27], self-interacting stochastic processes have been studied for many years from different aspects, and they have been an intensive research area. It has been considered from the modelling and asymptotic analysis of polymer to more abstract self-interaction processes.

4.3 The Self-interactive Process with External Factors for Personal Growth

In the field of social sciences, the subjects are individuals with different kinds of beliefs, emotions, and requirements. It is the differences of persons making the social systems more and more complex. How to describe the evolution of society, in [24], based on the idea of self-interactive systems, and also based on the dialectics of philosophy, the model of self-interactive systems with external factors was proposed as follows:

$$dx_t = -\left(\int_0^t f(x_t - x_s - Y_s)ds\right)dt + dB_t + du_t \tag{2}$$

where x_t is the state of the main index, it is determined by the beliefs, the pursuits, u_t is the interactive factors from the others, B_t refers to the Brownian motion, Y_t is the external factors to the index, for example, the political or natural constraints, moral rule's factors, and f is a proper function. Of course, the function f may be different for different persons or differnent groups.

The model proposed above is an original mathematical model on the the evolution of individuals based on the dialectical relationship between internal and external factors [29].

So, from the viewpoint of systems science, the social systems are composed by people and the environments. For a single person, the input is the belief, the effort, the political rule, the interactions with others, and restricted conditions from natural resources, etc. The desired inputs are the targets, pursuits, and the outputs are the certain comprehensive indexes. The error between the output and the desired input can be used as the feedback to regulate the system.

In fact, there are a lot of things for us to do, especially with the development of information technology, the consideration about the personality will becoming more possible and popular.

4.4 The New Model of Social Systems Based on the Self-interactive Process with External Factors

As to the evolution of social systems, it is more complex than that of a single person. For a group, it is the integrated model based on the models of single persons. At the same time, there are so many indexes needs to be considered.

Here, we propose a new model, which will contain the internal or external factors, for the evolution of the social systems.

4.4.1 The Essential Factors of Social Enterprises

As discussed above, the complexity of social systems, especially for companies, can be summarized as follows: there are contradictions between seizing opportunities and limited attentions; between pioneering or innovating and achieving the established goals; between focusing on self-interest and willingness to dedicate.

For a more detail description on the factors, we give an explanation as follows.

Competitive environment: in a competitive environment, creativity and initiative of employee are very important to the success of a business. New organizational forms and knowledge are competitive assets.

Goal: using adversarial forces to achieve a dynamic balance and getting the maximum return on management (ROM).

The law of balance: coordinating various constraints formed by organization, business strategy and human behavior. Restrictions must be balanced.

That is to say, for individuals or groups, the value and meaning of life can be realized by the intertwining of a series of characteristics, such as, targeted, organized, constantly updated, and the mixed emotions, such as, depression after defeat, joy after victory, etc.

So, we will consider the model for certain indexes from another point of view. That is to say, except for the conventional consideration of self-interactive processes, we also need to consider the possible organization forms, the targets, and the restrictions stated above for the evolution of the enterprises, and the influenced by certain impels from outside, for example, the natural environments or factors, the social rules, political rules, culture, economics, and so on.

And then we combine all the indexes together by the proper method.

4.4.2 The Evolution Model of Social Groups

As to the understanding of group interactions related to decision making, it may be advanced by considering a limited typography of communication acts [30].

As a matter of fact, the evolution of social groups can be compared with the organized sports teams. As a football team, for example, every player, whether he played or not, no matter how long he played, no matter how many goals he scored, is an indispensable part of the entire team. And, if the entire team is flawless, it should be the one with the features, such as, the excellence of each player, the effective combination of the whole team, and the creation of the leader. The players in each position must be on their own, facing the pressure from their opponents, and they must respond to teammates in their adjacent positions to form a team. Such an operational pattern is exactly the best model of human being's social operation after the post-industrial revolution.

The team's combat effectiveness will depend on the following features. Each of them has a division of labor and cooperation, i.e., each one should be not only sincere and united but also stand on his/her own. Each individual player has excellent special technology. At the same time, they and their surrounding companions can also produce an efficient response, and an efficient football team is based on complete personal capabilities. They have integrated almost everyone into the collective ones without obvious loopholes. This mode of operation is like a precision-running machine that roars and is invincible. For a football team, such kind of operation coincides with the working principle of human society after the industrial revolution.

As mentioned above, social groups are also with the hierarchical organization or multiscale structure, at the lower scale, the system is the one with the characteristics of self-interactive with external factors, and at the higher scale, the system is also the self-interactive with external factors. So the groups or companies are with the fractal structure. The difference with the former research is that the basic unit is the one with the characteristics of self-interactive system with external factors. The systems are more like self-interactive process with external factors iterated by self-interactive process with external factors. Such a structure forms a social pyramid structure.

As a group, a family or basic unit of society, in addition to the the close-loop of himself/herself as the basic elements, there are interactive among themselves. To an upper level, the interactive will be governed by the organization rule. All the above components will join together to form a social structure.

Finally, all of those structures integrated to form the whole society. That is the spatial model we propose.

Meanwhile, such structure is moving in a changing environment. And the evolution of whole society will be governed by natural environments, political rules, economical factors, or the general development speeds, behaviour or moral rules of individuals. They are the temporal part of the whole model.

For the more details on the model of special groups, such as, the estimation of f in (2), the infinitesimal generator of the diffusion processes, and so on, we will consider in the future.

5 Conclusions

In this chapter, some of the definitions and methods, such as, the defining on the social complexity, the hierarchical structure of social systems, the methods on multiscale analysis, networks method of the social factors, index methods, statistical methods, and so on, are summarized. The analysis on the complexity of social systems is considered from the original motivations or internal mechanism of the society. For modelling the evolution of the main social indexes, a novel spatial-temporal model, in which the spatial part is fractal structure model, which is iterated by self-interactive processes with the external factors, and the temporal part is a diffusion process, is proposed. It should be mentioned that, in this chapter, we just provide a structure or framework for the model on social evolution, it will be refined and improved in the future.

References

1. Kappeler, Peter M.: A framework for studying social complexity. Behav. Ecol. Sociobiol. **73**, 13 (2019)
2. Perc, M., Jordan, J.J., Rand, D.G., Wang, Z., Boccaletti, S., Szolnoki, A.: Statistical physics of human cooperation. Phys. Rep. **687**, 1–51 (2017)
3. Lukas, D., Clutton-Brock, T.: Social complexity and kinship in animal societies. Ecol. Lett. **21**, 1129–1134 (2018)
4. Peng, F., Zhang, Y.: The Measure Theory for Development: Chinese Development Practice and Reflection. Chinese Social Science Press, Beijing, China (2012)
5. Turchin, P., Currie, T.E., Turner, E.A.L., Gavrilets, S.: War, space, and the evolution of Old World complex societies, PNAS, October 8, vol. 110, no. 41 (2013)
6. Maynard Smith, J., Szathmáry, E.: The Major Transitions in Evolution. Oxford University Press, New York, NY (1997)
7. Wang, W.: The complex system's method on financial science and technology. In: Proceedings of The First International Forum on Financial Mathematics and Fianncial Computing, Suzhou, China (2019) (to appear)
8. Siegenfeld, A.F., Yaneer, B.-Y.: An Introduction to Complex Systems Science and its Applications (2019) (submitted)
9. Wang, Y., Zheng, S., Zhang, W., et al.: Fuzzy entropy complexity and multifractal behavior of statistical physics financial dynamics. Phys. A **506**, 486–498 (2018)
10. Wang, Y., Shang, P., Liu, Z.: Analysis of time series through complexity—entropy curves based on generalized fractional entropy. Nonlinear Dyn. **96**, 585–599 (2019)
11. de Waal, F.B.M., Tyack, P.L.: Animal Social Complexity: Intelligence, Culture, and Individualized Societies. Harvard University Press, Cambridge (2003)
12. Kwapiea, J., Droda, S.: Physical approach to complex systems. Phys. Rep. **515**, 115–226 (2012)
13. Oh, G.: Multifractals of investor behavior in stock market. J. Korean Phys. Soc. **71**, 19–27 (2017)
14. Falconer, K.: Fractal Geometry, Mathematical Foundations and Applications, 2nd edn. Wiley (2003)
15. Ahn, Y.-Y., Bagrow, J.P., Lehmann, S.: Link communities reveal multiscale complexity in networks. Nature **466**, 761–765 (2010)
16. Laurence, E., Doyon, N., Dubäe, Louis J., Desrosiers, P.: Spectral dimension reduction of complex dynamical networks. Phys. Rev. **X**(9), 011042 (2019)
17. Blumstein, D.T., Armitage, K.B.: Life history consequences of social complexity: a comparative study of ground-dwelling sciurids. Behav. Ecol. **9**, 8–19 (1998)
18. Avilés, L., Harwood, G.: A quantitative index of sociality and its application to group-living spiders and other social organisms. Ethology **118**, 1219–1229 (2012)
19. Bergman, T.J., Beehner, J.C.: Measuring social complexity. Anim. Behav. **103**, 203–209 (2015)
20. Fischer, J., Farnworth, M.S., Sennhenn-Reulen, H., Hammerschmidt, K.: Quantifying social complexity. Anim. Behav. **130**, 57–66 (2017)
21. Turchin, P., Currie, T.E., Whitehouse, H., Francois, P., Feeney, K., Mullins, D., Hoyer, D., Collins, C., Grohmann, S., Savage, P., Mendel-Gleason, G., Turner, E., Dupeyron, A., Cioni, E., Reddish, J., Levine, J., Jordan, G., Brandl, E., Williams, A., Cesaretti, R., Krueger, M., Ceccarelli, A., Figliulo-Rosswurm, J., Tuan, P.J., Peregrine, P., Marciniak, A., Preiser-Kapeller, J., Kradin, N., Korotayev, A., Palmisano, A., Baker, D., Bidmead, J., Bol, P., Christian, D., Cook, C., Covey, A., Feinman, G., Jlusson, D., Kristinsson, A., Miksic, J., Mostern, R., Petrie, C., Rudiak-Gould, P., ter Haar, B., Wallace, V., Mair, V., Xie, L., Baines, J., Bridges, E., Manning, J., Lockhart, B., Bogaard, A., Spencer, C.: Quantitative historical analysis uncovers a single dimension of complexity that structures global variation in human social organization. P Natl. Acad. Sci. USA **115**, E144–E151 (2018)
22. Arroba, P., Risco-Martn, J.L., Zapater, M., Moya, J.M., Ayala, J.L.: Enhancing regression models for complex systems using evolutionary techniques for feature engineering. J. Grid Comput. **13**, 409–423 (2015)

23. Hadfield, J.D., Nakagawa, S.: General quantitative genetic methods for comparative biology: phylogenies, taxonomies and multi-trait models for continuous and categorical characters. J. Evol. Biol. **23**, 494–508 (2010)
24. Wang, W.: The method on modifying the dynamic properties of self-interactive systems by external factors. Appl. Mech. Mater. **109**, 410–414 (2012)
25. Durrett, R.T., Rogers, L.C.G.: Asymptotic behavior of Brownian polymers. Prob. Theo. Rel. Fields **92**, 337–349 (1992)
26. Stroock, D.W., Varadhan, S.R.S.: Multidimensional Diffusion Processes. World Book Inc., Beijing, China (2009)
27. Norris, J.R., Rogers, L.C.G., Williams, D.: Self-avoiding random walk: a Brownian motion model with local time drift. Prob. Theory Rel. Fields **74**, 271–287 (1987)
28. Benaïm, M., Ledoux, M., Raimond, O.: Self-interacting diffusions. Prob. Theory Relat. Fields **122**, 1–41 (2002)
29. Wang, W.: The conceptual models for the growth of individuals based on the viewpoint of philosophy. In: Proceedings of the 2014 Asia-Pacific Humanities and Social Sciences Conference, Shanghai, China (2014)
30. Yaneer B.-Y., David, K.: A mathematical theory of interpersonal interactions and group behavior. NECSI Rep. (2018)

On Generalized Langevin Dynamics and the Modelling of Global Mean Temperature

Nicholas W. Watkins, Sandra C. Chapman, Aleksei Chechkin, Ian Ford, Rainer Klages, and David A. Stainforth

Abstract Climate science employs a hierarchy of models, trading the tractability of simplified energy balance models (EBMs) against the detail of Global Circulation Models. Since the pioneering work of Hasselmann, stochastic EBMs have allowed treatment of climate fluctuations and noise. However, it has recently been claimed that observations motivate heavy-tailed temporal response functions in global mean temperature to perturbations. Our complementary approach exploits the correspondence between Hasselmann's EBM and the original mean-reverting stochastic model in physics, Langevin's equation of 1908. We propose mapping a model well known in statistical mechanics, the Mori-Kubo Generalised Langevin Equation (GLE) to generalise the Hasselmann EBM. If present, long range memory then simplifies the GLE to a fractional Langevin equation (FLE). We describe the corresponding EBMs that map to the GLE and FLE, briefly discuss their solutions, and relate them to Lovejoy's new Fractional Energy Balance Model.

Keywords Climate sensitivity · Hasselmann model · Langevin equation · Fractional derivative · Long-range dependence

---

N. W. Watkins (✉) · S. C. Chapman · D. A. Stainforth
University of Warwick, Coventry CV4 7AL, UK
e-mail: nickwatkins62@fastmail.com
URL: http://warwick.ac.uk/fac/sci/physics/research/cfsa/people/watkins/

N. W. Watkins · D. A. Stainforth
London School of Economics and Political Science, London WC2A 2AE, UK

N. W. Watkins
The Open University, Milton Keynes MK7 6AA, UK

A. Chechkin
Akhiezer Institute for Theoretical Physics, Kharkov, Ukraine

Institute for Physics and Astronomy, University of Potsdam, Potsdam, Germany

I. Ford
University College London, London WC1E 6BT, UK

R. Klages
Queen Mary University of London, London E1 4NS, UK

D. Braha et al. (eds.), *Unifying Themes in Complex Systems X*, Springer Proceedings in Complexity, https://doi.org/10.1007/978-3-030-67318-5_29

1 Introduction and Aims of Paper

The importance of modelling the response of the Earth's climate to external perturbations, whether anthropogenic or solar, and internal fluctuations, is well recognised. It has been addressed by models with many complementary levels of complexity [1], from general circulation models to very simplified energy balance models [2]. Key steps included the first deterministic EBMs, and Hasselmann's development [3–5] of these into a stochastic differential equation with added white noise. The Hasselmann model was far from the last word in stochastic modelling for climate applications, however, and a review of progress in this area was given by the contributors to a recent book [6]. Of particular note has been the extensive study of multivariate stochastic models and statistical inference (e.g. [7]), and stochastic modelling [8, 9] in climate science using the Mori-Zwanzig formalism which provides a very general mathematical framework for the Langevin equation [10, 11]. Some of this work has benefitted from a formal equivalence between the Hasselmann model and the driven Langevin equation of statistical mechanics, and much of it has been done while still assuming that the noise terms are spectrally white. This greatly eases the mathematical tractability by linking the problem to the most mature parts of stochastic calculus (e.g. [11]). However there have long been physically-motivated arguments for making the driving noise term itself spectrally red rather than white. In particular Leith argued more than 25 years ago [12] that: "unfortunately there is evidence that [a model of the white noise-driven Langevin type] would be unsatisfactory to capture some of the low frequency phenomena observed in the atmosphere. This is referred to as the *infrared climate problem* and appears to be caused by nonlinear interactions of the chaotic internal weather frequencies that potentially induce a "piling-up" of extra variance at the low frequencies".

More recently evidence has been given for long-range dependence (LRD, [13]) in the climate system based on "1/f" power spectra and other diagnostics [14]. An alternative approach to using a "1/f" noise term in a stochastic EBM has been to propose non-exponential, long-tailed response functions in the time domain. These were introduced by Rypdal [15] in a fractional Brownian setting. This pioneering paper was followed by further work from the Tromso group, notably [16] and [17]. More recently several papers by Lovejoy and co-workers (e.g. [18]) have demonstrated the utility of a fractional Gaussian model (StocSIPS) for climate prediction on monthly and longer time scales.

As Leith's comments implied, however, the physical arguments for extending the Hasselmann formalism are much more general than those which motivate LRD, and there are many other well-known features of the climate system [6, 14] including periodicities and apparent spectral breaks that also need to be captured. So an ideal but still simple stochastic EBM would permit a more general range of dependency structures than either just the shortest possible range (Markovian) or longest range (LRD) behaviours, while allowing both as limiting cases. Just such a formalism has long been available in statistical mechanics [10, 19–22], the Mori-Kubo generalised

Langevin equation. This is a natural extension of the Langevin equation whereby an integral over a kernel replaces its constant damping term.

In this paper we propose extending the Hasselmann model by this GLE-inspired route, and show what the analogous generalised Hasselmann equation would be. We then recap how taking a power-law form for the damping kernel arrives at the fractional Langevin equation studied by researchers in topics including anomalous diffusion, and write down a corresponding fractional Hasselmann equation. Fractionally integrating both sides of this allows us to identify how Lovejoy et al.'s new [23, 24] fractional energy balance equation (FEBE) relates to our scheme.

2 Deterministic and Stochastic Energy Balance Models

The simplest energy balance models are deterministic. They describe the (assumed linear) change $\Delta T(t)$ in the Earth's global mean temperature $T(t)$ away from its radiative equilibrium value T_0 (i.e. the anomaly in GMT) due to changes in the net radiative forcing $F(t)$. ΔT is multiplied by the effective heat capacity C to give the change ΔQ in the heat content of the earth system, while the feedbacks in the climate system are lumped into a single parameter λ, to give a one-dimensional deterministic EBM:

$$\frac{d\Delta Q}{dt} = C\frac{d\Delta T}{dt} = -\lambda \Delta T(t) + F(t) \tag{1}$$

Hasselmann's model was a key step in improving the realism of this EBM, by incorporating delta correlated ("white") noise $\xi(t)$:

$$< \xi(t)\xi(t+t') >= \sigma_Q^2 \delta(t-t') \tag{2}$$

giving rise to what we will refer to as the driven Hasselmann equation (DHE):

$$(C\frac{d}{dt} + \lambda)\Delta T(t) = F(t) + \xi(t) \tag{3}$$

3 The Hasselmann-Langevin Correspondence

The DHE is the simplest possible Markovian stochastic linear EBM, and as such, isomorphic to the driven Langevin equation (DLE) of statistical mechanics [20, 25], for the velocity of a particle executing Brownian motion:

$$(M\frac{d}{dt} + \eta)V(t) = F(t) + \xi(t) \tag{4}$$

Table 1 A lookup table expressing the isomorphism between the driven Langevin equation (3 left columns) and the driven Hasselmann equation (3 right columns)

Quantity	Symbol	Units	Quantity	Symbol	Units
Velocity	$V(t)$	(m s^{-1})	GMT anomaly	$\Delta T(t)$	(K)
Initial V	v_0	(m s^{-1})	Initial ΔT	ΔT_0	(K)
Drift V	v_d	(m s^{-1})		F/λ	
Mass	M	(kg)	Heat capacity	C	(J K^{-1} m^{-2})
Damping rate	η/M	(s^{-1})		λ/C	(s^{-1})
Noise strength	σ_v/M	(m $s^{-3/2}$)		σ_Q/C	(K $s^{-1/2}$)

where we followed the convention of [25] that a stochastic variable is upper case while known values of the variable are lower case. The correspondence is well-known in climate science, and can be expressed via a lookup table (Table 1):

It should be noted that the Hasselmann model is sometimes used without a deterministic forcing term $F(t)$ (e.g. [26]). We will refer to this as the undriven Hasselmann equation (UHE):

$$(C\frac{d}{dt} + \lambda)\Delta T(t) = \xi(t) \tag{5}$$

and note that it is of course in correspondence with the undriven Langevin equation

$$(M\frac{d}{dt} + \eta)V(t) = \xi(t) \tag{6}$$

whose solution [11, 25, 27] is the well-known Ornstein-Uhlenbeck process.

In standard texts such as [27] it is shown that in the undriven $F = 0$ case, and for a solution started at $t = 0$ with velocity v_0, the velocity process of the ULE is

$$V(t) = v_0 e^{-(\eta/M)t} + \frac{1}{M}\int_0^t e^{-(\eta/M)(t-\tau)}\xi(\tau)d\tau \tag{7}$$

The equivalent solution for the temperature process in the UHE is well-known to be:

$$\Delta T(t) = \Delta T_0 e^{-(\lambda/C)t} + \frac{1}{C}\int_0^t e^{-(\lambda/C)(t-\tau)}\xi(\tau)d\tau \tag{8}$$

As noted by [27] the use of the alternative initial condition whereby the evolution starts at $t = -\infty$ means that the first term in the above can be dropped, giving

$$V(t) = \frac{1}{M} \int_{-\infty}^{t} e^{-(\eta/M)(t-\tau)} \xi(\tau) d\tau \tag{9}$$

and

$$\Delta T(t) = \frac{1}{C} \int_{-\infty}^{t} e^{-(\lambda/C)(t-\tau)} \xi(\tau) d\tau \tag{10}$$

respectively. The extension of the solution to the case with a nonzero deterministic component of forcing is discussed e.g. by [17].

4 Non-Markovian Models

We noted in the introduction, however, that in both condensed matter physics and climate science we may not always (e.g. [10, 28]) be entitled to make the (often very good) approximation that the noise term $\xi(t)$ is white and that the response function is exponential. Debate continues [1, 8] in climate research about when a fluctuation-dissipation theorem applies (or a priori could apply), while in condensed matter applications both internal and external noises are studied [21, 22], so will not prescribe the power spectrum of $\xi(t)$ in what follows. Since the 1960s one standard way to handle such cases in statistical mechanics has been to use the full GLE rather than the Langevin equation which approximates it. In use the initial time a is often be taken to be 0 or $-\infty$ as discussed by e.g. [10]. Following the notation of [29] but with $U'(x) = F(t)$ we have

$$M \frac{d}{dt} V(t) + M \int_{a}^{t} \gamma(t-\tau) V(\tau) d\tau = F(t) + \xi(t) \tag{11}$$

In the special case that the response is instantaneous and has no memory we have that the damping kernel $\gamma(t) = (\eta/M)\delta(t)$ and the ordinary DLE is retrieved.

By means of the correspondence used previously we can argue that a generalised Hasselmann equation (GHE) should be:

$$C \frac{d}{dt} \Delta T(t) + C \int_{a}^{t} \gamma(t-\tau) \Delta T(\tau) d\tau = F(t) + \xi(t) \tag{12}$$

which would become the familiar Hasselmann equation in the limit $\gamma(t) = (\lambda/C)\delta(t)$.

We are not aware of a general solution-beyond what is available in the Mori-Zwanzig approach-of the above equations for data-derived $F(t)$ of the type encountered in climate modelling, but sophisticated numerics are now available (e.g. [30]).

5 Fractional Non-Markovian Models

A special case of the GLE which has received interest (e.g. [20, 29, 31–33]) because of its relationship to anomalous diffusion is when the damping kernel $\gamma(t)$ is chosen to be a power law $\gamma(t) \sim t^{-\alpha}$ (e.g. [29, 31]). The integral operator term in the GLE can then be replaced (e.g. [29, 34]) by a Riemann-Liouville fractional derivative with respect to time:

$$\int_0^t d\tau \frac{1}{(t-\tau)^\alpha} \equiv \gamma_\alpha \frac{d^{\alpha-1}}{dt^{\alpha-1}} \tag{13}$$

where the gamma function in the definition of the derivative was absorbed into γ_α and again we follow the notational conventions of Lutz [29].

In this case the GLE becomes the fractional Langevin equation (FLE) (e.g. [20, 29, 31, 32, 35]) which in Lutz' notation is:

$$M\frac{d}{dt}V(t) + M\gamma_\alpha \frac{d^{\alpha-1}}{dt^{\alpha-1}}V(t) = F(t) + \xi(t) \tag{14}$$

while the GHE that we proposed above would become a fractional Hasselmann equation (FHE):

$$C\frac{d}{dt}\Delta T(t) + C\gamma_\alpha \frac{d^{\alpha-1}}{dt^{\alpha-1}}\Delta T(t) = F(t) + \xi(t) \tag{15}$$

The exponential functions in the solutions of the Langevin equation are known to be a special case of the solution of the FLE in terms of Mittag-Leffler functions (e.g. [20, 29, 31]), and the analogous expressions would follow for the FHE (see also [24]).

6 Fractionally Integrating the Standard Fractional Non-Markovian Models

The FHE has not so far been studied in the climate literature, but a closely related equation has [24]. To see this we can fractionally integrate both sides of the FLE and FHE with respect to $t^{\alpha-1}$. In the FLE case we may refer to the result as the fractionally integrated fractional Langevin equation (FIFLE):

$$M\frac{d^{2-\alpha}}{dt^{2-\alpha}}V(t) + M\gamma_\alpha V(t) = \frac{d^{1-\alpha}}{dt^{1-\alpha}}(F(t) + \xi(t)) \tag{16}$$

The corresponding fractionally integrated fractional Hasselmann equation (FIFHE) is

$$C\frac{d^{2-\alpha}}{dt^{2-\alpha}}\Delta T(t) + C\gamma_\alpha \Delta T(t) = \frac{d^{1-\alpha}}{dt^{1-\alpha}}(F(t) + \xi(t)) \tag{17}$$

To establish the relationship between the FIFHE and Lovejoy's FEBE we first note that the H defined in [24] is not that of Mandelbrot's fractional Brownian motion(e.g. [13, 36]), but is instead Mandelbrot's $H - 1/2$. We can use the well-known [36] relationship $H = 1/2 + d$ to identify Lovejoy's H as the memory parameter d used in much of the mathematical literature on stochastic processes which exhibit LRD (e.g. [13]).

If we then identify d with our $2 - \alpha$, define a time scale τ via $\tau^d = \gamma_\alpha^{-1}$, and multiply across by τ^d we see that the FIFHE becomes:

$$(\tau^d \frac{d^d}{dt^d} + 1)\Delta T(t) = \frac{\tau^d}{C}\frac{d^{d-1}}{dt^{d-1}}(F(t) + \xi(t)) \tag{18}$$

The left hand side is of the same form as Lovejoy's FEBE [24]. The right hand side differs from FEBE, being a fractional integral of the forcing term plus a fractionally integrated white noise, the details of which depend on those of the fractional derivative.

7 Conclusions

This paper is a first report on our project [37, 38] to use the well-known Mori-Kubo generalised Langevin equation of statistical mechanics to seek a flexible framework in which the Hasselmann equation and our FIFHE are respectively the short range and long range correlated limits. We intend to further explore the properties of the FHE and FIFHE and their relation to the FEBE in future work, as well as those of the more general GHE for climate.

Acknowledgements We acknowledge valuable interactions with Raphael Calel, Michael Ghil, Shaun Lovejoy, Nick Moloney, Cecile Penland, Kristoffer Rypdal and Martin Rypdal. NWW acknowledges support from the Max Planck Society, London Mathematical Laboratory, KLIMAFORSK project number 229754 and ONR grant NICOP – N62909-15-1-N143. NWW and SCC acknowledge the hospitality of Holger Kantz at MPIPKS, Ralf Metzler at the University of Potsdam, Juergen Kurths at PIK and Sid Redner at SFI during the development of this work. SCC acknowledges support from STFC grant ST/T000252/1 and AFOSR grant FA9550-17-1-0054. DAS acknowledges support from the Grantham Research Institute on Climate Change and the Environment, at the London School of Economics, and the ESRC Centre for Climate Change Economics and Policy (CCCEP) (ref. ES/R009708/1).

References

1. Ghil, M.: A century of nonlinearity in the geosciences. Earth Space Sci. **6**, 1007–1042 (2019). https://doi.org/10.1029/2019EA000599
2. North, G.R., Cahalan, R.F., Coakley Jr., J.A.: Energy balance climate models. Rev. Geophys. Space Phys. **19**(1), 91–121 (1981)
3. Hasselmann, K.: Stochastic climate models. Part I. Theory. Tellus **28**(6), 473–485 (1976)
4. Frankignoul, C., Hasselmann, K.: Stochastic climate models. Part II. Application to sea-surface temperature anomalies and thermocline variability. Tellus **29**(4), 289–305 (1977)
5. Lemke, P.: Stochastic climate models. Part III. Application to zonally averaged energy models. Tellus **29**(5), 385–392 (1977)
6. Franzke, C., O'Kane, T. (eds.): Nonlinear and Stochastic Climate Dynamics. Cambridge University Press, Cambridge (2017)
7. Penland, C.: Random forcing and forecasting using principal oscillation pattern analysis. Mon. Weather. Rev. **117**, 2165–2185 (1989)
8. Franzke, C.L.E., O'Kane, T.J., Berner, J., Williams, P.D., Lucarini, V.: Stochastic climate theory and modelling. WIREs Clim. Change **6**, 63–78 (2015). https://doi.org/10.1002/wcc.318
9. Gottwald, G., Crommelin, D., Franzke, C.: Stochastic climate theory. In: Franzke, C., O'Kane, T. (eds.) Nonlinear and Stochastic Climate Dynamics, pp. 209–240. Cambridge University Press, Cambridge (2017). https://doi.org/10.1017/9781316339251.009
10. Zwanzig, R.: Nonequilibrium Statistical Mechanics. Oxford University Press, Oxford (2001)
11. Chorin, A.J., Hald, O.H.: Stochastic Tools in Mathematical Science, 2nd edn. Springer, Berlin (2009)
12. Leith, C.: Climate signal and weather noise. In: Madden, R.A., Katz, R.W. (eds.) Applications of Statistics to Modeling the Earth's Climate System, Technical Note NCAR/TN-409+PROC. NCAR, Colorado (1994)
13. Beran, J.: Statistics for Long Memory Processes. Chapman and Hall/CRC, Boca Raton (1998)
14. Franzke, C.L.E., Barbosa, S., Blender, R., Fredriksen, H.-B., Laepple, T., Lambert, F., et al.: The structure of climate variability across scales. Rev. Geophys. **58**, e2019RG000657 (2020). https://doi.org/10.1029/2019RG000657
15. Rypdal, K.: Global temperature response to radiative forcing: solar cycle versus volcanic eruptions. J. Geophys. Res. Atmos. **117**, D06115 (2012)
16. Fredriksen, H.-B., Rypdal, M.: Long-range persistence in global surface temperatures explained by linear multibox energy balance models. J. Clim. **30**(18), 7157–7168 (2017)
17. Rypdal, M., Fredriksen, H.-B., Myrvoll-Nilsen, E., Rypdal, K., Sorbye, S.H.: Emergent scale invariance and climate sensitivity. Climate **6**(4), 93 (2018). https://doi.org/10.3390/cli6040093
18. Del Rio Amador, L., Lovejoy, S.: Predicting the global temperature with the Stochastic Seasonal to Interannual Prediction System (StocSIPS). Clim. Dyn. (2019). https://doi.org/10.1007/s00382-019-04791-4
19. Mori, H.: Transport, collective motion and Brownian motion. Prog. Theor. Phys. **33**(3), 423–455 (1965). https://doi.org/10.1143/PTP.33.423
20. Coffey, W.T., Kalmykov, Y.P.: The Langevin Equation: With Applications to Stochastic Problems in Physics, Chemistry and Electrical Engineering, 4th edn. World Scientific, Singapore (2017)
21. Chechkin, A.V., Klages, R.: Fluctuation relations for anomalous dynamics. J. Stat. Mech. L03002/1-11 (2009)
22. Chechkin, A.V., Lenz, F., Klages, R.: Normal and anomalous fluctuation relations for Gaussian stochastic dynamics. J. Stat. Mech. L11001/1-13 (2012)
23. Lovejoy, S.: Weather, Macroweather, and the Climate: Our Random Yet Predictable Atmosphere. Oxford University Press, Oxford (2019)
24. Lovejoy, S.: Fractional relaxation noises, motions and the fractional energy balance equation. Nonlinear Process. Geophys. Discuss. 1–52 (2019). https://doi.org/10.5194/npg-2019-39
25. Lemons, D.S.: An Introduction to Stochastic Processes in Physics. Johns Hopkins University Press, Baltimore (2002)

26. Cox, P.M., Huntingford, C., Williamson, M.S.: Emergent constraint on equilibrium climate sensitivity from global temperature variability. Nature **553**, 319–322 (2018). https://doi.org/10.1038/nature25450
27. Gardiner, C.W.: Handbook of Stochastic Methods: For Physics, Chemistry and the Natural Sciences, 3rd edn. Springer, Berlin (2004)
28. Hänggi, P., Jung, P: Colored noise in dynamical systems. In: Prigogine, I., Rice, S.A. (eds.)Advances in Chemical Physics, vol. LXXXIX. Wiley (1995)
29. Lutz, E.: Fractional Langevin equation. Phys. Rev. E **64**, 051106 (2001)
30. Darve, E., Solomon, J., Kia, A.: Computing generalized Langevin equations and generalized Fokker-Planck equations. Proc. Natl. Acad. Sci. U. S. A. **106**(27), 10884–10889 (2009)
31. Porrà, J.M., Wang, K.G., Masoliver, J.: Generalized Langevin equations: anomalous diffusion and probability distributions. Phys. Rev. E **53**(6), 2044–2056 (1996)
32. Molina-Garcia, D., Sandev, T., Safdari, H., Pagnini, G., Chechkin, A., Metzler, R.: Crossover from anomalous to normal diffusion: truncated power-law noise correlations and applications to dynamics in lipid bilayers. New J. Phys. **20**, 103027 (2018). https://doi.org/10.1088/1367-2630/aae4b2
33. Kobelev, V., Romanov, E.: Fractional Langevin equation to describe anomalous diffusion. Prog. Theor. Phys. Suppl. **139**, 470–476 (2000)
34. Metzler, R., Jeon, J., Cherstvy, A.G., Barkai, E.: Phys. Chem. Chem. Phys. **16**, 24128 (2014). https://doi.org/10.1039/C4CP03465A
35. Mainardi, F., Pironi, P.: The fractional Langevin equation: Brownian motion revisited. Extr. Math. **11**, 140–154 (1996)
36. Watkins, N.W.: Mandelbrot's stochastic time series models. Earth Space Sci. **6**, 2044–2056 (2019). https://doi.org/10.1029/2019EA000598
37. https://meetingorganizer.copernicus.org/EGU2019/presentations/watkins
38. https://doi.org/10.1002/essoar.10501367.1

Author Index

D. Braha et al. (eds.), *Unifying Themes in Complex Systems X*, Springer Proceedings
in Complexity, https://doi.org/10.1007/978-3-030-67318-5

The manufacturer's authorised representative in the EU is Springer Nature Customer Service Centre GmbH, Europaplatz 3, 69115 Heidelberg, Germany. If you have any concerns regarding our products, please contact ProductSafety@springernature.com

Printed and bound by CPI Group (UK) Ltd, Croydon, CR0 4YY
30/09/2026
02225594-0003